Analytical Chemistry: Concepts and Applied Principles

Analytical Chemistry: Concepts and Applied Principles

Editor: Jessica Carol

NY RESEARCH
P R E S S

New York

Published by NY Research Press
118-35 Queens Blvd., Suite 400,
Forest Hills, NY 11375, USA
www.nyresearchpress.com

Analytical Chemistry: Concepts and Applied Principles
Edited by Jessica Carol

International Standard Book Number: 978-1-63238-547-5 (Hardback)

Cataloging-in-Publication Data

Analytical chemistry : concepts and applied principles / edited by Jessica Carol.
   p. cm.
Includes bibliographical references and index.
ISBN 978-1-63238-547-5
1. Chemistry, Analytic. 2. Chemistry. 3. Metallurgical analysis. I. Carol, Jessica.
QD75.22 .A53 2017
543--dc23

Printed in the United States of America.

Contents

Preface IX

Chapter 1 **Nanomechanical DNA origami 'single-molecule beacons' directly imaged by atomic force microscopy** 1
Akinori Kuzuya, Yusuke Sakai, Takahiro Yamazaki, Yan Xu, Makoto Komiyama

Chapter 2 **Rapid *in vivo* detection of isoniazid-sensitive *Mycobacterium tuberculosis* by breath test** 9
Seong W. Choi, Mamoudou Maiga, Mariama C. Maiga, Viorel Atudorei,
Zachary D. Sharp, William R. Bishai, Graham S. Timmins

Chapter 3 **Attomolar DNA detection with chiral nanorod assemblies** 15
Wei Ma, Hua Kuang, Liguang Xu, Li Ding, Chuanlai Xu, Libing Wang,
Nicholas A. Kotov

Chapter 4 **Hydride ions in oxide hosts hidden by hydroxide ions** 23
Katsuro Hayashi, Peter V. Sushko, Yasuhiro Hashimoto, Alexander L. Shluger,
Hideo Hosono

Chapter 5 **Quantifying thiol–gold interactions towards the efficient strength control** 31
Yurui Xue, Xun Li, Hongbin Li, Wenke Zhang

Chapter 6 **First enantioseparation and circular dichroism spectra of Au_{38} clusters protected by achiral ligands** 40
Igor Dolamic, Stefan Knoppe, Amala Dass, Thomas Bürgi

Chapter 7 **All-dielectric metasurface analogue of electromagnetically induced transparency** 46
Yuanmu Yang, Ivan I. Kravchenko, Dayrl P. Briggs, Jason Valentine

Chapter 8 **Hydrochromic conjugated polymers for human sweat pore mapping** 53
Joosub Lee, Minkyeong Pyo, Sang-hwa Lee, Jaeyong Kim, Moonsoo Ra,
Whoi-Yul Kim, Bum Jun Park, Chan Woo Lee, Jong-Man Kim

Chapter 9 **Ultrasensitive and label-free molecular-level detection enabled by light phase control in magnetoplasmonic nanoantennas** 63
Nicolò Maccaferri, Keith E. Gregorczyk, Thales V.A.G. de Oliveira, Mikko Kataja,
Sebastiaan van Dijken, Zhaleh Pirzadeh, Alexandre Dmitriev,
Johan Åkerman, Mato Knez, Paolo Vavassori

Chapter 10 **Structural analysis and mapping of individual protein complexes by infrared nanospectroscopy** 71
Iban Amenabar, Simon Poly, Wiwat Nuansing, Elmar H. Hubrich,
Alexander A. Govyadinov, Florian Huth, Roman Krutokhvostov, Lianbing Zhang,
Mato Knez, Joachim Heberle, Alexander M. Bittner, Rainer Hillenbrand

Chapter 11   **Intracellular temperature mapping with a fluorescent polymeric thermometer and fluorescence lifetime imaging microscopy**.. 80
Kohki Okabe, Noriko Inada, Chie Gota, Yoshie Harada, Takashi Funatsu,
Seiichi Uchiyama

Chapter 12   **Direct mechanochemical cleavage of functional groups from graphene**............89
Jonathan R. Felts, Andrew J. Oyer, Sandra C. Hernández, Keith E. Whitener Jr,
Jeremy T. Robinson, Scott G. Walton, Paul E. Sheehan

Chapter 13   **Enantioselective recognition at mesoporous chiral metal surfaces**............. 96
Chularat Wattanakit, Yémima Bon Saint Côme, Veronique Lapeyre,
Philippe A. Bopp, Matthias Heim, Sudarat Yadnum, Somkiat Nokbin,
Chompunuch Warakulwit, Jumras Limtrakul, Alexander Kuhn

Chapter 14   **Solution-based circuits enable rapid and multiplexed pathogen detection**................. 104
Brian Lam, Jagotamoy Das, Richard D. Holmes, Ludovic Live, Andrew Sage,
Edward H. Sargent, Shana O. Kelley

Chapter 15   **Coherent anti-Stokes Raman scattering with single-molecule sensitivity using a plasmonic Fano resonance**... 112
Yu Zhang, Yu-Rong Zhen, Oara Neumann, Jared K. Day, Peter Nordlander,
Naomi J. Halas

Chapter 16   **A platform for designing hyperpolarized magnetic resonance chemical probes**............119
Hiroshi Nonaka, Ryunosuke Hata, Tomohiro Doura, Tatsuya Nishihara,
Keiko Kumagai, Mai Akakabe, Masashi Tsuda, Kazuhiro Ichikawa, Shinsuke Sando

Chapter 17   **Radiolysis as a solution for accelerated ageing studies of electrolytes in Lithium-ion batteries**...126
Daniel Ortiz, Vincent Steinmetz, Delphine Durand, Solène Legand, Vincent Dauvois,
Philippe Maître, Sophie Le Caër

Chapter 18   **Understanding silicate hydration from quantitative analyses of hydrating tricalcium silicates**.. 134
Elizaveta Pustovgar, Rahul P. Sangodkar, Andrey S. Andreev, Marta Palacios,
Bradley F. Chmelka, Robert J. Flatt, Jean Baptiste d'Espinose de Lacaillerie

Chapter 19   **Noble metal-comparable SERS enhancement from semiconducting metal oxides by making oxygen vacancies**.......................................143
Shan Cong, Yinyin Yuan, Zhigang Chen, Junyu Hou, Mei Yang, Yanli Su,
Yongyi Zhang, Liang Li, Qingwen Li, Fengxia Geng, Zhigang Zhao

Chapter 20   **Identification of phases, symmetries and defects through local crystallography**............150
Alex Belianinov, Qian He, Mikhail Kravchenko, Stephen Jesse, Albina Borisevich,
Sergei V. Kalinin

Chapter 21   **Gln40 deamidation blocks structural reconfiguration and activation of SCF ubiquitin ligase complex by Nedd8**...158
Clinton Yu, Haibin Mao, Eric J. Novitsky, Xiaobo Tang, Scott D. Rychnovsky,
Ning Zheng, Lan Huang

Chapter 22   **Hot electron-induced reduction of small molecules on photorecycling metal surfaces**... 170
Wei Xie, Sebastian Schlücker

Chapter 23   **A wearable chemical electrophysiological hybrid biosensing system for real-time health and fitness monitoring**.. 176
Somayeh Imani, Amay J. Bandodkar, A.M. Vinu Mohan, Rajan Kumar, Shengfei Yu,
Joseph Wang, Patrick P. Mercier

Chapter 24   **Lysosome triggered near-infrared fluorescence imaging of cellular trafficking processes in real time**.. 183
Marco Grossi, Marina Morgunova, Shane Cheung, Dimitri Scholz, Emer Conroy,
Marta Terrile, Angela Panarella, Jeremy C. Simpson, William M. Gallagher,
Donal F. O'Shea

Chapter 25   **Operando NMR spectroscopic analysis of proton transfer in heterogeneous photocatalytic reactions**.. 196
Xue Lu Wang, Wenqing Liu, Yan-Yan Yu, Yanhong Song, Wen Qi Fang, Daxiu Wei,
Xue-Qing Gong, Ye-Feng Yao, Hua Gui Yang

Permissions

List of Contributors

Index

Preface

The world is advancing at a fast pace like never before. Therefore, the need is to keep up with the latest developments. This book was an idea that came to fruition when the specialists in the area realized the need to coordinate together and document essential themes in the subject. That's when I was requested to be the editor. Editing this book has been an honour as it brings together diverse authors researching on different streams of the field. The book collates essential materials contributed by veterans in the area which can be utilized by students and researchers alike.

Analytical chemistry seeks to study the fundamental aspects of matter. It uses rigorous scientific methods to extract, study and analyse the objects of its research. This book on analytical chemistry explains the fundamental concepts of the field and the methods through which properties of matter are analysed and examined. The research that has been included in this book closely analyses the technical aspects of the field and enumerates the various applied principles and technology that have been developed within this discipline. The extensive content of this book provides the readers with a thorough understanding of the subject. Scientists and students actively engaged in this field will find this book full of crucial and unexplored concepts.

Each chapter is a sole-standing publication that reflects each author's interpretation. Thus, the book displays a multi-facetted picture of our current understanding of applications and diverse aspects of the field. I would like to thank the contributors of this book and my family for their endless support.

Editor

Nanomechanical DNA origami 'single-molecule beacons' directly imaged by atomic force microscopy

Akinori Kuzuya[1,†], Yusuke Sakai[1], Takahiro Yamazaki[1], Yan Xu[1,†] & Makoto Komiyama[1]

DNA origami involves the folding of long single-stranded DNA into designed structures with the aid of short staple strands; such structures may enable the development of useful nanomechanical DNA devices. Here we develop versatile sensing systems for a variety of chemical and biological targets at molecular resolution. We have designed functional nanomechanical DNA origami devices that can be used as 'single-molecule beacons', and function as pinching devices. Using 'DNA origami pliers' and 'DNA origami forceps', which consist of two levers ~170 nm long connected at a fulcrum, various single-molecule inorganic and organic targets ranging from metal ions to proteins can be visually detected using atomic force microscopy by a shape transition of the origami devices. Any detection mechanism suitable for the target of interest, pinching, zipping or unzipping, can be chosen and used orthogonally with differently shaped origami devices in the same mixture using a single platform.

[1] Research Center for Advanced Science and Technology, The University of Tokyo, 4-6-1 Komaba, Meguro, Tokyo 153-8904, Japan. †Present addresses: Department of Chemistry and Materials Engineering, Kansai University, 3-3-35 Yamate-cho, Suita, Osaka 564-8680, Japan (A.K.), Department of Medical Sciences, University of Miyazaki, 5200 Kihara, Kiyotake, Miyazaki 889-1692, Japan (Y.X.). Correspondence and requests for materials should be addressed to A.K. (email: kuzuya@kansai-u.ac.jp) or to M.K. (email: komiyama@mkomi.rcast.u-tokyo.ac.jp).

The rapid development of nanotechnology has enabled the precise manipulation of nanomaterials. However, typical analytical methods for chemical or biochemical targets are still based on spectroscopic principles, which reflect the average behaviour of a vast number of molecules. To analyse the behaviour of an individual molecule, nanomechanical devices that can work with target molecules in a single-molecule manner are required. Structural DNA nanotechnology[1–3], based on the programmed assembly of branched DNA helices, is a key technology that can provide such functional nanomechanical devices. DNA devices such as DNA tweezers[4–6], DNA scissors[7,8], and DNA walkers have been constructed[9–13] and used as molecular sensors for various targets[14,15], following the success of molecular beacons[16,17]. However, individual molecules of these devices are still too small to be easily analysed at molecular resolution using current microscopy techniques. DNA origami[18–21], in which long single-stranded DNA is folded into a designed nanostructure with the aid of many short staple strands, is a powerful new tool in structural DNA nanotechnology that provides robust and precise nanostructures in both 2D and 3D that are visible through atomic force microscopy (AFM) or electron microscopy. Taking advantage of their precise addressability at nanometre resolution, some custom nano-instruments for studying single-molecule interactions in biology and chemistry, using DNA origami assembly as a nanoscale stage, have recently been proposed[22–24]. Despite these properties, only a limited number of studies have been published on DNA origami as a building material for nanomechanical DNA devices[25,26]; some recent reports have discussed its usefulness as a scaffold on which nanomechanical DNA devices could be located[27–29].

Here we present functional nanomechanical DNA origami devices that can be used as versatile and visible 'single-molecule beacons' (Fig. 1). Using 'DNA origami pliers' and 'DNA origami forceps' (Fig. 2) that consist of two levers of ~170 nm in length connected at a fulcrum, various inorganic/organic targets, from metal ions to proteins, were visually detected in a single-molecule manner as a shape transition of DNA origami devices, using AFM.

Results

The design of nanomechanical DNA origami devices.
The DNA origami devices used in this study behave like nano-sized pinching devices (Fig. 1). Shape transitions of nanomechanical DNA origami devices, such as the closing or opening of pliers, are selectively triggered by interactions with target molecules and visualized using AFM imaging. The existence of the target is therefore determined at molecular resolution. This sensing method is applicable to versatile targets because a variety of interactions (for example, protein–ligand, DNA–RNA, or DNA–metal ions) can be employed.

Figure 2 shows the detailed structure of the DNA origami devices used in this study. The first origami device shown is the 'DNA origami pliers' (Fig. 2a), which consist of two ca. 170-nm lever domains, each of which is made of six antiparallel DNA helices from a part of M13 scaffold and 117 staple strands (ca. 20 nm wide). These levers are joined together at a fulcrum via two phosphodiester linkages in the M13 scaffold to form an immobile Holliday junction. In a Mg^{2+} solution, typically used in DNA origami preparation, the DNA four-way junction is known to be a right-handed, antiparallel stacked X-structure with a small angle of 60° (refs 30–32). DNA origami pliers in solution are thus expected to be twisted, as illustrated in Figure 2b. AFM measurements of DNA origami devices require deposition on a mica surface. Origami devices adhered to the 2D surface can adopt three forms. The most likely form is a 'cross', which most closely resembles the expected twisted structure of the origami pliers. The second feasible structure is an 'antiparallel form', in which two levers are aligned in parallel on a plane but point in opposite directions. Configuration of the M13 scaffold at the fulcrum in this form is in 'antiparallel' mode, which is rather relaxed, and is a typical configuration in DNA origami design. The third feasible

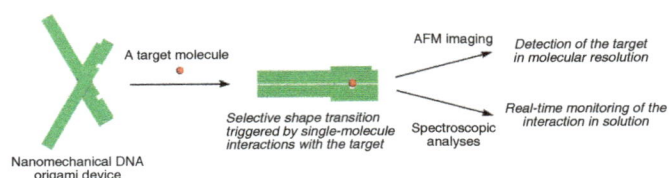

Figure 1 | Detection of various targets using DNA origami devices. The selective shape transitions of DNA origami devices triggered by intermolecular interactions with the target are visualized at molecular resolution using AFM on mica or monitored in real-time by fluorescence measurement in solution.

form is a 'parallel closed form' in which the levers are parallel. This configuration of the DNA four-way junction at the fulcrum is considered unfavourable compared with the antiparallel mode[30–32]. Thus, origami devices in the parallel closed form are expected to be the least populated species in the AFM images unless some further interaction rotates the fulcrum. Each of the levers has a small concavity two helical turns long and two helices wide (ca. 7 nm×7 nm). These concavities serve as the jaws of pliers that pinch the target molecule. When DNA pliers are in the parallel closed form, these concavities lie next to each other to form a larger cavity that accommodates the pinched target stably during AFM scanning, as shown in our previous studies[33,34].

The shape transition of DNA pliers in the presence of their targets is triggered by three independent mechanisms: 'pinching' (Fig. 3), 'zipping' (Fig. 4), and 'unzipping' (Fig. 5).

Detection of proteins by pinching mechanism.
Pinching is a process that detects single-target molecules that bind to multiple ligands each (Fig. 3a). For this purpose, two ligands are attached to each of the staple strands placed in the concavities (anchor strands, drawn in blue in Fig. 2a), and these ligands cooperatively capture a single-target molecule between the jaws. This intramolecular process triggers a shape transition in the origami device from the cross or antiparallel form to the parallel closed form, which can be visualized using AFM imaging or monitored in solution using spectroscopic analyses.

Figure 3b shows typical AFM images of selective, single-molecule pinching of a streptavidin tetramer (SA), which strongly binds to four biotin groups, using DNA origami pliers. Here the DNA pliers were modified with a biotin group in each of the jaws. The dominant species initially observed in 1×TAE/Mg^{2+} buffer solution without SA (left, i) was DNA pliers in cross (58% yield), which is consistent with the X-structure of the DNA four-way junction. The population of DNA pliers in antiparallel form was 16%; only 5% of DNA pliers were in the parallel closed form. The preference for the antiparallel over the parallel closed form was in accordance with previous observations. When SA was added to the solution, the populations were drastically altered. Almost 58% of DNA pliers were found in the parallel closed form in the presence of SA (middle). As shown in Figure 3c, a bright spot 5 nm in height, which corresponds to the expected diameter of the pinched SA molecule, was found in the jaws of most of the parallel pliers. The yields of antiparallel and cross pliers in the presence of SA were 5 and 23%, respectively.

Spectroscopic analyses have shown that the parallel pliers observed under AFM were formed in solution before deposition, triggered by SA addition. As shown in Figure 3e, the fluorescence spectra of doubly dye-labelled DNA pliers (see Fig. 2) showed a significant quenching of TXR emission (ca. 23% at 612 nm) and an enhancement of FAM emission (ca. 17% at 517 nm) immediately after the addition of one equivalent of SA to the solution. The only reasonable explanation for such a drastic shape change is that the

a

b

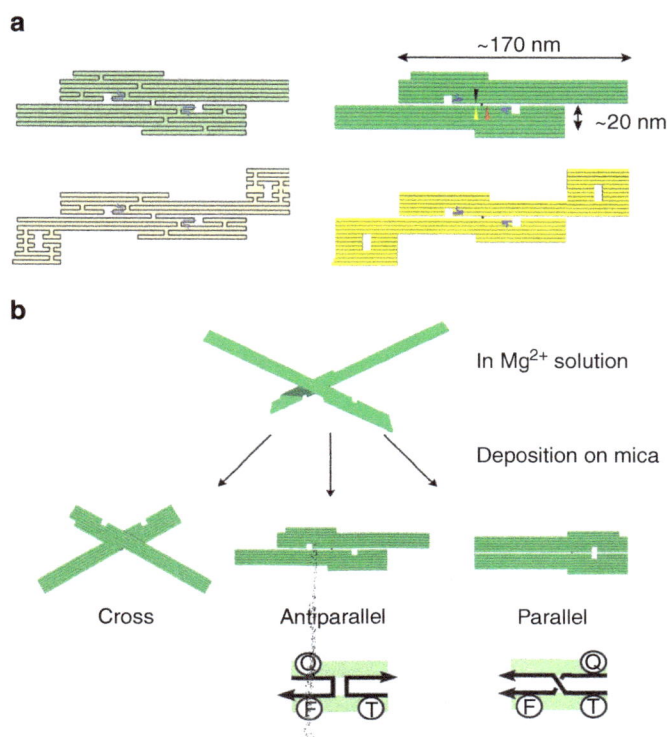

Figure 2 | Structure of nanomechanical DNA origami devices. (a) Folding patterns of M13 scaffold in DNA origami pliers and DNA origami forceps with extra finger-hole-like structure (lower left) for visual discrimination of coexisting targets in solution. Staple strands are omitted, except for the anchor strands with extra 8-nucleotide toeholds (presented in blue). Illustrations at right represent the approximate size of DNA helices, levers and concavities. The two levers are joined at a fulcrum through two phosphodiester bonds in the M13 scaffold so that an immobile Holliday junction is formed. Yellow, red, and black wedges indicate the positions of fluorescent dyes and a quencher (see below). **(b)** Expected solution structure of DNA pliers and three possible forms: cross, antiparallel, and parallel closed, observed on mica. The configuration of the M13 scaffold (drawn in black lines) at the fulcrum is shown below. To monitor the structure of origami devices in solution, one of the levers of the DNA pliers was fluorescently labelled with 6-fluoresceincarboxyamide (yellow wedge; F, FAM) and Texas Red (red wedge; T, TXR) at two sites one and a half helical turns away from the fulcrum. The other lever was modified with a Black hole quencher-2 (black wedge, Q, BHQ-2) group so that the fluorescence of FAM would be quenched by nearby BHQ-2 when DNA pliers were in antiparallel form. The shape transition of the pliers into parallel form brings BHQ-2 closer to TXR than FAM, resulting in the quenching of TXR and recovery of FAM emission.

single SA tetramer bridged the two levers by binding the attached biotins. The potential for nanomechanical DNA origami as a single-molecule detector for noncovalent and multidentate biochemical interactions has thus been shown.

To further prove that the shape transition to parallel pliers was achieved solely by biotin–SA–biotin bidentate binding, the biotinylated anchor strands were selectively detached from the DNA pliers after SA pinching using a DNA strand displacement technique with the aid of the extra 8-nucleotide toehold sequence on the anchor strands (right in Fig. 3b)[4]. After 2 h incubation at room temperature in the presence of 10 equivalents of unset strands complementary to the anchor strands, the population of parallel pliers decreased to 10%, and pliers in the cross form became the dominant species (53%) again as expected. Recovery of TXR emission to ca. 90% of the initial intensity and nearly complete re-quenching

of FAM emission was also consistent with the AFM measurements (Fig. 3e). As we recently showed through SA nanopatterning on a punched DNA origami scaffold[35], repetitive cycles of pinching and release are possible through the readdition of anchor and then unset strands.

The target molecules that can be pinched by DNA origami devices are not limited to SA. When FAM-modified anchor strands were used in place of biotinylated anchor strands, DNA pliers selectively pinched anti-fluorescein IgG in the jaws and adopted the parallel closed form (ca. 39% yield among clearly resolved motifs, Fig. 3f). According to the height profile analyses (Supplementary Fig. S1), objects slightly lower (3–4 nm) than SA were observed in the jaws, which is consistent with the expected thickness of IgG.

The closing of nanomechanical DNA origami is completely chemoselective, and, therefore, visual discrimination of coexisting proteins in a solution is possible using a mixture of differently shaped and modified origami devices. To demonstrate this concept, we prepared a second DNA origami device, 'DNA origami forceps', which have very similar jaws to DNA pliers but with an extra finger-hole-like structure at the other end of the levers (lower, Fig. 2a). Biotinylated DNA forceps (white arrows) and FAM-modified DNA pliers (yellow arrows) were separately annealed and mixed together in a 1:1 ratio (left, Fig. 3g). When SA was added to the mixture (right), biotinylated DNA forceps selectively pinched SA tetramers and adopted the parallel closed form (69% yield), whereas FAM-modified DNA pliers remained intact throughout the process.

Detection of metal ions by zipping mechanism. Although selective pinching and detection of a target molecule was successful using protein–ligand bindings, among the strongest of all biological interactions, single-molecule pinching of other targets that bind more weakly is usually difficult. A zipping mechanism involving multiple binding events is a second detection mechanism available from the present nanomechanical DNA origami devices and is appropriate for such targets (Fig. 4). Here multiple elements that bind together in the presence of the target are introduced to each of the levers, cooperatively triggering selective closure of the origami devices.

Na^+ ion sensing by DNA origami devices utilizes G-quadruplex formation (Fig. 4a), which has been popularly used in spectroscopic detection systems for metal ions[36]. A 12-nucleotide human telomere sequence (5'-TAGGGTTAGGGT-3'; a Telomere Element, TeloE) was attached to the 3'-ends of the staple strands placed at the four sites on each of the levers, spaced every three helical turns on the same side of the fulcrum (sites a, b, c, and d). This 12-nucleotide sequence is known to dimerize in the presence of Na^+ by forming an antiparallel G-quadruplex motif and thus acts as a zipper element[37]. The effects of salt on the form of the origami devices were first examined for DNA pliers without any TeloE modification (Fig. 4b). On addition of Na^+ to the solution (final concentration, 200 mM), the population of antiparallel DNA pliers significantly increased from 16% in the solution containing 12.5 mM Mg^{2+} alone (top left) to 46% (top right), whereas the percentage of DNA pliers in the cross form decreased (Fig. 4c). This change probably occurred because Mg^{2+}, which neutralizes phosphates and stabilizes the X-structure of the four-way junction, was partially replaced with the less efficient Na^+. The effect of Na^+ addition on DNA pliers bearing four TeloE (sites a, b, c, and d) on each lever was quite dramatic. In AFM measurements, almost 64% of the pliers were in parallel closed form, 19% were in antiparallel form, and only 6% were in cross form (bottom left in Fig. 4b). The yield of parallel closed form clearly correlated to the amount of introduced TeloE and the Na^+ concentration. The yields of parallel pliers for the motif with one TeloE at site a, with two TeloE at sites a and b, and with three TeloE at sites a, b, and c were 36%, 45%, and 54%, respectively (Fig. 4c). For the motif with four TeloE, the yield of parallel pliers correlated almost linearly with Na$^+$ concentration up to 100 mM and was approximately saturated thereafter (see Supplementary Fig. S2). Buffer exchange of the solution to remove Na$^+$ from the system completely opened the parallel closed DNA pliers, as expected (bottom right in Fig. 4b, iv). Fluorescence spectra have shown that Na$^+$-triggered zipping of DNA pliers through G-quadruplex formation proceeds efficiently in solution (Fig. 4d) but is rather slow compared with SA-biotin systems. It took 2–3 h to reach equilibrium (20% enhancement of FAM and 52% quenching of TXR emission) after Na$^+$ addition, according to the time course of the quenching of TXR emission (inset). Buffer

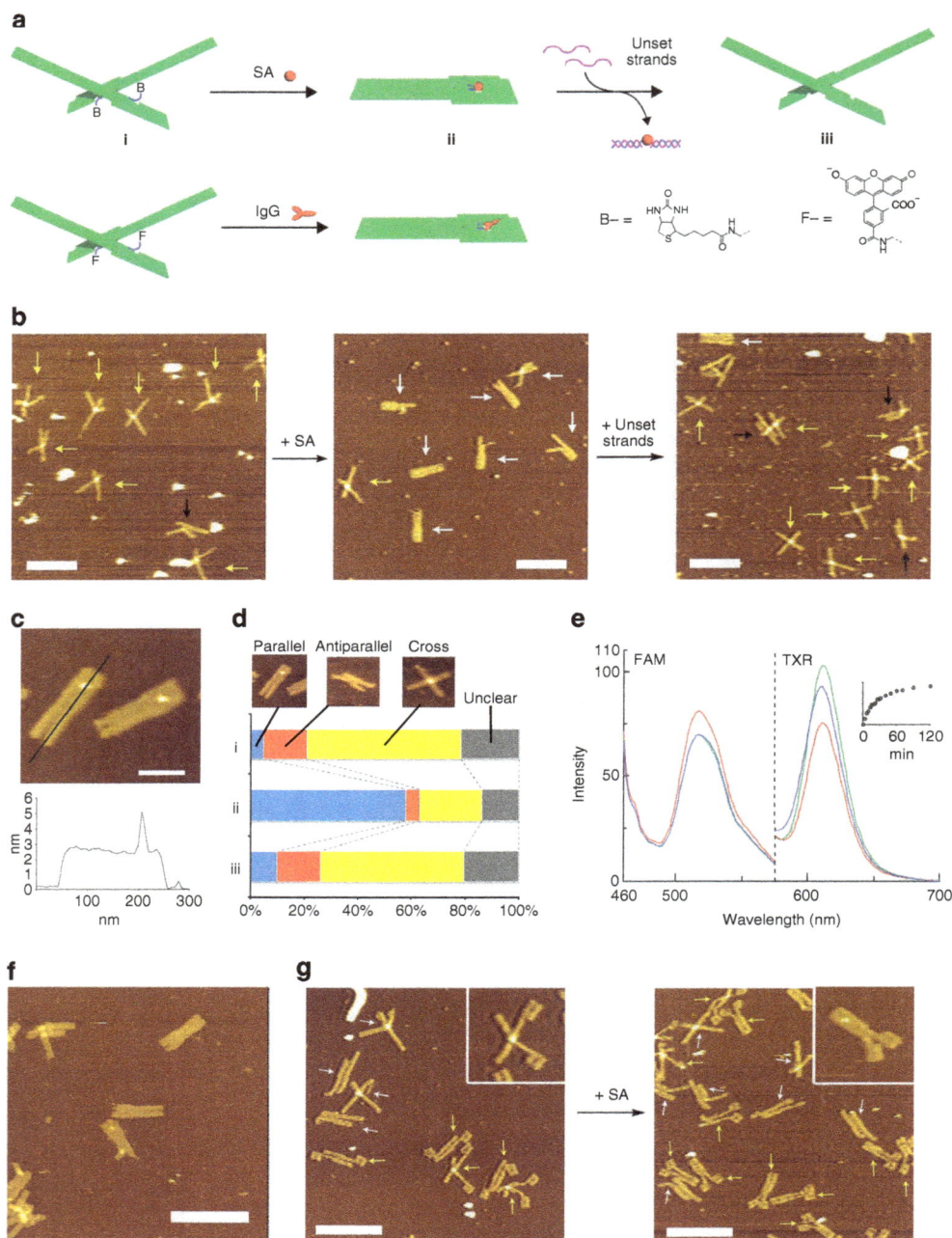

Figure 3 | Single-molecule pinching of proteins. (**a**) Schematic illustration of SA and IgG pinching. (**b**) AFM images for SA pinching and release by biotinylated DNA pliers. The scale bars are 300 nm. The dominant form of DNA pliers in Mg^{2+} solution before SA addition (left in **a**, i) was a cross (yellow arrows). After SA addition (middle in **a**, ii), DNA pliers selectively pinched exactly one SA tetramer and closed into the parallel closed form (white arrows). The population of DNA pliers in the cross form recovered significantly when the biotinylated anchor strands, together with SA, were selectively detached from the DNA pliers (right in **a**, iii). DNA pliers in the antiparallel form (black arrows) were always a minor species. (**c**) A height profile of parallel DNA pliers with SA (black line in the image). The scale bar is 100 nm. (**d**) Form distribution at each step in **a**, estimated by counting the motifs in AFM images. (**e**) Fluorescence spectra of doubly dye-labelled DNA pliers before (left in **a**, green) and 10 min after SA addition (middle in **a**, red) and 2 h after the addition of unset strands at room temperature (right in **a**, blue). The excitation wavelengths for 6-fluoresceincarboxyamide (FAM, 460–575 nm) and Texas Red, (TXR, 575–700 nm) spectra were 450 and 550 nm, respectively. Inset, time course of the recovery of TXR emission at 612 nm after the addition of unset strands (from the red spectrum to the blue one). (**f**) Single-molecule pinching of anti-fluorescein IgG by FAM-modified DNA pliers. The scale bar is 300 nm. Slightly lower objects than SA (3–4 nm) were observed in the jaws. (**g**) Selective pinching of SA by biotinylated DNA forceps (white arrows) in a mixture with FAM-labelled DNA pliers (yellow arrows). The scale bars are 300 nm. The insets show zoomed images of typical DNA forceps.

exchange using ultrafiltration resulted in backward transition of the emission to the initial level. The final FAM emission was 80% of the initial intensity, and that of TXR was 89%.

The TeloE sequence is known to dimerize in the presence of K[+] as well, although the dimer is a mixture of two species, parallel and antiparallel G-quadruplex motifs[38]. Selective closure of DNA

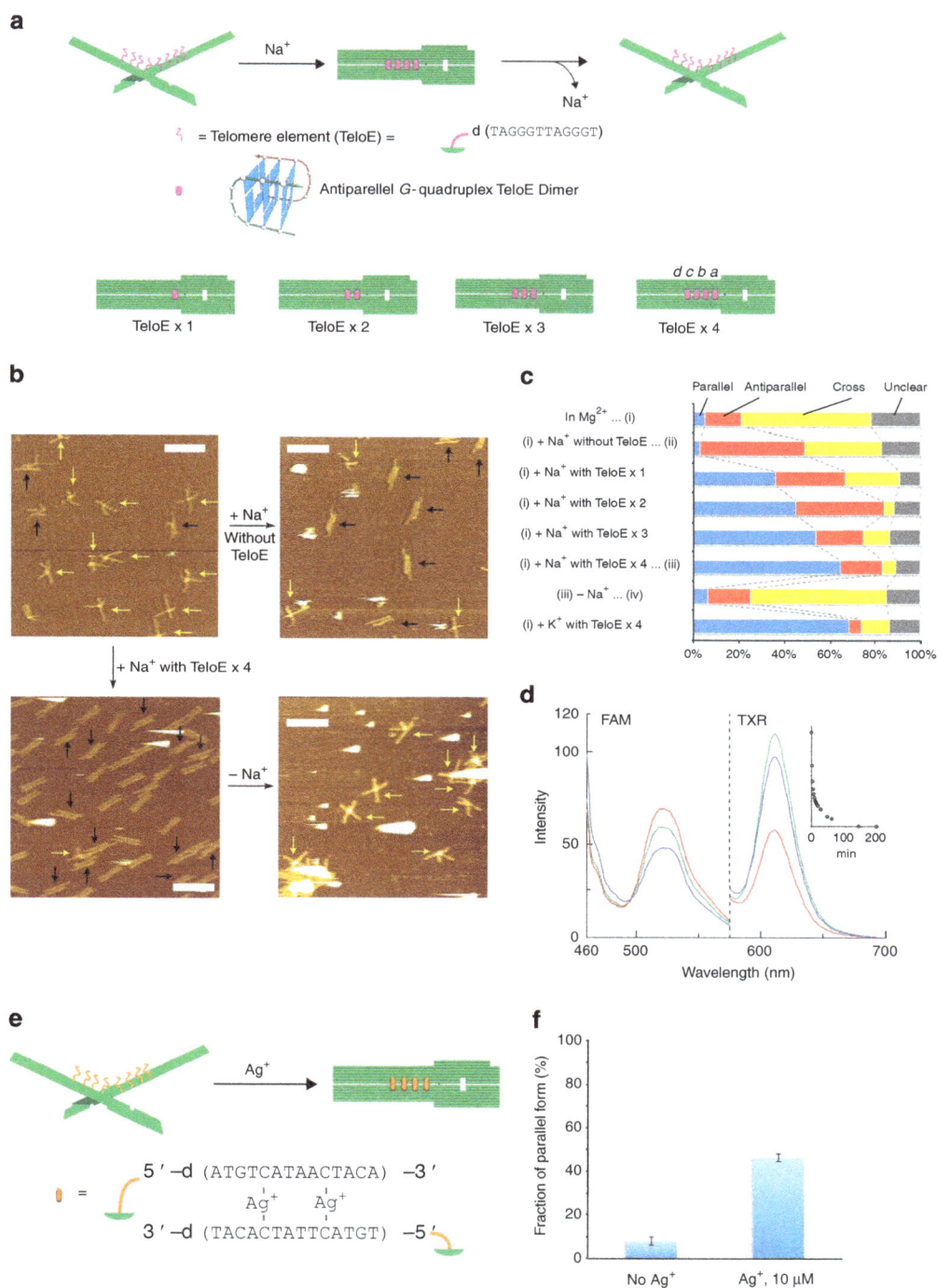

Figure 4 | Metal ion detection by zipping mechanism. (**a**) Schematic illustration of Na[+]-triggered zipping of DNA pliers bearing a G-quadruplex zipper. The four modified sites are indicated below. (**b**) AFM images for Na[+]-sensing. Un-indexed DNA pliers in the images are all in the parallel form. For the pliers without any TeloE, a significant increase in antiparallel DNA pliers (black arrows) was observed in a solution containing an extra 200 mM Na[+] (top right) compared with the solution containing 12.5 mM Mg[2+] alone (top left). For DNA pliers bearing four TeloE on the levers, on the other hand, most of the pliers were found in the parallel closed form, and few crosses (yellow arrows) were observed (bottom left). Most of the parallel closed pliers in opened up again when Na[+] was removed from the solution (bottom right). (**c**) Form distribution of DNA pliers estimated by counting the motifs in AFM images. Concentration of K[+] in the bottom was 100 mM. (**d**) Fluorescence spectra before (green) and 2 h after (red) Na[+] addition at room temperature and after Na[+] removal (blue). Inset, time course of the quenching of TXR emission at 612 nm after the addition of Na[+] to the solution (from the green spectrum to the red one). (**e**) Zipper elements involving C–C mismatches used for Ag[+] detection. (**f**) Ag[+] detection by DNA pliers bearing C–C mismatched zipper elements. Almost 47% of the clearly resolved DNA pliers ($n = 138$) were in the parallel closed form after the addition of Ag[+] to the solution (final concentration, 10 μM), whereas parallel pliers represented only 8% ($n = 111$) of DNA pliers in the solution without Ag[+]. The error bars indicate the standard errors.

pliers was also observed with K^+ addition (Fig. 4c). Representing the higher stability of the G-quadruplex containing K^+ (T_m values for TeloE dimer in the presence of 100 mM K^+ and Na^+ are 54 and 42 °C, respectively, data not shown), the yields of parallel pliers were slightly higher than in the systems with corresponding amounts of Na^+ (Supplementary Fig. S2).

In a similar manner to Na^+ or K^+, Ag^+ ions can be detected by a zipping mechanism by using Ag^+-mediated selective stabilization of C–C mismatches (Fig. 4e)[39]. Four pairs of complementary DNA elements that involve two C–C mismatches in a 14-mer sequence were attached to each of the levers, as with TeloE. As shown in Figure 4f, almost 47% of DNA pliers were in parallel closed form after the addition of Ag^+ to the solution.

Detection of small molecules by unzipping mechanism. Unzipping, which is the reverse process of the zipping mechanism, is the third detection mechanism of the present nanomechanical DNA origami devices (Fig. 5). The zipper elements for this mechanism were designed to bind together at the initial stage and selectively unbind in the presence of target molecules.

As shown in Figure 5a, the presence of human micro RNA can be clearly detected using origami devices by employing an unzipping mechanism. For example, 23-mer sequences complementary to miR20 were attached to the four sites for zipper elements in one of the levers of DNA pliers. The other lever was modified with partially complementary 10-mer sequences. DNA pliers bearing these elements were directly annealed into the parallel closed form in the absence of the target. The 13-mer sequences in the 3′ end of the 23-mer elements (red nucleotides in Fig. 5a) were left single-stranded to serve as a toehold to trigger selective strand displacement by the target. AFM measurements with increasing numbers of preclosing zipper-element pairs from one to four, just after the annealing of DNA pliers, revealed that at least three zipper-element pairs are necessary to obtain sufficient (around 80%) preclosing (Supplementary Fig. S3). Figure 5c shows selective detection of miR20 achieved using a 1:1 mixture of miR20-targeting DNA pliers and miR16-targeting DNA forceps with four preclosing zipper-element pairs. The addition of miR20 to the mixture selectively opened the DNA pliers (shown in blue bars), whereas the DNA forceps persisted in the parallel closed form (red bars). Selective detection of miR16 using the same 1:1 mixture was also possible (Supplementary Fig. S4). Unzipping of dual dye-labelled DNA pliers in solution was successfully monitored by fluorescence measurements, as shown in Figure 5d. Nearly 87% enhancement of TXR emission and 16% quenching of FAM emission were observed within 4 h after the addition of the target under room temperature.

A unique advantage of the unzipping mechanism is that different kinds of zipper elements can be introduced into one origami device to work together as an AND logic gate. For example, DNA pliers bearing two miR20-targeting elements (at sites *a* and *c*) and two miR16-targeting elements (at *b* and *d*) did not open until both targets (miR20 AND miR16) were added to the solution (Fig. 5e). Origami devices can therefore handle multiple targets within a single device molecule, which may not be easily achievable with simple DNA tweezers.

The unzipping mechanism is also appropriate for the detection of small molecules using zipper elements based on DNA (or RNA) aptamers (Fig. 5b). When ATP-aptamer-based elements, which bind to two ATP molecules by forming an intrastrand stem-loop structure[40,41], were attached to DNA pliers in combination with corresponding closer elements, 45% of the parallel closed pliers initially formed after annealing were unzipped after the addition of ATP to the solution and were imaged as cross or antiparallel forms (Fig. 5f). As expected, the addition of GTP to the solution did not trigger unzipping of the parallel closed pliers.

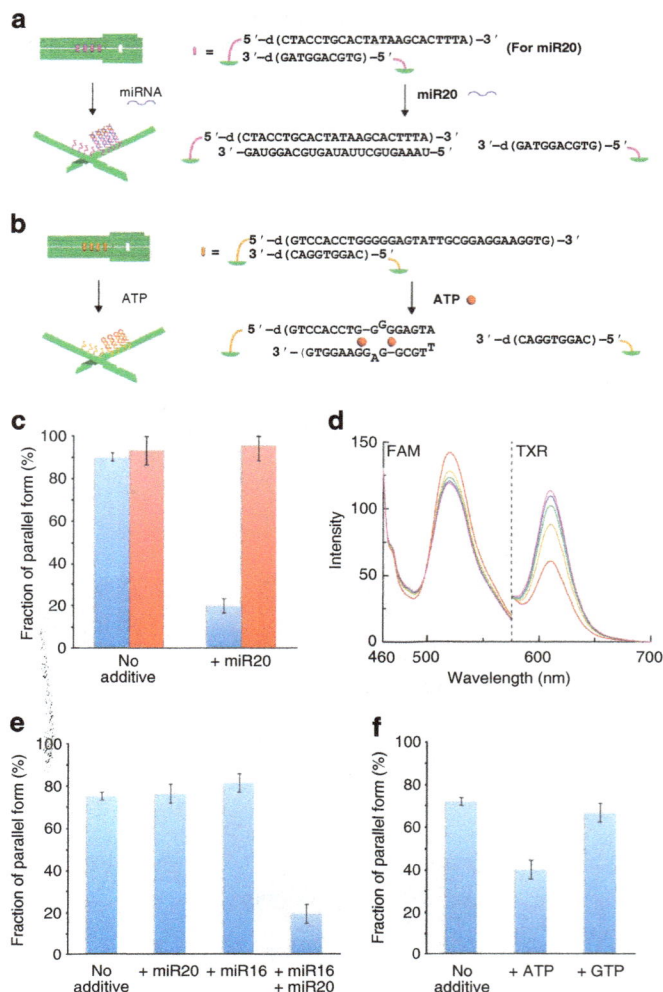

Figure 5 | Detection of small molecules by unzipping mechanism.
Detection of miRNA was performed using a selective strand displacement technique (**a**). ATP detection was done using aptamer-based elements (**b**). (**c**) Selective detection of human miR20 (final concentration, 200 nM) with miR20-targeting DNA pliers (blue bars) in a 1:1 mixture with miR16-targeting DNA forceps (red bars). The error bars indicate the standard errors ($n = 80$ and 55 for DNA pliers before and after miR20 addition, respectively, and $n = 44$ for DNA forceps in both of the systems). (**d**) Fluorescence spectra of dual-labelled miR20-targeting DNA pliers before (red) and 1 (orange), 2 (green), 3 (blue), and 4 (magenta) hours after miR20 addition at room temperature. (**e**) AND detection of two miRNA targets using miR20 and miR16 dual-targeting DNA pliers. Addition of only one of the two targets to the solution did not alter the fraction of parallel closed pliers ($n = 127$ for miR20 and 113 for miR16) from the value before addition ($n = 121$). Only when both targets were added was a significant decrease of the fraction observed ($n = 137$). The error bars indicate the standard errors. (**f**) ATP detection by DNA pliers bearing aptamer-based zipper elements. The initial yield of parallel pliers after annealing was 72% ($n = 199$). After ATP was added to the solution (final concentration, 1 mM), the yield decreased to 40% ($n = 217$), whereas the addition of GTP did not affect the yield significantly (66%, $n = 126$). The error bars indicate the standard errors.

Discussion

We have successfully applied nanomechanical DNA origami devices to construct visual 'single-molecule beacons' that can detect various targets of a wide range of molecular weights, from metal ions (a few tens of Da) to proteins (hundreds of kDa), at molecular resolution using the same platform. Each of the detection mechanisms we have described—pinching, zipping, or unzipping—for the targets of interest can be freely chosen and used orthogonally on differently shaped origami devices in a single mixture.

Comparison between the zipping experiments with Na^+ and TeloE (Fig. 4c) and preclosure of miR20-triggered unzipping systems (Supplementary Fig. S3) may hint at the border between targets appropriate for pinching and those require zipping. When only one zipper-element pair was introduced into DNA pliers, Na^+ with one TeloE gave ca. 40% closure of DNA pliers. In contrast, the yield of preclosed DNA pliers with only one preclosing zipper-elements pair was as low as 15% (whereas the parallel fraction without any zipper element was around 6%). Increasing the number of zipper-element pairs to two resulted in almost the same yield, around 50%, for both systems. Thus, the border should not be far from the zipper elements. The melting temperature of the TeloE dimer in the presence of 100 mM Na^+ is 42 °C, whereas the melting temperature for the 10-bp complementary part of the miR20 targeting elements is 31 °C under the conditions employed in the study (12.5 mM Mg^{2+} and 4 nM strand concentration, data not shown) or $\Delta G^{\circ}_{37} = -11.04$ kcal mol^{-1} (1 M Na^+)[42].

The maximum yield of parallel form in a solution containing a single kind of DNA origami device was ~80%, even for the preclosed DNA pliers used in unzipping systems, which were directly annealed into the closed form whereas the whole lever structure was in the folding process (Supplementary Fig. S3). This maximum yield is rather low compared with the ~90% folding yield reported for simple rectangular DNA origami[18], but it is still high considering the relatively small number of linkages, including an unfavourable parallel four-way junction, joining the two levers.

The observation of origami devices using AFM is a completely single-molecule method. Therefore, the theoretical detection limit of the systems should be exceedingly small, if the reaction volume is further reduced with the aid of microfluidics and MEMS technology. For example, a 4 nM solution of DNA origami devices, a typical condition used in the present study, corresponds to 1 origami device molecule in every 0.4 femtolitre ($= \mu m^3$), which contains 2 target molecules for SA and IgG detection, 50 molecules for 200 nM miRNA, or 2,500 atoms for 10 μM Ag^+.

Moreover, almost any kind of protein or small molecule can be a target of DNA origami devices by inverting the polarity of the pinching systems from an antigen-modified origami device/antibody combination to an antibody-modified origami device/antigen combination. Taking advantage of the rather large size of the origami devices, the capture and detection of whole virus capsids is feasible too[43]. The development of allosteric metaenzymes by attachment of another functional nanomaterial such as an enzyme to the other end of the levers, enabling the switching of their activity by mechanical movement, would be another interesting application of DNA pliers. Such metaenzymes may provide an extra detection pathway for the structural changes of DNA origami devices: chemical signal amplification, which is popularly employed today, for example, in enzyme immunoassays. The present system may be a first step toward powerful tools in future studies of various nano-biochemical interactions.

Methods

Materials. Staple DNA strands and dye-labelled staples were purchased from Sigma Genosys and used without further purification. Biotin-TEG- and FAM-modified anchor strands were chemically synthesized using the appropriate CPG columns (Glen Research) and were purified by reverse-phase HPLC. AFM imaging was performed on a SPA-300HV system (SII).

Preparation of nanomechanical DNA origami devices. Detailed structures of nanomechanical DNA origami devices used in the present study are shown in Supplementary Figures S5–S10, and the sequences of the staple strands are shown in Supplementary Tables S1–S3. The formation of DNA origami devices was accomplished with M13mp18 ssDNA (4 nM, Takara), staples, and anchor

strands or zipper elements (16 nM for each strand) in a solution containing Tris (40 mM), acetic acid (20 mM), EDTA (10 mM), and magnesium acetate (12.5 mM, 1×TAE/Mg buffer, 50 μl). This mixture was cooled from 90 to 25 °C at a rate of −1.0 °C min^{-1} using a PCR thermal cycler to anneal the strands.

Induction of the shape transitions. After excess staples and anchors were removed from the mixture using an ultrafiltration microtube (Amicon Ultra 0.5 ml–100 K, Millipore), 2 eq. (or 1 eq. for fluorescent measurements) of SA to DNA origami devices (final concentration 4 nM) was added to the solution and immediately subjected to AFM measurement. The selective release of SA via strand displacement was performed using 10 eq. of unset strands at room temperature for 2 h. Pinching of IgG was performed identically to SA pinching, except that the mixture was incubated at room temperature for 24 h before AFM imaging. The closing of DNA pliers using a G-quadruplex zipper was initiated by adding 1/10 volume of 2 M NaCl to the origami solution. The post-opening of DNA pliers by buffer exchange was achieved by concentrating the mixture with the ultrafiltration microtube and re-suspending it in 1×TAE/Mg^{2+} buffer three times over 1 h at room temperature (the calculated reduction factor of Na^+ was 1/11,000). The sensing of K^+ and Ag^+ was done in a similar manner, but 1X TA/Mg^{2+} buffer was used. The final concentration of KCl was 100 mM, and the concentration of $AgNO_3$ was 10 μM. The unzipping of preclosed miRNA-targeting DNA pliers was done in the presence of target miRNA (200 nM, 12.5 eq. to each zipper element) at room temperature overnight. The detection of ATP was performed using the appropriate zipper elements and 1 mM ATP. After ATP was added to the solution of DNA pliers, the mixture was cooled from 37 to 25 °C at a rate of −1.0 °C min^{-1} three times using a PCR thermal cycler.

AFM imaging. A mixture containing DNA origami devices and the target (3 μl) was deposited on freshly cleaved mica; extra 1×TAE/Mg^{2+} buffer (200 μl) was added; and imaging was performed in the fluid DFM scanning mode with a BL-AC40TS tip (Olympus). Typical zoom-out images are shown in Supplementary Figures S11–S18. DNA origami devices in an image were counted as the cross form, when both of the ends were clearly separated AND the levers around the fulcrum were clearly not laid in parallel. They were counted as the parallel form when at least one of the ends was clearly identified to be in head-to-head (the end of a lever close to the concavity) or tail-to-tail (the opposite end of the lever) contact, or into antiparallel form when head-to-tail contact was clearly observed for at least one of the ends. Devices not in any of the above conditions were counted as unclear motifs. The classification was confirmed by at least three persons. Counted numbers of the motifs for each experiment are shown in Supplementary Tables S4–S14.

Fluorescence measurements. A total of 500 μl of doubly dye-labelled DNA origami solution was annealed and concentrated to 50 μl using an ultrafiltration microtube. The final concentration of DNA pliers was 40 nM. The excitation wavelengths for the FAM and TXR spectra were 450 and 550 nm, respectively.

References

1. Seeman, N. C. Nucleic-acid junctions and lattices. *J. Theor. Biol.* **99**, 237–247 (1982).
2. Li, H., Carter, J. D. & LaBean, T. H. Nanofabrication by DNA self-assembly. *Mater. Today* **12**, 24–32 (2009).
3. Bath, J. & Turberfield, A. J. DNA nanomachines. *Nature Nanotech.* **2**, 275–284 (2007).
4. Yurke, B., Turberfield, A. J., Mills, A. P., Simmel, F. C. & Neumann, J. L. A DNA-fuelled molecular machine made of DNA. *Nature* **406**, 605–608 (2000).
5. Elbaz, J., Moshe, M. & Willner, I. Coherent activation of DNA layers: A 'SET-RESET' logic system. *Angew. Chem. Int. Ed.* **48**, 3834–3837 (2009).
6. Zhou, M., Liang, X., Mochizuki, T. & Asanuma, H. A light-driven DNA nanomachine for the efficient photoswitching of RNA digestion. *Angew. Chem. Int. Ed.* **49**, 2167–2170 (2010).
7. Shen, W. Q., Bruist, M. F., Goodman, S. D. & Seeman, N. C. A protein-driven DNA device that measures the excess binding energy of proteins that distort DNA. *Angew. Chem. Int. Ed.* **43**, 4750–4752 (2004).
8. Gu, H., Yang, W. & Seeman, N. C. DNA scissors device used to measure MutS binding to DNA mis-pairs. *J. Am. Chem. Soc.* **132**, 4352–4357 (2010).
9. Sherman, W. B. & Seeman, N. C. A precisely controlled DNA biped walking device. *Nano Lett.* **4**, 1203–1207 (2004).
10. Shin, J.- S. & Pierce, N. A. A synthetic DNA walker for molecular transport. *J. Am. Chem. Soc.* **126**, 10834–10835 (2004).
11. Yin, P., Yan, H., Daniell, X. G., Turberfield, A. J. & Reif, J. H. A unidirectional DNA walker that moves autonomously along a track. *Angew. Chem. Int. Ed.* **43**, 4906–4911 (2004).
12. Tian, Y., He, Y., Chen, Y., Yin, P. & Mao, C. D. Molecular devices—A DNAzyme that walks processively and autonomously along a one-dimensional track. *Angew. Chem. Int. Ed.* **44**, 4355–4358 (2005).
13. Omabegho, T., Sha, R. & Seeman, N. C. A bipedal DNA brownian motor with coordinated legs. *Science* **324**, 67–71 (2009).
14. Lee, J. B. *et al.* DNA-based nanostructures for molecular sensing. *Nanoscale* **2**, 188–197 (2010).

15. Li, D., Song, S. & Fan, C. Target-responsive structural switching for nucleic acid-based sensors. *Acc. Chem. Res.* **43**, 631–641 (2010).
16. Tyagi, S. & Kramer, F. R. Molecular beacons: probes that fluoresce upon hybridization. *Nat. Biotechnol.* **14**, 303–308 (1996).
17. Wang, K. *et al.* Molecular engineering of DNA: molecular beacons. *Angew. Chem. Int. Ed.* **48**, 856–870 (2009).
18. Rothemund, P. W. K. Folding DNA to create nanoscale shapes and patterns. *Nature* **440**, 297–302 (2006).
19. Kuzuya, A. & Komiyama, M. DNA origami: fold, stick, and beyond. *Nanoscale* **2**, 310–322 (2010).
20. Shih, W. M. & Lin, C. Knitting complex weaves with DNA origami. *Curr. Opin. Struct. Biol.* **20**, 276–282 (2010).
21. Nangreave, J., Han, D., Liu, Y. & Yan, H. DNA origami: a history and current perspective. *Curr. Opin. Chem. Biol.* **14**, 608–615 (2010).
22. Ke, Y., Lindsay, S., Chang, Y., Liu, Y. & Yan, H. Self-assembled water-soluble nucleic acid probe tiles for label-free RNA hybridization assays. *Science* **319**, 180–183 (2008).
23. Endo, M., Katsuda, Y., Hidaka, K. & Sugiyama, H. Regulation of DNA methylation using different tensions of double strands constructed in a defined DNA nanostructure. *J. Am. Chem. Soc.* **132**, 1592–1597 (2010).
24. Voigt, N. V. *et al.* Single-molecule chemical reactions on DNA origami. *Nature Nanotech.* **5**, 200–203 (2010).
25. Andersen, E. S. *et al.* Self-assembly of a nanoscale DNA box with a controllable lid. *Nature* **459**, 73–76 (2009).
26. Kuzuya, A. & Komiyama, M. Design and construction of a box-shaped 3D-DNA origami. *Chem. Commun.* 4182–4184 (2009).
27. Gu, H., Chao, J., Xiao, S.-J. & Seeman, N. C. A proximity-based programmable DNA nanoscale assembly line. *Nature* **465**, 202–205 (2010).
28. Lund, K. *et al.* Molecular robots guided by prescriptive landscapes. *Nature* **465**, 206–210 (2010).
29. Wickham, S. F. J. *et al.* Direct observation of stepwise movement of a synthetic molecular transporter. *Nature Nanotech.* **6**, 166–169 (2011).
30. Murchie, A. I. H. *et al.* Fluorescence energy transfer shows that the four-way DNA junction is a right-handed cross of antiparallel molecules. *Nature* **341**, 763–766 (1989).
31. Mao, C. D., Sun, W. Q. & Seeman, N. C. Designed two-dimensional DNA Holliday junction arrays visualized by atomic force microscopy. *J. Am. Chem. Soc.* **121**, 5437–5443 (1999).
32. Khuu, P. A., Voth, A. R., Hays, F. A. & Ho, P. S. The stacked-X DNA Holliday junction and protein recognition. *J. Mol. Recognit.* **19**, 234–242 (2006).
33. Kuzuya, A., Numajiri, K. & Komiyama, M. Accommodation of a single protein guest in nanometer-scale wells embedded in a (DNA Nanotape). *Angew. Chem. Int. Ed.* **47**, 3400–3402 (2008).
34. Kuzuya, A. *et al.* Precisely programmed and robust 2D streptavidin nanoarrays by using periodical nanometer-scale wells embedded in DNA origami assembly. *ChemBioChem* **10**, 1811–1815 (2009).
35. Numajiri, K., Kimura, M., Kuzuya, A. & Komiyama, M. Stepwise and reversible nanopatterning of proteins on a DNA origami scaffold. *Chem. Commun.* **46**, 5127–5129 (2010).
36. Kong, D.-M., Ma, Y.-E., Guo, J.-H., Yang, W. & Shen, H.-X. Fluorescent sensor for monitoring structural changes of G-quadruplexes and detection of potassium ion. *Anal. Chem.* **81**, 2678–2684 (2009).
37. Vorlíčková, M., Chládková, J., Kejnovská, I., Fialová, M. & Kypr, J. Guanine tetraplex topology of human telomere DNA is governed by the number of (TTAGGG) repeats. *Nucleic Acids Res.* **33**, 5851–5860 (2005).
38. Phan, A. T. & Patel, D. J. Two-repeat human telomeric d(TAGGGTTAGGGT) sequence forms interconverting parallel and antiparallel G-quadruplexes in solution: distinct topologies, thermodynamic properties, and folding/unfolding kinetics. *J. Am. Chem. Soc.* **125**, 15021–15027 (2003).
39. Ono, A. *et al.* Specific interactions between silver(I) ions and cytosine–cytosine pairs in DNA duplexes. *Chem. Commun.* 4825–4827 (2008).
40. Huizenga, D. E. & Szostak, J. W. A DNA aptamer that binds adenosine and ATP. *Biochemistry* **34**, 656–665 (1995).
41. Lin, C. H. & Patel, D. J. Structural basis of DNA folding and recognition in an AMP-DNA aptamer complex: distinct architectures but common recognition motifs for DNA and RNA aptamers complexed to AMP. *Chem. Biol.* **4**, 817–832 (1997).
42. SantaLucia, J. A unified view of polymer, dumbbell, and oligonucleotide DNA nearest-neighbor thermodynamics. *Proc. Natl Acad. Sci. USA* **95**, 1460–1465 (1998).
43. Stephanopoulos, N. *et al.* Immobilization and one-dimensional arrangement of virus capsids with nanoscale precision using DNA origami. *Nano Lett.* **10**, 2714–2720 (2010).

Acknowledgements

This work was partially supported by a Grant-in-Aid for Specially Promoted Scientific Research (22000007), a Grant-in-Aid for Scientific Research (S) (22220001), and a Grant-in-Aid for Young Scientists (B) (22750144) from the Ministry of Education, Science, Sports, Culture and Technology, Japan. Support from the Global COE Program for Chemistry Innovation and from the Association for the Progress of New Chemistry is also acknowledged.

Author contributions

A.K. and M.K. conceived the project and wrote the paper. A.K. and Y.X. designed the structures. A.K., Y.S. and T.Y. performed sample preparation and AFM imaging. Y.X. measured melting temperatures. All the authors analysed the data.

Additional information

2

Rapid *in vivo* detection of isoniazid-sensitive *Mycobacterium tuberculosis* by breath test

Seong W. Choi[1,*], Mamoudou Maiga[2,*], Mariama C. Maiga[2], Viorel Atudorei[3], Zachary D. Sharp[3], William R. Bishai[2] & Graham S. Timmins[1]

There is urgent need for rapid, point-of-care diagnostic tools for tuberculosis (TB) and drug sensitivity. Current methods based on *in vitro* growth take weeks, while DNA amplification can neither differentiate live from dead organisms nor determine phenotypic drug resistance. Here we show the development and evaluation of a rapid breath test for isoniazid (INH)-sensitive TB based on detection of labelled N_2 gas formed specifically from labelled INH by mycobacterial KatG enzyme. *In vitro* data show that the assay is specific, dependent on mycobacterial abundance and discriminates between INH-sensitive and INH-resistant (S315T mutant KatG) TB. *In vivo*, the assay is rapid with maximal detection of $^{15}N_2$ in exhaled breath of infected rabbits within 5–10 min. No increase in $^{15}N_2$ is detected in uninfected animals, and the increases in $^{15}N_2$ are dependent on infection dose. This test may allow rapid detection of INH-sensitive TB.

[1] Department of Pharmaceutical Sciences, College of Pharmacy, University of New Mexico, Albuquerque, New Mexico 87131, USA. [2] Department of Medicine, Center for Tuberculosis Research, Johns Hopkins University, Baltimore, Maryland 21231, USA. [3] Department of Earth and Planetary Sciences, College of Pharmacy, University of New Mexico, Albuquerque, New Mexico 87131, USA. * These author contributed equally to this work. Correspondence and requests for materials should be addressed to G.S.T. (email: gtimmins@salud.unm.edu).

Bacterially activated prodrugs are unusually well represented among the first- and second-line TB drugs. These include not only established drugs such as isoniazid (INH)[1], ethionamide[2] or pyrazinamide[3] but also newly approved and developing agents such as the nitroimidazoles delamanid[4] and PA824 (ref. 5). The selectivity of these agents arises from their specific activation by mycobacterial enzymes, usually to reactive intermediates, and is underlined by the major mode of resistance to these agents, with mutations in genes of their activating enzymes such as katG for INH[6], ethA for ethionamide[7], pncA for pyrazinamide[8] and ddn for nitroimidazoles[9]. Since gene inactivation may occur through a multiplicity of single-nucleotide polymorphisms (SNPs) or insertion/deletion (indel) events, nucleic acid amplification and SNP-indel detection approaches provide only partially predictive drug susceptibility data. Beyond single-gene mutational resistance, multiple other alleles[10–14] and other drugs[15,16] may influence enzymatic activity of prodrug conversion, factors that may also limit nucleic acid-based techniques for drug susceptibility testing. Despite the importance of prodrug activation, studies have been limited to *in vitro* samples or bacterial culture, and at present there are no POC techniques to directly measure prodrug conversion and enzymatic activity.

The mycobacterial enzyme KatG, which is responsible for INH activation, produces a range of INH-derived radicals that react with cellular components, especially the isonicotinoyl acyl radical that adds covalently to NAD^+ and $NADP^+$. The adducts formed by these radicals are potent inhibitors of the key mycobacterial targets. The first target of such inhibition to be elucidated was 2-trans-enoyl-acyl carrier protein reductase (InhA) that binds INacyl-NAD^+ adducts, tightly inhibiting mycolic acid synthesis[17]. Although other targets or reactive species may play roles, the importance of these alternative mechanisms compared with the widely accepted inhibition of InhA remains unclear[18].

The detection of degradation products of the INacyl-NAD^+ adduct, such as 4-isonicotinoylnicotinamide in urine or other fluids held great promise as a measure of INH prodrug conversion in TB, and hence determining KatG activity[19]. However, this appears to lack specificity for *Mycobacterium tuberculosis* as 4-isonicotinoylnicotinamide was found in urine of uninfected mice treated with INH, and in urine of TB patients even when they were culture-negative after treatment[19].

Mycobacterial KatG activates INH by oxidation to a hydrazyl radical that undergoes beta scission to form isonicotinoyl acyl radical. The other product of this beta-scission reaction, diazene, has received little to no attention in the literature. To study diazene production in KatG-expressing mycobacteria, we used doubly $^{15}N_2$-hydrazyl-labelled INH (1) to produce doubly labelled diazene (Fig. 1a). Under physiologic conditions, this diazene rapidly undergoes either oxidation by unsaturated bonds (Fig. 1b)[20] or bimolecular disproportionation (Fig. 1c) to produce $^{15}N_2$ (ref. 21). Diazene is widely used synthetically in the stereospecific reduction of a wide range of carbon–carbon double bonds[22].

This $^{15}N_2$ produced from INH-derived diazene may be readily detected by isotope ratio mass spectrometry (IRMS), and its abundance is reported as $\delta^{15}N_2$ where

$$\delta^{15}N_2 = 1{,}000 \times \left[\left(^{15}N^{15}N / ^{14}N^{14}N \right)_{sample} \right.$$
$$\left. - \left(^{15}N^{15}N / ^{14}N^{14}N \right)_{standard} \right] / \left(^{15}N^{15}N / ^{14}N^{14}N \right)_{standard}$$

Atmospheric ^{15}N is much lower in abundance than ^{14}N ($\sim 0.36\%$), hence $^{15}N_2$ is very low in abundance (~ 13 p.p.m.); therefore, even small amounts of $^{15}N_2$ generation may be detected through changes in $\delta^{15}N_2$. For example, an increase in the value

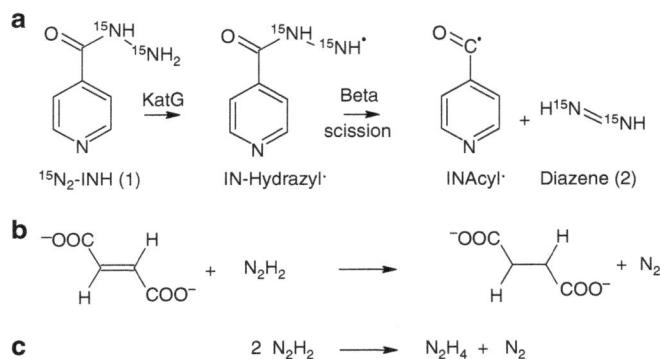

Figure 1 | Production of N₂ from KatG activation of isoniazid.
(**a**) Production of labelled diazene from $^{15}N_2$-hydrazyl-labelled INH; (**b**) oxidation of diazene to N_2 by reaction with unsaturated carbon bonds such as fumarate shown, rate constant $8 \times 10^2 M^{-1} s^{-1}$ (ref. 20); (**c**) disproportionation of diazene to N_2 and hydrazine rate constant $2.2 \times 10^4 M^{-1} s^{-1}$ (ref. 21).

of $\delta^{15}N_2$ of 250 would indicate a 25% increase in the absolute amount of $^{15}N_2$ in a sample. This same principle is exploited by other isotope ratio breath diagnostics including the urease breath test for *Helicobacter pylori* infection.

In this report, we describe the detection of $^{15}N_2$ products of INH activation that are specific for mycobacterial KatG, and test their specificity against other important lung bacterial pathogens that possess related peroxidase enzymes. By measuring the increase over baseline $\delta^{15}N_2$ upon addition of the $^{15}N_2$-hydrazyl INH (a method termed INH→N here), we hypothesized that IRMS detection of this $^{15}N_2$ may allow sensitive measurement of INH activation by KatG.

Results

In vitro cultures of *M. tuberculosis* H37Rv or *M. bovis* Bacillus Calmette–Guérin (BCG) were treated with $^{15}N_2$-hydrazyl INH in sealed tubes and portions of headspace gas collected, filtered and analysed. Treatment with 1 mg ml^{-1} $^{15}N_2$-hydrazyl INH resulted in marked increases in $\delta^{15}N_2$ that were dependent upon bacterial density (CFU ml^{-1}; Fig. 2a). Next, we determined the correlation between the accumulated $\delta^{15}N_2$ and the dose of $^{15}N_2$-hydrazyl INH administered (Fig. 2b), and these experiments showed sensitive IRMS detection of headspace $\delta^{15}N_2$ following $^{15}N_2$-hydrazyl INH doses of 0.1 mg ml^{-1}, a concentration we subsequently used throughout. The generation of headspace $\delta^{15}N_2$ occurred rapidly (Fig. 2c), and plateau levels were reached in ~ 1 h. Similar data were also observed with *M. bovis* BCG (Fig. 3), another KatG-expressing mycobacterial species, although generally lower levels of $^{15}N_2$ production were observed compared with *M. tuberculosis* H37Rv. These data confirmed our ability to measure mycobacterial KatG activity quantitatively with IRMS monitoring of conversion of $^{15}N_2$-hydrazyl INH to $^{15}N_2$ using *in vitro* cultures of mycobacteria.

We then evaluated the specificity of our $^{15}N_2$-hydrazyl INH to $^{15}N_2$ detection method for mycobacterial KatG activity. As may be seen in Fig. 4a, the common respiratory pathogens *Staphylococcus aureus*, *Pseudomonas aeruginosa* and *Escherichia coli* did not produce $^{15}N_2$ when treated with $^{15}N_2$-hydrazyl INH. To determine whether our $^{15}N_2$-hydrazyl INH to $^{15}N_2$ detection method for INH prodrug conversion was specific for the mycobacterial KatG, we tested the production of $^{15}N_2$ using an *M. tuberculosis* strain harbouring a mutated KatG. This strain possessed the *M. tuberculosis* katG-S315T mutation that is known to profoundly decrease INH activation and result in drug

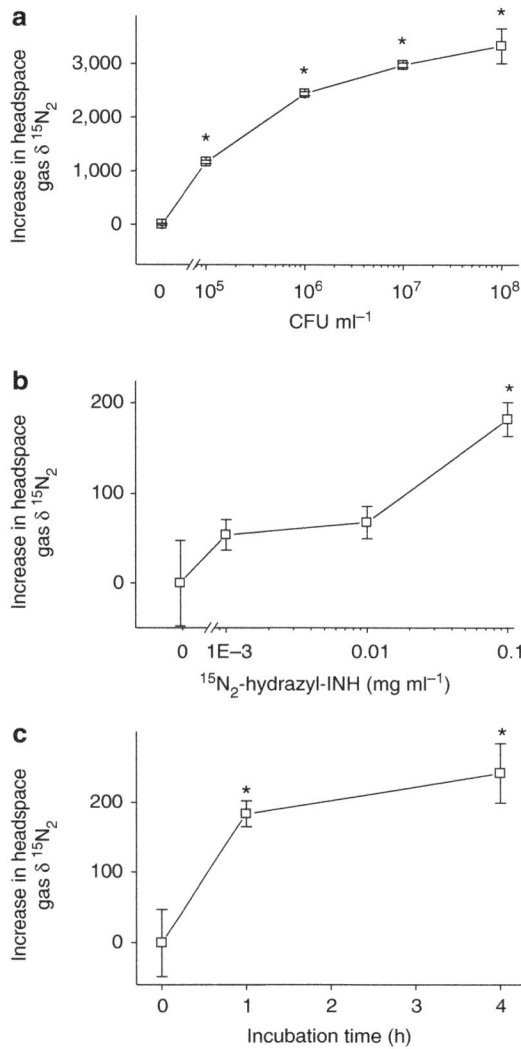

Figure 2 | CFU, dose and time dependence of $^{15}N_2$ production by M. tuberculosis H37Rv. Increased headspace $\delta^{15}N_2$ (mass 30) in $^{15}N_2$-hydrazyl INH-treated cultures. $^{15}N_2$ production was dependent upon (**a**) bacterial density of M. tuberculosis H37Rv (3 ml) incubated with $^{15}N_2$-hydrazyl INH (1 mg ml^{-1}) for 1h, *P < 0.001; (**b**) concentration of $^{15}N_2$-hydrazyl INH (H37Rv (10^8 CFU ml^{-1}, 3 ml) was incubated with $^{15}N_2$-hydrazyl INH at the indicated dose for 1h, *P < 0.001); (**c**) incubation time (H37Rv (10^8 CFU ml^{-1}, 3 ml) was incubated with $^{15}N_2$-hydrazyl INH (0.1 mg ml^{-1}) for the indicated time, *P < 0.001). Data represent mean ± s.d. of four separate biological replicates. One-way analysis of variance (ANOVA) with Tukey post hoc test.

Figure 3 | CFU, dose and time dependence of $^{15}N_2$ production by M. bovis BCG. Increased headspace δ^5N_2 (mass 30) in $^{15}N_2$-hydrazyl INH-treated cultures was dependent upon (**a**) bacterial density of M. bovis BCG (3 ml) incubated with 1 mg ml^{-1} of $^{15}N_2$-hydrazyl INH for 1h, *P < 0.001; (**b**) concentration of $^{15}N_2$-hydrazyl INH (M. bovis BCG (10^8 CFU ml^{-1}, 3 ml) was incubated with $^{15}N_2$-hydrazyl INH at the indicated dose for 1h, *P < 0.001); (**c**) incubation time (M. bovis BCG (10^8 CFU ml^{-1}, 3 ml) was incubated with $^{15}N_2$-hydrazyl INH (1 mg ml^{-1}) for the indicated time, *P < 0.001). Data represent mean ± s.d. of three separate biological replicates. One-way ANOVA with Tukey post hoc test.

resistance[6]. When compared with M. tuberculosis H37Rv, we found that the katG-S315T mutant did not produce any measurable $^{15}N_2$ (Fig. 4b).

These in vitro characteristics supported our hypothesis that the $^{15}N_2$-hydrazyl INH to $^{15}N_2$, INH → N detection method might be used to detect KatG activation of INH in vivo, using a breath test approach[23,24]. Rabbits were infected with high dose (10^4 CFU) or low dose (10^3 CFU) M. tuberculosis H37Rv using an inhalation exposure system (Glas-col) as previously described[25]. After a 6-week incubation period, rabbits were treated with 10 mg of $^{15}N_2$-hydrazyl INH instilled bronchoscopically. Direct delivery to the lung was chosen to rapidly expose lung bacteria to $^{15}N_2$-hydrazyl INH in order to allow rapid assay, as opposed to an oral dosage form which would require absorption and

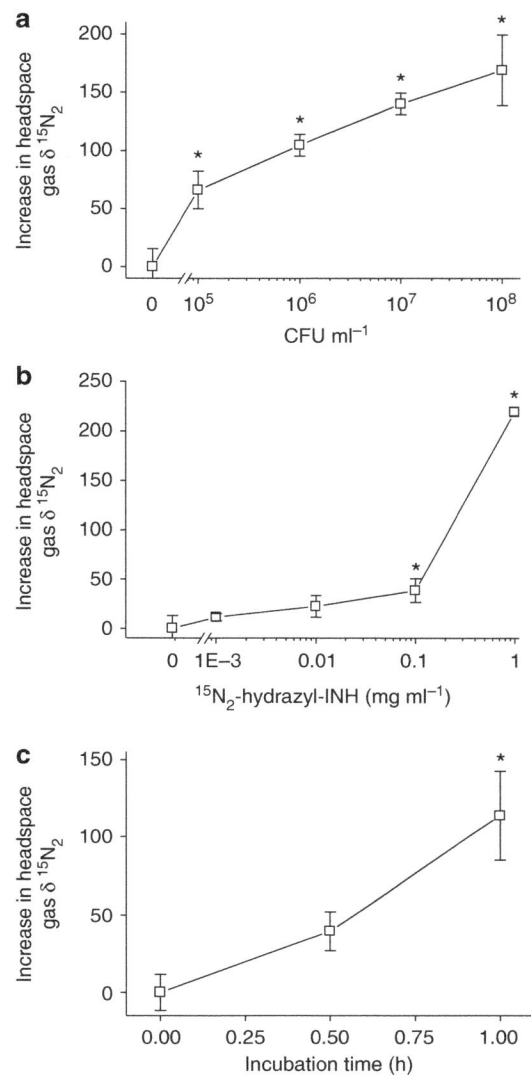

redistribution. Inhaled INH has been used clinically in humans[26]. Breath samples were collected before dosing, and then at 5, 10 and 20 min post dose. Four non-infected rabbits were used as control group.

It was seen that $\delta^{15}N_2$ increased rapidly in breath of all infected animals, with no observed increase in breath $\delta^{15}N_2$ of the four uninfected controls (Fig. 5a,b). The lack of signal in uninfected animals, together with significant signals in all infected animals, suggests that a high degree of sensitivity and specificity is inherent in this assay. Breath $\delta^{15}N_2$ reached a maximum after 5–10 min, and then variably decreased, likely because of differential distribution and absorption of $^{15}N_2$-hydrazyl INH from the lung into systemic circulation from the more focal pattern of delivery arising from instillation. A relationship between peak levels of

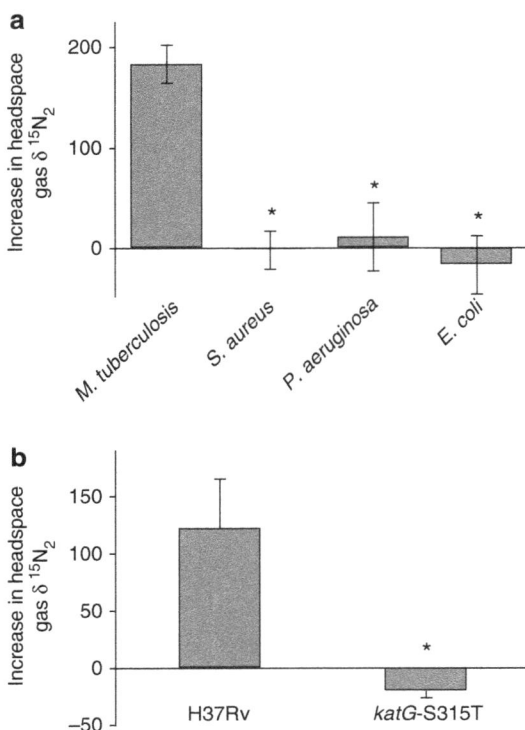

Figure 4 | Specificity of $^{15}N_2$ production. (a) Increased headspace $\delta\ ^{15}N_2$ in $^{15}N_2$-hydrazyl INH-treated overnight cultures of *S. aureus*, *P. aeruginosa* and *E. coli* compared with *M. tuberculosis* H37Rv. Bacterial culture (10^8 CFU ml^{-1}, 3 ml) was incubated with ^{15}N-INH (0.1 mg ml^{-1}) for 1 h. Data represent mean ± s.d. ($n = 3$ biological replicates). Student's *t*-test, *$P < 0.001$. (b) Comparison in headspace $\delta^{15}N_2$ in $^{15}N_2$-hydrazyl INH-treated drug-sensitive *M. tuberculosis* H37Rv, and an INH-resistant KatG mutant strain (*katG*-S315T). H37Rv or *katG*-S315T strains (10^8 CFU ml^{-1}, 3 ml) were incubated with $^{15}N_2$-hydrazyl INH (0.1 mg ml^{-1}) for 1 h. Data represent mean ± s.d. ($n = 4$ biological replicates). Student's *t*-test, *$P < 0.005$.

$\delta^{15}N_2$ and lung CFU was observed (Fig. 6a,b reflecting $\delta^{15}N_2$ as a function of lung CFU at the time of killing and initial infective CFU, respectively). This suggests that the approach might be sensitive to the amount of lung mycobacteria present, although significant further work is needed to delineate this relationship. Repetitive use of the technique is also likely to be complicated by the highly bactericidal nature of inhaled INH, and for monitoring of bacterial load other techniques such as urease breath tests[23] or sputum CFU may be more useful.

Discussion

The INH→N detection method for mycobacterial KatG activity described here is capable of discriminating between INH-susceptible and -resistant *M. tuberculosis* and between KatG-expressing mycobacteria and other common lung pathogens *in vitro*. It is also capable of rapidly discriminating between controls and animals infected with INH-susceptible TB. Potential advantages of the INH→N test are the rapid non-radioactive breath test approach, based upon detecting prodrug activation, and that samples the entire lung. The readout of this test, $^{15}N_2$, is detected using IRMS, and portable MS detection devices are available and under development[27], supporting eventual development into a POC technology. Residual gas analyser MS, a technique with great potential for portability, has recently been shown to be effective in clinical IRMS[28], and represents one

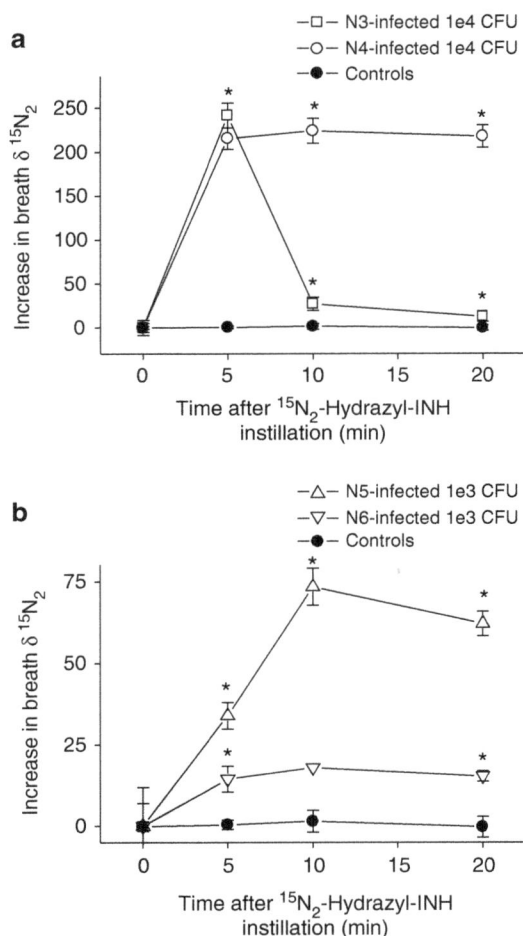

Figure 5 | *In vivo* $^{15}N_2$ production in TB-infected and control rabbits. Increased breath in (a) high-dose- and (b) low-dose TB-infected rabbits. Rabbits were infected with high or low doses of *M. tuberculosis* H37Rv, instilled with 10 mg $^{15}N_2$-hydrazyl INH and breath-collected. Rabbits (pathogen-free outbred New Zealand White) were infected with the indicated CFU by aerosol. At week 6, rabbits were anaesthetized with ketamine (15–25 mg kg^{-1}) and xylazine (5–10 mg kg^{-1}), and treated with $^{15}N_2$-hydrazyl INH (10 mg kg^{-1} in 0.4 ml PBS) by intratracheal intubation. Exhaled breath gas (12 ml) was collected into Helium gas-flushed tubes at 0, 5, 10 and 20 min post $^{15}N_2$-hydrazyl INH administration. Data represent mean ± s.d. ($n = 3$ repeats for each rabbit N3 through N6, $n = 12$ for four uninfected control rabbits with three repeats for each). Two-way mixed ANOVA with Bonferroni *post hoc* test. *$P < 0.001$. See detailed rabbit data in Table 1.

avenue forward. As with any new potential diagnostic approach, ultimate clinical usage and utility must be determined in trials.

Clinically, high-level INH resistance is strongly correlated to *katG*-S315T mutations with greatly lowered INH-activating (and INH→N) activity, whereas lower-level resistance is associated with *inhA* promoter mutations that will likely not be differentiated from INH-sensitive strains by the INH→N test[29]. However, INH→N assay would allow rapid point-of-care detection of *katG*-S315T and other katG mutations as part of a diagnostic approach, to enable rapid and optimal therapy. The potential for the INH→N method to report as a rapid and specific biomarker of mycobacterial load may provide useful tool for monitoring clinical trials and therapeutic efficacy. INH→N may also prove useful in diagnosis of some non-tuberculous mycobacteria, such as INH-sensitive *M. kansasii*[30] that can otherwise be challenging. However, since some peroxidases other than mycobacterial KatG enzymes bind INH (such as lactoperoxidase[31]), further studies of specificity are planned.

Similar approaches may also be extended to other TB prodrug classes so that effective and rapid detection of drug sensitivity/ resistance through prodrug conversion can guide therapy. One example would be Delamanid and PA824 that are activated to bactericidal NO by mycobacterial Ddn[5]: using ^{15}N-nitro-PA824 would result in ^{15}NO $\cdot$ that could be directly detected in breath, or as ^{15}N-nitrate/nitrite in other samples such as blood or urine. This could provide rapid detection of drug activation (and hence sensitivity) in patients when conventional techniques such as MS detection of des-nitro-PA824 are difficult (Clif Barry, personal communication). This would allow optimal use of these drugs in therapy of multidrug resistant and extensively drug-resistant disease.

More generally, while pathogen genotypes are rapidly determined without culture, the study of bacterial phenotypes in the host (as opposed to culture in which it can greatly change) is extremely challenging. However, the broad importance of phenotype and phenotype variance in pathogenesis is becoming increasingly appreciated, with specific examples of both growth phase-dependent[32] and stochastic[14] isoniazid-resistant phenotypes being recently elucidated. The ability to determine bacterial phenotypes through stable isotope detection of specific bacterial metabolic pathways without requiring culture could prove broadly valuable in complementing genomic approaches in studying microbiomes. Finally, it is worth noting that yet another reactive species from mycobacterial KatG activation of INH, in this case diazene, could play a role in INH action through reducing the key unsaturated mycobacterial molecules.

Methods

Bacterial cultures. *M. tuberculosis* H37Rv (H37Rv), *M. bovis* BCG, *E. coli* DH5α and *P. aeruginosa* PAO1 were gifts from Professor Vojo Deretic[33,34], *M. tuberculosis katG*-S315T (*katG*-S315T) was a gift from Professor Alex Pym[35]. *S. aureus* USA300 LAC was a gift from Professor Pamela Hall[36]. All bacterial cultures were grown at 37 °C with shaking. Mycobacterium cultures were prepared by thawing frozen stock aliquots: H37Rv and *katG*-S315T were grown in 7H9 Middlebrook liquid medium supplemented with oleic acid, albumin, dextrose and catalase (Becton Dickinson Inc., Sparks, MD, USA), 0.5% glycerol and 0.05% Tween 80. *BCG* was grown in the same culture medium omitting oleic acid. *E. coli* DH5α was grown overnight in LB broth (Becton Dickinson), *P. aeruginosa* strain PAO1 was grown overnight in LB broth supplemented with 1.76% NaCl and 1% glycerol, and *S. aureus* USA300 LAC was grown overnight in BBL Trypticase soybroth (Becton Dickinson).

In vitro KatG assay. Three millilitres of mycobacterial cultures (BCG, H37Rv or *katG*-S315T) were diluted as appropriate from week-old cultures, while other bacterial cultures (*P. aeruginosa*, *E. coli* or *S. aureus*) were diluted from overnight cultures. The 3-ml cultures were shaken aerobically and then were incubated with ^{15}N$_2$-hydrazyl INH (at 0.1 mg ml^{-1} unless noted) in 12 ml Exetainer vials (Labco Ltd., Ceredigion, UK) for 1 h at 37 °C with shaking at 250 r.p.m. unless otherwise indicated. Collected headspace gas (1 ml) was filtered through 0.25-µ syringe filters and transferred into Helium-flushed Exetainers.

Measurement of ^{15}N$_2$ conversion. Sampled gas was analysed for ^{15}N enrichment in headspace N$_2$ by gas IRMS (Delta^{plus}XL, Thermo Scientific Inc., Waltham, MA, USA). Samples were separated by GC immediately upstream of their inlet into the IRMS using a 30-m column packed with 5 Å molecular sieves operating at 60 °C and using ultrahigh purity helium as carrier gas. IRMS of the N$_2$ peak measured relative ratio of mass 30 ^{15}N$_2$ versus mass 28 ^{14}N$_2$. Nitrogen gas of purity > 99.99% (Matheson Tri-Gas, Albuquerque, NM, USA) was used as the reference gas.

Animal experiments. These consisted of four uninfected control rabbits, two rabbits infected at high dose (N3 and N4) and two rabbits infected at low dose (N5 and N6). Rabbits (females, 16–20 weeks old, 3.5–4 kg pathogen-free outbred New Zealand White, Robinson Services Inc., Mocksville, NC, USA) were aerosol-infected with *M. tuberculosis* H37Rv at either 10^3 or 10^4 CFUs using an inhalation exposure system as previously described[37,38] (Glas-col, Terre Haut, IN, USA). At week 6, rabbits were anaesthetized with ketamine 15–25 mg kg^{-1} and xylazine 5–10 mg kg^{-1}, and then 10 mg ^{15}N$_2$-hydrazyl INH in 0.4-ml saline was instilled using intratracheal insertion through an endotracheal tube. To collect breath gas, a 14-French feeding tube connected to a 30-ml syringe was introduced through the endotracheal tube into the level of the carina to aspirate the exhaled air when the rabbit is breathing out. Breath gas (12 ml) was filtered with a 0.35-µ filter into Helium-flushed tubes before and after ^{15}N$_2$-hydrazyl INH treatment at 0, 5, 10 and 20 min. ^{15}N$_2$ enrichment in breath gas was measured by IRMS. Immediately after breath testing, the animals were killed, and lung weight and CFU measured (Table 1. Rabbits were killed with intravenous euthasol (Virbac Corporation, Fort Worth, TX, USA). The rabbit model was chosen as it is the smallest model that enables ready endoscopic infection, instillation of INH and collection of breath.

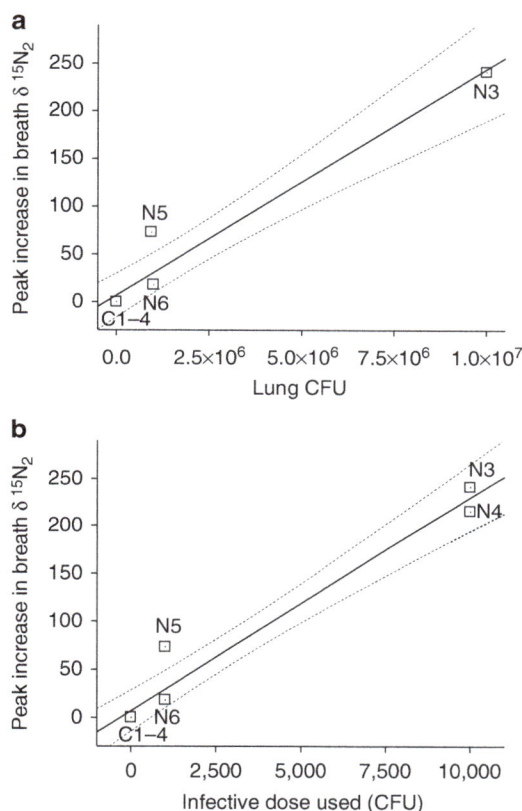

Figure 6 | Dependence of peak increase in breath δ ^{15}N$_2$ upon infection level. The maximal increase in breath δ ^{15}N$_2$ after ^{15}N$_2$-hydrazyl INH delivery in Fig. 5 is plotted here as a function of (**a**) lung CFU determined at the time of killing, and (**b**) initial infective dose delivered. Rabbit identity numbers are shown, and 95% confidence limits are presented as dashed lines.

Table 1 | Details of TB-infected rabbits.

| | Rabbit ID no. | | | |
	N3	**N4**	**N5**	**N6**
Inoculum size	10^4 CFU	10^4 CFU	10^3 CFU	10^3 CFU
CFU per g lung at the time of killing	1.8 × 10^5	ND	7.6 × 10^4	7.4 × 10^4
Lung weight (g)	58	ND	12.5	14.1
Total lung CFU	1 × 10^7	ND	9.4 × 10^5	1 × 10^6

CFU, colony-forming unit; ND, not performed; TB, tuberculosis.

Ethic statement. Animal work in this study was carried out in strict accordance with the recommendations in the Guide for the Care and Use of Laboratory Animals of the National Institutes of Health, the Animal Welfare Act and US federal law. The protocol was approved by the Institutional Animal Care and Use Committees at Johns Hopkins University (RB11M466).

Statistical analysis. All statistical analyses were performed using SPSS version 19 (SPSS Inc., Chicago, IL, USA). P values were determined using analysis of variance (ANOVA) and Student's t-test, and values <0.05 were considered statistically significant.

References

1. Zhang, Y., Heym, B., Allen, B., Young, D. & Cole, S. The catalase-peroxidase gene and isoniazid resistance of *Mycobacterium tuberculosis*. *Nature* **358**, 591–593 (1992).
2. Baulard, A. R. *et al.* Activation of the pro-drug ethionamide is regulated in mycobacteria. *J. Biol. Chem.* **275**, 28326–28331 (2000).
3. Zhang, Y. & Mitchison, D. The curious characteristics of pyrazinamide: a review. *Int. J. Tuberc. Lung Dis.* **7**, 6–21 (2003).
4. Matsumoto, M. *et al.* OPC-67683, a nitro-dihydro-imidazooxazole derivative with promising action against tuberculosis *in vitro* and in mice. *PLoS Med.* **3**, e466 (2006).
5. Singh, R. *et al.* PA-824 kills nonreplicating Mycobacterium tuberculosis by intracellular NO release. *Science* **322**, 1392–1395 (2008).
6. Heym, B., Alzari, P. M., Honore, N. & Cole, S. T. Missense mutations in the catalase-peroxidase gene, katG, are associated with isoniazid resistance in *Mycobacterium tuberculosis*. *Mol. Microbiol.* **15**, 235–245 (1995).
7. Morlock, G. P., Metchock, B., Sikes, D., Crawford, J. T. & Cooksey, R. C. ethA, inhA, and katG loci of ethionamide-resistant clinical *Mycobacterium tuberculosis* isolates. *Antimicrob. Agents Chemother.* **47**, 3799–3805 (2003).
8. Scorpio, A. & Zhang, Y. Mutations in pncA, a gene encoding pyrazinamidase/nicotinamidase, cause resistance to the antituberculous drug pyrazinamide in tubercle bacillus. *Nat. Med.* **2**, 662–667 (1996).
9. Manjunatha, U. H. *et al.* Identification of a nitroimidazo-oxazine-specific protein involved in PA-824 resistance in *Mycobacterium tuberculosis*. *Proc. Natl Acad. Sci. USA* **103**, 431–436 (2006).
10. Zahrt, T. C., Song, J., Siple, J. & Deretic, V. Mycobacterial FurA is a negative regulator of catalase-peroxidase gene katG. *Mol. Microbiol.* **39**, 1174–1185 (2001).
11. Pagan-Ramos, E., Song, J., McFalone, M., Mudd, M. H. & Deretic, V. Oxidative stress response and characterization of the oxyR-ahpC and furA-katG loci in *Mycobacterium marinum*. *J. Bacteriol.* **180**, 4856–4864 (1998).
12. Lucarelli, D., Vasil, M. L., Meyer-Klaucke, W. & Pohl, E. The metal-dependent regulators FurA and FurB from *Mycobacterium tuberculosis*. *Int. J. Mol. Sci.* **9**, 1548–1560 (2008).
13. Master, S., Zahrt, T. C., Song, J. & Deretic, V. Mapping of *Mycobacterium tuberculosis* katG promoters and their differential expression in infected macrophages. *J. Bacteriol.* **183**, 4033–4039 (2001).
14. Wakamoto, Y. *et al.* Dynamic persistence of antibiotic-stressed mycobacteria. *Science* **339**, 91–95 (2013).
15. Flipo, M. *et al.* Ethionamide boosters. 2. Combining bioisosteric replacement and structure-based drug design to solve pharmacokinetic issues in a series of potent 1,2,4-oxadiazole EthR inhibitors. *J. Med. Chem.* **55**, 68–83 (2012).
16. Frenois, F., Engohang-Ndong, J., Locht, C., Baulard, A. R. & Villeret, V. Structure of EthR in a ligand bound conformation reveals therapeutic perspectives against tuberculosis. *Mol. Cell* **16**, 301–307 (2004).
17. Rozwarski, D., Grant, G., Barton, D., Jacobs, W. & Sacchettini, J. Modification of the NADH of the isoniazid target (InhA) from *Mycobacterium tuberculosis*. *Science* **279**, 98–102 (1998).
18. Wang, F. *et al.* Mycobacterium tuberculosis dihydrofolate reductase is not a target relevant to the antitubercular activity of isoniazid. *Antimicrob. Agents Chemother.* **54**, 3776–3782 (2010).
19. Mahapatra, S. *et al.* A novel metabolite of antituberculosis therapy demonstrates host activation of isoniazid and formation of the isoniazid-NAD+ adduct. *Antimicrob. Agents Chemother.* **56**, 28–35 (2012).
20. Tang, H. R., Mckee, M. L. & Stanbury, D. M. Absolute rate constants in the concerted reduction of olefins by diazene. *J. Am. Chem. Soc.* **117**, 8967–8973 (1995).
21. Tang, H. R. & Stanbury, D. M. Direct-detection of aqueous diazene - its uv spectrum and concerted dismutation. *Inorg. Chem.* **33**, 1388–1391 (1994).
22. Pasto, D. J. & Taylor, R. T. Reduction with diimide. *Org. React.* **40**, 91–155 (1991).
23. Jassal, M. S. *et al.* 13[C]-urea breath test as a novel point-of-care biomarker for tuberculosis treatment and diagnosis. *PLoS ONE* **5**, e12451 (2010).
24. Maiga, M. *et al. In vitro* and *in vivo* studies of a rapid and selective breath test for tuberculosis based upon mycobacterial CO dehydrogenase. *mBio* **5**, e00990 (2014).
25. Be, N. A., Klinkenberg, L. G., Bishai, W. R., Karakousis, P. C. & Jain, S. K. Strain-dependent CNS dissemination in guinea pigs after *Mycobacterium tuberculosis* aerosol challenge. *Tuberculosis* **91**, 386–389 (2011).
26. Yokota, S. & Miki, K. [Effects of INH (Isoniazid) inhalation in patients with endobronchial tuberculosis (EBTB)]. *Kekkaku: [Tuberculosis]* **74**, 873–877 (1999).
27. Ouyang, Z., Noll, R. J. & Cooks, R. G. Handheld miniature ion trap mass spectrometers. *Anal. Chem.* **81**, 2421–2425 (2009).
28. Maity, A. *et al.* Residual gas analyzer mass spectrometry for human breath analysis: a new tool for the non-invasive diagnosis of *Helicobacter pylori* infection. *J. Breath Res.* **8**, 016005 (2014).
29. Dantes, R. *et al.* Impact of isoniazid resistance-conferring mutations on the clinical presentation of isoniazid monoresistant tuberculosis. *PLoS ONE* **7**, e37956 (2012).
30. Griffith, D. E. *et al.* An official ATS/IDSA statement: diagnosis, treatment, and prevention of nontuberculous mycobacterial diseases. *Am. J. Respir. Crit. Care Med.* **175**, 367–416 (2007).
31. Singh, A. K. *et al.* Mode of binding of the tuberculosis prodrug isoniazid to heme peroxidases: binding studies and crystal structure of bovine lactoperoxidase with isoniazid at 2.7 A resolution. *J. Biol. Chem.* **285**, 1569–1576 (2010).
32. Niki, M. *et al.* A novel mechanism of growth phase-dependent tolerance to isoniazid in mycobacteria. *J. Biol. Chem.* **287**, 27743–27752 (2012).
33. Curcic, R., Dhandayuthapani, S. & Deretic, V. Gene expression in mycobacteria: transcriptional fusions based on xylE and analysis of the promoter region of the response regulator mtrA from *Mycobacterium tuberculosis*. *Mol. Microbiol.* **13**, 1057–1064 (1994).
34. Schurr, M. J., Martin, D. W., Mudd, M. H. & Deretic, V. Gene cluster controlling conversion to alginate-overproducing phenotype in *Pseudomonas aeruginosa*: functional analysis in a heterologous host and role in the instability of mucoidy. *J. Bacteriol.* **176**, 3375–3382 (1994).
35. Pym, A. S., Saint-Joanis, B. & Cole, S. T. Effect of katG mutations on the virulence of *Mycobacterium tuberculosis* and the implication for transmission in humans. *Infect. Immun.* **70**, 4955–4960 (2002).
36. Hall, P. R. *et al.* Nox2 modification of LDL is essential for optimal apolipoprotein B-mediated control of agr type III *Staphylococcus aureus* quorum-sensing. *PLoS Pathog.* **9**, e1003166 (2013).
37. Converse, P. J. *et al.* The impact of mouse passaging of *Mycobacterium tuberculosis* strains prior to virulence testing in the mouse and guinea pig aerosol models. *PLoS ONE* **5**, e10289 (2010).
38. Barry, C. E. *et al.* The spectrum of latent tuberculosis: rethinking the biology and intervention strategies. *Nat. Rev. Microbiol.* **7**, 845–855 (2009).

Acknowledgements

This study was funded by NIH Grants AI064386 and AI081015 (G.S.T.) and AI36973 and AI37856 (W.R.B.).

Author contributions

S.W.C. developed techniques, performed *in vitro* bacteriology and isotope ratio experiments, analysed IRMS data and wrote the manuscript. M.M. developed techniques, performed *in vivo* experiment, analysed data and wrote the manuscript. M.C.M. performed *in vivo* experiments and analysed data. V.A. developed techniques, performed and analysed IRMS analysis. Z.D.S. developed techniques, performed and analysed IRMS analysis and wrote the manuscript. W.R.B. designed experiments, analysed *in vivo* data and wrote the manuscript. G.S.T. synthesized chemicals, designed experiments, analysed *in vivo* data and wrote the manuscript.

Additional information

Competing financial interests: G.S.T. acts as the Chief Science Officer of, and W.R.B. consults for, Avisa Pharma (a company that is developing a clinical-stage ^{13}C urea-based stable isotope breath test for detection of TB and other urease-producing lung infections). The remaining of the authors declare no competing financial interest.

3

Attomolar DNA detection with chiral nanorod assemblies

Wei Ma[1,*], Hua Kuang[1,*], Liguang Xu[1], Li Ding[2], Chuanlai Xu[1], Libing Wang[1,2] & Nicholas A. Kotov[3,4,5,6]

Nanoscale plasmonic assemblies display exceptionally strong chiral optical activity. So far, their structural design was primarily driven by challenges related to metamaterials whose practical applications are remote. Here we demonstrate that gold nanorods assembled by the polymerase chain reaction into DNA-bridged chiral systems have promising analytical applications. The chiroplasmonic activity of side-by-side assembled patterns is attributed to a 7–9 degree twist between the nanorod axes. This results in a strong polarization rotation that matches theoretical expectations. The amplitude of the bisignate 'wave' in the circular dichroism spectra of side-by-side assemblies demonstrates excellent linearity with the amount of target DNA. The limit of detection for DNA using side-by-side assemblies is as low as 3.7 aM. The chiroplasmonic method may be particularly useful for biological analytes larger than 2–5 nm which are difficult to detect by methods based on plasmon coupling and 'hot spots'. Circular polarization increases for inter-nanorod gaps between 2 and 20 nm when plasmonic coupling rapidly decreases. Reaching the attomolar limit of detection for simple and reliable bioanalysis of oligonucleotides may have a crucial role in DNA biomarker detection for early diagnostics of different diseases, forensics and environmental monitoring.

[1] State Key Lab of Food Science and Technology, School of Food Science and Technology, Jiangnan University, Wuxi, Jiangsu 214122, China. [2] State Key Lab of Food Safety Test (Hunan), Changsha, Hunan 410004, China. [3] Department of Chemical Engineering, University of Michigan, Ann Arbor, Michigan 48109, USA. [4] Department of Materials Science, University of Michigan, Ann Arbor, Michigan 48109, USA. [5] Department of Biomedical Engineering, University of Michigan, Ann Arbor, Michigan 48109, USA. [6] Biointerface Institute, University of Michigan, Ann Arbor, Michigan 48109, USA. * These authors contributed equally to this work. Correspondence and requests for materials should be addressed to C.X. (email: xcl@jiangnan.edu.cn) or to N.A.K. (email: kotov@umich.edu).

Chirality of nanoparticles (NPs) and their assemblies have recently attracted substantial interest among materials scientists[1-5]. There now exist examples of NP assemblies with pyramidal[3,6,7] and helical morphologies[2,8], and strong chiral responses are predicted for other geometries[9,10]. The chiro-optical properties of nanomaterials originate from the atomic-scale chirality of the inorganic core of NPs[11], the chiral arrangement of the thiolates on their surfaces[12], the electronic 'imprint' of chirality due to adsorbed chiral organic molecules on NPs surface (for example, DNA and peptides)[13,14], and from the intrinsic chiral geometry of NPs or their assemblies at nano- and submicron-scales[2,3,6-8]. Currently, chiral nanostructures are prepared using chiral templates, for instance, DNA including the origami approach[2], helical fibres[15], twisted nanoribbons[8] or by lithography[16], and the primary motivation for the development of chiral nanomaterials is the possibility of creating chiral metamaterials with negative refractive indices[17]. Optical devices utilizing these phenomena are intriguing, but fundamental challenges remain for the practically relevant infrared and visible range.

In this study, we pursue the bioanalytical potential of self-assembled chiral nanoscale superstructures. We demonstrate that the limit of DNA detection reached by side-by-side (SBS) assemblies of Au nanorods (NRs) using chiral bisignate plasmonic signals could be markedly lower than those reported for other widely discussed optical methods employing ultraviolet–visible absorption[18] of coupled plasmons, fluorescence tagging[19] and surface-enhanced Raman scattering (SERS)[20]. In addition, these results compete well for single molecular detection of analytes[21], due to an alternative dependence of optical polarization effects between interdistant plasmonic particles.

Results

NR assemblies by PCR. In this study, chiral assemblies of gold NRs were made using a polymerase chain reaction (PCR) (Fig. 1a)[22]. The use of PCR allowed for the controlled growth of NP and NR assemblies connected by DNA, where the number of

thermal cycles determined the lengths and complexity of the resulting superstructures (Fig. 1). The mode of attachment of NRs to each other followed either an end-to-end (ETE) or a SBS assembly pattern, controlled by the placement of the PCR primers (Fig. 1).

The gold NRs had lengths and diameters of 62 nm and 22 nm, respectively, with an aspect ratio of 2.9. Preferential binding of thiol-terminated primers to the end facets of the NRs allowed for the ETE growth mode for the NR chains (Fig. 1b)[5]. To obtain SBS assemblies, Au NRs were modified by dithiothreitol (DTT) binding to the end sites and thiol polyethylene glycol. These modifications make NRs stable for a wide range of solution conditions and protects them from excessive modification by thiols[22]. Subsequent addition of the thiolated primer resulted in preferential attachment to the sides of the NRs. Once introduced to the PCR replication system, NRs modified with DNA strands either at their sides or ends acted as 'monomers' for the PCR assembly (Fig. 1c) and 'building blocks' for the resulting nanoscale assemblies. Variations in the placement of the primers allowed for finely controlled synthesis of extended NR 'chains' and 'ladders'. The number of PCR cycles, n, regulated the length of the assemblies (Figs 2a–f and 3a–f, Supplementary Figs S1–S11). The NR attachment patterns were retained until $n = 20$ and 30 for ETE and SBS assemblies, respectively (Supplementary Figs S6, S11). Modified Au NRs without PCR ($n = 0$) organized sporadically (Figs 2a and 3a). The consistent elongation of NR 'chains' and 'ladders' with increasing n was confirmed by dynamic light scattering measurements of the hydrodynamic diameter (D_h). For the ETE assembly, D_h increased from 102 ± 2.1 to 701 ± 15 nm as n increased from 2 to 20 cycles. For the same values of n, D_h increased from 88 ± 3.1 to 408 ± 23.2 nm in the case of SBS assembly (Supplementary Fig. S12). This change in D_h correlates well with the increase of statistically averaged number of NRs in ETE and SBS assemblies for different n values (Supplementary Figs S13–S17). As expected, the length of ETE is greater than those for SBS assemblies because of the sideway attachment of NRs in

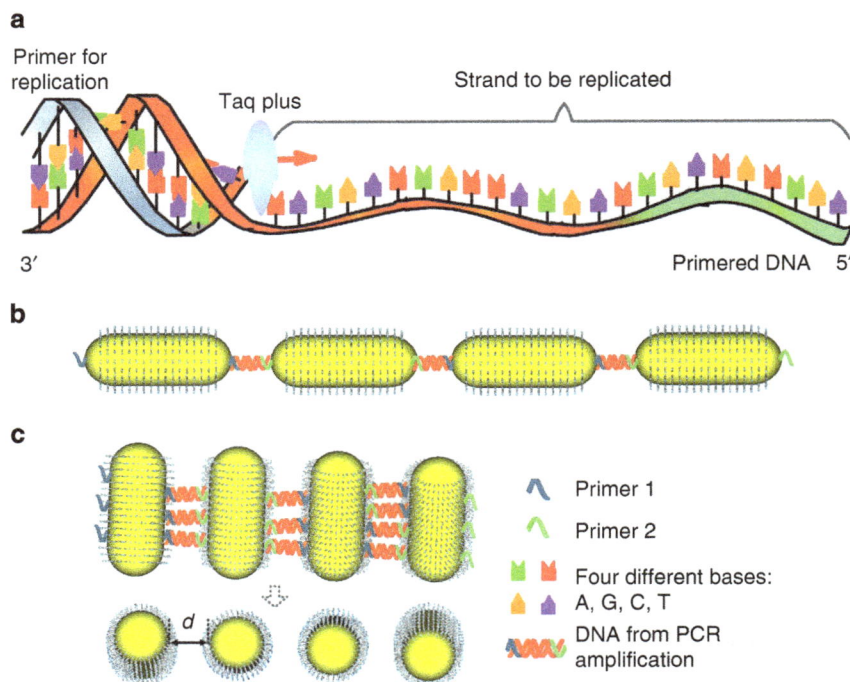

Figure 1 | Schematics for PCR assembly of Au NRs. (**a**) PCR replication procedure in which a DNA strand can be ampilified using primer, template DNA, taq plus polymerase and four different DNA bases. (**b**) PCR-based gold NRs ETE assembly. (**c**) PCR-based gold NRs SBS assembly with inter-NR gap d; in the bottom part of the panel the DNA chains were removed for clarity.

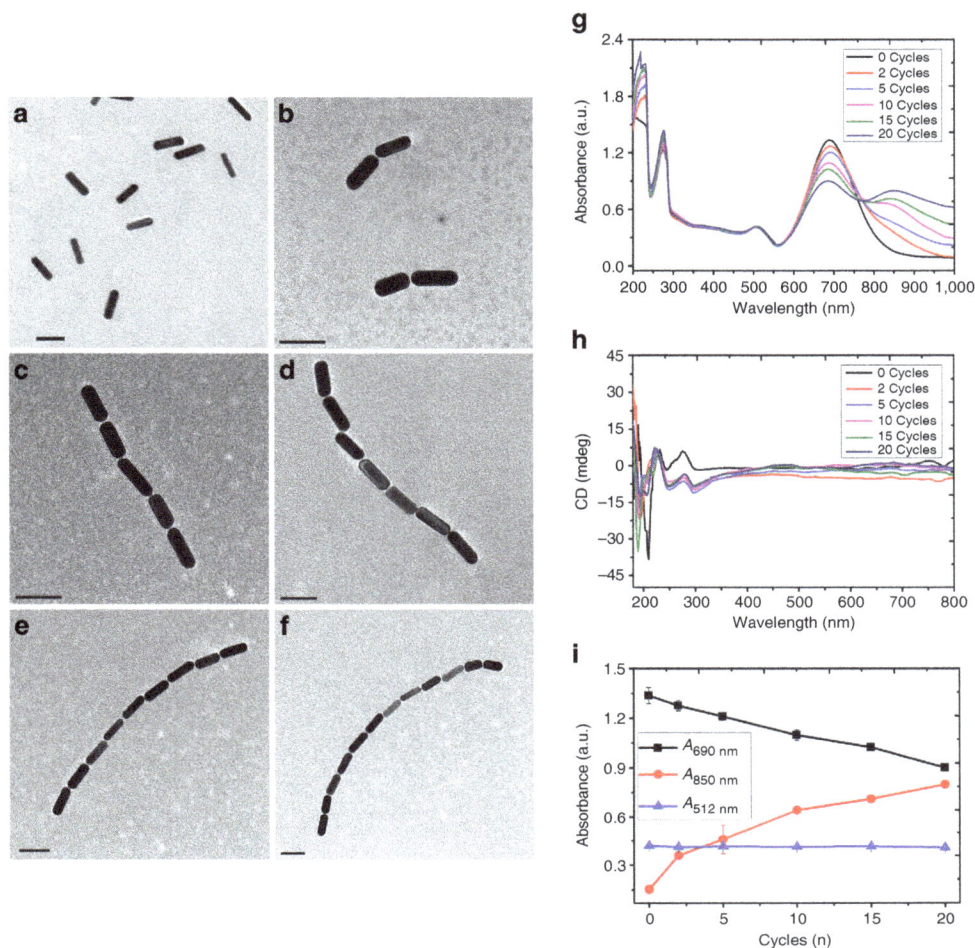

Figure 2 | Structure and optical properties of ETE assemblies of Au NRs. (**a–f**) Representative TEM images for ETE assembly obtained after different number of PCR cycles, $n = 0$ (**a**), 2 (**b**), 5 (**c**), 10 (**d**),15 (**e**) and 20 (**f**); scale bar, 50 nm. (**g,h**) Ultraviolet–visible (**g**) and CD spectra (**h**) for ETE assembly obtained for different n. (**i**) Intensity of absorption maxima for ETE assemblies obtained for different n. The error bars represent the standard deviation of sample measurements.

the latter case. In concert with previous studies[23], the transverse plasmon for the ETE assembly changed very little with n (Fig. 2), whereas the longitudinal peak shifted to the red (Fig. 2g). For the SBS assemblies (Fig. 3a–f), the longitudinal plasmon band (λ_L) experienced a blue shift by 17 nm from $n = 0$ to $n = 30$ (Fig. 3i,k).

Chiral properties of the NR assemblies. The chiroptical properties of the NR assemblies and evaluation of their prospects for bioanalysis were the primary foci of this work. A distinct CD signal was seen in the ultraviolet part of the spectrum from 180 to 250 nm for both ETE and SBS assemblies. This should be attributed to DNA ligands and is expected for this system. A strong bisignate CD wave was seen in the plasmonic λ_L part of the spectrum between 500 and 800 nm. For the Cotton effect, the 620–800 nm spectral window was negative, whereas for 500–620 nm wavelengths it was positive, which has definitive analogies in molecular systems[24] although displaying greater CD intensity and g-factors (Fig. 3g, Supplementary Fig. S18). Chirality of the assemblies was observed only for SBS and not for ETE assemblies or single DNA-modified NRs (Figs 2h and 3j). The strong chiroplasmonic response was observed for n as few as two (Fig. 3j) when assemblies of 2–4 NRs were dominant (Supplementary Fig. S14a). We saw progressively stronger CD

signals as n increased, indicating that chiral geometries of NR assemblies persisted along with increasing the length of SBS assemblies.

Appearance of the CD signal should be attributed to the twisted structure of the SBS assemblies[22]. The three-dimensional (3D) images of the assemblies obtained with state-of-the-art cryo transmission electron microscopy (TEM) tomography showed a distinct and consistent twist between two adjacent NRs in the SBS assemblies (Fig. 3g,h)[22]. Note these images reflect the conformation of the assemblies as they exist in solution and are not affected by high-vacuum conditions and drying. The negative values of the dihedral angle (θ) between the adjacent NRs in SBS assemblies corresponding to the right rotating enantiomers (Supplementary Fig. S16) persist throughout the PCR assembly process[22]. These dihedral angles for dimer, trimer, tetramer and pentamer were consistently negative and equal to -9.0, -7.1, -8.0 and -7.0 degrees, respectively (Fig. 3g,h, Supplementary Figs S16, S17). The preference for one enantiomer as opposed to another is related to symmetry breaking of the parallel NR due to twisting of the connected DNA bridges and the general preference of non-parallel orientation of charged nanoscale rods as the conformation with minimal energy with multiple examples in biomolecules[22]. The consistency of the sign of θ leads to strong chirality of SBS-assembled NRs. ETE assemblies do not exhibit CD response (Fig. 2h) because the torsional force of chiral DNA

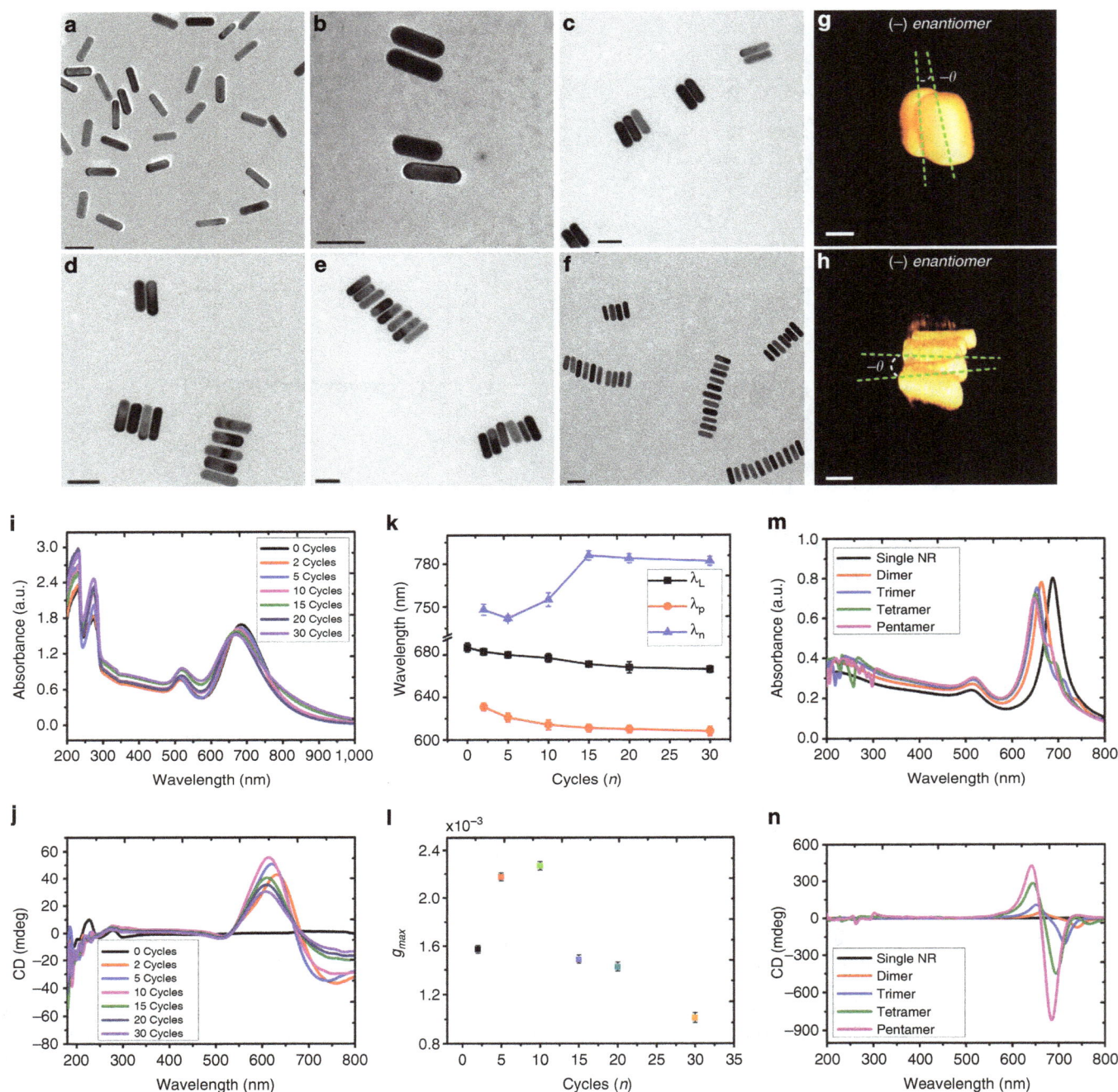

Figure 3 | Structure and optical properties of SBS assemblies of Au NRs. (**a–f**) Representative TEM images for SBS assemblies obtained after different number of PCR cycles, $n = 0$ (**a**), 2 (**b,c**), 5 (**d**), 10 (**e**) and 15 (**f**); scale bar, 50 nm. (**g,h**) Cryo TEM tomography images for NRs SBS assembled trimer (**g**) and pentamer (**h**); scale bar, 25 nm. (**i, j**) Experimental ultraviolet–visible (**i**) and CD spectra (**j**) for SBS assemblies with $n = 0$–30. (**k**) Evolution of spectral features of SBS NR assemblies represented by λ_L (longitudinal peak maximum in ultraviolet–visible spectra), λ_p (positive peak in plasmonic part of the CD spectra) and λ_n (negative peak in plasmonic part of the CD spectra) with increasing n. (**l**) Dependence of the maximum of chiral anisotropy factor g_{max} on n. (**m,n**) Calculated absorption (**m**) and CD spectra (**n**) for NRs SBS assemblies. The number of NRs, n, ranged from 1 to 5. The error bars in **i** and **l** represent the standard deviation of sample measurements.

oligomers connecting one end of the rod to another is substantially smaller than in SBS assemblies and therefore the enantiomer distribution is equilibrated. Chirality of the twisted NR assemblies was predicted theoretically[10] but was never observed experimentally[5] until recently[22]. Unlike earlier studies by Kadowala and coworkers on chiroplasmonic shifts on chiral coatings[25], the prospects for analytical and other applications of their chiroplasmonic properties of twisted assemblies in solutions

were not considered theoretically or experimentally. We hypothesized it to be a promising research direction due to the intensity of the chiral signals in these assemblies, their bisignate nature, stronger polarization rotation in solution than in thin films and continuous increase of the CD signal for $0 < n < 10$. The peak values of the anisotropy factor (g-factor) increased from 1.6×10^{-3} to 2.3×10^{-3} with 2–10 cycles (Fig. 3l, Supplementary Fig. S18).

As the number of PCR cycles increased, both positive (λ_p) and negative (λ_n) CD bands exhibited spectral shifts. The λ_p CD band moved to the blue part of the spectrum by 24 nm as PCR cycles increased from 2 to 20 cycles (Fig. 3k), whereas λ_n shifted by 16 nm from 2 to 5 cycles and became broader after 5 cycles. Also important is that as n exceeds a certain threshold value, the amount of disorder increases and the CD intensity of the bands stops growing. The larger aggregates do not have the consistency in the signs of their dihedral angles and therefore their CD signals decrease (Fig. 3i). For $n = 20$, complex agglomerates formed (Supplementary Fig. S11) and racemization of the assemblies occurred.

The absorbance (Fig. 3n) and CD (Fig. 3m) spectra calculated for SBS assemblies of 2–5 NRs showed excellent agreement with experimental results. The simulated electric fields on gold surface (E-fields) under right/left circular polarized (RCP/LCP) light excitation at λ_p and λ_n bands showed that the twisted SBS assemblies were indeed chiral: the coupling efficiencies to RCP and LCP (Supplementary Fig. S19) differ considerably. Agreeing with experimental results, the ETE-assembled structure did not show any CD signal (Fig. 2h) in the plasmonic region. The simulations of ETE NR trimer performed according to the 3D geometry-based state-of-the-art cryo TEM tomography and indeed exhibited a very weak CD signal (Supplementary Fig. S20).

NR assemblies for attomolar DNA biosensing. Chirality of SBS assemblies employing different reactant DNA templates can be potentially used for biosensing of oligonucleotides. The possibility of improving the limit of detection (LOD) by taking advantage of the bisignate wave-shape of the CD signals can be one of the advantages of the method. As n increases, λ_p becomes more positive, whereas λ_n becomes more negative. These spectroscopic changes occur in synch with each other, which essentially increases the detected signal and improves the signal-to-noise ratio. Note also that nanoscale assemblies amplify the chiral adsorption compared with the atomic-scale chirality in organic molecules. In addition to amplification of signal due to bisignate nature of the CD wave and nanoscale dimentions of chiral chromophores, there are also other factors related to dependence of intensity of CD signal on inter-NR gap, d. This dependence is conducive to ultrasensitive detection of biomacromolecules and represents one of the key advantages of this methods compared to others based on plasmon coupling.

We evaluated CD and other spectral optical responses for different amounts of target DNA oligomers. The calculated LOD was found to be 3.9 aM, 8.1 aM and 3.7 aM based on CD intensity of $C(\lambda_p)$, $C(\lambda_n)$ and $C(\lambda_p)-C(\lambda_n)$, respectively (Fig. 4). The

LOD values above characterizes this techniques as substantially more sensitive than typical PCR methods with or without NPs that give LOD = 0.1 fM (ref. 26). To validate the bioanalysis by chiroplasmonic effects, the widely used dilution method (see Methods) with *a priori* known concentrations of DNA was adopted (Supplementary Fig. S21). The concentrations returned by the chiroplasmonic method matched those expected for the specific dilution. The uncertainty coefficient, u, was as low as 0.0367 (see Methods) and indicative of the high accuracy of the method. Such u is associated, among other factors, with PCR amplification, high g, and improved signal-to-noise ratio due to bisignate nature of the chiroplasmonic spectra.

The use of CD spectroscopy is not uncommon for biosensing and sometimes offer better levels of detection compared with ultraviolet–visible[18], fluorescence spectroscopy[27], and electrochemistry-based methods[28], with the best LODs equal to 10,200 and 500 fM, respectively, especially in combination with plasmonic substrates[29,30]. However it never reached the level of attomolar (10^{-18} M) level of detection. The femtomolar detection level (10^{-15} M) is not sufficient to meet the demands of biomedical and environmental applications for a variety of clinical tasks especially for reliable early diagnosis of diseases using DNA biomarkers in complex biological samples.

It is relevant to compare analytical capabilities of chiroplasmonic method with its parent/related techniques, for instance, PCR and SERS that do not use CD spectra modulated by plasmonic particles. In some cases methods based on polaron coupling were able to reach nearly single-molecule detection capabilities[20] and high clinical relevance for detection of prostate cancer[21]. We compared chiroplasmonic method with well-established and highly sensitive reverse transcription PCR (RT-PCR) used for clinical and forensic purposes. LOD for the same DNA strands as in Fig. 4 was found to be 156 aM for RT-PCR (Supplementary Fig. S22). The NR assembly method was, therefore, *ca.* 40 times more sensitive than RT-PCR. The parallel experiments with chiroplasmonic method yielded identical concentrations with the chiroplasmonic bioanalysis.

Analytical capabilities of SERS are being rapidly advanced in many laboratories including successful SERS analysis using ETE and SBS assemblies of Au NRs[23,31] featuring high-intensity E-fields between the rods, and therefore, deserve special attention in this study. One of the best case scenarios for DNA detection would probably be the use a high-intensity SERS tag, such as 4-aminothiophenol (4-ATP) with Raman-active transitions υ(C–S) at 1,083 cm^{-1} and υ(C–C) at 1,590 cm^{-1}. Its SERS signal is further enhanced by E-fields on gold surface amplified after bridging of NRs by DNA. The Raman intensity of 4-ATP increased with increasing n for ETE assemblies with addition of

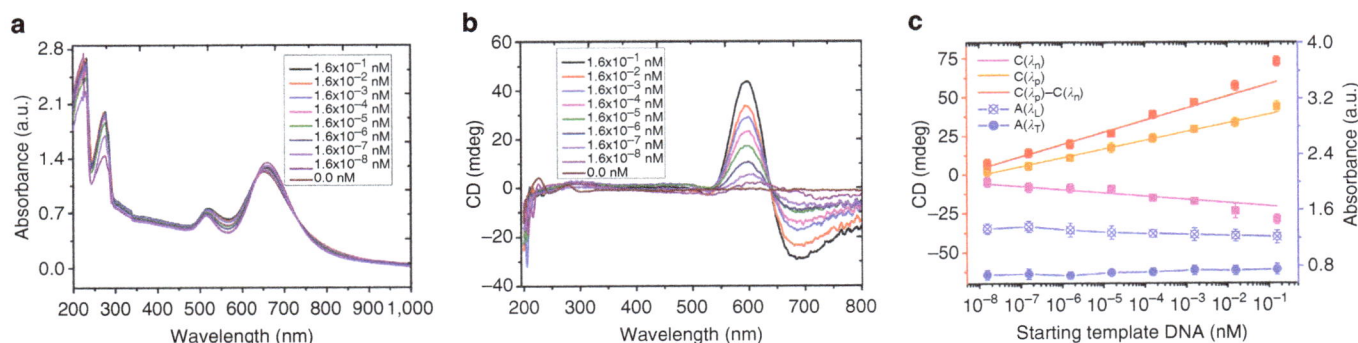

Figure 4 | DNA analysis with SBS NRs assemblies. (a,b) Experimental ultraviolet–visible (**a**) and CD spectra (**b**) for NR assemblies obtained for different DNA concentrations starting from 0.156 nM with stepwise 10 × dilution. (**c**) Calibration curves obtained using CD and ultraviolet–visible spectra of SBS assemblies. The error bars represent the standard deviation of sample measurements.

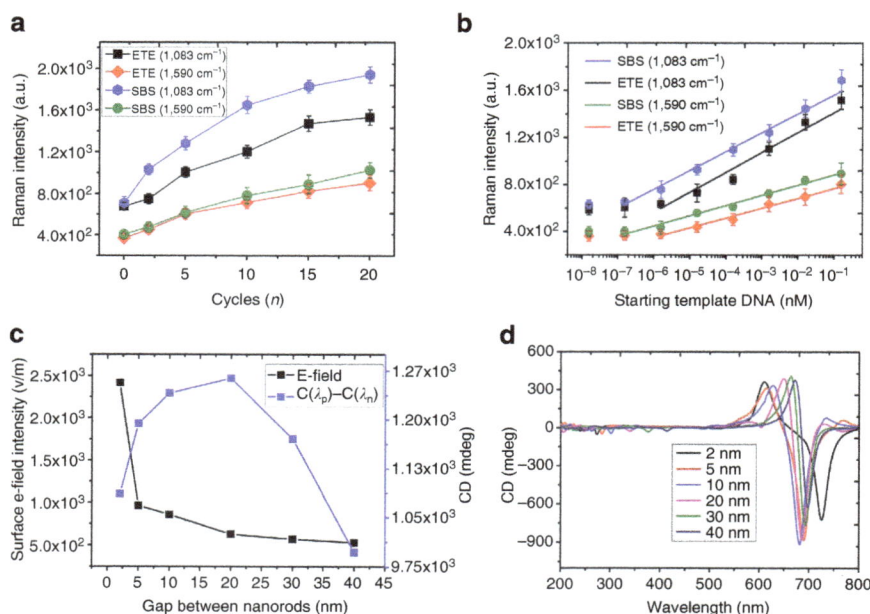

Figure 5 | DNA analysis using SERS capabilities. (**a,b**) SERS intensity of 4-ATP tag at 1,083 cm^{-1} and 1,590 cm^{-1} for ETE and SBS NR assemblies obtained after different n (**a**) and different starting DNA concentrations (**b**). (**c**) Calculated dependence of the intensity of the surface E-field and CD intensity for a NR pentamers with gaps of 2, 5, 10, 20, 30 and 40 nm. (**d**) Calculated CD spectra for NR pentamers with gaps of 2, 5, 10, 20, 30 and 40 nm. The error bars represent the standard deviation of sample measurements.

new NRs (Fig. 5a, Supplementary Figs S23–S25). Note that Lee et al.[31] observed a continuous decrease of SERS and E-field intensity with increasing numbers of NRs in SBS assemblies, whereas CD signal in DNA-bridged ETE assemblies display the opposite trend and increase for $n < 10$ which contributes to improved sensitivity of the chiroplasmonic method. LOD using the strongest SERS line of 4-ATP at 1,083 cm^{-1} was 1.14 and 1.58 fM for SBS and ETE assemblies (Fig. 5b), respectively, which is 290 and 403 times higher than for chiroplasmonic method for the identical DNA strand. Importantly, detection of long DNA is inherently suboptimal using SERS because formation of hotspots critical for its success requires the gaps between NRs to be small, that is, *ca.* 0.5–2 nm. The same is also true for SERS detection of all high molecular weight compounds that are larger than the size of optimal hotspots and similar spectroscopic methods based on strong polaron coupling, such as surface-enhanced Raman optical activity[24]. Molecules with diameters > 2 nm are particularly common in bioanalysis. As indicated by the calculations in Fig. 5c,d, the read-out parameter used in chiroplasmonic method, $C(\lambda_p)-C(\lambda_n)$, is much less sensitive to the distance between the NRs than the surface E-field necessary to enhance scattering from 4-ATP or other Raman tags. In fact, it initially increases for DNA-relevant size regime rather than decreasing as E-field intensity. Polarization rotation reaches maximum for the NR gap of *ca.* 20 nm, whereas the intensity of the surface E-field generated by NRs drops off dramatically when gap reaches 5 nm (Fig. 5c, Supplementary Fig. S26).

Discussion

We showed that the SBS assemblies of plasmonic NRs with strong polarization rotation make possible detection of DNA markers with unusually low LOD that is greatly needed for medical diagnostics, forensics and environmental needs. The physical phenomena behind this capability include enhancement of polarization rotation by plasmonic structures, chiral symmetry breaking for SBS assemblies and bisignate nature of CD spectra.

Chiroplasmonic method of detection has sensitivity advantage for the analysis of biomolecules > 2 nm, while other plasmonic methods could be preferred for smaller analytes, unless steps for narrowing the gap by depositing additional layers of plasmonic material are taken. In perspective, the high sensitivity of the CD signal to geometry of the twisted NR assembly allows for experimental observation of the torsional dynamics of helical systems in solutions and better understanding of the 3D geometry of plasmonic assemblies. The strong polarization rotation in DNA-bridged SBS assemblies can also be utilized in intracellular monitoring of low-occurrence markers and signaling molecules.[39]

Methods

Au nanorod preparation. Au NRs were synthesized by Au seeds growth method[32]. Synthesis of Au seeds: hydrogen tetrachloroaurate trihydrate ($HAuCl_4 \cdot 3H_2O$) was dissolved in 2.5 ml, 0.2 M hexadecyltrimethylammonium bromide (CTAB) solution, added by 0.3 ml pre-cooled 300 µl, 0.01 M sodium borohydride ($NaBH_4$) and quickly mixed for 2 min and left to reaction at 25 °C for 2 h. The growth of Au NR, 0.15 ml of 0.004 M $AgNO_3$ (NR): 70 µl of 0.079 M (ascorbic acid, Vc) was added to 5 ml, 0.001 M, $HAuCl_4$ was added to 5 ml, 0.2 M CTAB solution and left to reduce for 2 min, finally 12 µl of prepared Au seeds were added, strongly stirred for 20 s and left at 25 °C. The concentration of gold NRs was 0.25 nM.

NR modification. Modification of Au NRs included primer on end, side and 4-ATP modifications.

End modification. A 1 ml aliquot of synthesized Au NRs was first centrifuged (10,000 g, 10 min). The precipitate was then dissolved in 200 µl of 0.005 M CTAB solution (five times concentrated). The Au NRs modification of the primer was carried out at 25 °C for 12 h with a reaction ratio of 80:1 between the Au NRs and the primer. Then unreacted primer was removed by centrifuging (7,000 g, 10 min) and dissolved in 200 µl of 0.005 M CTAB solution.

Side modification. A 1 ml aliquot of synthesized Au NRs was five times concentrated into 200 µl of 0.005 M CTAB solution by centrifugation. Then the end sites were modified with DTT for 8 h with a molar ratio of 10:1 between the Au NRs and the primer. The DTT-modified Au NRs were centrifuged again (7,000 g, 10 min) and dissolved in 200 µl of 0.005 M CTAB solution. Then the Au NRs were modified by the addition of polyethylene glycol with a molar ratio of 120:1.

Finally, the side modification of the primer was reacted for 10 h at a molar ratio of 400:1 between the primers and the Au NRs. The Au NRs were then centrifuged (70,000 g, 10 min) and stored in 200 μl of 0.005 M CTAB solution.

SERS tag modification. For SERS measurement, 4-ATP was used to modify the Au NR after being modified by primers. The 4-ATP was modified after the primer was attached with the same modification ratio for end and side modifications. Generally, 4-ATP modification employed similar procedures with the same reaction ratio as that of the NRs modified with primer. The molar ratio between the Au NRs and the 4-ATP was set at 400:1 and reacted for 8 h both for the NRs used for ETE and SBS assemblies. The Au NRs were then centrifuged at 70,000 r.p.m. for 10 min and finally dispersed in 200 μl of 0.005 M CTAB solution. After 4-ATP modification, the as-prepared NRs were employed in PCR-based assemblies with different cycles and the assembled products were monitored in the liquid phase.

PCR conditions and Au NR assembly. Using Lambda DNA as a template[23], the PCR reaction was performed in a 100 μl amplification system. The reaction mixture contained 10 μl of PCR buffer, 2 μl of dNTPs (1 mM), 1 μl of template DNA, 1 μl (5 U) of Taq Plus polymerase and 10 μl of each Au NR-forward/reverse-primer (F/R-primer) conjugates (Au NR-F/R-primer); finally, amplification mixture was set to 100 μl by adding 66 μl of ultrapure water. The $5 \times$ PCR buffer contained of 50 M KCl, 10 mM tris–HCl, pH 9.0 at 25 °C, 0.1% TritonX-100 and 1.5 mM MgCl$_2$. Each of the Au NR-F/R-primer conjugates and Au NR-reverse-primer (R-primer) conjugates were concentrated to 1.25 nmol. The thermal cycling protocol began with a 3 min predenaturation step at 94 °C, followed by varying cycles of 94 °C denaturation (30 s), 60 °C annealing (30 s) and 72 °C extension (1 min) steps, followed by a further extension for 10 min at 72 °C. Last, the PCR system was held at 4 °C for ca. 10 min before use. Sequences for F-primer and R-primer were listed as follows: F-primer: TGGCTGACCCTGATGAGTTCG; R-primer: GGGCCATG ATTACGCCAGTT. The assembly parameter of PCR cycle (n) and starting template DNA was set as follows: the n was set as 0, 2, 5, 10, 15, 20 and 30 cycles with starting template DNA concentration of 0.156 nM. The template DNA concentration was set by 10 times stepwise dilution of the starting material, 0.156 nM to 0.0156 fM.

Analytical calculations, experimental statistics and uncertainty analysis. The LOD was calculated according to the high sensitivity analysis. The calibration curve was plotted as

$$y = a + b \cdot x \tag{1}$$

where a and b are the variable obtained via least-square root linear regression for the signal–concentration curve for variable y representing the CD signal (mdeg) at DNA concentration of x (nM).

When

$$y = C_{blank} + 3\text{s.d.} \tag{2}$$

where s.d. is the standard deviation and C_{blank} is the CD of signal of blank sample (without DNA).

The LOD was calculated as follows:

$$\text{LOD} = 10^{\frac{(C_{blank} + 3\text{s.d.}) - a}{b}}. \tag{3}$$

The uncertainty for LOD of our assay was calculated using statistical analysis methods based on standard deviations of different concentration points (A-type uncertainty). The standard deviation (s.d.) was calculated according to the well-known formula:

$$\text{s.d.} = \sqrt{\frac{1}{n_r - 1} \cdot \sum_{i=1}^{N_r} \left(X_i - X_{avg} \right)^2} \tag{4}$$

where n_r is the total number of the samples. X_i is the ith sample of the series of measurements. X_{avg} is the average value of the CD (or other) signals obtained for the specific series of identical samples repeated n_r times.

The uncertainty coefficient, u was calculated as follows:

$$u = \frac{\text{s.d.}}{\sqrt{n_r}}. \tag{5}$$

The sample repeat number, n_r was $n_r = 9$ for all of the experiments. For DNA detection following the chiroplasmonic protocol presented below, the uncertainty coefficient was 0.0367.

According to the methods of uncertainty analysis accepted in analytical chemistry[33], this value of uncertainty coefficient corresponds to a high accuracy method.

Calibration protocol. Progressive dilution method with known concentrations of analytes is used for calibration of a wide variety of analytical methods including those for exceptionally small concentrations and single-molecule detection limits. These methods can be based on ultraviolet–visible[18], fluorescence spectroscopy[27], or electrochemistry[28]. The reliability of this calibration technique stems from the high accuracy of volume measurements when a sample of known (high) concentration is diluted according to the power law.

The template DNA concentrations was acquired by stepwise $10 \times$ dilution, following the procedure accepted in the field of bioanalysis and other branches of analytical chemistry[18,34]. For each dilution, 5 μl DNA solution (by pipette with 1–10 μl measurement range) was added to 45 μl (by pipette with 5–50 μl measurement range) ultra-pure water. The pipette tips were replaced and discarded every time. Analogous procedure was carried out for the blank experiment without starting DNA analyte and was used in equation (2) as C_{blank}. One microlitre of as-diluted DNA solution for a specific concentration was added into 100 μl PCR system for amplification and subsequent analysis by CD spectroscopy.

RT-PCR detection protocol. The RT-PCR was performed by standard procedure according to the process described by Ma et al.[35] RT-PCR was carried out on CFX-96 real-time PCR system with 25 ml amplification volume. The PCR amplification mixture was composed of 1.5 ml $20 \times$ EvaGreen dye, 2.5 ml of 1 mM dNTP, 2.5 ml of $10 \times$ PCR buffer, 0.25 ml of 5 U Taq DNA polymerase and 0.5 ml of 2.5 mM upstream and downstream primer, respectively, and finally ultra-pure water was added to have a volume of 25 ml. The PCR cycling parameters were set as 94 °C denaturation (30 s), 60 °C annealing (30 s) and 72 °C extension (1 min) steps, followed by a further extension for 10 min at 72 °C. Fluorescence measurements were taken after each annealing step. The standard curve for RT-PCR is plotted in Supplementary Fig. S22 according to the number of the threshold cycle, $n_{threshold}$, defined as the number of PCR cycles when the fluorescence reached the value of 900 (a.u.) for specific concentration of DNA (exponential amplification stage).

Computer simulations. Computer simulations ETE and SBS assemblies were performed using CST Microwave Studio[36,37]. The geometry of the NR ladders and chains was defined by d, surface-to-surface gap between Au NRs, the angle between the long axes of the two NRs. The propagation of excitation beam was defined by φ_x and φ_z, the angles between excitation beam and x axis, z axis, respectively. The surface of gold NR is composed of two semispherical surfaces and one side cylinder surface. Surface E-field simulations by RCP and LCP were carried out at λp and λn, respectively. Surface E-field enhancement simulations were carried out using linearly polarized beam with E-field vector parallel and perpendicular with longitudinal direction of NR. The total E-field enhancement was the sum of these two fields. Simulations of CD and absorbance spectra were accomplished by parameter sweep of φ_z and φ_x from 0 to 2π with step of $\pi/6$ (30°), which was similar to theoretical methods previously explored[9,14,38].

Instrumentation. The ultraviolet–visible spectra were measured using a ultraviolet–visible spectrometer (200–1,000 nm) in a quartz cell. The CD spectra were performed on a Bio-Logic MOS-450 CD spectrometer. TEM micrographs were collected on a JEOL-2010 microscope operated at 120 kV. The 3D reconstruction of electron tomography was carried out using a Tecnai Spirit 120 kV TEM. Dynamic light scattering data were obtained using a Malvern Zetasizer ZS instrument with a 632.8 nm laser source and a backscattering detector at 173°. Raman scattering spectra were measured in liquid cell using a LabRam-HR800 Micro-Raman spectrometer with Lab-spec 5.0 software. The slit and pinhole were set at 100 and 400 mm, respectively, an air-cooled He-Ne laser for 632.8 nm excitation with a power of ~8 mW.

References

1. Sheikholeslami, S., Jun, Y. W., Jain, P. K. & Alivisatos, A. P. Coupling of optical resonances in a compositionally asymmetric plasmonic nanoparticle dimer. *Nano Lett.* **10**, 2655–2660 (2010).
2. Kuzyk, A. et al. DNA-based self-assembly of chiral plasmonic nanostructures with tailored optical response. *Nature* **483**, 311–314 (2012).
3. Chen, W. et al. Nanoparticle superstructures made by polymerase chain reaction: collective interactions of nanoparticles and a new principle for chiral materials. *Nano Lett.* **9**, 2153–2159 (2009).
4. Xia, Y. H., Zhou, Y. L. & Tang, Z. Y. Chiral inorganic nanoparticles: origin, optical properties and bioapplications. *Nanoscale* **3**, 1374–1382 (2011).
5. Nie, Z., Petukhova, A. & Kumacheva, E. Properties and emerging applications of self-assembled structures made from inorganic nanoparticles. *Nat. Nanotech.* **5**, 15–25 (2010).
6. Mastroianni, A. J., Claridge, S. A. & Alivisatos, A. P. Pyramidal and chiral groupings of gold nanocrystals assembled using DNA scaffolds. *J. Am. Chem. Soc.* **131**, 8455–8459 (2009).
7. Yan, W. J. et al. Self-assembly of chiral nanoparticle pyramids with strong R/S optical activity. *J. Am. Chem. Soc.* **134**, 15114–15121 (2012).
8. Srivastava, S. et al. Light-controlled self-assembly of semiconductor nanoparticles into twisted ribbons. *Science* **327**, 1355–1359 (2010).
9. Govorov, A. O., Fan, Z., Hernandez, P., Slocik, J. M. & Naik, R. R. Theory of circular dichroism of nanomaterials comprising chiral molecules and nanocrystals: plasmon enhancement, dipole interactions, and dielectric effects. *Nano Lett.* **10**, 1374–1382 (2010).

10. Auguié, B., Alonso-Gómez, J. L., Guerrero-Martínez, A. s. & Liz-Marzán, L. M. Fingers crossed: optical activity of a chiral dimer of plasmonic nanorods. *J. Phys. Chem. Lett.* **2**, 846–851 (2011).

11. Dolamic, I., Knoppe, S., Dass, A. & Burgi, T. First enantioseparation and circular dichroism spectra of Au-38 clusters protected by achiral ligands. *Nat. Commun.* **3**, 798 (2012).

12. Bovet, N., McMillan, N., Gadegaard, N. & Kadodwala, M. Supramolecular assembly facilitating adsorbate-induced chiral electronic states in a metal surface. *J. Phys. Chem. B* **111**, 10005–10011 (2007).

13. Lu, F. *et al.* Discrete nanocubes as plasmonic reporters of molecular chirality. *Nano Lett.* **13**, 3145–3151 (2013).

14. Slocik, J. M., Govorov, A. O. & Naik, R. R. Plasmonic circular dichroism of peptide-functionalized gold nanoparticles. *Nano Lett.* **11**, 701–705 (2011).

15. Guerrero-Martínez, A., Alonso-Gómez, J. L., Auguié, B., Cid, M. M. & Liz-Marzán, L. M. From individual to collective chirality in metal nanoparticles. *Nano Today* **6**, 381–400 (2011).

16. Hentschel, M., Schäferling, M., Weiss, T., Liu, N. & Giessen, H. Three-dimensional chiral plasmonic oligomers. *Nano Lett.* **12**, 2542–2547 (2012).

17. Soukoulis, C. M., Linden, S. & Wegener, M. Negative refractive index at optical wavelengths. *Science* **315**, 47–49 (2007).

18. Elghanian, R., Storhoff, J. J., Mucic, R. C., Letsinger, R. L. & Mirkin, C. A. Selective colorimetric detection of polynucleotides based on the distance-dependent optical properties of gold nanoparticles. *Science* **277**, 1078–1081 (1997).

19. Zhang, Z., Sharon, E., Freeman, R., Liu, X. & Willner, I. Fluorescence detection of DNA, adenosine-5'-triphosphate (ATP), and telomerase activity by zinc(II)-protoporphyrin IX/G-quadruplex labels. *Anal. Chem.* **84**, 4789–4797 (2012).

20. Lim, D.-K., Jeon, K.-S., Kim, H. M., Nam, J.-M. & Suh, Y. D. Nanogap-engineerable Raman-active nanodumbbells for single-molecule detection. *Nat. Mater.* **9**, 60–67 (2010).

21. Rodríguez-Lorenzo, L., de la Rica, R., Álvarez-Puebla, R. A., Liz-Marzán, L. M. & Stevens, M. M. Plasmonic nanosensors with inverse sensitivity by means of enzyme-guided crystal growth. *Nat. Mater.* **11**, 604–607 (2012).

22. Ma, W. *et al.* Chiral plasmonics of self-assembled nanorod dimers. *Sci. Rep.* **3**, 1934 (2013).

23. Lee, A. *et al.* Probing dynamic generation of hot-spots in self-assembled chains of gold nanorods by surface-enhanced raman scattering. *J. Am. Chem. Soc.* **133**, 7563–7570 (2011).

24. Berova, N., Bari, L. D. & Pescitelli, G. Application of electronic circular dichroism in configurational and conformational analysis of organic compounds. *Chem. Soc. Rev.* **36**, 914–931 (2007).

25. Mulligan, A. *et al.* Going beyond the physical: instilling chirality onto the electronic structure of a metal. *Angew. Chem.* **117**, 1864–1867 (2005).

26. Deng, H. *et al.* Gold nanoparticles with asymmetric polymerase chain reaction for colorimetric detection of DNA sequence. *Anal. Chem.* **84**, 1253–1258 (2012).

27. Cui, D. *et al.* Self-assembly of quantum dots and carbon nanotubes for ultrasensitive DNA and antigen detection. *Anal. Chem.* **80**, 7996–8001 (2008).

28. Park, S. J., Taton, T. A. & Mirkin, C. A. Array-based electrical detection of DNA with nanoparticle probes. *Science* **295**, 1503–1506 (2002).

29. Kravets, V. G. *et al.* Singular phase nano-optics in plasmonic metamaterials for label-free single-molecule detection. *Nat. Mater.* **12**, 304–309 (2013).

30. Hendry, E. *et al.* Ultrasensitive detection and characterization of biomolecules using superchiral fields. *Nat. Nanotech.* **5**, 783–787 (2010).

31. Lee, A. *et al.* Side-by-side assembly of gold nanorods reduces ensemble-averaged SERS intensity. *J. Phys. Chem. C* **116**, 5538–5545 (2012).

32. Nikoobakht, B. & El-Sayed, M. A. Preparation and growth mechanism of gold nanorods (NRs) using seed-mediated growth method. *Chem. Mater.* **15**, 1957–1962 (2003).

33. Ellison, S. L. R. (Eds) A.W. Eurachem/CITAC guide: Quantifying Uncertainty in Analytical Measurement, Analytical Measurement, Third edition (2012).

34. Ma, W. *et al.* A PCR based magnetic assembled sensor for ultrasensitive DNA detection. *Chem. Commun.* **49**, 5369–5371 (2013).

35. Wang, L. *et al.* Side-by-Side and End-to-End Gold Nanorod Assemblies for Environmental Toxin Sensing. *Angew. Chemie Int. Ed.* **49**, 5472–5475 (2010).

36. Alvarez-Puebla, R. A. *et al.* Gold nanorods 3D-supercrystals as surface enhanced Raman scattering spectroscopy substrates for the rapid detection of scrambled prions. *Proc. Natl Acad. Sci. USA* **108**, 8157–8161 (2011).

37. Lilly, G. D., Agarwal, A., Srivastava, S. & Kotov, N. A. Helical assemblies of gold nanoparticles. *Small* **7**, 2004–2009 (2011).

38. Zhukovsky, S. V., Kremers, C. & Chigrin, D. N. Plasmonic rod dimers as elementary planar chiral meta-atoms. *Opt. Lett.* **36**, 2278–2280 (2011).

39. Xu, L. *et al.* Regiospecific Plasmonic Assemblies for *in-situ* Raman Spectroscopy in Live Cells. *J. Am. Chem. Soc.* **134**, 1699–1709 (2012).

Acknowledgements

This material is based upon the work partially supported by the Center for Solar and Thermal Energy Conversion, an Energy Frontier Research Center funded by the U.S. Department of Energy, Office of Science, Office of Basic Energy Sciences under Award Number DE-SC0000957. We acknowledge support from NSF under grants ECS-0601345, EFRI-BSBA 0938019, CBET 0933384, CBET 0932823 and CBET 1036672. The work is also partially supported by ARO MURI W911NF-12-1-0407 'Coherent Effects in Hybrid Nanostructures for Lineshape Engineering of Electromagnetic Media' (N.A.K.). This work is financially supported by the National Natural Science Foundation of China (21071066, 91027038, 21101079 and 21175034), the Key Programs from MOST (2012BAC01B07, 2012BAD29B05, 2012AA06A303 and 2012BAD29B04) and grants from Jiangsu Province, MOF and MOE (NCET-12-0879, BE2011626, BK2010001, BK2010141 and JUSRP51308A) (C.L.X.). The authors thank the NSF for grant no. DMR-9871177 for funding of the JEOL 2010F analytical electron microscope used in this work.

Author contributions

C.X., L.W. and N.A.K. designed the experiments, interpreted and analyzed the data, conceptualized the findings and co-wrote the paper. W.M., H.K., L.X., L.D. performed the experiments, characterizations, prepared the samples and carried out computer calculations. W.M. and H.K. analyzed the data and carried out three-dimensional reconstruction of electron tomography.

Additional information

4

Hydride ions in oxide hosts hidden by hydroxide ions

Katsuro Hayashi[1,†], Peter V. Sushko[2,†], Yasuhiro Hashimoto[3], Alexander L. Shluger[2] & Hideo Hosono[4]

The true oxidation state of formally 'H$^-$' ions incorporated in an oxide host is frequently discussed in connection with chemical shifts of ^{1}H nuclear magnetic resonance spectroscopy, as they can exhibit values typically attributed to H$^+$. Here we systematically investigate the link between geometrical structure and chemical shift of H$^-$ ions in an oxide host, mayenite, with a combination of experimental and *ab initio* approaches, in an attempt to resolve this issue. We demonstrate that the electron density near the hydrogen nucleus in an OH$^-$ ion (formally H$^+$ state) exceeds that in an H$^-$ ion. This behaviour is the opposite to that expected from formal valences. We deduce a relationship between the chemical shift of H$^-$ and the distance from the H$^-$ ion to the coordinating electropositive cation. This relationship is pivotal for resolving H$^-$ species that are masked by various states of H$^+$ ions.

[1] Center for Secure Materials, Materials and Structures Laboratory, Tokyo Institute of Technology, R3-34, 4259 Nagatsuta, Yokohama 226-8503, Japan. [2] Department of Physics and Astronomy, University College London, London WC1E 6BT, UK. [3] Department of New Business Development, Asahi Kasei Corporation, 1-105 Kanda-Jinbocho, Tokyo 101-8101, Japan. [4] Frontier Research Center, Tokyo Institute of Technology, S2-13, 4259 Nagatsuta, Yokohama 226-8503, Japan. †Present addresses: Department of Applied Chemistry, Kyushu University, Fukuoka 819-0395, Japan (K.H.), Fundamental and Computational Sciences Directorate, Pacific Northwest National Laboratory, Richland, Washington 99352, USA (P.V.S.). Correspondence and requests for materials should be addressed to K.H. (email: k.hayashi@cstf.kyushu-u.ac.jp) or to P.V.S. (email: peter.sushko@pnnl.gov).

Hydrogen is a ubiquitous element that exhibits various types of bonding in materials[1-3]. For example, hydrogen reacts with electropositive metal elements to form metal hydrides, in which hydrogen is treated formally as H^- ions, whereas in most metal oxides, hydrogen forms OH^- ions and is formally in the H^+ state. Both H^- and H^+ can be doped into polar semiconductor hosts and the stability of each species is controlled by Fermi level[3]. Recently, incorporating H^- ions into metal oxides and mixed-anion compounds has attracted much attention as a new approach to modify the physical and chemical properties of inorganic materials[4-12] including high-temperature superconductors[9,10]. Such materials are accessible by new chemical processes using metal hydride as a strong reducing reagent and/or hydrogen source. The effects of H^- incorporation include antiferromagnetic coupling of two neighbouring cations[5,6], tuning of ionic size of halide ion sites by hydride ion-substitution[9,10], stabilization of high-symmetry polymorphic phases[11] and hosts for fast ion diffusion[11,12]. Furthermore, the large negative redox potential of H_2/H^- (-2.2 V with respect to H^+/H_2 (SHE) and 0.3–0.5 V with respect to Li^+/Li (ref. 13)) may not only be useful for employing H^- as a reductant in organic and electrochemical reactions[14] but may also facilitate redox reactions inside materials containing H^- ions. An example of the latter has been reported for H^- ion-doped mayenite[15-26], in which H^- ions are photochemically converted to protons by releasing carrier electrons[15-17]. A similar photochemical conversion process is also found on the surface of MgO (ref. 27).

Nevertheless, whether 'H^-' ions can be routinely incorporated in oxide hosts and, if so, whether they indeed possess the chemical state of H^- ions, is still debated. These questions frequently arise in connection to the chemical shifts of ^1H nuclear magnetic resonance (NMR) spectroscopy in non-metals, which exhibit values that are typical for H^+, for example, $\sim +5$ p.p.m. with respect to tetramethylsilane (TMS)[11,17,18], whereas 'metallic' hydrides, such as TiH_2, exhibit large negative shifts[28].

To resolve this controversy regarding the chemical shifts of H^- ions in ^1H NMR spectra, here we systematically investigate the effect of the local environment of H^- and OH^- ions in an oxide host, mayenite, on chemical shift with a combination of experimental and *ab initio* approaches as well as re-examining published data. We deduce that the distance from the H^- ion to the coordinating cation is a simple but essential parameter that determines the chemical shift of H^- via a linear relationship. This relationship allows the overlapping NMR signatures of H^- and H^+ ions to be separated from each other, which we demonstrate using H^- ions diluted in calcium phosphate apatite as an example.

Results

Incorporation of H^- and OH^- ions into mayenites. The structural formula of mayenite is $[M_{24}Al_{28}O_{64}]^{4+} \cdot 4X^-$, where the brackets correspond to the lattice framework of a unit cell with a space group $I\bar{4}3d$, and M is Ca (refs 15–25) or Sr (ref. 26). A framework of AlO_4 tetrahedra, together with Ca^{2+} or Sr^{2+} ions embedded in it, forms cages that can accommodate various 'extraframework' anions indicated by $4X^-$ (see Fig. 1). Both Ca- and Sr-mayenites (described as C12A7 and S12A7, respectively) can theoretically accommodate up to four H^- or OH^- ions ($4X^-$ hereafter) per 12 cages of the unit cell. Five mayenite samples incorporating different extraframework anionic species ($X^- = OH^-$, H^- and e^-) with concentrations close to theoretical maxima were prepared. The chemical compositions of these samples were determined using diffractometries, NMR spectroscopy and supporting analysis. (see the Methods section)

Figure 2 displays the results obtained from Rietveld analysis and electron density analysis with the maximum entropy method

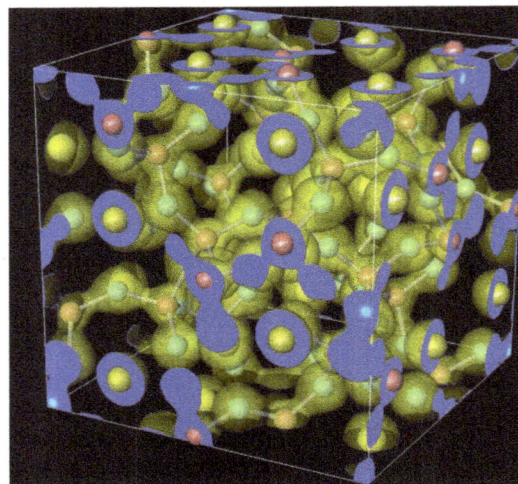

Figure 1 | Crystal structure of C12A7. The electron density isosurface (0.5×10^{-6} $e \cdot pm^{-3}$) in $[Ca_{24}Al_{28}O_{64}]^{4+} \cdot 4e^-$ obtained by maximum entropy method analysis and superposed on the atomic structure of the unit cell determined by Rietveld analysis (Ca: yellow; Al: red; O: cyan). The framework charge can also be compensated by incorporation of anions into cages, whose occupancy is one-sixth in the case of divalent anions such as O^{2-} and S^{2-}, and one-third in the case of monovalent anions such as F^-, OH^- and H^-.

(MEM) of synchrotron X-ray diffraction (XRD) data measured for the five samples. When comparing the three electron density maps for the C12A7 (Fig. 2a–c), only OH^- ions are clearly observed in the cages, whereas the cage centres in samples containing extraframework e^- and H^- appear to be empty. However, the presence of H^- is manifested by another feature in the density maps. In particular, the electron density for Ca^{2+} ions is spherical in the C12A7:e^- sample, whereas it has an ellipsoidal shape elongated along the S_4 axis in the samples containing H^- and OH^- ions. This elongation is interpreted as a superposition of the densities of Ca^{2+} ions located at the poles of the vacant cages and those occupied by H^- or OH^- anions. In both cases, the concentration ratio of the vacant cages to the occupied ones is very close to the ideal ratio of 2:1 (refs 22,24,26; see Supplementary Tables 1–3, and Supplementary Note 1), which allows us to calculate the displacement of two Ca^{2+} ions at the cage poles inwards along the S_4 axis to be ~ 40 pm with respect to their sites in an empty cage in both C12A7:H^- and C12A7:OH^-. A similar effect is observed in S12A7 (Fig. 2d,e), where the extraframework H^- and OH^- ions induce the displacement of the pole Sr^{2+} ions inwards by ~ 30 pm (see Supplementary Tables 4 and 5, and Supplementary Note 2). Again the MEM analysis directly resolves OH^- ions in the cage centre, but not H^- ions. However, neutron powder diffraction (NPD) of the S12A7:H^- sample implied almost theoretical maximum occupation of hydrogen in cages (see Supplementary Table 6 and Supplementary Note 3), confirming the presence of H^- in the cages.

The distances between the pole cations in C12A7:OH^- and S12A7:OH^- differ by 33 pm, which is consistent with the difference between ionic diameters defined by Shannon[29]: 242 and 212 pm for Sr^{2+} (VII) and Ca^{2+} (VII) (in this case, each cation is coordinated with six framework O^{2-} ions and one OH^- ion), respectively. This agreement indicates that an OH^- ion is sandwiched by the two pole cations in both C12A7 and S12A7 and its effective ionic radius is 132 ± 2 pm, which, again, coincides with the value defined for OH^- (II)[29]. Similarly, using the distance between the metal cation and hydride ion

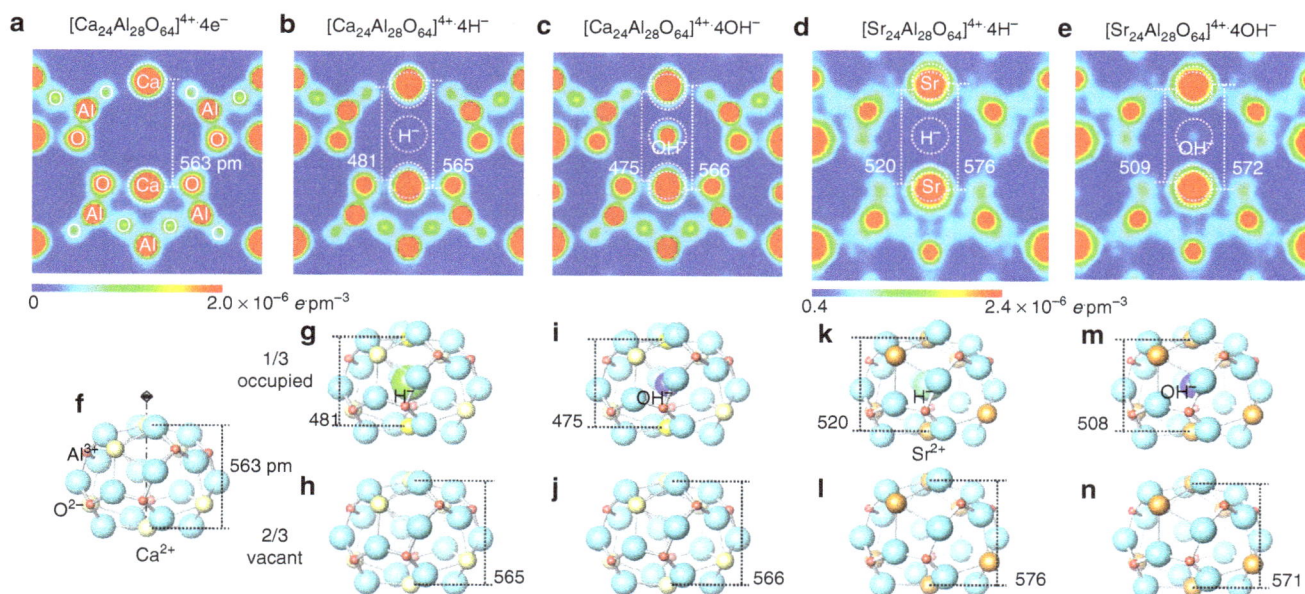

Figure 2 | Incorporation of H⁻ and OH⁻ anions into mayenites observed by X-ray diffraction. Upper row shows maximum entropy method electron density maps in the {400} planes for C12A7:e⁻ (**a**), C12A7:H⁻ (**b**), C12A7:OH⁻ (**c**), S12A7:OH⁻ (**d**) and S12A7:OH⁻ (**e**) samples. Distances (pm) between the two pole Ca^{2+} ions on the S_4 axes, as determined by Rietveld analysis, are indicated on the maps. The lower row shows the geometrical structures of the cages (Ca^{2+}: yellow; Sr^{2+}: orange; Al^{3+}: red; O^{2-}: cyan; H⁻: green; OH⁻: blue) and the corresponding Ca–Ca and Sr–Sr distances. Structures of the occupied and vacant cages are shown separately for C12A7:H⁻ (**g**: occupied; **h**: vacant) and C12A7:OH⁻ (**i**: occupied; **j**: vacant), S12A7:OH⁻ (**k**: occupied; **l**: vacant) and S12A7:OH⁻ (**m**: occupied; **n**: vacant) systems.

($d_{Ca-H} = 240$ pm and $d_{Sr-H} = 266$ pm), the ionic radius of an H⁻ ion is 134 ± 2 pm, which is slightly larger than the radius of F⁻ (II) (128.5 pm) and almost the same as the radius of O^{2-} (II) (135 pm). This evaluation agrees with the generally accepted trend that the radius of an H⁻ ion is nearly the same or slightly larger than that of an F⁻ ion and increases with increasing electropositivity of the coordinating cation[29].

Another important geometrical parameter is the distance between the oxygen atom of the OH⁻ ion and an O^{2-} ion ($d_{O-H...O}$), which forms a hydrogen bond[1,30,31]. Taking into account the average displacements of OH⁻ ions from the cage centre as estimated from the thermal parameters of Rietveld analysis, these distances are evaluated to be 313 and 322 pm in C12A7 and S12A7, respectively. These $d_{O-H...O}$ values are much larger than those in various other host materials, suggesting that these OH⁻ ions have only weak hydrogen bonds. Nevertheless, these $d_{O-H...O}$ values are consistent with an empirical relation[1] linking $d_{O-H...O}$ and O–H stretching mode frequency (3,560 cm⁻¹ for C12A7 (ref. 25) and 3,650 cm⁻¹ for S12A7 (ref. 26)). Important structural parameters of the samples are summarized in Table 1.

NMR chemical shifts. The magic-angle-spinning (MAS) technique is indispensable for NMR characterization of a solid material. However, information about the anisotropy of the electron density around the nucleus and dipole–dipole interactions is sacrificed to improve spectrum resolution. Thus, the isotropic chemical shift, δ_{iso}, is a principal parameter to characterize the chemical states of hydrogen[32,33]. Figure 3 shows ¹H-MAS-NMR spectra measured for H⁻ and H⁺ (extraframework OH⁻ ion) in C12A7 and S12A7. The H⁻ ion has a peak at +5.1 p.p.m. with respect to TMS reference, whereas H⁺ of the OH⁻ ion appears at −0.8 p.p.m. in C12A7. Nearly, the same areal intensities of the spectra further confirm full incorporation of H⁻ and OH⁻ ions. In S12A7, the difference between the values

of δ_{iso} for H⁻ (+6.1 p.p.m.) and OH⁻ ions (−1.3 p.p.m.) expands with cage size. Overall, δ_{iso} for H⁻, $\delta_{iso}(H^-)$, is thought to become more positive as the cage size increases and, at the same time, δ_{iso} for H⁺, $\delta_{iso}(H^+)$, becomes more negative. Interestingly, the two values of $\delta_{iso}(H^-)$ are outside their typical ranges and apparently reversed compared with the typical positions for protons and metallic hydrides. Indeed, usually $\delta_{iso}(H^+)$ varies from +20 to 0 p.p.m. (ref. 32), whereas metallic hydrides, such as TiH_2, typically exhibit $\delta_{iso}(H^-)$ below 0 p.p.m. (ref. 28). Our data suggest that electron density should be larger around H⁺ than around H⁻ nuclei, provided that the chemical shift is predominantly determined by the chemical shielding effect. This relationship contradicts the formal number of electrons associated with H⁺ (no electrons) and H⁻ (two electrons). Furthermore, our results for hydrogen show opposite behaviour to the NMR spectra of alkali elements. According to the NMR spectra of alkali-organic complexant alkalides, alkali ions such as Na^+ and K^+ have positive chemical shifts, whereas alkalide ions such as Na^- and K^- exhibit relatively negative shifts[34,35]. This relationship can be easily understood from the difference in the occupation numbers of the outer s shell, for example, $[Ne]3s^0$ for Na^+ and $[Ne]3s^2$ for Na^-. Clearly, the same logic does not apply to H⁺ and H⁻ ions.

These experimental results were corroborated with *ab initio* calculations using the periodic and embedded cluster approaches[16,21,36,37]. Geometrical parameters and NMR chemical shifts were calculated for 12 configurations of H⁻ and OH⁻ ions in C12A7 and S12A7 (for details, see Supplementary Table 7 and Supplementary Note 4). The calculated geometrical parameters are very close to those determined experimentally. In particular, the distances between the pole cations agree with those evaluated from XRD data within 1–16 pm, whereas the calculated $d_{O-H...O}$ distance is within 10 pm of the corresponding experimental values. The chemical shifts were calculated using a procedure suggested by Cheeseman *et al.*[38] that has produced reliable values for the chemical shifts of

Table 1 | Structural and electronic data.

	Experimental method			Theoretical method/basis set			
	MEM/Rietveld	NMR		GGA PBE	HF/6-311 + G(2d,p)		
	Distance (pm)	δ_{iso} (p.p.m. w.r.t. TMS)	Splitting (p.p.m.)	Distance (pm)	Electron density (10^{-6} e pm^{-3})	δ_{iso} (p.p.m. w.r.t. TMS)	Splitting (p.p.m.)
$C12A7:OH^-$	$d_{O-H\ldots O} = 313$	−0.8	5.9	$d_{O-H\ldots O} = 309\ (+4, -7)$	2.72	+4.9	5.2
$C12A7:H^-$	$d_{M-H} = 240$	5.1		$d_{M-H} = 247$	2.28	−0.3	
$S12A7:OH^-$	$d_{O-H\ldots O} = 322$	−1.3	7.4	$d_{O-H\ldots O} = 318\ (+9, -10)$	2.73	+5.2	6.0
$S12A7:H^-$	$d_{M-H} = 257$	6.1		$d_{M-H} = 261$	2.26	−0.8	

GGA, the generalised gradient approximation; HF, Hartree-Fock; MEM, maximum entropy method; NMR, nuclear magnetic resonance; PBE, the density functional by Perdew, Burke and Ernzerhof; TMS, tetramethylsilane; w.r.t, with respect to.
For detail, see Supplementary Notes 5 and 6.

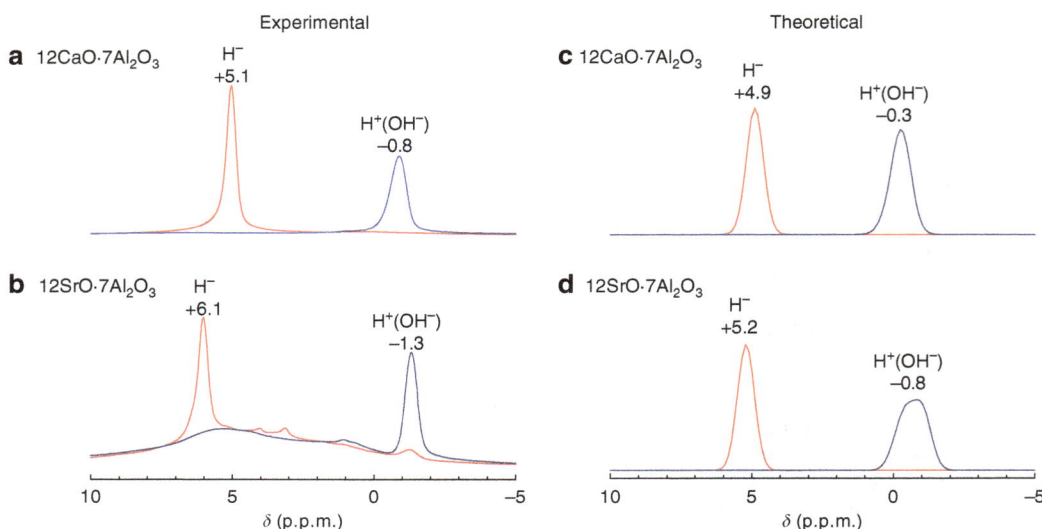

Figure 3 | ^{1}H nuclear magnetic resonance (NMR) spectroscopy. Experimental ^{1}H-magic-angle-spinning NMR spectra for (**a**) C12A7:H$^-$ (red line) and C12A7:OH$^-$ (blue line), and (**b**) S12A7:H$^-$ (red line) and S12A7:OH$^-$ (blue line) samples, and corresponding theoretical spectra for (**c**) C12A7 and (**d**) S12A7 systems. Broad signals between +7 and −2 p.p.m. in the experimental spectra of S12A7 are attributed to adsorbed water and hydroxide impurities.

molecular species. The calculated chemical shifts are in a very good agreement with the experimental data (Table 1 and Fig. 3) and, importantly, they confirm that $\delta(H^-) > \delta(H^+)$. The latter is attributed to the difference of the electron density on the hydrogen nucleus: $\rho(H^-) < \rho(H^+)$ in both C12A7 and S12A7.

The interaction of a spin-1/2 nucleus (such as ^{1}H) with its surroundings can be described using the Hamiltonian, $H_{int} = H_D + H_{CS} + H_P + H_K + \cdots$, which includes the magnetic dipolar interaction, H_D; the chemical shielding effect, H_{CS}; the paramagnetic spin interaction, H_P, and the Knight shift effect, H_K (refs 28,33). The negative δ_{iso} of metallic hydrides is principally ascribed to the Knight shift effect, because of the interaction of nuclear spin with conduction electrons, which introduce an 'extra' effective field at the nuclei. Its typical value ranges from 0 to −250 p.p.m. for ^{1}H (ref. 28). Both C12A7:H$^-$ and S12A7:H$^-$ samples have measurable electrical conductivities because of residual electrons. However, the primary conductivity mechanism is thermally activated polaron hopping[21], and hence the Knight shift effect can be ruled out. Contribution from H_P is also negligible, because the chemical shift is barely affected by the concentration of residual electrons: shifts of less than 0.1 p.p.m. were observed for the electron concentration range of $2 \times 10^{19} - 8 \times 10^{20}$ cm^{-3} (ref. 18; Supplementary Fig 1). Thus, chemical shielding has the dominant effect on the $\delta_{iso}(H^-)$ value.

It follows that the electron density around the nucleus is higher for H$^+$ than H$^-$ ions in mayenites.

Effect of structural parameters on chemical shift. It is well established that the $\delta_{iso}(H^+)$ value has a close relationship with the $d_{O-H\ldots O}$ value[30,31,39]. The values of $\delta_{iso}(H^+)$ observed for the two mayenites also follow this general trend, as shown in Fig. 4. The fact that $\delta_{iso}(H^+)$ in mayenites are negative with respect to TMS is consistent with exceptionally large $d_{O-H\ldots O}$ values. Similarly, we found a correlation between $\delta_{iso}(H^-)$ and the distance between the H$^-$ ion and its nearest coordinating cation, d_{M-H}. To focus on the electron shielding effect on $\delta_{iso}(H^-)$ only, we consider ionic hydrides[40–42] and mixed-anion hydrides[7,8], in which H$^-$ ions are coordinated with only saline (alkaline-earth or alkali metals) cations. Their $\delta_{iso}(H^-)$ values are plotted against d_{M-H} in Fig. 4, and are approximated to a linear function. Taking account of the scattering of data, this relationship can be described as:

$$\delta_{iso}(H^-)(ppm) = 0.070\ d_{M-H}(pm) - 11.5(\pm 2). \quad (1)$$

To rationalize the observed results, we considered two model systems (Fig. 5a,b): (i) a cluster comprising two planar Mg_2O_2 blocks, sandwiching an OH$^-$ ion, was used to investigate the dependence of $\delta_{iso}(H^+)$ on $d_{O-H\ldots O}$ and (ii) an H$^-$ ion occupying anion site in the bulk of a rock-salt lattice, such as that of MgO, was used to investigate the dependence of $\delta_{iso}(H^-)$ on d_{M-H}. The results of these calculations summarized in Fig. 5e reproduce well the general trends observed for the H^- and OH^- ions experimentally.

Here, let us consider how $d_{O-H\cdots O}$ affects the $\delta_{iso}(H^+)$ value. Figure 5c shows profiles of calculated electron density with varying $d_{O-H\cdots O}$. The electron density around the proton nuclei increases with increasing distance. When the distance is short, the proton is attracted to the host oxygen, while the negative charge

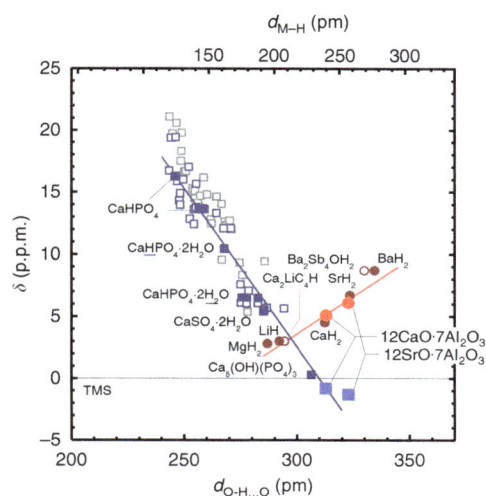

Figure 4 | Dependence of δ_{iso} on $d_{O-H\cdots O}$ and d_{M-H}. Data for mayenites are co-plotted with literature data: filled squares for CaO-based compounds[31]; blue open squares for silicate glasses[39]; grey open squares for some solid salts[30]; filled circles for binary saline hydrides with alkaline-earth metals[40], Mg[41] and Li[42]; open circles for Ca_3LiC_3H (ref. 7) and $Ba_8Sb_4OH_2$ (ref. 8). Horizontal axes of $d_{O-H\cdots O}$ and d_{M-H} were adjusted so that each of the two sets of mayenite data were aligned.

of the counter oxygen pushes the bonding electrons away from the proton. As a result, the electron density around the proton decreases. As $d_{O-H\cdots O}$ increases, the chemical bond of the OH^- ion becomes less polarized and its bond length, d_{O-H}, decreases concurrently. Both factors lead to higher electron density at the hydrogen site, in turn leading to negative shift. The model used to calculate $\delta_{iso}(H^+)$ (Fig. 5a) reproduces the structure of mayenite cages well, in which two pole cations sandwich an OH^- ion. In general, the polarization of an O–H bond is affected not only by the distance $d_{O-H\cdots O}$ but also by ligand ions. When cations are located near the proton, they tend to shorten the O–H bond, which in turn causes an additional negative shift; the effect is opposite for anions. Different ligands from each crystal structure may influence the relationship between $d_{O-H\cdots O}$ and $\delta_{iso}(H^+)$. This is discussed in Supplementary Note 5.

As for the H^- ion, Fig. 5d demonstrates that as d_{M-H} increases, the electron density around the hydrogen nucleus decreases, inducing the positive chemical shift. The sensitivity of the electron density to d_{M-H} is ascribed to the large polarizability of the H^- ion. In general, the site volume for an H^- ion increases markedly as the bond character in metal hydrides changes from covalent to ionic[29]. In contrast, saline cations become more electropositive as their radius increases, leading to formation of a bond with more ionic character to the H^- ion[29,43]. Therefore, the general dependence of $\delta_{iso}(H^-)$ on d_{M-H} (Fig. 4) can be explained from the fact that a longer d_{M-H} gives the H^- ion more space, which in turn decreases the electron density around the nucleus and moves $\delta_{iso}(H^-)$ to a more positive value.

The dependence of the δ_{iso} values on the lattice structural parameters discussed above (Fig. 4) can be used to predict the δ_{iso} values for OH^- and H^- ions. Because OH^- and H^- ions can occupy equivalent lattice sites, as is the case in mayenites, the ability to make such predictions has practical importance. When both OH^- and H^- ions are incorporated at equivalent sites in an oxide host, the NMR signal of H^- is in the typical range for hydrogen-bonded H^+ ions. Hence, it may be difficult to

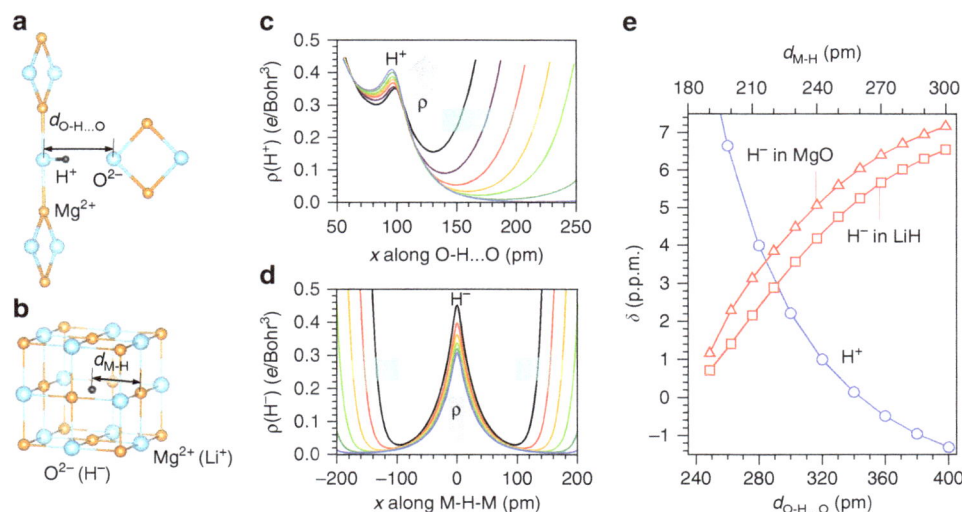

Figure 5 | Dependence of $\delta_{iso}(H^+)$ and $\delta_{iso}(H^-)$ on the characteristic parameters of proton local environment. Model systems are: (**a**) OH^- ion sandwiched between two Mg_2O_2 clusters and forming a hydrogen bond with another Mg_2O_2 cluster, (**b**) an H^- ion at the anion site in the bulk of a rock-salt ionic crystal, such as MgO or LiH. (**c**) Electron density along the O–H bond of an OH^- ion for several values of $d_{O-H\cdots O}$ plotted against distance from O atom of the OH^- ion. Black to blue lines correspond to $d_{O-H\cdots O}$ values of 220, 240, 260, 280, 300, 340 and 400 pm. (**d**) Electron density along the Mg–H bond of an H^- ion in MgO for several values of d_{M-H}. Black to blue lines correspond to d_{M-H} values of 200, 220, 240, 260, 280 and 300 pm. (**e**) Dependences of $\delta_{iso}(H^+)$ and $\delta_{iso}(H^-)$ on $d_{O-H\cdots O}$ (circles) and d_{M-H} (triangles for M = Mg and squares for M = Li), respectively. Light-green arrows in (**c**) and (**d**) indicate increasing volume associated with hydrogen species, while grey arrows indicate the corresponding change of the charge density magnitude at the hydrogen site.

distinguish H^+ and H^- ions solely from their δ_{iso} value. Generally, hydrogen is ubiquitous and forms various configurations of H^+ in oxides, possibly masking the weak NMR signal of diluted H^- ions. This suggests that there may be H^- ions present in materials that have been overlooked in NMR spectra. In materials with a site volume similar to that of mayenite, the apparent reversal of chemical shift takes place, facilitating identification of the H^- signal.

Scouting for H^- ions among OH^- ions. To demonstrate that the relationship between d_{M-H} and δ_{iso} is useful for revealing the existence of H^- ions and predicting their chemical shifts and to establish a methodology for identifying H^- out of various states of H^+ that 'hide' the H^- ions, we considered calcium phosphate apatite[31,44-49]. As described below, this material hosts similar sets of anions in its special crystallographic sites as mayenite does. Calcium phosphate apatite generally crystallizes in a hexagonal (space group $P6_3/m$) or pseudo-hexagonal (more exactly monoclinic with a space group $P2_1/b$) system and is characterized by an isolated tetrahedron of PO_4^{3-} and two sites with Ca^{2+} ions. One of these (the Ca2 site) forms an array of triangles that run along the c-axis. Its inner space is often called a 'channel' and is occupied partially or completely by various kinds of anions. The general formula of the unit cell is $[Ca_{10}(PO_4)_6]^{2+} \cdot 2Y^-$, where brackets refer to a positively charged host lattice that incorporates guest anions: $2Y^- = 2F^-$, $2Cl^-$, $2OH^-$ or O^{2-}. F^- (ref. 44) and O^{2-} (ref. 46) ions are located at the centre of Ca_3 triangles, whereas

OH^- (refs 31,46) and Cl^- (ref. 44) ions are positioned on an off-triangle plane along the c-axis (see Fig. 6a). Although there have been no reports of incorporation of H^- ions in the channel of apatite, it is expected from its analogy to mayenite that H^- will be formed when apatite is severely reduced in the presence of hydrogen. Because an H^- ion has an ionic radius comparable with F^- and O^{2-} ions, the most feasible site in the apatite crystal to accommodate an H^- ion is the centre of the calcium triangle (see Fig. 6a). A chemical shift of $+4.9 \pm 2$ p.p.m. is predicted from the triangle centre to calcium distance of 234 pm according to the relationship given by equation (1). Similarly, the $O-H\ldots O$ distance of 307 pm for the OH^- in the channel gives a chemical shift of $+1.0 \pm 2$ p.p.m., which agrees with experimental values[31,45,47] (cf. Fig. 4).

Then, we prepared several apatite samples annealed in wet air, dry oxygen and with titanium hydride (see the Methods section). NMR and infrared transmission spectra measured for these samples are displayed in Fig. 6b,c, respectively. The signal around 0 p.p.m. is well known to arise from OH^- ions in the channel. However, in contrast to mayenite that exhibits only two distinct signals for H^- and H^+, many peaks appear in the series of apatite samples. Most of them are associated with O-H stretching bands in the infrared spectra, originating from HPO_4^{2-} groups (indicated iii and iv in Fig. 6c)[31,47,48] in the lattice framework of apatite, complex defects of OH^- in CaO precipitates (ii and iii)[49] in some apatite samples as well as channel OH^- ions (i)[31,45,47]. This is consistent with a general relationship that correlates the strength of the hydrogen bond with the shift of the O-H stretching frequency to lower wavenumber[1]. At the same time,

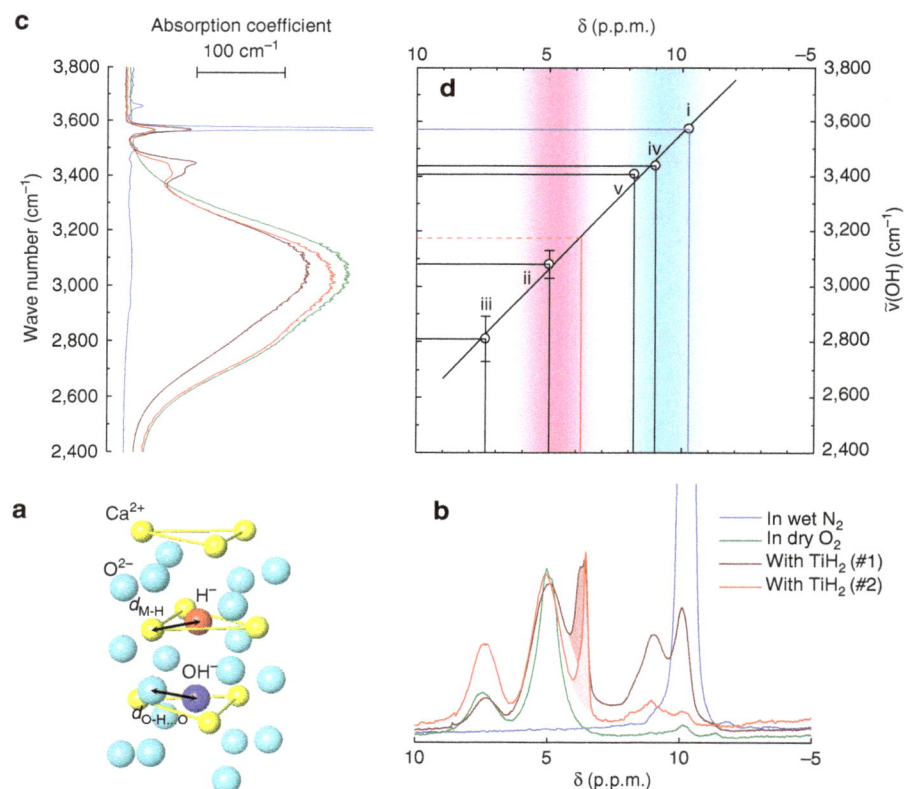

Figure 6 | H^- ions in apatites. (**a**) Local structure around the channel in apatite, showing the coordinating ions with OH^- and H^- ions. (**b**) ^1H magic-angle-spinning nuclear magnetic resonance (NMR) spectra collected in apatite samples annealed in wet N_2 (indicated by blue lines), in dry O_2 (orange) and with TiH_2 (brown and red). Filled areas highlight the signals assigned to H^- ions. (**c**) Infrared transmission spectra measured for the same set of samples. (**d**) Correlation between chemical shifts (δ_{iso}) in the NMR spectra and wavenumbers in the infrared spectra. Blue and red shaded areas indicate chemical shift ranges predicted using the d_{M-H} and $d_{O-H\ldots O}$ correlations with δ_{iso} (H^-) and δ_{iso} (H^+), respectively, given in Fig. 4 and equation 1.

the chemical shift moves to lower magnetic field because of the decrease in the electron density around the hydrogen nucleus. An unassigned signal remains at 3.6 p.p.m. (shaded in Fig. 6b), which is in the range of the prediction for the H^- ion in the channel. This signal has not been experimentally reported to date but appears only in the severely reduced samples prepared here, and thus is unambiguously assigned to H^- ions. Their concentration is estimated to be one-thirtieth of the theoretical maximum, which makes them very difficult to detect and characterize by diffraction techniques.

Beyond mayenites and apatite, many oxide-based crystals possess crystallographic sites for OH^- or halide ions, whose exchange with H^- ions is expected to be possible and would lead to novel material functions. Identification of such novel H^- ions both fully occupying a certain crystallographic site and doped at a dilute level will be now possible using NMR spectroscopy.

Methods

Mayenite sample preparation and characterization. The C12A7:OH^- sample was prepared by annealing stoichiometric C12A7 powder in a wet N_2 atmosphere at 900 °C. The OH^- ion concentration was controlled to a maximum value of 2.3×10^{21} cm^{-3} (ref. 25). The C12A7:H^- sample was obtained by annealing a Czochralski-refined C12A7 crystal sealed together with TiH_2 in a silica tube at 1,100 °C. The reacted layer formed on the crystal was mechanically removed. Electron spin resonance spectra revealed that the samples do not contain any spin-active species except residual (2×10^{19} cm^{-3}) extraframework electrons. The C12A7:e^- sample was obtained by chemical reduction of a C12A7 single crystal using metallic Ti as a reducing agent, as described in ref. 19. The S12A7:OH^- sample was prepared by solid-state reaction of $Sr(OH)_2 \cdot 8H_2O$ and γ-Al_2O_3 powders at 800 °C in wet N_2 atmosphere as described in ref. 26. Thermogravimetry-mass spectrometry measurements showed weight loss of 0.9% upon H_2O desorption suggesting that the concentration of OH^- ions is close to the maximum possible. The S12A7:H^- sample was prepared by thermally annealing pressed pellets of S12A7:OH^- powder mixed with CaH_2 powder, which were encapsulated in a stainless steel tube with Sweagelok plugs, at 500 °C for 16 days. Residual CaH_2 and side product Ca compounds were removed by dissolving the sample in a solution of 0.1 M NH_4Cl in methanol. The sample powder was washed several times with pure methanol. Residual extraframework electron concentration was evaluated to be $\sim 1 \times 10^{19}$ cm^{-3} by electron spin resonance. The concentration of H^- ions in this sample was evaluated to be $2.0 \pm 0.1 \times 10^{21}$ cm^{-3} by volumetry[18], suggesting nearly full incorporation of H^- ions (see Supplementary Note 6).

Apatite sample preparation and characterization. Commercial high-density hydroxyapatite ceramic (Cell yard, Pentax/Asahi techno glass/Hoya, purity of > 99%, Ca/P ratio of 1.667) was used as a starting material. The pellets were annealed in dry O_2 atmosphere at 1,200 °C for 4 days (dry oxygen sample) or in wet N_2 at 900 °C for 6 days using a tube furnace (wet N_2 sample). In other batches, the apatite pellets were put in an Ti-foil envelope containing TiH_2 powder, and then the envelope was sealed in an evacuated silica glass capsule. Heat treatment was carried out at 1,000 °C for 4 days (TiH_2 (#1) sample) or 20 days (TiH_2 (#2) sample). After that, the reaction layer formed on the surface was removed by mechanical grinding. infrared transmission spectra were measured with a Perkin-Elmer Spectrum One spectrometer (Perkin-Elmer) using mirror-finished samples with thicknesses of 40–200 μm.

Diffractometry. Powder XRD patterns for C12A7 samples were collected at 300 K with a diffraction angle range of 4–70° and an incident X-ray wavelength of 0.5016 Å using a Debye-Scherrer camera at beam line BL02B2, SPring-8. The Enigma programme was used for Rietveld/MEM analysis. The powder XRD patterns for S12A7 samples were collected at 298 K with a diffraction angle range of 10–140° using a Rigaku Rint-2000 XRD (Rigaku; Bragg-Brentano geometry, Cu $K\alpha$ source). The Rietan-2000/Prima programme was used for Rietveld/MEM analysis. NPD data were collected for the S12A7:H^- sample at 290 K with a diffraction angle range of 10–165° at a wavelength of 1.495 Å using a high-resolution powder diffractometer for thermal neutrons (HRPT, Paul Scherrer Institut). Rietveld analysis was performed using the Rietan-FP programme. The Vesta programme was used to plot all structural models and electron density maps.

NMR spectroscopy. ^1H NMR spectroscopic measurements were performed on a Bruker Biospin DSX-400 spectrometer (Bruker) operating at a resonance frequency of 400.13 MHz equipped with a high-speed MAS probe with a diameter of 5 mm. Each granulated sample was weighed (about 70 mg) to obtain a quantitative result and sealed in a zirconia rotor. The rotation frequency was 30 kHz. Spectra were acquired with 90° pulses with a duration of 2.5 μs and an interval of 10–100 s.

Chloroform ($+ 7.25$ p.p.m. with respect to TMS) was used as a secondary reference for the chemical shifts.

Theoretical calculations. A quantum-mechanically treated cluster, which included a framework cage hosting extraframework H^- and OH^- ions, was embedded into the rest of C12A7 and S12A7 lattices considered classically. The total energy of these systems was minimized with respect to coordinates of ~ 500 atoms including the quantum-mechanically treated cluster and its surrounding. Density functional theory and the hybrid B3LYP density functional, as implemented in the Gaussian 03 package[37], together with the 6-31 G(d) basis set, were used for the geometry optimization. Then, chemical shifts were calculated using two hybrid density functionals, the Hartree–Fock method and the 6-311 + G(2d,p) basis set[38] (see Supplementary Notes 5 and 6.).

References

1. Steiner, T. The hydrogen bond in the solid state. *Angew. Chem. Int. Ed.* **41**, 48–76 (2002).
2. Norby, T., Widerøe, M., Glöckner, R. & Larring, Y. Hydrogen in oxide. *Dalton Trans.* **19**, 3012–3018 (2004).
3. Janotti, A. & Van de Walle, C. G. Hydrogen multicentre bonds. *Nat. Mater.* **6**, 44–47 (2007).
4. Huang, B. & Corbett, J. D. Ba_3AlO_4H: synthesis and structure of a new hydrogen-stabilized phase. *J. Solid State Chem.* **141**, 570–575 (1998).
5. Heyward, M. A. et al. The hydride anion in an extended transition metal oxide array: $LaSrCoO_3H_{0.7}$. *Science* **295**, 1882–1884 (2002).
6. Helps, R. M., Rees, N. H. & Hayward, M. A. $Sr_3Co_2O_{4.33}H_{0.84}$: an extended transition metal oxide-hydride. *Inorg. Chem.* **49**, 11062–11068 (2010).
7. Lang, D. A., Zaikina, J. V., Lovingood, D. D., Gedris, T. E. & Latturner, S. E. Ca_2LiC_3H: A new complex carbide hydride phase grown in metal flux. *J. Am. Chem. Soc.* **132**, 17523–17530 (2010).
8. Boss, M., Petri, D., Pickhard, F., Zönnchen, P. & Röhr, C. New barium antimonide oxides containing zintl ions $[Sb]^{3-}$, $[Sb_2]^{4-}$, and $\frac{1}{\infty}[Sb_n]^{n-}$. *Z. Anorg. Allg. Chem.* **631**, 1181–1190 (2005).
9. Hanna, T. et al. Hydrogen in layered iron arsenides: indirect electron doping to induce superconductivity. *Phys. Rev. B* **84**, 024521 (2011).
10. Iimura, S. et al. Two-dome structure in electron-doped iron arsenide superconductors. *Nat. Commun.* **3**, 943 (2012).
11. Kobayashi, Y. et al. An oxyhydride of $BaTiO_3$ exhibiting hydride exchange and electronic conductivity. *Nat. Mater.* **11**, 507–511 (2012).
12. Oumellal, Y., Rougier, A., Nazri, G. A., Tarascon, J.-M. & Aymard, L. Metal hydrides for lithium-ion batteries. *Nat. Mater.* **7**, 916–921 (2008).
13. Shriver, D. F. & Atkins, P. W. *Inorganic Chemistry*, 3 edn 253–282 (Freeman, 1999).
14. Ito, H., Hasegawa, Y. & Ito, Y. Li–H_2 cells with molten alkali chlorides electrolyte. *J. Appl. Electrochem.* **35**, 507–512 (2005).
15. Hayashi, K., Matsuishi, S., Kamiya, T., Hirano, M. & Hosono, H. Light-induced conversion of an insulating refractory oxide into a persistent electronic conductor. *Nature* **419**, 462–655 (2002).
16. Hayashi, K., Sushko, P. V., Shluger, A. L., Hirano, M. & Hosono, H. Hydride ion as a two electron donor in a nanoporous crystalline semiconductor $12CaO \cdot 7Al_2O_3$. *J. Phys. Chem. B* **109**, 23836–23842 (2005).
17. Hayashi, K. Heavy doping of H^- Ion in $12CaO \cdot 7Al_2O_3$. *J. Solid State Chem.* **184**, 1428–1432 (2011).
18. Yoshizumi, T., Kobayashi, Y., Kageyama, H. & Hayashi, K. Simultaneous quantification of hydride ions and electrons incorporated in $12CaO \cdot 7Al_2O_3$ cages by deuterium-labeled volumetric analysis. *J. Phys. Chem. C* **116**, 8747–8752 (2012).
19. Kim, S.-W. et al. Metallic state in a lime-alumina compound with nanoporous structure. *Nano Lett.* **7**, 1138–1143 (2007).
20. Toda, Y. et al. Work function of a room-temperature stable electride $Ca_{24}Al_{28}O_{64}^{4+}(e^-)_4$. *Adv. Mater.* **19**, 3564–3569 (2007).
21. Sushko, P. V., Shulger, A. L., Hayashi, K., Hirano, M. & Hosono, H. Electron localization and a confined electron gas in a nanoporous inorganic electrides. *Phys. Rev. Lett.* **19**, 126401 (2003).
22. Nomura, T. et al. Anion incorporation-induced cage deformation in $12CaO \cdot 7Al_2O_3$ crystal. *Chem. Lett.* **36**, 902–903 (2007).
23. Lerch, M. et al. Oxide nitrides: from oxides to solids with mobile nitrogen ions. *Prog. Solid State Chem.* **37**, 81–131 (2009).
24. Palacios, L. et al. Structure and electrons in mayenite electrides. *Inorg. Chem.* **47**, 2661–2667 (2008).
25. Hayashi, K., Hirano, M. & Hosono, H. Thermodynamics and kinetics of hydroxide ion formation in $12CaO \cdot 7Al_2O_3$. *J. Phys. Chem. B* **109**, 11900–11906 (2005).
26. Hayashi, K. et al. Solid state syntheses of nanoporous crystal $12SrO \cdot 7Al_2O_3$ and formation of high density oxygen radicals. *Chem. Mater.* **20**, 5987–5996 (2008).
27. Chiesa, M. et al. Excess electrons stabilized on ionic oxide surfaces. *Acc. Chem. Res.* **39**, 861–867 (2006).

28. Bowman, R. C. NMR studies of electronic structure and hydrogen diffusion in transition metal hydrides. *Hyperfine Interact.* **24–26**, 583–606 (1985).

29. Shannon, R. D. Revised effective ionic radii and systematic studies of interatomic distances in halides and chalcogenides. *Acta Cryst.* **A32**, 751–767 (1976).

30. Berglund, B. & Vaughn, W. Correlations between proton chemical shift tensors, deuterium quadrupole couplings, and bond distances for hydrogen bonds in solids. *J. Chem. Phys.* **73**, 2037–2043 (1980).

31. Yesinowski, J. P. & Eckert, H. Hydrogen environments in calcium phosphates: ^{1}H MAS NMR at high spinning speeds. *J. Am. Chem. Soc.* **109**, 6274–6282 (1987).

32. Akitt, J. W. *NMR and Chemistry*, 3rd edn (Chapman & Hall, 1992).

33. MacKenzie, K. J. D. & Smith, M. E. *Multinuclear Solid-State NMR of Inorganic Materials* (Pergamon, 2002).

34. Ellaboudy, A., Tinkham, M. L., Van Eck, B., Dye, J. L. & Smith, P. B. Magic-angle spinning sodim-23 nuclear magnetic resonance studies of crystalline sodides. *J. Phys. Chem.* **88**, 3852–3855 (1984).

35. Tinkham, M. L. & Dye, J. L. First observation by ^{39}K NMR of K$^-$ in solution and in crystalline potassides. *J. Am. Chem. Soc.* **107**, 6129–6130 (1985).

36. Sushko, P. V., Shluger, A. L. & Catlow, C. R. A. Relative energies of surface and defect states: ab initio calculations for the MgO (001) surface. *Surf. Sci.* **450**, 153–170 (2000).

37. Frisch, M. J. *et al. Gaussian 03, Revision C.02* (Gaussian Inc., 2004).

38. Cheeseman, J. R., Trucks, G. W., Keith, T. A. & Frisch, M. J. A comparison of models for calculating nuclear magnetic resonance shielding tensors. *J. Chem. Phys.* **104**, 5497–5509 (1996).

39. Eckert, H., Yesinowski, J. P., Silver, L. A. & Stolper, E. M. Water in silicate glasses: quantification and structural studies by ^{1}H solid echo and MAS-NMR methods. *J. Phys. Chem.* **92**, 2055–2064 (1988).

40. Nicol, A. T. & Vaughan, R. W. Proton chemical shift tensors of alkaline earth hydrides. *J. Chem. Phys.* **69**, 5211–5213 (1978).

41. Magusin, P. C. M. M., Kalisvaart, W. P., Notten, P. H. L. & van Santen, R. A. Hydrogen sites and dynamics in light-weight hydrogen-storage materials magnesium-scandium hydride investigated with ^{1}H and ^{2}H NMR. *Chem. Phys. Lett.* **456**, 55–58 (2008).

42. Bowman, R. C., Hwang, S.-J., Ahn, C. C. & Vajo, J. J. NMR and X-ray diffraction studies of phases in the destabilized LiH-Si system. *Mater. Res. Soc. Symp. Proc.* **837**, N3.6.1–N3.6.6 (2005).

43. Sanderson, R. T. Electronegativity and bond energy. *J. Am. Chem. Soc.* **105**, 2259–2261 (1983).

44. Mengoet, M., Bartam, R. H. & Gilliam, O. R. Paramagnetic holelike defect in irradiated calcium hydroxyapatite single crystals. *Phys. Rev. B* **11**, 4110–4124 (1975).

45. Hartmann, P., Jäger, C., Barth, St., Vogel, J. & Meyer, K. Solid state NMR, X-ray diffraction, and infrared characterization of local structure in heat-treated oxyhydroxyapatite microcrystals: an analog of the thermal decomposition of hydroxyapatite during plasma-spray prodedure. *J. Solid State Chem.* **160**, 460–468 (2001).

46. de Leeuew, N. H., Bowe, J. R. & Rabone, J. A. A computational investigation of stoichiometric and calcium-deficient oxy- and hydroxy-apatites. *Faraday Discuss.* **134**, 195–214 (2007).

47. Isobe, T., Nakamura, S., Nemoto, R., Senna, M. & Sfihi, H. Solid-state double nuclear magnetic resonance study of the local structure of calcium phosphate nanoparticles synthesized by a wet-mechanochameical reaction. *J. Phys. Chem. B* **106**, 5169–5176 (2002).

48. Zahn, D. & Hochrein, O. On the composition and atomic arrangement of calcium-deficient hydroxyapatite: An ab-initio analysis. *J. Solid State Chem.* **181**, 1712–1716 (2008).

49. Freund, F. & Wengeler, H. The infrared spectrum of OH-compensated defect sites in C-doped MgO and CaO single crystals. *J. Phys. Chem. Solids* **42**, 129–145 (1982).

Acknowledgements

This research was supported by an Elements Strategy Initiative Project to Form Core Research Centers, from the Ministry of Education, Culture, Sports, Science and Technology of Japan. P.V.S. acknowledges supports by the Royal Society. We thank Y. Kubota, T. Nomura and H. Tanaka for Rietveld/MEM analysis, M. Yoshida and S.-W. Kim for sample preparation, M. Nayuki for NMR measurement, H. Nozaki and V. Pomjakushin for NPD measurements, Y. Kubota, M. Takata and J.-E. Kim for synchrotron XRD measurements.

Author contributions

H.H., K.H., and P.V.S. directed the entire project and co-wrote the manuscript. K.H. synthesized and characterized materials. K.H. carried out diffractometry analyses. Y.H. performed NMR measurements. P.V.S. and A.L.S. carried out *ab initio* calculations.

Additional information

5

Quantifying thiol–gold interactions towards the efficient strength control

Yurui Xue[1], Xun Li[1], Hongbin Li[2] & Wenke Zhang[1]

The strength of the thiol–gold interactions provides the basis to fabricate robust self-assembled monolayers for diverse applications. Investigation on the stability of thiol–gold interactions has thus become a hot topic. Here we use atomic force microscopy to quantify the stability of individual thiol–gold contacts formed both by isolated single thiols and in self-assembled monolayers on gold surface. Our results show that the oxidized gold surface can enhance greatly the stability of gold–thiol contacts. In addition, the shift of binding modes from a coordinate bond to a covalent bond with the change in environmental pH and interaction time has been observed experimentally. Furthermore, isolated thiol–gold contact is found to be more stable than that in self-assembled monolayers. Our findings revealed mechanisms to control the strength of thiol–gold contacts and will help guide the design of thiol–gold contacts for a variety of practical applications.

[1] State Key Laboratory of Supramolecular Structure and Materials, College of Chemistry, Jilin University, 2699 Qianjin Street, Changchun 130012, China.
[2] Department of Chemistry, University of British Columbia, Vancouver, British Columbia, Canada V6T 1Z1. Correspondence and requests for materials should be addressed to W.Z. (email: zhangwk@jlu.edu.cn).

Self-assembled monolayers (SAMs) of thiols on gold surfaces are one of the most popular model systems for the study of the self-assembly of organic molecules on metal surfaces. SAMs that are based on thiol–gold chemistry have been widely employed in the fields of chemistry, physics, molecular biology, pharmaceutical engineering and materials science[1–3]. Considerable interests in such systems have increased, owing to their versatile applications, including the fabrication of nano-patterning[4,5], molecular-scale devices[6,7], optical materials[8,9], formulation of biosurfaces[10] and support for cell culture[2,11]. The strength of the gold–sulphur (Au–S) interaction formed between thiols and gold surfaces provides the basis to fabricate robust SAMs for diverse applications. The study on the nature of Au–S interaction and the stability of SAMs formed on gold surfaces under various conditions, such as different surface properties of gold, solution pH and types of solvents, is thus important. One of the effective ways for such study is to break the individual Au–S contact by external force. Given the limited availability of detection methods, such experiments are challenging.

The advent of several single-molecule manipulation techniques, including optical tweezers, magnetic tweezers, biomembrane force probe and atomic force microscopy (AFM)-based single-molecule force spectroscopy (SMFS), offers new and powerful tools to investigate the intra- or intermolecular interactions of both natural (biological) and synthetic macromolecules[12–16]. These interactions are unbinding forces that originated from specific interactions in biological systems (that is, the unfolding of proteins, melting of DNA strands, antigen–antibody interactions, ligand–receptor interactions and protein–nucleic acid interactions), entropic elasticity, basic supramolecular interactions (that is, hydrogen bonds, coordinated bonds, π–π interactions, hydrophobic interactions and so on) and even the strength of a single covalent bond[13,17–28]. The investigations of molecular interactions in complicated systems, such as on live bacterial surfaces, in intact virus particle, and in condensed polymer materials, have also been performed successfully via AFM-based SMFS[21,22,29].

For thiol–gold interactions, numerous investigations focusing on the exploration of the rupture mechanism of Au–S interaction have been carried out in the past few years by using both experimental and theoretical methods. Gaub et al.[24] pioneered in conducting SMFS experiments to detect the strength of a single covalent bond around Au–S binding sites. They determined a rupture force of 1.4 ± 0.3 nN in their experiments. However, they did not answer unambiguously the origin of the measured rupture force (that is, the rupture of the Au–S bond or the extraction of the anchored gold atoms from substrates). Ab initio molecular dynamics simulations[30,31] in the framework of 'virtual AFM experiments' demonstrate that pulling a thiol attached to the gold surface could result in the breakage of a Au–Au bond with rupture force of ~ 1.2 nN. Mechanically controlled break-junction experiments[32,33] that are slightly different from the AFM-based force experiments have demonstrated that the breaking of the molecular junctions takes place at Au–Au bonds near the molecule–electrode contact and produces a similar rupture force of 1.5 nN. However, an AFM-based force experiment conducted by Skulason et al.[25] showed a small force value of 0.1 nN, which was also demonstrated to be the abstraction of thiolate-complexed gold atoms from the tip surface. Recently, Venkataraman et al.[34] measured the breaking force in thiol junctions and showed a rupture force of ~ 1.0 nN. These results show that the forces required to break the Au–Au bonds obtained in their respective experiments are substantially different from each other, although the underlying rupture mechanism is thought to be identical in these different

experiments. A careful comparison of these studies shows that the experimental conditions, gold surface properties and sample preparation conditions (solvent and pH) are actually different. This observation implies that significant variations in rupture forces between these measurements may originate from different experimental conditions. This finding also suggests that the strength of a single Au–Au bond around Au–S binding sites can be influenced by external factors.

Investigating the factors that affect the strength of thiol–gold contact is necessary to clarify these controversial results. This approach can also facilitate a deep and direct understanding of the rupture mechanisms and dynamic processes of a single bond formed between thiol and gold surfaces, as well as the rational design of stable novel nanomaterials and nanodevices based on Au–S-specific interaction. However, to the best of our knowledge, detailed information on the fundamental factors affecting the strength of single bonds formed between Au and S atoms of thiol for both isolated single molecules and individual molecules in SAMs, especially at the single molecular level, remains unavailable.

In this paper, AFM-based SMFS is employed to study the effects of experimental conditions on the strength of single bonds at the thiol–gold interface formed by isolated thiols or in SAMs. Our results show that thiolate-bound gold atoms could be extracted from the gold surface with the retraction of the AFM tip, leading to the breakage of Au–Au bonds near the Au–S binding sites for both isolated thiols and SAMs. For isolated thiols, the strength of single bonds formed between thiols and Au surfaces is strongly affected by the properties of the gold surface, the solution pH and the interacting time. We also observe a shifting of the bond from coordinate to covalent bond upon increasing the environmental pH and the interaction time. However, the strength of individual Au–Au bonds around Au–S binding sites for SAMs is affected only by the properties of the gold surface and the reaction time. Our results also show a large difference in the rupture forces of Au–Au covalent bonds obtained from isolated thiols and individual thiols in SAM. The significance of our findings is discussed.

Results

Effect of surface properties on thiol–gold contacts. Proper cleaning of the gold surface is necessary to obtain high-quality thiol–gold-based SAM. For this purpose, rational methods for preparing highly reproducible gold surfaces, including the oxidative[35–38] and reductive[36,37] pretreatments, have been proposed. Briefly, gold substrates could be oxidized to a positive charge state via conventional methods, such as ultraviolet/ozone, oxygen plasma, electrochemical oxidation and piranha solution oxidation. The freshly prepared oxidized gold surfaces can be chemically reduced to zero state (metallic gold) after they were immersed in ethanol. Given that these pretreatment methods could produce different states (valence) of gold surfaces, correlating the gold properties with the stability of thiol–gold contact is necessary. To the best of our knowledge, no such study at the single molecule level is currently available. In this section, the effects of the properties of gold surfaces on the strength of single thiol–gold contacts at pH 8.0 are discussed.

As described above, oxidized and reduced gold surfaces were prepared by employing piranha solution and ethanol, respectively, to study this effect. In accordance with the phenomena reported in literature, the oxidized gold substrates were hydrophilic, whereas the reduced gold substrates were hydrophobic[39,40]. X-ray photoelectron spectroscopy (XPS) was used to precisely determine the oxidization states of the freshly prepared gold surfaces. As seen in Fig. 1, the Au $4f_{7/2}$ bands with

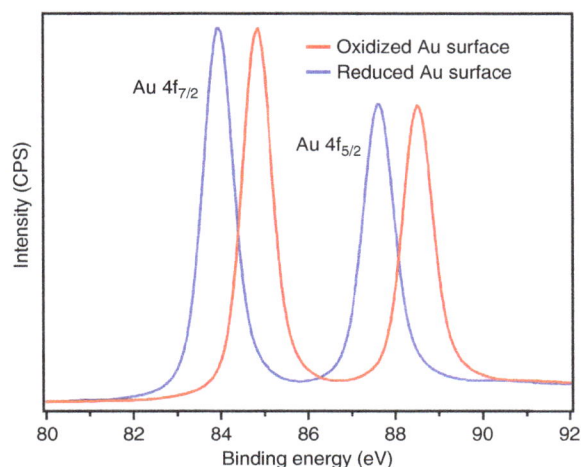

Figure 1 | XPS evidence for the oxidization and reduction of gold surfaces. The spectra represent the two sets of doublet ($4f_{7/2}$ and $4f_{5/2}$) of Au 4f core-level spectra of oxidized (red line) and reduced (blue line) gold surfaces.

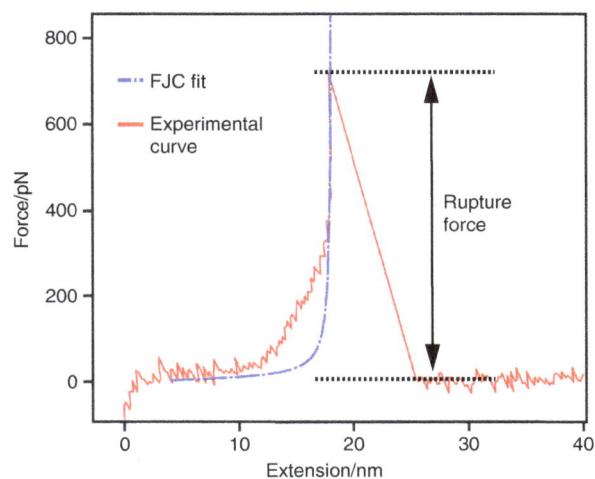

Figure 3 | Typical force-extension curve with the freely jointed chain (FJC) fit. The rupture forces obtained on such single molecule stretching curves were employed to construct the force histograms. The deviation between the experimental curve and FJC fit is due to the PEG linker[66].

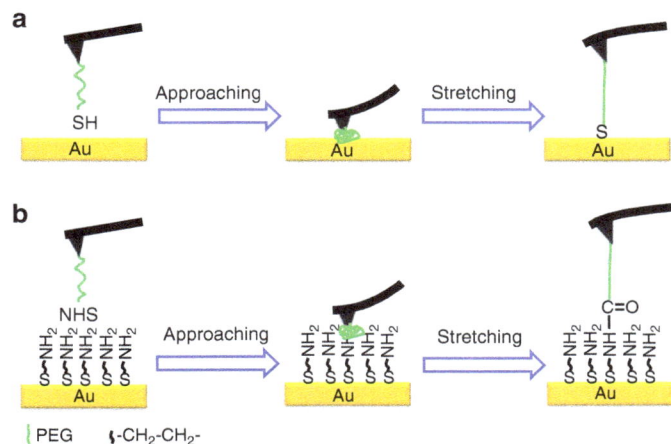

Figure 2 | Illustration of AFM-based SMFS study of the gold–sulphur interactions. (**a**) Illustration of the formation and breakage of isolated thiol–gold contact *in situ*. (**b**) Illustration of the detection of thiol–gold interactions in amino-terminated, self-assembled monolayers.

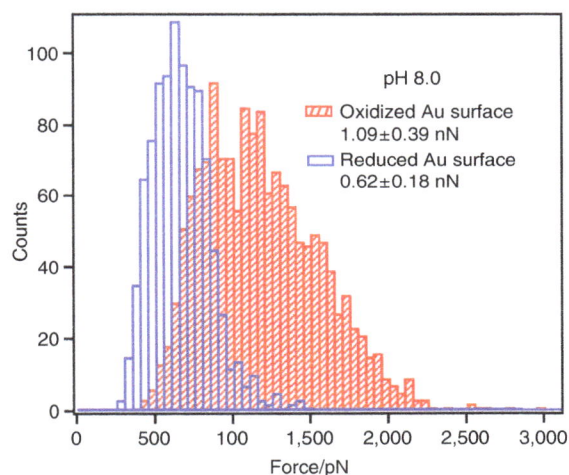

Figure 4 | Effect of surface properties on the strength of thiol–gold contacts at pH 8.0. Histograms of the rupture forces obtained by unbinding the isolated thiol from the oxidized (red) ($n=1{,}510$, n is number of events) and reduced (blue) ($n=970$) gold surfaces at pH 8.0.

binding energies of 84.85 and 83.95 eV correspond to oxidized and reduced gold surfaces, respectively, which indicate that the ethanol treatment changed the gold surface from the oxidized to the reduced state. The narrow scans in the Au 4f region of oxidized gold surface shifted to higher binding energy compared with the reduced surface[36,41,42], indicating a remarkable energy difference between each doublet.

α-Thiol-ω-carboxy-terminated poly (ethylene glycol) (HS-PEG-COOH) was attached to the amino-group-modified AFM tip via the amide bond, leaving the thiol ends exposed for attachment to the gold surface. During the contact of thiol-labelled AFM tip with the gold substrate (oxidized or reduced), a molecular bridge formed between the AFM tip and the gold substrate via the thiol–gold chemistry (Fig. 2a). During the separation of the AFM tip with the gold surface, the weakest part of the bridge structure broke, resulting in a rupture event. The resulting rupture force was measured (Fig. 3). Statistical analysis on the rupture forces was performed.

Figure 4 shows the histograms of rupture forces obtained on oxidized and reduced gold surfaces at pH 8.0, respectively. For the

oxidized gold surface, the most probable rupture force is 1.09 ± 0.39 nN. For the reduced gold surface, the force histogram indicates a most probable rupture force of 0.62 ± 0.18 nN. Previous studies[5,27,30–32,34,43,44] strongly indicated that the breakage takes place at the Au–Au bond around the Au–S binding sites, which was the weakest among the covalent bonds (that is, Si–O, Si–C, C–N, C–C, C–O, Au–S and Au–Au) in the linkage, leaving one or more gold atoms at the terminal of the tethered linker. Control experiments, in which the thiol-terminated PEG was replaced with sulphur-free methoxyl-terminated PEG, were performed to confirm that the rupture force obtained above originated from the interaction between the sulphydryl group and gold surfaces. The histograms of the resulted rupture forces obtained from the interaction between PEG-OCH$_3$ linker and gold surfaces gave force peaks of 43 pN at pH 7.4 and 34 pN at pH 8.0, respectively (see Supplementary Fig. 1).

Another control experiment on sulphydryl-terminated silicon surface (Fig. 5) indicates that the gold atoms have been extracted

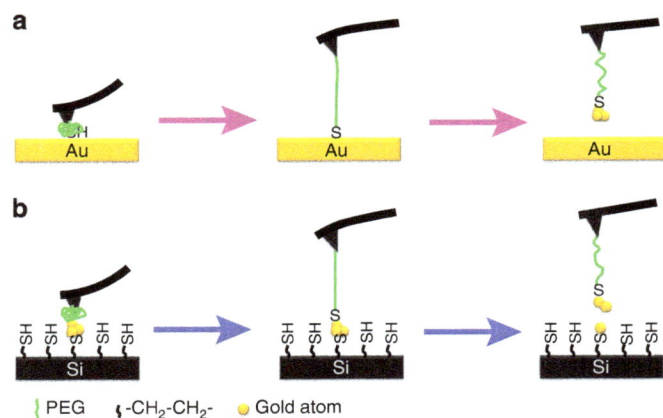

Figure 5 | Illustration of AFM-based SMFS study of the breakage of isolated Au–Au bonds. (**a**) Illustration of the abstraction of gold atoms from the gold surfaces. (**b**) Same AFM tip employed in **a** was brought to interact with the sulphydryl-terminated silicon surfaces.

from the gold surface and became attached to the AFM tip during the pulling experiment (see Fig. 5a and Supplementary Fig. 2). Based on these facts, we deem that the obtained rupture force shown in Fig. 4 can be ascribed to the cleavage of single Au–Au bonds.

Although the rupture of the same bonds on stretching is very likely to be in both cases, the most probable rupture force obtained on oxidized gold surfaces is larger than that on the reduced surfaces. As proposed by Ron et al.[37], thiols may directly react with the oxidative gold surface to form Au–S bonds via the oxidation–reduction reaction involving the reduction of the gold oxide and the direct adsorption of thiols onto the reduced gold surface. The formed species on both the oxidized and reduced gold surfaces are gold (I) thiolates. Different variation in the degree of valence may have distinct effects on the precise rearrangement behaviour of the gold atoms beneath the surface, which in turn could result in the variation of the strength of the Au–Au bond. Recently, *ab initio* investigations[45] on the formation of alkanethiol on gold surfaces suggested the existence of a noticeable perturbation of the position of the gold atoms around the Au–S binding sites, which affect the electronic structures of the neighbouring gold atoms. The difference of the unbinding force in our current system may thus come from the different rearrangement of the surrounding gold atoms on oxidized and reduced gold surfaces during the binding process.

Effect of pH on the strength of single thiol–gold contacts. The adsorption of thiol onto gold surface starts with physisorption, during which the H atom favourably remains on the S atom, followed by a chemisorption including the breaking of S–H bond and the formation of a Au–S bond as a result of the deprotonation of thiols and formation of thiyl radicals[45–51]. The existence of the dissociated hydrogen could affect the pH value of the microenvironment around the Au–S binding sites. Thus, the environmental pH also affects the formation of the Au–S bond, because an acidic environment inhibits the dissociation of S–H bonds, whereas an alkaline environment favours this dissociation. To the best of our knowledge, no detailed/direct information exists on the effect of environmental pH on dynamic processes for the formation of Au–S covalent bonds. AFM-based SMFS method was employed to quantify the effect of pH on the mechanical strength of an isolated thiol–gold contact.

Figure 6 shows the distributions of rupture forces of thiol–gold contacts obtained on oxidized (Fig. 6a) and reduced (Fig. 6b) gold

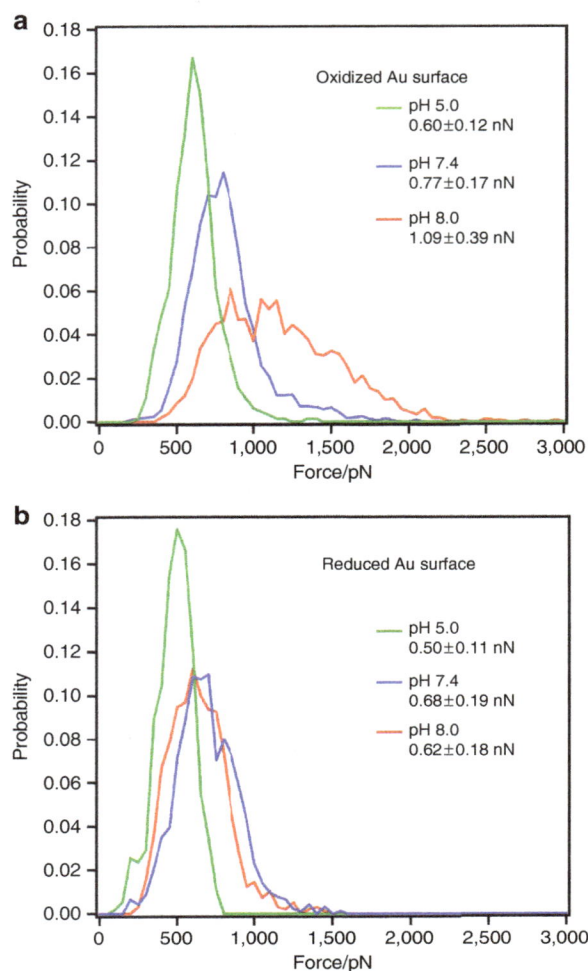

Figure 6 | pH effects on the strength of thiol–gold contacts. Histograms of the forces required to rupture single bonds formed between isolated thiols and (**a**) oxidized and (**b**) reduced Au surfaces at pH 5.0 (green line), 7.4 (blue line) and 8.0 (red line), respectively.

surfaces in aqueous solutions at different pH values, respectively. From the data, we can see that the bonds formed between isolated individual thiols and oxidized gold surfaces break with a most probable unbinding force of ~0.60, 0.77 and 1.09 nN at pH 5.0, 7.4 and 8.0, respectively. The most probable corresponding rupture forces obtained on reduced gold surfaces are ~0.50, 0.68 and 0.62 nN, respectively. The most probable rupture force obtained on oxidized gold surfaces increases with increasing pH value. The rupture forces obtained on reduced gold surfaces show a minimum force value at pH 5.0 and similar force values at pH 7.4 and 8.0, respectively.

During the formation of a Au–S covalent bond, multiple bonding scenarios exist depending on the location of the H atom, which could lead to drastic changes in the rupture force[44,52], and a possibility of forming weaker coordinate bonds between the protonated SH groups and gold surfaces[53]. Recently, a comprehensive investigation on the elongation process of molecular junctions formed by octanedithiol molecule and Au electrodes indicated that the bonds that are easiest to break are the coordinated Au-linker contacts (~0.6 nN), which are smaller than the covalent thiolate–gold junctions (from ~1.5 to ~2.2 nN)[52]. These findings indicate that the fraction of the intact –SH group could have great influence on the rupture force obtained in SMFS experiments. Considering the reaction mechanisms between thiols and gold surfaces, a higher proportion of intact –SH groups, which could form Au–SH coordination bonds, can most probably exist at lower pH values. On the contrary, more –SH groups deprotonated and the percentage of Au–S covalent bonds increased significantly at higher pH conditions, leading to a larger rupture force. As a result, the increase in pH value would shift the bond type of thiol–gold contact from a coordinate bond to a covalent bond. To further prove our hypothesis, both XPS and Raman spectroscopy have been used to study thiol–gold interactions and the main findings are in good agreement with our SMFS results (see Supplementary Figs 3 and 4).

The higher rupture force observed on oxidized gold surface may be ascribed to the fact that the presence of gold oxide (for example, the –OH groups on the surface) promotes the conversion of the protonated SH group to water in the reaction, whereas the released proton remains in the form of hydrogen ions for the reduced gold surface.

Effect of interaction time on thiol–gold contacts. As discussed above, the coordinate bond could actually form between the –SH group and gold surface in the initial stage, followed by the dissociation of S–H bond that results in a thiyl radical, which would finally form a gold–thiolate covalent bond at the gold–sulfur interface[45–51,54–58]. These findings indicate that the formation of Au–S covalent bonds needs a certain time to complete. Experimental evidence[58] also suggests that the loss of hydrogen can be prevented to some extent as long as no reaction occurs for hydrogen removal, which provides a way to control the bonding strength between thiols and gold.

By changing the waiting/reaction time during the contact of the thiol group and gold surfaces, we have investigated the effects of interaction time on the strength of thiol–gold interaction at the single molecule level.

Figure 7 shows histograms of the rupture forces obtained between thiols and gold surfaces at pH 8.0 with different interaction times. The most probable rupture forces of 0.69 nN at 1.0 s, 0.76 nN at 2.0 s, 1.09 nN at 3.0 s and 1.01 nN at 8.0 s have been observed. The rupture forces increased with the increase of interaction time from 1.0 to 3.0 s. This finding indicates that an apparent shifting of bond types from coordinate to covalent with the increase in the interaction time might have occurred.

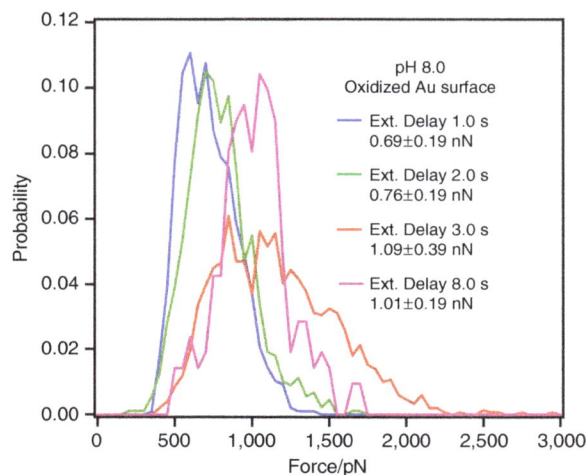

The rupture force obtained at the interaction time of 8.0 s is also similar to that obtained at 3.0 s, which indicates that after a reaction time of ≥ 3.0 s, the majority of the thiol–gold contact will form a covalent bond at pH 8.0. The quantitative information obtained on the stability of isolated thiol–gold contacts by using AFM-based SMFS is useful for sample systems where the strength of individual thiol–gold contact plays important roles, such as the picking up (or immobilization) and manipulation of thiol-labelled polyethylene oxide chain (or polyprotein)[22,59,60]. An *in situ* waiting time of ≥ 3.0 s for the formation of covalent S–Au bond is necessary to stabilize the thiol–gold chemistry-based molecular linkages.

Strength of individual thiol–gold contacts in SAMs. From the discussion above, we know that experimental conditions can significantly affect the stability of the isolated thiol–gold contacts. How much will these conditions affect the stability of the thiol–gold contacts in SAMs? To answer this question, SAMs of amino-terminated thiol have been prepared on oxidized and reduced gold surfaces, respectively; environmental conditions, such as surface properties, pH and reaction time, on its stability were also studied by using the SMFS method (Fig. 2b). It needs to be pointed out that we focus on the quantitative comparison of the stability of the thiol–gold contact in isolated state and in SAMs in this study. To realize this, we need to minimize the effect of neighbouring/adjacent molecules in SAMs on the apparent thiol–gold interactions (or the apparent rupture forces obtained by AFM). Thus, we have chosen cysteamine, which contain short carbon chain, for the preparation of SAMs, since thiol molecule with long alkane chain will further stabilize the SAMs[2,60].

A survey on the published papers shows that the commonly used reaction time for the preparation of SAMs of thiols on gold surfaces ranged from a few minutes to days. In accordance with frequently used reaction conditions published in literature, SAMs employed in our experiments were prepared by immersing oxidized or reduced Au substrates in 100 mM of cysteamine PBS (pH 7.4) solution for 0.5, 5.0 and 24 h, respectively. Subsequently, the mechanical stability of SAMs was studied by using AFM-based SMFS at pH 8.0.

Figure 8 shows the most probable rupture forces of 0.56 nN for 0.5 h, 0.54 nN for 5.0 h and 0.57 nN for 24.0 h for SAMs formed

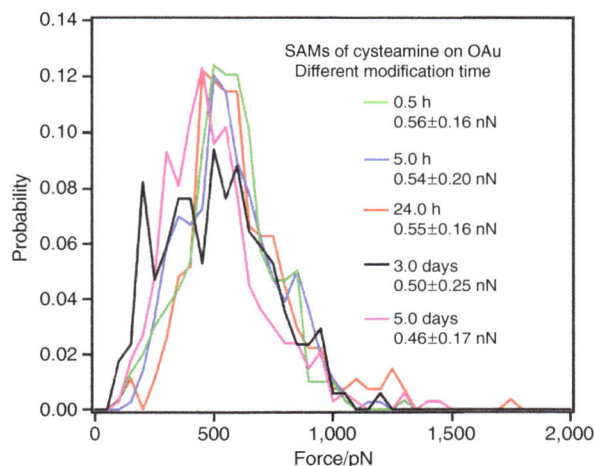

Figure 7 | Interaction time dependence of the strength of thiol–gold contacts. Histograms of the forces required to rupture single bonds formed between thiol and oxidized gold surfaces at pH 8.0 with increase of surface waiting time from 1.0 s to 2.0, 3.0 and 8.0 s.

Figure 8 | Effect of reaction time on stability of thiol–gold contacts in SAMs. Histograms of the rupture forces obtained from the SAMs, which were produced by immersing oxidized Au substrates in 100 mM of cysteamine solutions for 0.5 (green line), 5.0 (blue line), 24.0 h (red line), 3 days (black line) and 5 days (pink line). SMFS experiments were carried out at pH 8.0.

on oxidized gold surface. No significant variation of the rupture forces was observed, suggesting similar mechanical stability for them. A further prolonged reaction time, for example, 5 days, can even weaken the thiol–gold interactions so that the rupture force can reach ∼0.46 nN, which is close to that obtained on reduced gold surface, as shown in Supplementary Fig. 5. This result indicates that from the stability point of view, 0.5 h is enough for the formation of SAMs, and very long reaction time is not helpful. The information is useful for the immobilization of sensitive thiol-containing proteins (or other biological samples) on gold substrate[18].

By using a similar experimental strategy, we also investigated pH effects on the stability of the SAMs formed on oxidized and reduced gold surface, respectively. The most probable rupture forces for the breaking of individual molecules in SAMs on oxidized Au surfaces are 0.55 nN at pH 5.0, 0.53 nN at pH 7.4, and 0.57 nN at pH 8.0, respectively (see Supplementary Fig. 6a). The force profile corresponding to the rupturing force of individual molecules in SAMs on reduced gold surfaces indicates the most probable rupture forces of 0.49 nN at pH 5.0, 0.48 nN at pH 7.4, and 0.46 nN at pH 8.0 (see Supplementary Fig. 6b). Our results reveal that the rupture force obtained on an oxidized gold surface is larger than that obtained on a reduced gold surface at the same pH value. Another important phenomenon is that the most probable bond rupture force does not change with the change of the solution pH on respective gold surfaces. Considering the fact that the SAMs were formed by 5-h reaction at pH 7.4, the cleavage of S–H bond and the formation of Au–S covalent bond should be completed. As a result, the solution pH will not affect the binding (between the thiol and gold) process any more. Control experiments, in which the same AFM tip that has been used to break the thiol–gold contacts in SAMs (Fig. 2b and Supplementary Fig. 7), were used to perform the force measurement on sulphydryl-terminated silicon surface again to prove our hypothesis. The result shows that the Au–Au bond has been broken, and gold atoms were attached to the AFM tip during the force measurement on SAMs (see Supplementary Fig. 7, top row). In addition, it is interesting to note that the rupture force obtained in the control experiment (∼1.2 nN) (Step II) is bigger than that obtained in Step I (∼0.57 nN) (see Supplementary Fig. 7). This phenomenon may indicate that the arranging modes of Au atoms on the gold substrate are different from that attached to the AFM tip (more like gold nanocluster). Furthermore, it has been recognized that Au(111)/SAM interface can contain a layer of Au adatoms that are covalently bonded with the neighbouring thiol(ate) molecules, that is, the first atomic layer above a metallic Au(0) surface contains a layer of Au(I) and S. This bonding network is covalent $(-(S-Au)_n-)$[1,61]. Hence, when the AFM tip is contacted to a thiol(ate) that is a part of the SAM layer, a possible scenario where a complex molecular wire was peeled off from the gold substrate and get attached to the AFM tip would happen, as shown in the bottom row of Supplementary Fig. 7a. The breakage of such kind of molecular wire in Step II can happen either at the Au–Au or the Au–S sites, depending on the binding mode between thiol group (on the silicon surface) and the molecular wire of gold (see Supplementary Fig. 7b,c). As a result, the histogram may shift towards higher force values and get broadened. Our results also indicate that after the formation of SAMs, environmental pH will not affect the stability of the SAMs. However, during the formation of SAMs, the solution pH can affect the efficiency for the conversion of thiol–gold contacts from non-covalent (that is, coordinate) to covalent bonds (see Supplementary Figs 3 and 4). These findings are useful for the design of gold nanoparticle and thiol–gold chemistry-based drug (or gene) delivery systems.

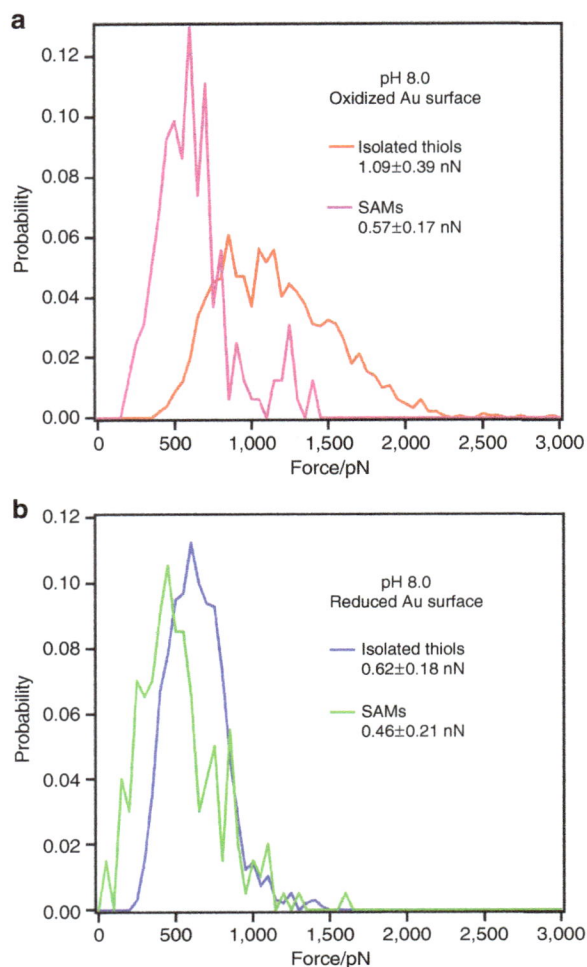

Figure 9 | Comparison of the strength of isolated thiol–gold contacts and that in SAMs. Histograms of rupture forces of the single thiol–gold contact formed by isolated molecules and in SAMs on (**a**) oxidized and (**b**) reduced gold surfaces at pH 8.0, respectively.

Comparison of isolated thiol–gold contacts and that in SAMs. Figure 9 shows the direct comparison of the mechanical strength of single thiol–gold contacts formed by isolated single thiols and in SAMs on oxidized (Fig. 9a) and reduced (Fig. 9b) surfaces at pH 8.0, respectively. Interestingly, our results show that the rupture forces of isolated single molecules immobilized on both oxidized and reduced gold substrates are larger than that obtained on SAMs formed on respective substrates. For oxidized gold surfaces (Fig. 9a), the most probable rupture force obtained from an isolated molecule (∼1.09 nN) is significantly larger than that obtained in SAMs (∼0.57 nN). For reduced gold surfaces (Fig. 9b), the most probable rupture force obtained from an isolated molecule (∼0.62 nN) is also larger than that obtained in SAMs (∼0.46 nN).

The formation of a Au–S covalent bond is a complicated process, involving the dissociation of the S–H bond and followed by the formation of the Au–S covalent bond[45–51]. The dissociated hydrogen atoms could either adsorb on gold surfaces or release from the surface in the form of H_2. Ensemble experiments also indicated that the loss of hydrogen could undergo a very slow process[56,58], even as long as several minutes, which is longer than the interaction time (3.0 s) between isolated thiols and gold surfaces in our SMFS experiments for isolated thiols. The released H located on the gold surface could lead to drastic changes in rupture forces[45,52]. These facts have caused the dependence of the unbinding force on pH and waiting time (within the time scale of 3.0 s) for isolated thiol–gold contacts. However, the SAMs adopted in our experiments were prepared by immersing the freshly prepared oxidized or reduced gold substrates into 100 mM of cysteamine aqueous or alcoholic solutions for 5 h at room temperature, respectively. The dissociated H from the SH group would have detached from the gold surfaces and have no effect on the rupture force of Au–Au bond in our experiments.

The large difference of rupture forces of the thiol–gold contacts formed by isolated and self-assembled thiols may be ascribed to the following reasons: (1) the large amount of cysteamine (100 mM) can act as a reduction reagent, and the originally oxidized gold surface has been reduced during the formation of SAMs (see Supplementary Fig. 8). As discussed above, the reduced gold surface can weaken the Au–Au bond; (2) the formation of SAMs on gold surfaces may weaken the interactions between gold atoms in the top and lower layers because earlier scanning tunnelling microscopy studies have shown the mobility of Au thiolates within the SAMs of alkanethiols[62,63].

Discussion

We have quantified the effects of experimental conditions, such as surface properties, solution pH/composition and reaction time, on the strength of chemical bonds formed between isolated thiols and gold surfaces or those in SAMs by using AFM-based SMFS. Our results unambiguously demonstrate that an oxidized gold surface can significantly enhance the stability of thiol–gold contacts in both isolated and self-assembled systems. However, for the later case, if the reaction time during the formation of SAMs on oxidized gold is too long (for example, longer than 5 days), the enhancement of stability will decrease greatly (see Fig. 8). This may indicate that the final molecular structures of the Au/SAM interface on oxidized and reduced gold are similar, as the XPS spectra of these two system are very similar (see Supplementary Fig. 8). When ethanol is adopted as the solvent during the preparation of SAMs on oxidized gold surface, the strength of thiol–gold contacts can be weakened because of the reduction effect of ethanol on gold. The formation of covalent thiol–gold contacts shows pH dependence: at lower pH, coordinate bonds dominate in the contacts; at higher pH, the bond type becomes covalent. Our results also show that the *in situ* formation of Au–S covalent bond requires a minimal interaction time of around 3.0 s. This observation means that to obtain a stable molecular bridge in between a gold-coated AFM tip and the substrate via the thiol–gold chemistry, a contact time of 3.0 s is necessary. Although the direct contact of thiol-modified AFM tip with gold substrate can speed up the reaction between the thiol and gold (as compared with the reaction in free solution) due to the shorter mass transport time, normally higher indentation force (>200 pN) will not speed up the reaction further, and indentation force that is too high may even destroy the thiol group or the PEG linker. Considering the fact that the time for mass transport in our AFM experiment is nearly zero, the 3.0 s interaction time necessary for strong interaction is the formation of covalent Au–S bonds[25,55]. However, the reaction time for the formation of SAMs can be different. Depending on the type of thiol molecules (for example, with long or short alkyl chain), the concentration as well as the cleanness of the gold surface, the formation time for a 'full monolayer' can vary from seconds to minutes, up to hours and days[2,55,64]. The weakened strength of the thiol–gold contact in SAMs compared with that of isolated thiols can be ascribed to both the reduction effect of the thiols and the further weakening effect of the self-assembled thiols on the Au–Au interactions.

Our findings on the stability of thiol–gold contacts are useful for practical applications. For example, to stabilize the anchor of single thiol-labelled molecules, we can use an oxidized gold surface and perform the reaction in aqueous (rather than in ethanolic) solution at higher pH with appropriate reaction time (for example, ~ 3.0 s for isolated thiol–gold contacts, while less than 1 day for SAMs). For the single-molecule pulling experiment, in which a thiol-labelled molecule needs to be attached to the gold substrate and the strength of the thiol–gold contact is crucial to the experiment[18,60], the molecule of interest needs to be immobilized onto the gold surface under a more dilute solution; the utilization of small thiol molecules as a 'dilute agent' is also avoided to co-assemble the target molecule[13].

Methods

Chemicals and reagents. HS-PEG-COOH (Mw = 3,400 Da, PDI = 1.4) and ω-carboxy-terminated poly (ethylene glycol) methyl ether (Mw = 20,000 Da, PDI = 1.05) were purchased from Polymer Source Inc. α, ω-dicarboxyl-terminated poly (ethylene glycol) (HOOC-PEG-COOH, Mw = 3,500 Da, PDI = 1.03) was purchased from JenKem Technology Co. 3-Aminopropyldimethylmethoxysilane was obtained from Fluorochem (UK). (3-Mercaptopropyl)trimethoxysilane, N-hydroxy-succinimide (NHS), 1-(3-dimethylaminopropyl)-3-ethylcarbodiimide hydrochloride (EDC) and cysteamine were purchased from Sigma-Aldrich. The PBS solution (pH 7.4) was prepared by dissolving one PBS tablet (Sigma) in 200 ml of deionized water and filtered. All other chemical reagents were of analytical reagent grade and were used as received without further purification. High-purity deionized water (dH$_2$O > 18 MΩ cm) purified with a Millipore System was adopted to prepare all aqueous solutions in this work.

Silanization of AFM tips. Silicon nitride AFM tips (Veeco Instruments, now Bruker Nano, Santa Barbara, CA; MSCT) with different functionalizations were employed for the SMFS experiments. Before modification, the AFM tips were treated with piranha solution (H$_2$SO$_4$ (98%)/H$_2$O$_2$ (30%) = 7:3 in volume), followed by thorough rinsing with high-purity deionized water, and then drying in an oven at 115 °C for 90 min to remove any remaining water. (*Caution: Piranha solution that may result in explosion or skin burns is a very hazardous oxidant. This solution must be handled with extreme care.*) The vapour-phase deposition method was introduced to silanize the cleaned AFM tips by suspending them in the atmosphere of the 3-aminopropyldimethylmethoxysilane in a dry nitrogen-purged desiccator for 1 h at 20 °C. After rinsing thrice with methanol, the silanized tips were subsequently placed in a 110 °C oven for 10 min.

Thiol derivatization of AFM tips. The thiol-terminated tips were prepared by introducing a covalent attachment of thiol-bearing polymer (HS-PEG-COOH) to amino-terminated AFM tips using standard EDC/NHS chemistry. Briefly, the carboxyl ends of the PEG (0.1 mM) were activated by reacting with EDC (6.0 mM) and NHS (10 mM) in PBS at pH 7.4 to introduce succinimide-reactive groups. The activated polymer was incubated with the amino-functionalized tips for 1 h and was then rinsed thrice with PBS (pH 7.4) to remove unanchored molecules.

NHS activation of AFM tips. HOOC-PEG-COOH was activated by employing EDC/NHS chemistry as described above before use. Afterwards, the amino-silanized tips were incubated with the resulting 4.2 mM α, ω-di(NHS)-PEG in PBS buffer for 30 min. The tips were rinsed thrice with the same buffer to remove unanchored molecules.

Preparation of oxidized gold surfaces. Gold surfaces (with ~50 nm of chromium and 200 nm of gold on glass substrate) were degreased for 5 min in anhydrous ethanol and then placed in freshly prepared piranha solution (H$_2$SO$_4$ (98%)/H$_2$O$_2$ (30%) = 7:3 in volume) for at least 10 min. The gold surfaces were rinsed with high-purity deionized water exhaustively and dried with high-purity nitrogen gas. The freshly prepared gold substrates were termed oxidized gold substrates.

Preparation of reduced gold surfaces. The reduced gold substrates were obtained by immersing the freshly prepared oxidized gold substrates in pure anhydrous ethanol for 2 h at room temperature.

Preparation of amino-terminated SAMs. Amino-terminated SAMs on oxidized gold surfaces were prepared by immersion of the oxidized gold substrate in 100 mM of cysteamine PBS (7.4) solution for 5 h. The SAMs on reduced gold surfaces were prepared by immersion of the reduced gold sample in 100 mM of ethanolic solution of cysteamine for 5 h. Afterwards, the samples were carefully rinsed with deionized water and dried with high purity nitrogen.

Preparation of sulphydryl-terminated silicon substrate. Silicon wafers were first treated with freshly prepared piranha solution (H_2SO_4 (98%)/H_2O_2 (30%)) = 7:3 in volume), followed by thorough rinsing with high-purity deionized water, and drying in an oven at 115 °C for 90 min to remove any remaining water. (*Caution: Piranha solution that may result in explosion or skin burns is a very hazardous oxidant. This solution must be handled with extreme care.*) The vapour-phase deposition method was introduced to silanize the cleaned silicon substrates by suspending them in the atmosphere of (3-mercaptopropyl)trimethoxysilane in a dry nitrogen-purged desiccator at 25 °C for 120 min. Subsequently, wafers were rinsed thrice with methanol, followed by drying under a stream of nitrogen gas.

AFM-based SMFS. Force spectroscopy experiments on both isolated single molecules (Fig. 2a) and individual molecules in SAMs (Fig. 2b) on gold were carried out on a NanoWizardII BioAFM (JPK Instrument AG, Berlin, Germany) in contact mode. Detailed descriptions on the operation of the AFM-based SMFS have been reported elsewhere[13,21,22]. Functionalized Si_3N_4 AFM tips were adopted in this work, and the spring constants of AFM cantilevers were calibrated by the thermal noise method, producing spring constants of 0.02–0.03 N m^{-1} (ref. 65). All measurements were carried out with freshly prepared AFM tips and samples in PBS buffer at room temperature. In each approach–retraction cycle, a modified AFM tip was first brought into contact with the gold surface at a constant indentation force of 200 pN and then held on the gold surface for 3.0 s, unless stated otherwise before retraction. During the separation of the AFM tip from substrate, the formed connective bridge in-between can be stretched and eventually broken, and the rupture force can be recorded. The pulling speed was kept constant at 1.0 μm s^{-1} for all experiments.

Analysis of SMFS data. Considering that PEG spacer was employed during the pulling experiment, force-extension curves that show characteristic mechanical properties of a single PEG chain (that is, those that show only a single rupture event, can be fitted by the modified freely jointed chain model below 100 pN and above 300 pN region with a fixed Kuhn length of 7 Å, and with the typical kink at around 300 pN[66]) were analysed (Fig. 3). Histograms of the rupture forces were fitted by a Gaussian function to obtain the most probable unbinding forces. All data analyses were performed by using custom software written in Igor Pro. (Wavemetrics).

X-ray photoelectron spectroscopy. XPS spectra were collected by using an electron spectrometer (ESCALAB 250) equipped with monochromatized Al Kα radiation source with pass energy of 30 eV. The binding energies were corrected by referencing the C (1s) 284.6 eV.

References

1. Häkkinen, H. The gold–sulfur interface at the nanoscale. *Nat. Chem.* **4**, 443–455 (2012).
2. Love, J. C., Estroff, L. A., Kriebel, J. K., Nuzzo, R. G. & Whitesides, G. M. Self-assembled monolayers of thiolates on metals as a form of nanotechnology. *Chem. Rev.* **105**, 1103–1169 (2005).
3. Boisselier, E. & Astruc, D. Gold nanoparticles in nanomedicine: preparations, imaging, diagnostics, therapies and toxicity. *Chem. Soc. Rev.* **38**, 1759–1782 (2009).
4. Kumar, A., Biebuyck, H. A. & Whitesides, G. M. Patterning self-assembled monolayers: applications in materials science. *Langmuir* **10**, 1498–1511 (1994).
5. Liao, W. S. *et al.* Subtractive patterning *via* chemical lift-off lithography. *Science* **337**, 1517–1521 (2012).
6. Motesharei, K. & Myles, D. C. Molecular recognition on functionalized self-assembled monolayers of alkanethiols on gold. *J. Am. Chem. Soc.* **120**, 7328–7336 (1998).
7. Schliwa, M. & Woehlke, G. Molecular motors. *Nature* **422**, 759–765 (2003).
8. Yao, H., Miki, K., Nishida, N., Sasaki, A. & Kimura, K. Large optical activity of gold nanocluster enantiomers induced by a pair of optically active penicillamines. *J. Am. Chem. Soc.* **127**, 15536–15543 (2005).
9. Gautier, C. & Bürgi, T. Chiral N-isobutyryl-cysteine protected gold nanoparticles: preparation, size selection, and optical activity in the UV – vis and infrared. *J. Am. Chem. Soc.* **128**, 11079–11087 (2006).
10. Ferretti, S., Paynter, S., Russell, D. A., Sapsford, K. E. & Richardson, D. J. Self-assembled monolayers: a versatile tool for the formulation of bio-surfaces. *Trends Anal. Chem.* **19**, 530–540 (2000).
11. Hudalla, G. A. & Murphy, W. L. Chemically well-defined self-assembled monolayers for cell culture: toward mimicking the natural ECM. *Soft Matter* **7**, 9561–9571 (2011).
12. Merkel, R. Force spectroscopy on single passive biomolecules and single biomolecular bonds. *Phys. Rep.* **346**, 343–385 (2001).
13. Zhang, W. K. & Zhang, X. Single molecule mechanochemistry of macromolecules. *Prog. Polym. Sci.* **28**, 1271–1295 (2003).
14. Neuman, K. C. & Nagy, A. Single-molecule force spectroscopy: optical tweezers, magnetic tweezers and atomic force microscopy. *Nat. Methods* **5**, 491–505 (2008).
15. Hao, X. *et al.* Direct measurement and modulation of single-molecule coordinative bonding forces in a transition metal complex. *Nat. Commun.* **4**, 2121 (2013).
16. Geisler, M., Balzer, B. N. & Hugel, T. Polymer adhesion at the solid-liquid interface probed by a single-molecule force sensor. *Small* **5**, 2864–2869 (2009).
17. Florin, E.-L., Moy, V. T. & Gaub, H. E. Adhesion forces between individual ligand-receptor pairs. *Science* **264**, 415–417 (1994).
18. Li, H. B. *et al.* Reverse engineering of the giant muscle protein titin. *Nature* **418**, 998–1002 (2002).
19. Fang, J. *et al.* Forced protein unfolding leads to highly elastic and tough protein hydrogels. *Nat. Commun.* **4**, 2974 (2013).
20. Cui, S. X., Yu, J., Kühner, F., Schulten, K. & Gaub, H. E. Double stranded DNA dissociates into single strands when dragged into a poor solvent. *J. Am. Chem. Soc.* **129**, 14710–14716 (2007).
21. Liu, N. N. *et al.* Pulling genetic RNA out of tobacco mosaic virus using single-molecule force spectroscopy. *J. Am. Chem. Soc.* **132**, 11036–11038 (2010).
22. Liu, K. *et al.* Extracting a single polyethylene oxide chain from a single crystal by a combination of atomic force microscopy imaging and single-molecule force spectroscopy: toward the investigation of molecular interactions in their condensed states. *J. Am. Chem. Soc.* **133**, 3226–3229 (2011).
23. Garnier, L., Gauthier-Manuel, B., van der Vegte, E. W., Snijders, J. & Hadziioannou, G. Covalent bond force profile and cleavage in a single polymer chain. *J. Chem. Phys.* **113**, 2497–2503 (2000).
24. Grandbois, M., Beyer, M., Rief, M., Clausen-Schaumann, H. & Gaub, H. E. How strong is a covalent bond? *Science* **283**, 1727–1730 (1999).
25. Skulason, H. & Frisbie, C. D. Detection of discrete interactions upon rupture of au microcontacts to self-assembled monolayers terminated with -S(CO)CH₃ or -SH. *J. Am. Chem. Soc.* **122**, 9750–9760 (2000).
26. Langry, K. C., Ratto, T. V., Rudd, R. E. & McElfresh, M. W. The AFM measured force required to rupture the dithiolate linkage of thioctic acid to gold is less than the rupture force of a simple gold-alkyl thiolate bond. *Langmuir* **21**, 12064–12067 (2005).
27. Ribas-Arino, J., Shiga, M. & Marx, D. Understanding covalent mechanochemistry. *Angew. Chem. Int. Ed.* **48**, 4190–4193 (2009).
28. Beyer, M. K. & Clausen-Schaumann, H. Mechanochemistry: the mechanical activation of covalent bonds. *Chem. Rev.* **105**, 2921–2948 (2005).
29. Alsteens, D. *et al.* Controlled manipulation of bacteriophages using single-virus force spectroscopy. *ACS Nano* **3**, 3063–3068 (2009).
30. Krüger, D., Fuchs, H., Rousseau, R., Marx, D. & Parrinello, M. Pulling monatomic gold wires with single molecules: an *ab initio* simulation. *Phys. Rev. Lett.* **89**, 186402 (2002).
31. Krüger, D., Rousseau, R., Fuchs, H. & Marx, D. Towards "mechanochemistry": mechanically induced isomerizations of thiolate-gold clusters. *Angew. Chem. Int. Ed.* **42**, 2251–2253 (2003).
32. Xu, B. Q., Xiao, X. Y. & Tao, N. J. Measurements of single-molecule electromechanical properties. *J. Am. Chem. Soc.* **125**, 16164–16165 (2003).
33. Huang, Z., Chen, F., Bennett, P. A. & Tao, N. J. Single molecule junctions formed *via* Au–thiol contact: stability and breakdown mechanism. *J. Am. Chem. Soc.* **129**, 13225–13231 (2007).
34. Frei, M., Aradhya, S. V., Hybertsen, M. S. & Venkataraman, L. Linker dependent bond rupture force measurements in single-molecule junctions. *J. Am. Chem. Soc.* **134**, 4003–4006 (2012).
35. Finklea, H. O., Avery, S., Lynch, M. & Furtsch, T. Blocking oriented monolayers of alkyl mercaptans on gold electrodes. *Langmuir* **3**, 409–413 (1987).
36. Ron, H. & Rubinstein, I. Alkanethiol monolayers on preoxidized gold. encapsulation of gold oxide under an organic monolayer. *Langmuir* **10**, 4566–4573 (1994).
37. Ron, H. & Rubinstein, I. Self-assembled monolayers on oxidized metals. 3. Alkylthiol and dialkyl disulfide assembly on gold under electrochemical conditions. *J. Am. Chem. Soc.* **120**, 13444–13452 (1998).
38. Carvalhal, R. F., Freire, R. S. & Kubota, L. T. Polycrystalline gold electrodes: a comparative study of pretreatment procedures used for cleaning and thiol self-assembly monolayer formation. *Electroanalysis* **17**, 1251–1259 (2005).
39. White, M. L. The wetting of gold surfaces by water. *J. Phys. Chem.* **68**, 3083–3085 (1964).
40. Smith, T. The hydrophilic nature of a clean gold surface. *J. Colloid Interface Sci.* **75**, 51–55 (1980).
41. Raiber, K., Terfort, A., Benndorf, C., Krings, N. & Strehblow, H.-H. Removal of self-assembled monolayers of alkanethiolates on gold by plasma cleaning. *Surf. Sci.* **595**, 56–63 (2005).
42. Fuchs, P. Low-pressure plasma cleaning of Au and PtIr noble metal surfaces. *Appl. Surf. Sci.* **256**, 1382–1390 (2009).
43. Konôpka, M., Rousseau, R., Štich, I. & Marx, D. Detaching thiolates from copper and gold clusters: which bonds to break? *J. Am. Chem. Soc.* **126**, 12103–12111 (2004).

44. Wang, G. M., Sandberg, W. C. & Kenny, S. D. Density functional study of a typical thiol tethered on a gold surface: ruptures under normal or parallel stretch. *Nanotechnology* **17**, 4819–4824 (2006).

45. Tielens, F. & Santos, E. AuS and SH bond formation/breaking during the formation of alkanethiol SAMs on Au(111): a theoretical study. *J. Phys. Chem. C* **114**, 9444–9452 (2010).

46. Poirier, G. E. & Pylantt, E. D. The self-assembly mechanism of alkanethiols on Au(1 1 1). *Science* **272**, 1145–1148 (1996).

47. Schreiber, F. *et al.* Adsorption mechanisms, structures, and growth regimes of an archetypal self-assembling system: Decanethiol on Au (111). *Phys. Rev. B* **57**, 12476–12481 (1998).

48. Xu, S. *et al. In situ* studies of thiol self-assembly on gold from solution using atomic force microscopy. *J. Chem. Phys.* **108**, 5002–5012 (1998).

49. Yamada, R. & Uosaki, K. *In situ* scanning tunneling microscopy observation of the self-assembly process of alkanethiols on gold(111) in solution. *Langmuir* **14**, 855–861 (1998).

50. Kankate, L., Turchanin, A. & Gölzhäuser, A. On the release of hydrogen from the S-H groups in the formation of self-assembled monolayers of thiols. *Langmuir* **25**, 10435–10438 (2009).

51. Grönbeck, H., Curioni, A. & Andreoni, W. Thiols and disulfides on the Au(111) surface: the headgroup-gold interaction. *J. Am. Chem. Soc.* **122**, 3839–3842 (2000).

52. Qi, Y., Qin, J., Zhang, G. & Zhang, T. Breaking mechanism of single molecular junctions formed by octanedithiol molecules and Au electrodes. *J. Am. Chem. Soc.* **131**, 16418–16422 (2009).

53. Basch, H., Cohen, R. & Ratner, M. A. Interface geometry and molecular junction conductance: geometric fluctuation and stochastic switching. *Nano Lett.* **5**, 1668–1675 (2005).

54. Petroski, J., Chou, M. & Creutz, C. The coordination chemistry of gold surfaces: formation and far-infrared spectra of alkanethiolate-capped gold nanoparticles. *J. Organomet. Chem.* **694**, 1138–1143 (2009).

55. Rouhana, L. L., Moussallem, M. D. & Schlenoff, J. B. Adsorption of short-chain thiols and disulfides onto gold under defined mass transport conditions: coverage, kinetics, and mechanism. *J. Am. Chem. Soc.* **133**, 16080–16091 (2011).

56. Matthiesen, J. E., Jose, D., Sorensen, C. M. & Klabunde, K. J. Loss of hydrogen upon exposure of thiol to gold clusters at low temperature. *J. Am. Chem. Soc.* **134**, 9376–9379 (2012).

57. Sashuk, V. Thiolate-protected nanoparticles *via* organic xanthates: mechanism and implications. *ACS Nano* **6**, 10855–10861 (2012).

58. Hasan, M., Bethell, D. & Brust, M. The fate of sulfur-bound hydrogen on formation of self-assembled thiol monolayers on gold: ^{1}H NMR spectroscopic evidence from solutions of gold clusters. *J. Am. Chem. Soc.* **124**, 1132–1133 (2002).

59. Oberhauser, A. F., Marszalek, P. E., Erickson, H. P. & Fernandez, J. M. The molecular elasticity of the extracellular matrix protein tenascin. *Nature* **393**, 181–185 (1998).

60. Popa, I. *et al.* Nanomechanics of HaloTag tethers. *J. Am. Chem. Soc.* **135**, 12762–12771 (2013).

61. Whetten, R. L. & Price, R. C. Nano-golden order. *Science* **318**, 407–408 (2007).

62. Sondag-Huethorst, J. A. M., Schonenberger, C. & Fokkink, L. G. J. Formation of holes in alkanethiol monolayers on gold. *J. Phys. Chem.* **98**, 6826–6834 (1994).

63. Stranick, S. J., Parikh, A. N., Allara, D. L. & Weiss, P. S. A new mechanism for surface diffusion: motion of a substrate-adsorbate complex. *J. Phys. Chem.* **98**, 11136–11142 (1994).

64. Peterlinz, K. A. & Georgiadis, R. *In situ* kinetics of self-assembly by surface plasmon resonance spectroscopy. *Langmuir* **12**, 4731–4740 (1996).

65. Butt, H. J. & Jaschke, M. Calculation of thermal noise in atomic force microscopy. *Nanotechnology* **6**, 1–7 (1995).

66. Oesterhelt, F., Rief, M. & Gaub, H. E. Single molecule force spectroscopy by AFM indicates helical structure of poly(ethylene-glycol) in water. *New J. Phys.* **1**, 6.1–6.11 (1999).

Acknowledgements

This work was supported by NSFC (grant numbers 91127031, 20921003 and 21221063), the National Basic Research Program (2013CB834503), the SRF for ROCS (SEM) and the Program for New Century Excellent Talents in University (NCET). We thank Cuicui Fu and Professor Weiqing Xu for their kind help with SERS measurements.

Author contributions

W.Z. conceived the project and designed the experiments. Y.X. performed experiments and data analysis. X.L. performed partial AFM experiments. Y.X., H.L. and W.Z. jointly wrote the manuscript. All authors edited the manuscript.

Additional information

6

First enantioseparation and circular dichroism spectra of Au$_{38}$ clusters protected by achiral ligands

Igor Dolamic[1,*], Stefan Knoppe[1,*], Amala Dass[2] & Thomas Bürgi[1]

Bestowing chirality to metals is central in fields such as heterogeneous catalysis and modern optics. Although the bulk phase of metals is symmetric, their surfaces can become chiral through adsorption of molecules. Interestingly, even achiral molecules can lead to locally chiral, though globally racemic, surfaces. A similar situation can be obtained for metal particles or clusters. Here we report the first separation of the enantiomers of a gold cluster protected by achiral thiolates, Au$_{38}$(SCH$_2$CH$_2$Ph)$_{24}$, achieved by chiral high-performance liquid chromatography. The chirality of the nanocluster arises from the chiral arrangement of the thiolates on its surface, forming 'staple motifs'. The enantiomers show mirror-image circular dichroism responses and large anisotropy factors of up to 4×10^{-3}. Comparison with reported circular dichroism spectra of other Au$_{38}$ clusters reveals that the influence of the ligand on the chiroptical properties is minor.

[1] Département de Chimie Physique, Université de Genève, 30 Quai Ernest-Ansermet, 1211 Genève 4, Switzerland. [2] Department of Chemistry and Biochemistry, University of Mississippi, 352 Coulter Hall, University, Mississippi 38677, USA. *These authors contributed equally to this work. Correspondence and requests for materials should be addressed to T.B. (email: Thomas.Buergi@unige.ch).

Chirality is ubiquitous in nature and has tremendous impact on biology, medicine, and pharmaceutical sciences. Whereas the origin of homochirality on earth is still unclear, it is now evident that many biological macromolecules are built from chiral building blocks. However, chiral assemblies can also emerge from achiral constituents. For example, achiral molecules may turn chiral on adsorption on a surface, even if the latter itself is not chiral[1–3]. In addition, achiral molecules can form chiral patterns on achiral surfaces[4]. This emergence of chirality is due to the restriction of the molecules to two-dimensional space on adsorption. This may lead to a reduced symmetry of the adsorbate complex or to chiral distortions of the molecule owing to its interaction with the surface. If chirality arises in such a way through the bonding of achiral constituents, a racemic mixture is obtained. Such phenomena have been studied on metal surfaces, where an adsorbate lattice can destroy the reflection symmetry of the metal surface underneath. Chiral domains are then formed on the surface with equal abundance of left- and right-handedness. This local chirality can be observed by scanning tunnelling microscopy[4].

An analogous situation was recently discovered on thiolate-protected gold particles or clusters. Jadzinsky *et al.* determined the structure of the gold nanocluster $Au_{102}(p\text{-MBA})_{44}$ (p-MBA: *para*-mercaptobenzoic acid) by X-ray crystallography[5]. An unusual bridged binding motif between gold and sulphur ('staple motif') was evidenced in which the sulphur atoms become chiral centres on adsorption. Moreover, it was found that the arrangement of the staples on the cluster surface forms a chiral pattern[5]. Because the p-MBA ligand used is achiral, both enantiomers are observed in the unit cell of the crystal. A similar situation is found for $Au_{38}(SR)_{24}$ clusters (see below)[6,7].

The chirality of gold nanoparticles has recently become an intensively studied field of modern nanoscience as it opens new possibilities in catalysis and sensing applications[8–10]. The use of plasmon resonances in chiral metamaterials has been discussed and employed in several examples[11–15]. As gold nanoparticles exhibit localized surface plasmon resonances at diameters above ca 2 nm, their smaller analogues (up to ca 200 Au atoms) show interesting, molecular properties[16]. Among these small nanoparticles (in the following referred to as nanoclusters), thiolate-protected systems of the general formula $[Au_n(SR)_m]^z$ (SR: thiolate; z: charge) have evolved as the most studied class, because of their extraordinary stability.

Optical activity in Au:thiolate nanoclusters has first been observed by Schaaff and Whetten in 1998 (refs 17,18). Since then, numerous examples of more or less defined systems have been reported[19]. Protecting ligands include several derivatives of cysteine[10,20–23] as well as 'artificial' ligands such as binaphthyl systems[24–26] or other small organic thiolates[27]. Besides circular dichroism (CD) studies on the electronic transitions, the conformational analysis of the stabilizing ligand was demonstrated, using vibrational circular dichroism in the infrared[20,21,25]. Chiroptical properties in the ultraviolet-visible were found to be strongly size-dependent[24]. Also, it was shown that only a small fraction of enantiopure ligands in a mixed ligand system is sufficient to induce significant optical activity to the clusters[28]. Optical activity can result from a number of effects. Several models have been proposed to explain its origin in gold clusters, including the trivial case of using a chiral ligand (in this, the electrons in the gold core are trapped in dissymmetric electric fields)[29]; a chiral footprint model (in analogy to classic surface chemistry, adsorption of a chiral ligand on the cluster surface perturbs the surface atoms in a chiral fashion)[20,30] and intrinsic core chirality[31,32], as it was proposed that the equilibrium geometry of the core atoms is asymmetric. The importance of these different mechanisms for the optical activity observed in protected metal nanoclusters is difficult to assess, because, up to now, only for nanoclusters containing chiral enantiopure (or at least enantioenriched) ligands optical activity was reported.

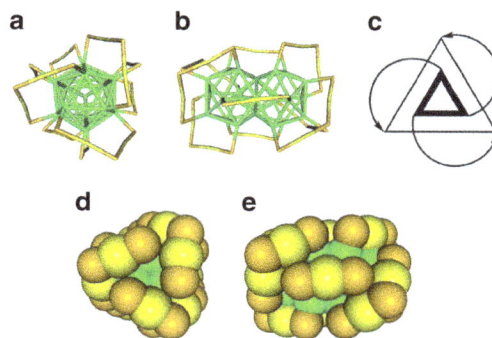

Figure 1 | Crystal structure of the left-handed enantiomer of $Au_{38}(SCH_2CH_2Ph)_{24}$. For clarity, the -$CH_2CH_2Ph$ units were removed; yellow, gold adatoms; green, core atoms (Au); orange, sulphur. (**a**) Top view of the cluster; (**b**) side-view; (**c**) schematic representation highlighting the handedness of the cluster. The inner triangle represents the top three core atoms binding to the long staples. The arrows represent long staples and the outer triangle represent the core Au atoms binding to the 'end' of the staple. This representation is a top view along the C_3 axis, and the two triangles are not in one plane. (**d**) Top-view in space-filling representation mode; (**e**) side-view in space-filling representation mode. The structures were created using the crystallographic data provided in ref. 7.

The above-mentioned staple motifs, which have been proposed earlier[33,34], are an essential part of cluster structures. They can be thought as [thiolate-Au(I)]$_x$-thiolate (x = 1, 2) oligomers that bind in a bidentate fashion to the gold atoms of the cluster core. Staple-type binding was also identified for extended surfaces (self-assembled monolayers of benzenethiol and methylthiol on Au(111))[35,36]. Such staples can be the source of chirality, as outlined above. The staple motif was also found in the crystal structure of $[Au_{25}(SCH_2CH_2Ph)_{18}]^{-1}$ and $Au_{38}(SCH_2CH_2Ph)_{24}$ (refs 7,37,38). Similar to Au_{102}, Au_{38} shows intrinsic chirality by the arrangement of the staple motifs on the cluster surface. In contrast, this is not the case for Au_{25}. A chiral arrangement of staples has also been proposed for $Au_{144}(SR)_{60}$ clusters[39].

$Au_{38}(SCH_2CH_2Ph)_{24}$ is of prolate shape, containing a face-fused biicosahedral Au_{23} core and is protected by 3 short $Au(SR)_2$ and 6 long $Au_2(SR)_3$ staples (Fig. 1)[6,7,40]. The bare core can be idealized as of D_{3h} symmetry (in reality, slight distortions are found), which is lowered by the protecting staples to adopt a D_3 symmetry. The staples are arranged in a chiral fashion: the long staples are arranged in a staggered configuration of two triblade fans (composed of three staples), that either rotate clockwise or anti-clockwise (but both in the same sense, within one enantiomer), depending on the enantiomer. Moreover, the short staples at the equator of the cluster are slightly tilted with respect to the threefold axis, following the handedness of the long staples.

In this contribution, we demonstrate for the first time that it is possible to separate the enantiomers of this Au_{38} cluster covered with achiral thiolates (2-phenylethylthiolate) by a high-performance liquid chromatography (HPLC) column. A major prerequisite is the isolation of pure Au_{38} cluster from the crude reaction product, as the employed thermal etching method usually yields polydisperse clusters[41]. This was achieved by gel permeation chromatography (GPC, or size exclusion chromatography, SEC). Successful enantioseparation enables us to measure optical activity for an enantiopure thiolate-protected gold cluster. The optical activity arising from the chiral arrangement of the staples is large. The anisotropy factors ($\Delta A/A$) are the largest reported so far for thiolate-protected gold clusters, indicating the importance of the chiral pattern for the chiroptical response of such systems.

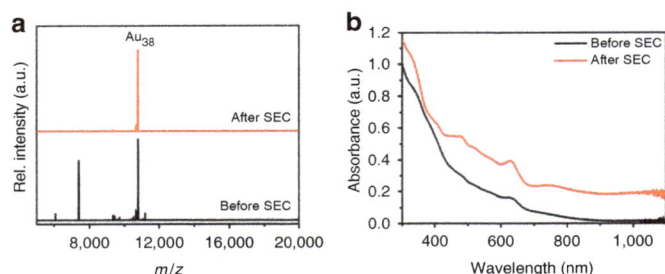

Figure 2 | Characterization of *rac*-Au₃₈(SCH₂CH₂Ph)₂₄. (**a**) MALDI mass spectra of Au₃₈(SCH₂CH₂Ph)₂₄ before (black) and after (red) size selection. The signals for Au₄₀(SCH₂CH₂Ph)₂₄ (11,173 Da) and Au₂₅(SCH₂CH₂Ph)₁₈ (7,391 Da) disappeared from the spectrum, indicating successful size exclusion. For a detailed description of this process, see ref. 41. (**b**) Ultraviolet-visible spectra of Au₃₈(SCH₂CH₂Ph)₂₄ before (black) and after (red) size selection. The absorption features of Au₃₈ are drastically enhanced.

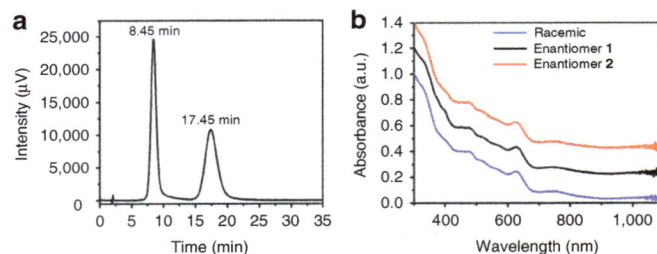

Figure 3 | HPLC-separation of *rac*-Au₃₈(SCH₂CH₂Ph)₂₄. (**a**) HPLC-chromatogram of the enantioseparation of *rac*-Au₃₈(SCH₂CH₂Ph)₂₄ with the ultraviolet-visible detector at 380 nm. The peak at 8.45 min corresponds to enantiomer **1**; the second peak at 17.45 corresponds to enantiomer **2**. (**b**) Ultraviolet-visible spectra of enantiomers **1** (black) and **2** (red) and of the racemate (blue). The spectra were normalized at 300 nm and off-set for clarity. The well-known ultraviolet-visible signature of Au₃₈ is perfectly reproduced in all spectra, showing that the two collected fractions are composed of Au₃₈(SCH₂CH₂Ph)₂₄.

Results

Isolation and characterization of *rac*-Au₃₈(SCH₂CH₂Ph)₂₄. Racemic Au₃₈(SCH₂CH₂Ph)₂₄ was prepared and purified according to previously reported protocols[28,41]. Briefly, tetrachloroauric acid and L-glutathione were co-dissolved in water and methanol and reduced by sodium borohydride. The resulting precursor material was then dissolved in water and a mixture of acetone and 2-phenylethylthiol was added. A mixture of Au$_n$(SCH₂CH₂Ph)$_m$ ($n = 25$–144, $m = 18$–60), containing Au₃₈(SCH₂H₂Ph)₂₄, as major component was gained by heating the system to 80 °C. Excess thiol was removed by extensive methanol washing and the crude clusters were size-selected by gel permeation chromatography.

The monodisperse racemic clusters were characterized by ultraviolet-visible spectroscopy and MALDI mass spectrometry (Fig. 2)[42,43]; the spectra are in agreement with previously reported data[41]. A single peak at 10,778 Da (calc: 10,778.08) and a characteristic fragmentation pattern in the mass spectrum indicates monodispersity based on the sensitivity of MALDI spectrometry. The unit cell of the crystal structure bears both enantiomers of Au₃₈(SCH₂CH₂Ph)₂₄, but the synthesis of the clusters involves the use of homochiral L-glutathione, and a chiral induction that might lead to an enantiomeric excess in Au₃₈ cannot be fully excluded. A CD spectrum was recorded showing no significant signals, suggesting that the Au₃₈(SR)₂₄ nanoclusters used are a truly racemic mixture.

HPLC separation of *rac*-Au₃₈(SCH₂CH₂Ph)₂₄. The racemic clusters were separated at room temperature using a chiral cellulose-based analytical HPLC column and hexane/isopropanol (80:20) as eluent. The eluting solutions were monitored by an ultraviolet detector at 380 nm. Two peaks well separated were observed at 8.45 and 17.45 min (enantiomers **1** and **2** according to increasing elution times, Fig. 3a). The second peak is broadened and less intense compared with the first one, but integration gives identical peak areas to within the accuracy of the measurement. Runs at different temperatures (in the range of 10–25 °C) showed that separation is slightly better at lower temperatures. The ultraviolet-visible spectra of both peaks clearly show the distinct signature of Au₃₈(SR)₂₄ clusters (Fig. 3b).

Circular dichroism of Au₃₈(SCH₂CH₂Ph)₂₄. To confirm the separation of enantiomers, we collected the fractions according to the peaks over several HPLC runs and concentrated the combined solutions. CD spectra of these concentrated solutions were measured (Fig. 4a). The ultraviolet-visible spectra include the Au₃₈-specific

electronic transition at 629 nm (Fig. 3b). The CD spectra give perfect mirror images and eleven clear signals are observable between 230 and 900 nm (245 (+), 255 (+), 308 (+), 345 (−), 393 (−), 440 (+), 479 (−) 564 (+), 629 (−) and 747 (+) nm; signs are given for enantiomer **1**, that is, the first enantiomer eluting from the HPLC column. Compared with the ordinary absorption spectrum of Au₃₈(SR)₂₄, which is highly structured showing several peaks and shoulders, more distinct transitions can be identified in the CD. The identified peaks are in good agreement with those reported for absorption spectra at low temperatures[44]. Anisotropy factors $g = \Delta A/A = \theta[\text{mdeg}]/(32980 \times A)$ were calculated over the spectral range (Fig. 4b). Surprisingly, these are quite strong with values between 1×10^{-3} and up to 4×10^{-3}; notably, the anisotropy factor increases with increasing wavelength (decreasing energy). The maximum anisotropy factor of gold nanoclusters protected with chiral thiols was reported to be up to 4×10^{-3} (ref. 24). This indicates that intrinsic chirality due to ligand arrangement can contribute significantly to the net optical activity.

Discussion

Comparison of the chiroptical properties of the intrinsically chiral Au₃₈ with those reported for Au₃₈ clusters with chiral thiols is self-evident, as it gives direct insight into the contribution of a chiral ligand to the shape and strength of the CD spectra. Schaaff and Whetten reported a series of glutathionate-protected Au clusters that were separated by gel electrophoresis[18]. The absorption features of compound **3** in ref. 18 are in good agreement with those of Au₃₈ clusters, although the assignment is not made in the report. Interestingly, the energies and signs of the peak maxima are in very good agreement with those of enantiomer **2** of Au₃₈(SCH₂CH₂Ph)₂ (Table 1). Minor differences occur for ultraviolet transitions (below 300 nm). In this region, the glutathionate-ligand should contribute to the CD spectrum. Of note, the maximum anisotropy factors of Au₃₈(SCH₂CH₂Ph)₂₄ (4×10^{-3}) exceed those of Au₃₈(SG)₂₄ (1.3×10^{-3}), but this could be due to parameters such as solvent or sample purity after gel electrophoresis. However, for most of the spectral range, the anisotropy factors are similar. This indicates that (in this case) the chiral ligand does not have the dominant influence on the chiroptical properties of Au₃₈ clusters. We doubt that this finding can be generalized as chiral ligands can induce optical activity to clusters that are not intrinsically chiral (such as Au₂₅(SR)₁₈)[17,27]. The good agreement between the CD spectra of Au₃₈(SCH₂CH₂Ph)₂₄ and Au₃₈(L-SG)₂₄ strongly indicates that, in the latter case, one-handedness of the cluster (which dominated the CD spectra in the Au₃₈ case) is favoured over the other due to the presence of the chiral glutathionate-ligand. In other words, there is a strong diastereoselectivity during the formation of $Au_{38}(L\text{-}SG)_{24}$.

In a recent article, Lopez-Acevedo et al. simulated the structure, and predicted the intrinsic chirality of $Au_{38}(SR)_{24}$ (ref. 6). Moreover, the CD spectra of the cluster were computed for the right-handed enantiomer (structure **1** in ref. 6 was identified as being the 'correct' structure). Comparison of the experimental spectra with those computed (with methylthiolate as model ligand) shows a good match for the spectral range from 800 to 500 nm available for comparison with transitions at 747, 629 and 564 nm (a comparison of a wider range is not possible as it was either not measured (lower energies) or not presented in the figures (higher energies)). Moreover, the experimental and calculated anisotropy factors are of similar magnitudes (Table 1). As the sign of the calculated spectrum agrees with the experimental spectrum of enantiomer **2**, we tentatively assign the latter as the right-handed one. Comparison of the spectra of $Au_{38}(SCH_2CH_2Ph)_{24}$ with $Au_{38}(SG)_{24}$ and (simulated) $Au(SMe)_{24}$ reveals a minor influence of the ligand to the shape of the CD spectra as all three spectra are in good agreement concerning energies, sign and magnitude of anisotropy.

As it is possible to separate the enantiomers and measure CD spectra of good quality, no racemization is thought to occur over several hours in solution (the sum of the anisotropy factor plots gives a zero line, indicating the same enantiomeric excess in both fractions). The collected solutions of the two chromatographic fractions were concentrated to dryness and stored at $-5\,^\circ C$ over 3 days. After that time, the anisotropy factors of the enantiomers

were identical to the one measured directly after chromatography, indicating that racemization does not take place under these conditions.

In this report, we present the first successful enantioseparation of racemic $Au_{38}(SR)_{24}$ nanoclusters. Moreover, the studied clusters are protected by achiral ligands ($-SCH_2CH_2Ph$) and the observed chirality is an intrinsic structural property of the cluster. The observed optical activity is the first spectroscopic evidence of chirality stemming only from the asymmetric arrangement of achiral adsorbates on a surface. This type of chirality has been identified by X-ray diffraction and, in the case of extended surfaces, by microscopy studies[1–4]. The observed anisotropy factors of up to 4×10^{-3} are surprisingly strong, considering the fact that no chiral ligands are present. Comparison with glutathionate-protected Au_{38} clusters only shows a minor influence of the chiral ligand to the spectrum (indicating that one enantiomer is selectively formed when glutathione is used)[18]. Moreover, the spectrum is in good qualitative and quantitative agreement with those of computed structures[6].

Methods

General. All chemicals were used as received, if not mentioned otherwise. Tetrachloroauric acid trihydrate (Aldrich, 99.9 + %), reduced L-glutathione (Sigma-Aldrich, >99%), sodium borohydride (Fluka, >96%), 2-phenylethylthiol (Aldrich, 98%), anhydrous sodium sulfate (Reactolab, Servion/CH), [3-(4-tert-butylphenyl)-2-methyl-2-propenylidene]malononitrile (Aldrich, >98%), methanol (VWR, >99.8%), acetone (Fluka, >99.5%), methylene chloride (Sigma-Aldrich, >99.9%), tetrahydrofuran (Acros, p.A.), hexane (Sigma-Aldrich, HPLC grade), isopropanol (Sigma-Aldrich, HPLC grade), regenerated cellulose membranes (0.2 μm, Sartorius), PTFE syringe filters (0.2 μm, Carl Roth) and Bio Beads SX-1 (Bio-Rad) were used as received, if not mentioned otherwise. Tetrahydrofuran was dried over sodium sulfate and stored under nitrogen. Nanopure water (>18 MΩ) was used.

Synthesis and isolation of rac-$Au_{38}(SCH_2CH_2Ph)_{24}$. Step 1. Preparation of L-glutathionate-protected clusters. Tetrachloroauric acid trihydrate (1 g, 2.54 mmol) was dissolved in methanol (200 ml); L-glutathione (3.1 g, 10.18 mmol) was dissolved in water (100 ml). The solutions were combined and stirred at room temperature for 30 min. During this, a yellow-brown suspension was formed. A freshly prepared, ice-cooled solution of sodium borohydride (1.1 g, 30 mmol) in water (60 ml) was added all at once. Immediately, the reaction mixture turned dark-brown to black. The solution was stirred at room temperature for 90 min, during which the clusters precipitated. The solvent was decanted and the crude material was washed with methanol several times.

Step 2. Thermal etching towards rac-$Au_{38}(SCH_2CH_2Ph)_{24}$. The L-glutathionate-protected clusters from Step 1 (ca 550 mg) were dissolved in 10 ml of water and 10 ml of acetone and 15 ml of 2-phenylethylthiol were added. The mixture was stirred at 80 °C for 3 h, during which the aqueous phase discoloured. Some insoluble white material formed. The crude reaction mixture was diluted with water and extracted with methylene chloride. The aqueous phase was discarded. The solvent was removed from the organic phase and the clusters were extensively washed with methanol to remove excess thiol and other byproducts and filtered over a regenerated cellulose filter (0.2 μm). Clusters were redissolved in methylene chloride and

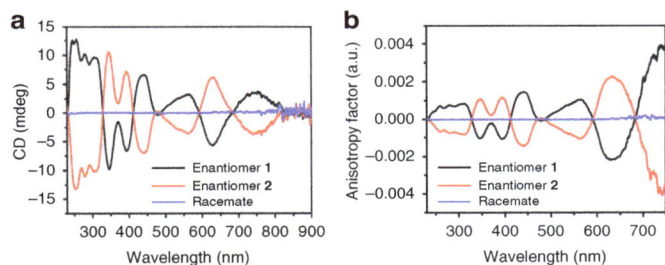

Figure 4 | CD spectra and anisotropy factors of $Au_{38}(SCH_2CH_2Ph)_{24}$. (**a**) CD spectra of isolated enantiomers **1** (black) and **2** (red) and the racemic $Au_{38}(SCH_2CH_2Ph)_{24}$ (blue) before separation; (**b**) corresponding anisotropy factors of enantiomers **1** and **2** and of the racemate. The spectra exhibit excellent mirror-image relationships and anisotropy factors $g = \Delta A/A$ of up to 4×10^{-3}.

Table 1 | Wavelengths, anisotropy factors, and signs of enantiomer 2 of $Au_{38}(SCH_2CH_2Ph)_{24}$, $Au_{38}(SG)_{24}$ (ref. 18) and $Au_{38}(SMe)_{24}$ (ref. 6).

Enantiomer 2		$Au_{38}(SG)_{24}$		$Au_{38}(SMe)_{24}$	
Wavelength (nm)	g (a.u.)	Wavelength (nm)	g (a.u.)	Wavelength (nm)	g (a.u.).
				ca 779	ca -3.3×10^{-3}
747	ca -4×10^{-3}	ca 747	ca -1.3×10^{-3}	ca 729	ca -0.28×10^{-3}
629	ca $+2\times10^{-3}$	ca 620	ca $+1.3\times10^{-3}$	ca 629	ca $+6.2\times10^{-3}$
564	ca -1×10^{-3}	ca 568	ca -1.2×10^{-4}	ca 568	ca -4.3×10^{-4}
479	ca 1×10^{-4}	ca 512	ca $+4\times10^{-4}$		
440	ca -1.4×10^{-3}	ca 449	ca -1.2×10^{-4}		
393	ca $+1\times10^{-3}$	ca 385	ca $+6\times10^{-4}$		
345	ca $+1\times10^{-3}$	ca 354	ca $+6\times10^{-4}$		
308	ca -8×10^{-4}	ca 296	ca -1.2×10^{-4}		
255	ca -6×10^{-4}				
245	ca -4×10^{-4}	ca 239	ca $+4\times10^{-4}$		

methanol precipitation and washing was repeated. Overall, five washing cycles were applied. Eventually, the clusters were dissolved in methylene chloride and passed through a PTFE syringe filter (0.2 μm) to remove insoluble byproducts. After this, ultraviolet-visible and MALDI mass spectra were recorded.

Step 3—Size-selection of rac-Au$_{38}$(SCH$_2$CH$_2$Ph)$_{24}$. A weight of 45 g of Bio-Rad BioBeads SX-1 was suspended in about 7 times the bed volume of tetrahydrofuran. The beads were allowed to swell overnight and given into a glass column (100 cm in length and 2.5 cm in diameter) equipped with a glass frit (G4) and inert gas inlet. The beads were allowed to settle (90 cm bed height) under a gentle stream of N$_2$ and washed extensively with tetrahydrofuran (ca 500 ml). The crude clusters from Step 2 were dissolved in a minimum amount of tetrahydrofuran and repeatedly eluted, using tetrahydrofuran as mobile phase (ca 1 ml min^{-1}) until the eluting clusters were purely composed of Au$_{38}$(SCH$_2$CH$_2$Ph)$_{24}$ (the eluting band overlaps with Au$_{40}$(SCH$_2$CH$_2$Ph)$_{24}$; therefore, repeated chromatographic separations are necessary.). The collected fractions were characterized by ultraviolet-visible spectroscopy, until no further change was observed. The fraction identified as rac-Au$_{38}$(SCH$_2$CH$_2$Ph)$_{24}$ was washed with methanol and passed over a PTFE syringe filter, as described in Step 2 before characterization with ultraviolet-visible and CD spectroscopy as well as MALDI mass spectrometry.

Ultraviolet-visible spectroscopy. Ultraviolet-visible spectra were recorded on a Varian Cary 50 spectrophotometer, using a quartz cuvette of 10 and 5 mm path length. Spectra were measured in methylene chloride and normalized at 300 nm.

CD spectroscopy. CD spectra were recorded on a JASCO J-815 CD-spectrometer using a quartz cuvette of 5 mm path length. The spectra were recorded in diluted solutions of methylene chloride and the signal of the blank solvent was subtracted. For each spectrum, eight scans at a scanning speed of 100 nm/min at a data pitch of 0.1 nm were averaged. The spectra were recorded at 20 °C; for temperature control, a JACSO PFD-350S Peltier element was used. Anisotropy factors $g = \theta$[mdeg]/ $(32980 \times A)$ were calculated using the ultraviolet-visible spectrum provided by the CD spectrometer.

MALDI analysis. Mass spectra were obtained using a Bruker Autoflex mass spectrometer equipped with a nitrogen laser at near threshold laser fluence in positive linear mode. [3-(4-tert-Butylphenyl)-2-methyl-2-propenylidene]malononitrile was used as the matrix with a 1:1,000 analyte : matrix ratio[42]. A volume of 2 μl of the analyte/matrix mixture was applied to the target and air-dried.

HPLC. Chromatographic separation of the enantiomers was achieved on a JASCO 20XX HPLC system equipped with a Phenomenex Lux-Cellulose-1 column (5 μm, 250 mm×4.6 mm). For detection, a JASCO 2070plus ultraviolet-visible detector was used. Path length was 10 mm and the wavelength was set to 380 nm. The analytes were eluted at a flow rate of 2 ml min^{-1} using hexane:isopropanol (80:20). For separation at different temperatures, a Thermasphere TS-430 HPLC column chiller/heater was used.

References

1. Böhringer, M., Morgenstern, K., Schneider, W.- D. & Berndt, R. Separation of a Racemic Mixture of Two-Dimensional Molecular Clusters by Scanning Tunneling Microscopy. *Angew. Chem. Int. Ed.* **38**, 821–823 (1999).
2. Chen, Q., Frankel, D. J. & Richardson, N. V. Chemisorption induced chirality: glycine on Cu{110}. *Surf. Sci.* **497**, 37–46 (2002).
3. Parschau, M., Romer, S. & Ernst, K. H. Induction of homochirality in achiral enantiomorphous monolayers. *J. Am. Chem. Soc.* **126**, 15398–15399 (2004).
4. Lorenzo, M. O., Baddeley, C. J., Muryn, C. & Raval, R. Extended surface chirality from supramolecular assemblies of adsorbed chiral molecules. *Nature* **404**, 376–379 (2000).
5. Jadzinsky, P. D., Calero, G., Ackerson, C. J., Bushnell, D. A. & Kornberg, R. D. Structure of a thiol monolayer-protected gold nanoparticle at 1.1 A resolution. *Science* **318**, 430–433 (2007).
6. Lopez-Acevedo, O., Tsunoyama, H., Tsukuda, T., Hakkinen, H. & Aikens, C. M. Chirality and electronic structure of the thiolate-protected Au38 nanocluster. *J. Am. Chem. Soc.* **132**, 8210–8218 (2010).
7. Qian, H., Eckenhoff, W. T., Zhu, Y., Pintauer, T. & Jin, R. Total structure determination of thiolate-protected Au38 nanoparticles. *J. Am. Chem. Soc.* **132**, 8280–8281 (2010).
8. Zhu, Y., Qian, H. F. & Jin, R. C. Catalysis opportunities of atomically precise gold nanoclusters. *J. Mater. Chem.* **21**, 6793–6799 (2011).
9. Kang, Y. J., Oh, J. W., Kim, Y. R., Kim, J. S. & Kim, M. Chiral gold nanoparticle-based electrochemical sensor for enantioselective recognition of 3,4-dihydroxyphenylalanine. *Chem. Commun. (Camb.)* **46**, 5665–5667 (2010).
10. Shukla, N., Bartel, M. A. & Gellman, A. J. Enantioselective separation on chiral Au nanoparticles. *J. Am. Chem. Soc.* **132**, 8575–8580 (2010).
11. Slocik, J. M., Govorov, A. O. & Naik, R. R. Plasmonic circular dichroism of Peptide-functionalized gold nanoparticles. *Nano Lett.* **11**, 701–705 (2011).
12. Hendry, E. *et al.* Ultrasensitive detection and characterization of biomolecules using superchiral fields. *Nature Nanotech.* **5**, 783–787 (2010).
13. Oh, H. S. *et al.* Chiral Poly(fluorene-alt-benzothiadiazole) (PFBT) and Nanocomposites with Gold Nanoparticles: Plasmonically and Structurally Enhanced Chirality. *J. Am. Chem. Soc.* **132**, 17346–17348 (2010).
14. Lilly, G. D., Agarwal, A., Srivastava, S. & Kotov, N. A. Helical assemblies of gold nanoparticles. *Small* **7**, 2004–2009 (2011).
15. Guerrero-Martínez, A., Alonso-Gómez, J. L., Auguié, B., Cid, M. M. & Liz-Marzán, L. M. From individual to collective chirality in metal nanoparticles. *Nano Today* **6**, 381–400 (2011).
16. Dass, A. Faradaurate nanomolecules: a superstable plasmonic 76.3 kDa cluster. *J. Am. Chem. Soc.* **133**, 19259–19261 (2011).
17. Schaaff, T. G., Knight, G., Shafigullin, M. N., Borkman, R. F. & Whetten, R. L. Isolation and selected properties of a 10.4 kDa Gold: Glutathione cluster compound. *J. Phys. Chem. B* **102**, 10643–10646 (1998).
18. Schaaff, T. G. & Whetten, R. L. Giant gold-glutathione cluster compounds: Intense optical activity in metal-based transitions. *J. Phys. Chem. B* **104**, 2630–2641 (2000).
19. Gautier, C. & Burgi, T. Chiral gold nanoparticles. *ChemPhysChem* **10**, 483–492 (2009).
20. Gautier, C. & Burgi, T. Chiral N-isobutyryl-cysteine protected gold nanoparticles: preparation, size selection, and optical activity in the UV-vis and infrared. *J. Am. Chem. Soc.* **128**, 11079–11087 (2006).
21. Gautier, C. & Burgi, T. Vibrational circular dichroism of N-acetyl-l-cysteine protected gold nanoparticles. *Chem. Commun. (Camb.)*, 5393–5395 (2005).
22. Yao, H., Miki, K., Nishida, N., Sasaki, A. & Kimura, K. Large optical activity of gold nanocluster enantiomers induced by a pair of optically active penicillamines. *J. Am. Chem. Soc.* **127**, 15536–15543 (2005).
23. Yao, H., Fukui, T. & Kimura, K. Chiroptical responses of D-/L-penicillamine-capped gold clusters under perturbations of temperature change and phase transfer. *J. Phys. Chem. C* **111**, 14968–14976 (2007).
24. Gautier, C., Taras, R., Gladiali, S. & Burgi, T. Chiral 1,1′-binaphthyl-2,2′-dithiol-stabilized gold clusters: size separation and optical activity in the UV-vis. *Chirality* **20**, 486–493 (2008).
25. Gautier, C. & Burgi, T. Vibrational Circular Dichroism of Adsorbed Molecules: BINAS on Gold Nanoparticles. *J. Phys. Chem. C* **114**, 15897–15902 (2010).
26. Tamura, M. & Fujihara, H. Chiral bisphosphine BINAP-stabilized gold and palladium nanoparticles with small size and their palladium nano-particle-catalyzed asymmetric reaction. *J. Am. Chem. Soc.* **125**, 15742–15743 (2003).
27. Zhu, M. *et al.* Chiral Au nanospheres and nanorods: synthesis and insight into the origin of chirality. *Nano Lett.* **11**, 3963–3969 (2011).
28. Knoppe, S., Dharmaratne, A. C., Schreiner, E., Dass, A. & Burgi, T. Ligand exchange reactions on Au(38) and Au(40) clusters: a combined circular dichroism and mass spectrometry study. *J. Am. Chem. Soc.* **132**, 16783–16789 (2010).
29. Goldsmith, M. R. *et al.* The chiroptical signature of achiral metal clusters induced by dissymmetric adsorbates. *Phys. Chem. Chem. Phys.* **8**, 63–67 (2006).
30. Humblot, V., Haq, S., Muryn, C., Hofer, W. A. & Raval, R. From local adsorption stresses to chiral surfaces: (R,R)-tartaric acid on Ni(110). *J. Am. Chem. Soc.* **124**, 503–510 (2002).
31. Garzón, I. L. *et al.* Chirality, defects, and disorder in gold clusters. *Eur. Phys. J. D* **24**, 105–109 (2003).
32. Garzón, I. L. *et al.* Chirality in bare and passivated gold nanoclusters. *Phys. Rev. B* **66**, 073403 (2002).
33. Hakkinen, H., Walter, M. & Gronbeck, H. Divide and protect: capping gold nanoclusters with molecular gold-thiolate rings. *J. Phys. Chem. B* **110**, 9927–9931 (2006).
34. Walter, M. *et al.* A unified view of ligand-protected gold clusters as superatom complexes. *Proc. Natl Acad. Sci. USA* **105**, 9157–9162 (2008).
35. Maksymovych, P. & Yates, J. T. Jr. Au adatoms in self-assembly of benzenethiol on the Au(111) surface. *J. Am. Chem. Soc.* **130**, 7518–7519 (2008).
36. Voznyy, O., Dubowski, J. J., Yates, J. T. & Maksymovych, P. The role of gold adatoms and stereochemistry in self-assembly of methylthiolate on Au(111). *J. Am. Chem. Soc.* **131**, 12989–12993 (2009).
37. Heaven, M. W., Dass, A., White, P. S., Holt, K. M. & Murray, R. W. Crystal structure of the gold nanoparticle [N(C$_8$H$_{17}$)$_4$][Au$_{25}$(SCH$_2$CH$_2$Ph)$_{18}$]. *J. Am. Chem. Soc.* **130**, 3754–3755 (2008).
38. Zhu, M., Aikens, C. M., Hollander, F. J., Schatz, G. C. & Jin, R. Correlating the crystal structure of a thiol-protected Au25 cluster and optical properties. *J. Am. Chem. Soc.* **130**, 5883–5885 (2008).
39. Lopez-Acevedo, O., Akola, J., Whetten, R. L., Gronbeck, H. & Hakkinen, H. Structure and bonding in the ubiquitous icosahedral metallic gold cluster Au144(SR)60. *J. Phys. Chem. C* **113**, 5035–5038 (2009).
40. Pei, Y., Gao, Y. & Zeng, X. C. Structural prediction of thiolate-protected Au38: a face-fused bi-icosahedral Au core. *J. Am. Chem. Soc.* **130**, 7830–7832 (2008).
41. Knoppe, S., Boudon, J., Dolamic, I., Dass, A. & Burgi, T. Size exclusion chromatography for semipreparative scale separation of Au38(SR)24 and Au40(SR)24 and larger clusters. *Anal. Chem.* **83**, 5056–5061 (2011).

42. Dass, A., Stevenson, A., Dubay, G. R., Tracy, J. B. & Murray, R. W. Nanoparticle MALDI-TOF mass spectrometry without fragmentation: $Au_{25}(SCH_2CH_2Ph)_{18}$ and mixed monolayer $Au_{25}(SCH_2CH_2Ph)_{18-x}(L)_x$. *J. Am. Chem. Soc.* **130,** 5940–5946 (2008).

43. Harkness, K. M., Cliffel, D. E. & McLean, J. A. Characterization of thiolate-protected gold nanoparticles by mass spectrometry. *Analyst* **135,** 868–874 (2010).

44. Devadas, M. S. *et al.* Temperature-Dependent Optical Absorption Properties of Monolayer-Protected Au25and Au38Clusters. *J. Phys. Chem. Lett.* **2,** 2752–2758 (2011).

Acknowledgements

We gratefully acknowledge financial support from the University of Geneva, the Swiss National Science Foundation, NSF 0903787 (AD) and the University of Mississippi (AD). Prof. Stefan Matile (University of Geneva) is acknowledged for providing the CD spectrometer.

Author contributions

I.D. developed and performed the HPLC work; S.K. prepared and selected the analytes and measured CD and ultraviolet-visible spectra; both authors equally contributed to the writing of the manuscript and are co-first authors. Mass spectra were measured by A.D. T.B. designed the concept of the work. A.D. and T.B. supervised the project.

Additional information

7

All-dielectric metasurface analogue of electromagnetically induced transparency

Yuanmu Yang[1], Ivan I. Kravchenko[2], Dayrl P. Briggs[2] & Jason Valentine[3]

Metasurface analogues of electromagnetically induced transparency (EIT) have been a focus of the nanophotonics field in recent years, due to their ability to produce high-quality factor (Q-factor) resonances. Such resonances are expected to be useful for applications such as low-loss slow-light devices and highly sensitive optical sensors. However, ohmic losses limit the achievable Q-factors in conventional plasmonic EIT metasurfaces to values $< \sim 10$, significantly hampering device performance. Here we experimentally demonstrate a classical analogue of EIT using all-dielectric silicon-based metasurfaces. Due to extremely low absorption loss and coherent interaction of neighbouring meta-atoms, a Q-factor of 483 is observed, leading to a refractive index sensor with a figure-of-merit of 103. Furthermore, we show that the dielectric metasurfaces can be engineered to confine the optical field in either the silicon resonator or the environment, allowing one to tailor light–matter interaction at the nanoscale.

[1] Interdisciplinary Materials Science Program, Vanderbilt University, Nashville, Tennessee 37212, USA. [2] Center for Nanophase Materials Sciences, Oak Ridge National Laboratory, Oak Ridge, Tennessee 37831, USA. [3] Department of Mechanical Engineering, Vanderbilt University, Nashville, Tennessee 37212, USA. Correspondence and requests for materials should be addressed to J.V. (email: jason.g.valentine@vanderbilt.edu).

Electromagnetically induced transparency (EIT) is a concept originally observed in atomic physics and arises due to quantum interference, resulting in a narrowband transparency window for light propagating through an originally opaque medium[1]. This concept was later extended to classical optical systems using plasmonic metamaterials[2–10], among others[11,12], allowing experimental implementation with incoherent light and operation at room temperature. The transparent and highly dispersive nature of EIT offers a potential solution to the long-standing issue of loss in metamaterials as well as the creation of ultra-high-quality-factor (Q-factor) resonances, which are critical for realizing low-loss slow-light devices[2,3,6,10], optical sensors[13,14] and enhancing nonlinear interactions[15].

The classical analogue of EIT in plasmonic metamaterials relies on a Fano-type interference[16,17] between a broadband 'bright'-mode resonator, which is accessible from free space, and a narrowband 'dark' mode resonator, which is less-accessible, or inaccessible, from free space. If these two resonances are brought in close proximity in both the spatial and frequency domains, they can interfere resulting in an extremely narrow reflection or transmission window. Due to the low radiative loss of the dark mode, the Fano resonance can be extremely sharp, resulting in complete transmission, analogous to EIT[2–10], or complete reflection[13], from the sample across a very narrow bandwidth. However, the main limitation of metal-based Fano-resonant systems is the large non-radiative loss due to ohmic damping, which limits the achievable Q-factor[17] to $< \sim 10$.

High-refractive-index dielectric particles offer a potential solution to the issue of material (non-radiative) loss. Such particles exhibit magnetic and electric dipole, and higher order, Mie resonances while suffering from minimal absorption loss, provided the illumination energy is sub-bandgap[18–22]. For instance, dielectric Fano-resonant structures based on oligomer antennae[23], silicon (Si) nanostripe[24] and asymmetric cut-wire metamaterials[25,26] have been demonstrated with Q-factors up to 127.

In this article, we describe the development of Si-based metasurfaces possessing sharp EIT-like resonances with a Q-factor of 483 in the near-infrared regime. The high-Q resonance is accomplished by employing Fano-resonant unit cells in which both radiative and non-radiative damping are minimized through coherent interaction among the resonators combined with the reduction of absorption loss. Combining the narrow resonance linewidth with strong near-field confinement, we demonstrate an optical refractive index sensor with a figure-of-merit (FOM) of 103. In addition, we demonstrate unit-cell designs consisting of double-gap split-ring resonators that possess narrow feed-gaps in which the electric field can be further enhanced in the surrounding medium, allowing interaction with emitters such as quantum dots.

Results

Design and characterization. The schematic of the designed dielectric metasurface is shown in Fig. 1a. The structure is formed from a periodic lattice made of a rectangular bar resonator and a ring resonator, both formed from Si. The rectangular bar resonator serves as an electric dipole antenna, which couples strongly to free-space excitation with the incident E-field oriented along the x axis. The collective oscillations of the bar resonators form the 'bright' mode resonance. The ring supports a magnetic dipole mode wherein the electric field is directed along the azimuthal direction, rotating around the ring's axis. The magnetic dipole mode in the ring cannot be directly excited by light at normal incidence, as the magnetic arm of the incident wave is perpendicular to the dipole axis; however, it can couple to the bright-mode bar resonator. Furthermore, the ring resonators interact through near-field coupling, resulting in collective oscillation of the resonators and suppression of radiative loss, forming the 'dark' mode of the system.

The interference between the collective bright and dark modes form a typical 3-level Fano-resonant system[17], as illustrated in

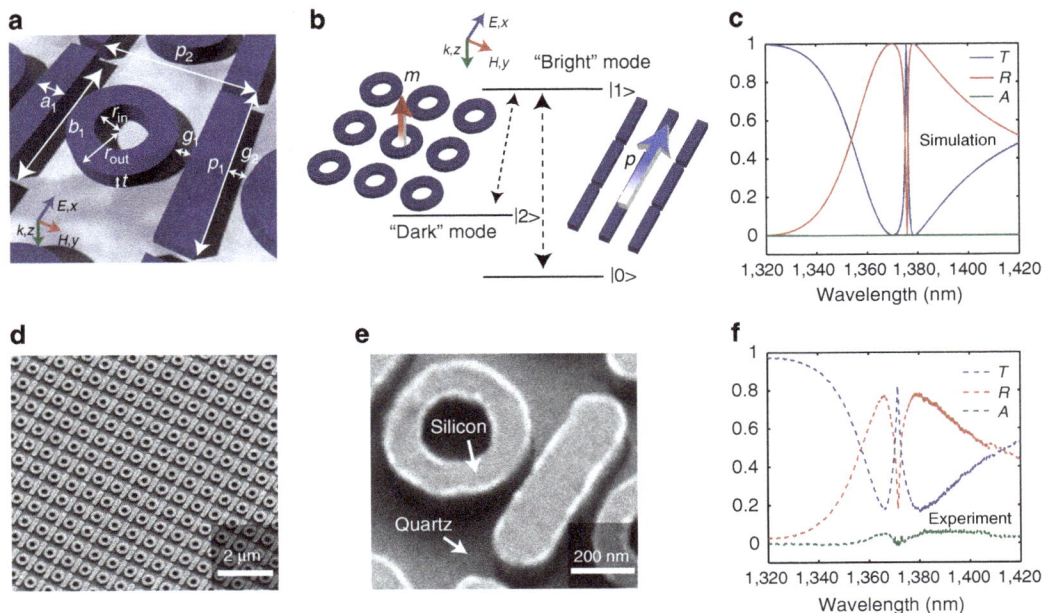

Figure 1 | Configuration and performance of the EIT metasurface. (a) Diagram of the metasurface. The geometrical parameters are: $a_1 = 150$ nm, $b_1 = 720$ nm, $r_{in} = 110$ nm, $r_{out} = 225$ nm, $g_1 = 70$ nm, $g_2 = 80$ nm, $t = 110$ nm, $p_1 = 750$ nm and $p_2 = 750$ nm. (b) Schematic of interference between the bright- and dark-mode resonators. (c) Simulated transmittance (blue curve), reflectance (red curve) and absorption (green curve) spectra of the metasurface. (d) Oblique scanning electron microscope image of the fabricated metasurface. (e) Enlarged image of a single unit cell. (f) Experimentally measured transmittance, reflectance and absorption spectra of the metasurface.

Fig. 1b. The response of the dielectric metasurface is similar to the collective modes found in asymmetric double-gap split-ring resonators, in which the asymmetry in the rings yields a finite electric dipole moment that can couple the out-of-plane magnetic dipole mode to free space[27–32]. In our case, coupling to free space is provided by the bright-mode resonators, which are placed in close proximity to the symmetric dark-mode resonators. Numerical simulations of the structure were carried out using commercially available software (CST Microwave Studio) using the finite-element frequency-domain solver (see Methods). The simulated transmittance, reflectance and absorption spectra of the designed structure are shown in Fig. 1c, where a distinct EIT-like peak can be observed at a wavelength of 1,376 nm. In these simulations, the resonators are sitting on a quartz substrate and embedded in a medium with a refractive index of $n = 1.44$, matching the experiments described below. We have also used the dielectric function of Si as determined using ellipsometry and assumed an infinitely large array, resulting in a Q-factor that reaches 1,176 (see Methods for details on the extraction of the Q-factor). Furthermore, the peak of the transparency window approaches unity, demonstrating the potential to realize highly dispersive, yet lossless, 'slow-light' devices.

The designed structure was fabricated by starting with a 110-nm-thick polycrystalline Si (refractive index $n = 3.7$)-thin film deposited on a quartz (SiO$_2$) ($n = 1.48$) wafer. The structure was patterned using electron-beam lithography (EBL) for mask formation followed by reactive-ion etching (RIE). A scanning electron microscope (SEM) image of a fabricated sample is shown in Fig. 1d,e. Before optical measurements, the resonator array was immersed in a refractive-index-matching oil ($n = 1.44$) within a polydimethylsiloxane flow cell (see Methods for details). The

experimentally measured transmittance, reflectance and absorption spectra, plotted in Fig. 1f, were acquired by illuminating the sample with normal-incident white light with the electric field oriented along the long axis of the bar resonator. A peak transmittance of 82% was observed at a wavelength of 1,371 nm with a Q-factor of 483, as determined by fitting the dark-mode resonance to a Fano line shape (see Methods and Supplementary Fig. 1). The shape of the measured spectra have good agreement with the simulation, although the Q-factor and peak transmittance are reduced. This is most likely due to imperfections within the fabricated sample, which introduce scattering loss and break coherence among the resonators. The role of coherence will be further addressed below. Furthermore, additional loss in the Si arising from surface states created during reactive-ion etching leads to slightly increased absorption in the array compared with theory. It should also be noted that the use of index-matching oil is not needed to achieve such high Q-factors. Designs comprised of the resonators placed on a quartz substrate and surrounded by air are presented in Supplementary Fig. 2; Supplementary Note 1.

Theoretical treatment. The response of the metasurface can be qualitatively understood by applying the widely used coupled harmonic oscillator model[33,34], described by the following equations,

$$\dot{x}_1 - j(\omega_0 + j\gamma_1)x_1 + j\kappa x_2 = gE_0 e^{j\omega t},$$
$$\dot{x}_2 - j(\omega_0 + \delta + j\gamma_2)x_2 + j\kappa x_1 = 0, \tag{1}$$

where x_1 and x_2 represent the amplitudes of the collective modes supported by oscillators 1 (bright mode) and 2 (dark mode), respectively. γ_1 and γ_2 are the damping rates, given by

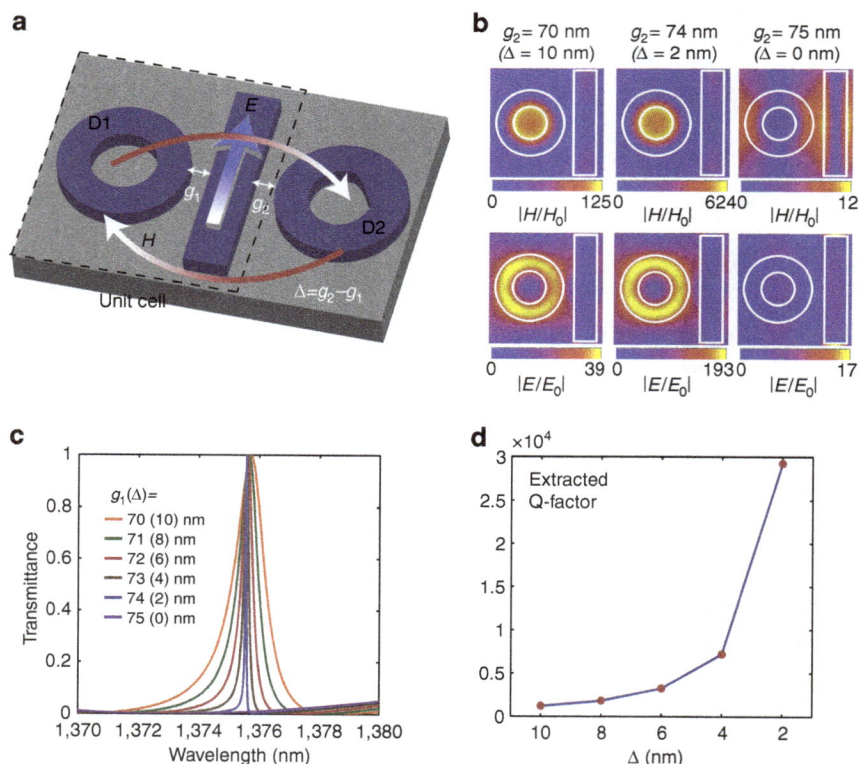

Figure 2 | Dependence of the Q-factor on the bright- and dark-mode resonator spacing. (a) Schematic showing the coupling of the bright-mode resonator to the neighbouring dark-mode resonator. Here, we schematically provide the fields arising due to the bright-mode resonance. **(b)** Simulated magnetic and electric field amplitudes as a function of resonator spacing. **(c)** Transmittance curve for the metasurface with varying values of g_1 obtained through finite-element frequency-domain simulation. $\Delta = g_2 - g_1$ is used to track the difference in spacing between the bright-mode resonator and the adjacent dark-mode resonators. **(d)** The extracted Q-factors of the EIT resonance as a function of Δ.

$\gamma = \gamma_R + \gamma_{NR}$ where γ_R and γ_{NR} are the radiative and non-radiative decay rates, respectively. ω_0 is the central resonant frequency of oscillator 1, δ is the detuning of resonance frequency of oscillator 1 and 2 and g is the bright-mode dipole-coupling strength to the incident electric field E_0. κ is the coupling coefficient between oscillators 1 and 2 and given the close proximity of the resonators, it should be considered an effective coupling coefficient that takes into account interaction between the bright-mode atom and its two nearest neighbours.

Of particular interest is the value of γ_2 that reflects the damping of the collective dark mode and plays a large role in dictating the linewidth of the EIT-like resonance. The radiative damping term, γ_{R2}, is minimized due to the collective oscillations of the array, mediated by near-field coupling between the unit cells. One consequence of utilizing collective modes is that the value of γ_{R2} is inversely proportional to the size of the array due to the fact that the discontinuity at the edges of the array allows for light leakage to free space[27,28]. While collective modes have been utilized in past demonstrations of plasmonic EIT metamaterials to realize extremely small γ_{R2} values, the Q-factors have intrinsically been limited by the ohmic loss in the metal. By replacing the resonator constituents with lossless dielectrics, the non-radiative damping term γ_{NR2} can also be minimized, resulting in the large increase in the Q-factor and peak transmittance.

Equally important is the role of the coupling coefficient, κ, in realizing a high-Q-factor resonance for this system. It has previously been shown that the slope of dispersion is inversely proportional to κ^2 (ref. 2). In the limit of $\gamma_2 \to 0$, reduction of κ will result in a monotonic increase of the Q-factor until $\kappa = 0$ is reached, at which point the Fano resonance will vanish, leaving only the bright-mode resonance. In our system, the magnetic field from the bright-mode resonator is inducing the dark-mode resonance, as illustrated in Fig. 2a. Thus, as the spacing between the resonators (g_1) is increased the coupling coefficient decreases, resulting in an increase in the Q-factor (Fig. 2c,d) and strong magnetic and electric field localization (Fig. 2b). At a spacing of $g_1 = 74$ nm, the Q-factor reaches a value of $\sim 30{,}000$ which is indicative of the fact that both radiative and non-radiative losses in the system are minimal, resulting in a situation wherein $\gamma_2 \sim 0$. However, losses in the Si primarily arising from surface states created during etching will ultimately limit the Q-factor, as has been demonstrated in microcavities[35]. Such losses are not included in these calculations. It is also important to realize that the bright-mode resonator is also interacting with the dark-mode resonator in the adjacent unit cell (D2), inducing a magnetic field that is 180° out of phase with the excitation from the bright-mode resonator within its own unit cell. Therefore, when the gap between the bright-mode resonator and adjacent dark-mode resonators are equal ($g_1 = 75$ nm, $\Delta = g_2 - g_1 = 0$) the fields destructively interfere and κ goes to zero, resulting in elimination of the Fano resonance, as can be observed in Fig. 2c and in the magnetic and electric field profiles in Fig. 2b.

At the transmission peak the energy in the array is concentrated in the collective dark mode with the magnetic dipoles in each of the split-rings oscillating coherently as shown in Fig. 3a. The array size thus becomes an important factor in the overall Q-factor of the metasurface due to the fact that lattice perturbations at the array's edge break the coherence[27,28], leading to strong scattering of light into free space and broadening of the resonance peak. To characterize this effect, we fabricated arrays of five different sizes, consisting of 400 to 90,000 unit cells (SEM image of the test samples are shown in Fig. 3b). In the measurement, we used a $\times 5$ objective for illumination and a $\times 50$ objective for collection with an aperture in the conjugate image plane on the transmission side to confine the collection area to $15\,\mu m \times 15\,\mu m$. The measured spectra are shown in

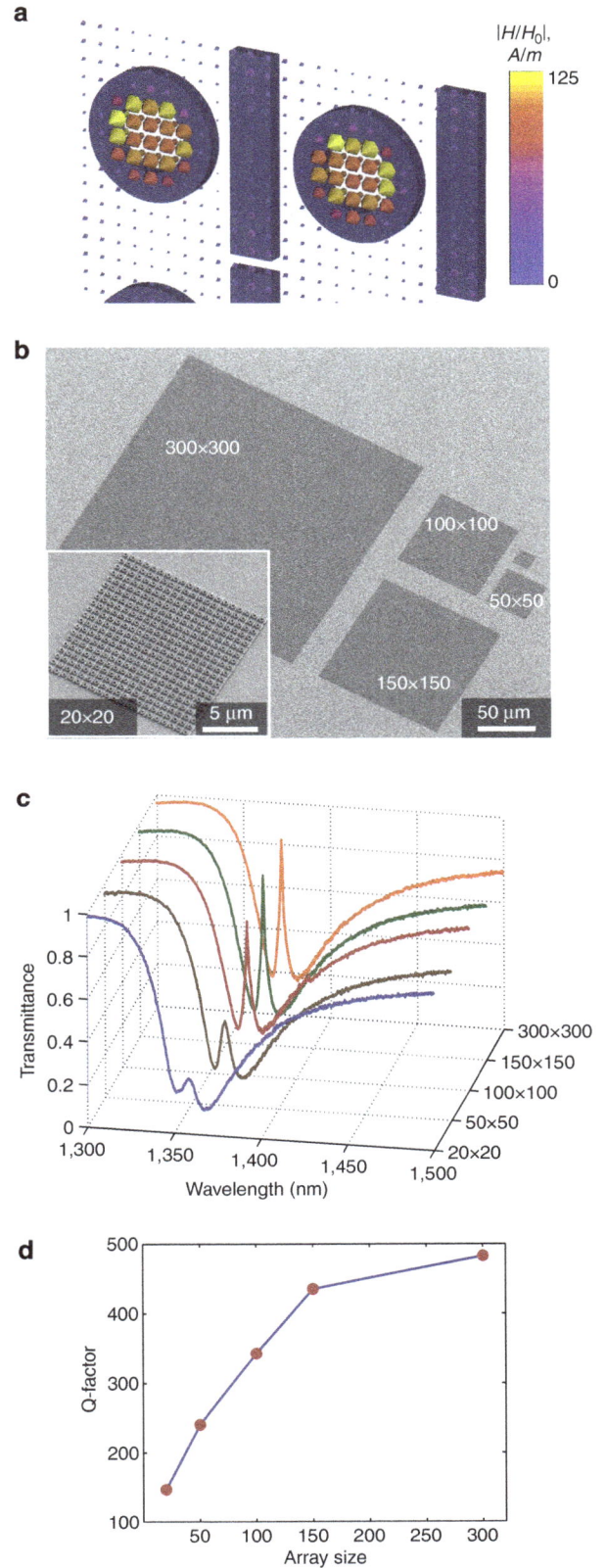

Figure 3 | Dependence of the Q-factor on the array size of the metasurface. (**a**) Vector plot of the magnetic field showing the coherent excitation of the magnetic dipoles within the dark-mode resonators. (**b**) Scanning electron microscope images of the dielectric metasurface with different array sizes. The inset shows the magnified view of a 20 × 20 array. (**c**) Transmittance spectra of arrays with different sizes. (**d**) The extracted Q-factors of the metasurface as a function of array size.

Fig. 3c and we observe that the Q-factor increases with increasing array size, saturating as we approach 90,000 unit cells, as is shown in Fig. 3d. The saturation of the Q-factor indicates the limit imposed by both incoherent radiative loss arising from inhomogeneities and non-radiative loss arising due to finite absorption in the Si. As expected, the required unit-cell number needed to achieve spectral convergence is larger than that reported in coherent plasmonic metamaterials due to the reduction in the non-radiative component.

Photonic crystal cavities possessing higher Q-factors have been reported[36], although they lack the spatial homogeneity present in the metasurface outlined here. Diffractive-guided mode structures can also exhibit extremely high Q-factors[37], although in our case we have the additional freedom of engineering the local field enhancement through modification of the dark-mode resonators. For instance, this can be done by placing symmetric feed-gaps in the resonator forming a double-gap split-ring, as shown in Fig. 4a. In this case, the azimuthally oriented field located in the gap is enhanced by a factor of $\varepsilon_d/\varepsilon_s$ (Fig. 4b), where ε_d and ε_s are the permittivities of the resonator and surrounding medium, respectively. For the design depicted in Fig. 4, this results in an electric field enhancement of 44 in the gap region. Having the advantages of both a sharp spectral response and a strongly enhanced field in the surrounding substance allows the metasurface to serve as an ideal platform for enhancing interaction with the surrounding medium. To investigate the response of such structures, samples were made using the same processes as described above and measured in $n = 1.44$ index-matching oil. An SEM image of the sample is shown in Fig. 4c. The simulated and experimentally measured transmittance curves

are shown in Fig. 4d and result in theoretical and experimentally measured Q-factors of 374 and 129, respectively. The lower theoretical Q-factor, compared with the ring geometry, is due to the fact that the electric fields are not equal in the two arms of the split-ring due to the asymmetry of the unit cell. This results in radiation loss to free space as the electric dipole moments do not fully cancel one another. The decrease in the experimental Q-factor, compared with theory, is attributed to more imperfections in the sample arising from the inclusion of the gaps, ultimately resulting in increased scattering from the structure. Single-gap split-ring dark-mode resonators were also explored and are described in Supplementary Fig. 3; Supplementary Note 2. More details regarding the field profiles of each of the metasurface variations can be found in Supplementary Fig. 4.

Refractive-index sensing. Due to their narrow linewidths, one interesting application of these metasurfaces is optical sensing. Optically resonant sensors are characterized by both the linewidth of the resonance ($\Delta\lambda$) as well as the shift in the resonance per refractive-index-unit change (S). These two values comprise the FOM that is given by FOM $= S/\Delta\lambda$ (ref. 38). The highest-demonstrated FOMs for Fano-resonant-localized surface plasmon resonance (LSPR) sensors are on the order of 20 (refs 13,38,39). Here, we examine the FOM of the ring-resonator metasurfaces and to further increase the sensitivity of our metasurface to local index changes, we etched a 100-nm-tall quartz pillar below the Si resonators such that the field overlap in the surrounding dielectric can be further promoted, as illustrated in Fig. 5a.

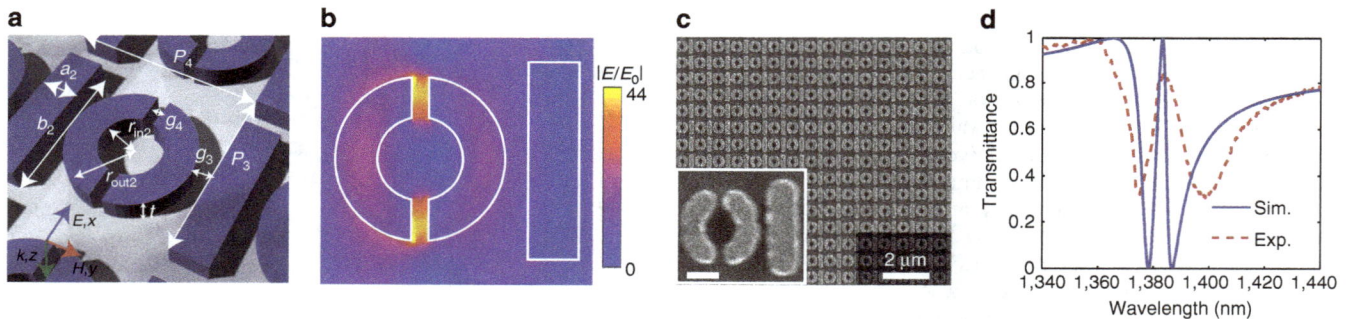

Figure 4 | Double-gap split-ring for enhancing E-field in the surrounding environment. (**a**) Diagram of the metasurface. The geometrical parameters are: $a_2 = 150$ nm, $b_2 = 600$ nm, $r_{in2} = 120$ nm, $r_{out2} = 250$ nm, $g_3 = 70$ nm, $g_4 = 50$ nm, $t = 110$ nm, $p_3 = 750$ nm and $p_4 = 800$ nm. (**b**) Simulated electric-field-amplitude distribution. (**c**) Scanning electron microscope image of the fabricated metasurface. The inset shows a single unit cell (scale bar, 200 nm). (**d**) Simulated (blue solid line) and measured (red dashed line) transmittance spectrum of the metasurface.

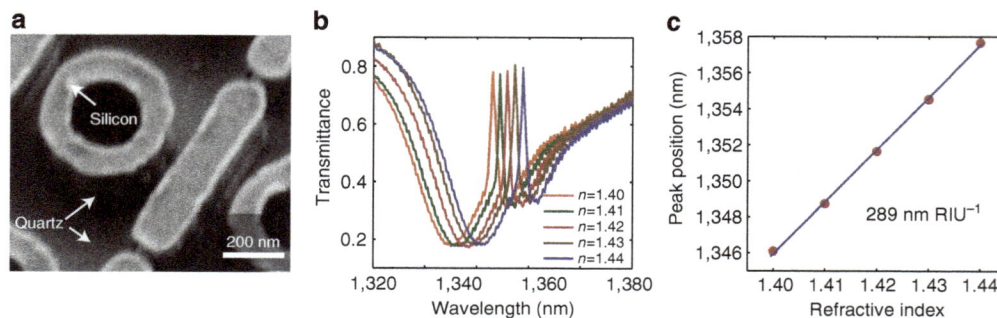

Figure 5 | Refractive index sensing. (**a**) SEM image of the sample with the quartz substrate etched by 100 nm. (**b**) Measured transmittance spectra of the EIT metasurface when immersed in oil with refractive index ranging from 1.40 to 1.44. (**c**) The red circles show the experimentally measured resonance peak position as a function of the background refractive index. The blue curve is a linear fit to the measured data, which was used to determine the sensitivity as 289 nm RIU^{-1}.

The transmittance spectra of the metasurface when immersed in oil with different refractive indices, from 1.40 to 1.44, are presented in Fig. 5b, and it can be seen that a substantial movement in the peak position is realized despite the small index change. A linear fit of the shift in the resonance peak (Fig. 5c) leads to a sensitivity of $S = 289\,\text{nm}\,\text{RIU}^{-1}$, which is comparable, but slightly lower, than the best Fano-resonant plasmonic sensors. The slight decrease in sensitivity is due to the fact that the index contrast, and corresponding field enhancement, at the dielectric metasurface–oil interface is still lower than that found in plasmonic structures. However, when combining the sensitivity with an average resonance peak linewidth $\Delta\lambda$ of only 2.8 nm, we arrive at an FOM of 103, which far exceeds the current record for Fano-resonant LSPR sensors. With further optimization of fabrication and more stringent control of the resonator coupling, FOMs on the order of 1,000 should be within reach. We also characterized the performance of the double-gap split-ring metasurface, finding it has a higher sensitivity ($S = 379\,\text{nm}\,\text{RIU}^{-1}$) due to the increased modal overlap with the surrounding material. However, the experimentally measured FOM is reduced to 37 due to the broader linewidth. More details can be found in Supplementary Fig. 5; Supplementary Note 3.

In conclusion, dielectric metasurfaces can be used to significantly improve upon the performance of their plasmonic counterparts in mimicking EIT due to their greatly reduced absorption loss, resulting in sensing FOMs that far exceed previously demonstrated LSPR sensors. With proper design, such metasurfaces can also confine the optical field to nanoscale regions, opening the possibility of using such metasurfaces for a wide range of applications including bio/chemical sensing, enhancing emission rates, optical modulation and low-loss slow-light devices.

Methods

Simulations. The numerical simulations (transmittance/reflectance and field distributions) presented in the manuscript were carried out using a commercially available software (CST Microwave Studio 2013) using the finite-element frequency-domain solver with periodic boundary conditions for each unit cell. The dimensions and material properties of the metasurface and the background are specified in the main text. Both normal-incident TE00 (polarization along the long axis of the bar antenna) and TM00 (polarization perpendicular to the long axis of the bar antenna) modes were defined at the incident and receiving ports to ensure that any polarization conversion is accounted for. At least 10 mesh steps per wavelength were used to ensure the accuracy of the calculated results. In addition, we reproduced the results using another commercially available full-wave simulation tool (Lumerical) using a finite-difference time-domain solver.

Q-factor extraction. We used a Fano model to extract the Q-factor of the EIT resonance from the dielectric metasurface. The experimentally measured transmittance T_{exp} was fitted to a Fano line shape given by $T_{\text{Fano}} = |a_1 + ja_2 + \frac{b}{\omega - \omega_0 + j\gamma}|^2$, where a_1, a_2 and b are constant real numbers; ω_0 is the central resonant frequency; γ is the overall damping rate of the resonance. The experimental Q-factor was then determined by $Q = (\omega_0/2\gamma)$. The fit was performed over the wavelength range from 1,367 to 1,380 nm as this model only encompasses the Fano line shape and does not include the fit to the background resonance provided by the bright mode. We used a fitting method identical to the one used in a recent demonstration of Si-based chiral metasurface[25], therefore rendering the reported Q-factors in both cases comparable. The fit is shown in Supplementary Fig. 1.

Sample fabrication. Polycrystalline Si was chosen as the resonator material due to its high index and small absorption in the near-infrared frequencies. The Si was deposited on a 4-inch quartz wafer using a low-pressure chemical vapour deposition horizontal tube furnace. The metasurfaces were created by a sequential process of EBL, mask deposition, lift-off and RIE. In the EBL process, we employed a 10-nm chromium charge-dissipation layer on top of the PMMA resist. The total pattern size was $\sim 225\,\mu\text{m} \times 225\,\mu\text{m}$, written using a JEOL 9300FS 100 kV EBL tool. A fluorine-based inductively coupled plasma RIE recipe involving C_4F_8, SF_6, O_2 and Ar gas flows was used to etch the poly-Si layer creating the metasurface layer. Polydimethylsiloxane flow cells were used to immerse the metasurface in the refractive-index-matching oils (Cargille Labs) for all the spectroscopy and sensing measurements. The samples were washed thoroughly after each measurement with isopropyl alcohol followed by placing them in a vacuumed desiccator for 1 h to ensure that no residual oil was left after each measurement.

Optical measurements. A custom-built free-space infrared microscope was used to measure the transmittance, reflectance and absorption spectra of the metasurfaces. Light from a tungsten/halogen lamp was focused through a long-working distance objective (Mitutoyo, $\times 5$, 0.14 numerical aperture) with the numerical aperture of the objective further cut down to 0.025 by an aperture at its back focal plane to restrict the incident angle to near-normal. The transmittance spectrum (T) was normalized with the transmittance of the white light through an unpatterned quartz substrate. The reflectance spectrum (R) was normalized to reflectance of a gold mirror, and the absorption of the sample is determined as $A = 1 - T - R$.

References

1. Harris, S. E. Electromagnetically induced transparency. *Phys. Today* **50**, 36 (1997).
2. Zhang, S., Genov, D. A., Wang, Y., Liu, M. & Zhang, X. Plasmon-induced transparency in metamaterials. *Phys. Rev. Lett.* **101**, 047401 (2008).
3. Papasimakis, N., Fedotov, V., Zheludev, N. & Prosvirnin, S. Metamaterial analog of electromagnetically induced transparency. *Phys. Rev. Lett.* **101**, 253903 (2008).
4. Verellen, N. *et al.* Fano resonances in individual coherent plasmonic nanocavities. *Nano Lett.* **9**, 1663–1667 (2009).
5. Singh, R., Rockstuhl, C., Lederer, F. & Zhang, W. Coupling between a dark and a bright eigenmode in a terahertz metamaterial. *Phys. Rev. B* **79**, 085111 (2009).
6. Tassin, P., Zhang, L., Koschny, T., Economou, E. & Soukoulis, C. Low-loss metamaterials based on classical electromagnetically induced transparency. *Phys. Rev. Lett.* **102**, 053901 (2009).
7. Liu, N. *et al.* Plasmonic analogue of electromagnetically induced transparency at the Drude damping limit. *Nat. Mater.* **8**, 758–762 (2009).
8. Liu, N., Hentschel, M., Weiss, T., Alivisatos, A. P. & Giessen, H. Three-dimensional plasmon rulers. *Science* **332**, 1407–1410 (2011).
9. Liu, X. *et al.* Electromagnetically induced transparency in terahertz plasmonic metamaterials via dual excitation pathways of the dark mode. *Appl. Phys. Lett.* **100**, 131101 (2012).
10. Gu, J. *et al.* Active control of electromagnetically induced transparency analogue in terahertz metamaterials. *Nat. Commun.* **3**, 1151 (2012).
11. Xu, Q. *et al.* Experimental realization of an on-chip all-optical analogue to electromagnetically induced transparency. *Phys. Rev. Lett.* **96**, 123901 (2006).
12. Yang, X., Yu, M., Kwong, D.-L. & Wong, C. All-optical analog to electromagnetically induced transparency in multiple coupled photonic crystal cavities. *Phys. Rev. Lett.* **102**, 173902 (2009).
13. Liu, N. *et al.* Planar metamaterial analogue of electromagnetically induced transparency for plasmonic sensing. *Nano Lett.* **10**, 1103–1107 (2010).
14. Dong, Z.-G. *et al.* Enhanced sensing performance by the plasmonic analog of electromagnetically induced transparency in active metamaterials. *Appl. Phys. Lett.* **97**, 114101 (2010).
15. Sun, Y. *et al.* Electromagnetic diode based on nonlinear electromagnetically induced transparency in metamaterials. *Appl. Phys. Lett.* **103**, 091904 (2013).
16. Fano, U. Effects of configuration interaction on intensities and phase shifts. *Phys. Rev.* **124**, 1866–1878 (1961).
17. Luk'yanchuk, B. *et al.* The Fano resonance in plasmonic nanostructures and metamaterials. *Nat. Mater.* **9**, 707–715 (2010).
18. Lewin, L. The electrical constants of a material loaded with spherical particles. *Proc. Inst. Elec. Eng. Part 3* **94**, 65–68 (1947).
19. O'Brien, S. & Pendry, J. B. Photonic band-gap effects and magnetic activity in dielectric composites. *J. Phys. Condens. Matter* **14**, 4035–4044 (2002).
20. Peng, L. *et al.* Experimental observation of left-handed behavior in an array of standard dielectric resonators. *Phys. Rev. Lett.* **98**, 157403 (2007).
21. Moitra, P. *et al.* Realization of an all-dielectric zero-index optical metamaterial. *Nat. Photonics* **7**, 791–795 (2013).
22. Yang, Y. *et al.* Dielectric meta-reflectarray for broadband linear polarization conversion and optical vortex generation. *Nano Lett.* **14**, 1394–1399 (2014).
23. Miroshnichenko, A. E. & Kivshar, Y. S. Fano resonances in all-dielectric oligomers. *Nano Lett.* **12**, 6459–6463 (2012).
24. Fan, P., Yu, Z., Fan, S. & Brongersma, M. L. Optical Fano resonance of an individual semiconductor nanostructure. *Nat. Mater.* **13**, 471–475 (2014).
25. Wu, C. *et al.* Spectrally selective chiral silicon metasurfaces based on infrared Fano resonances. *Nat. Commun.* **5**, 3892 (2014).
26. Zhang, J., MacDonald, K. F. & Zheludev, N. I. Near-infrared trapped mode magnetic resonance in an all-dielectric metamaterial. *Opt. Express* **21**, 26721–26728 (2013).
27. Fedotov, V. a. *et al.* Spectral collapse in ensembles of metamolecules. *Phys. Rev. Lett.* **104**, 223901 (2010).
28. Jenkins, S. D. & Ruostekoski, J. Metamaterial transparency induced by cooperative electromagnetic interactions. *Phys. Rev. Lett.* **111**, 147401 (2013).
29. Kao, T. S., Jenkins, S. D., Ruostekoski, J. & Zheludev, N. I. Coherent control of nanoscale light localization in metamaterial: creating and positioning isolated subwavelength energy hot spots. *Phys. Rev. Lett.* **106**, 085501 (2011).
30. Jenkins, S. D. & Ruostekoski, J. Theoretical formalism for collective electromagnetic response of discrete metamaterial systems. *Phys. Rev. B* **86**, 085116 (2012).

31. Adamo, G. *et al.* Electron-beam-driven collective-mode metamaterial light source. *Phys. Rev. Lett.* **109,** 217401 (2012).

32. Burns, M. M., Fournier, J.-M. & Golovchenko, J. A. Optical binding. *Phys. Rev. Lett.* **63,** 1233–1236 (1989).

33. Haus, H. A. *Waves and Fields in Optoelectronics* (Prentice Hall, 1983).

34. Fan, S., Suh, W. & Joannopoulos, J. D. Temporal coupled-mode theory for the Fano resonance in optical resonators. *J. Opt. Soc. Am. A* **20,** 569 (2003).

35. Vahala, K. J. Optical microcavities. *Nature* **424,** 839–846 (2003).

36. Akahane, Y., Asano, T., Song, B.-S. & Noda, S. High-Q photonic nanocavity in a two-dimensional photonic crystal. *Nature* **425,** 944–947 (2003).

37. Fan, S. & Joannopoulos, J. Analysis of guided resonances in photonic crystal slabs. *Phys. Rev. B* **65,** 235112 (2002).

38. Anker, J. N. *et al.* Biosensing with plasmonic nanosensors. *Nat. Mater.* **7,** 442–453 (2008).

39. Zhang, S., Bao, K., Halas, N. J., Xu, H. & Nordlander, P. Substrate-induced Fano resonances of a plasmonic nanocube: a route to increased-sensitivity localized surface plasmon resonance sensors revealed. *Nano Lett.* **11,** 1657–1663 (2011).

Acknowledgements

This work was funded by the Office of Naval Research under programmes N00014-12-1-0571 and N00014-14-1-0475. A portion of this research was conducted at the Center for Nanophase Materials Sciences, which is sponsored at Oak Ridge National Laboratory by the Scientific User Facilities Division, Office of Basic Energy Sciences, US Department of Energy. A portion of this work was also performed at the Vanderbilt Institute of Nanoscale Science and Engineering, we thank the staff for their support.

Author contributions

Y.Y. and J.V. conceived the idea. Y.Y. designed the structure, fabricated the sample and conducted the optical characterization. I.I.K. assisted in developing the electron-beam lithography and RIE processes and D.P.B. performed the low-pressure chemical vapour deposition. All authors discussed the results, and Y.Y. and J.V. prepared the manuscript. J.V. supervised the project.

Additional information

Competing financial interests: The authors declare no competing financial interests.

8

Hydrochromic conjugated polymers for human sweat pore mapping

Joosub Lee[1], Minkyeong Pyo[1], Sang-hwa Lee[2], Jaeyong Kim[2,3], Moonsoo Ra[4], Whoi-Yul Kim[4], Bum Jun Park[5], Chan Woo Lee[3] & Jong-Man Kim[1,3]

Hydrochromic materials have been actively investigated in the context of humidity sensing and measuring water contents in organic solvents. Here we report a sensor system that undergoes a brilliant blue-to-red colour transition as well as 'Turn-On' fluorescence upon exposure to water. Introduction of a hygroscopic element into a supramolecularly assembled polydiacetylene results in a hydrochromic conjugated polymer that is rapidly responsive (<20 µs), spin-coatable and inkjet-compatible. Importantly, the hydrochromic sensor is found to be suitable for mapping human sweat pores. The exceedingly small quantities (sub-nanolitre) of water secreted from sweat pores are sufficient to promote an instantaneous colorimetric transition of the polymer. As a result, the sensor can be used to construct a precise map of active sweat pores on fingertips. The sensor technology, developed in this study, has the potential of serving as new method for fingerprint analysis and for the clinical diagnosis of malfunctioning sweat pores.

[1] Department of Chemical Engineering, Hanyang University, Seoul 133-791, Korea. [2] Department of Physics, Hanyang University, Seoul 133-791, Korea. [3] Institute of Nano Science and Technology, Hanyang University, Seoul 133-791, Korea. [4] Department of Electronic Engineering, Hanyang University, Seoul 133-791, Korea. [5] Department of Chemical Engineering, Kyung Hee University, Youngin-Si, Gyeonggi-do 446-701, Korea. Correspondence and requests for materials should be addressed to J.-M.K. (email: jmk@hanyang.ac.kr) or to C.W.L. (email: lcw@hanyang.ac.kr).

Materials that undergo colour changes in response to physical (heat, light, pressure, current and so on) and chemical/biochemical (solvent, pH, ligand–receptor interactions and so on) inputs serve as the key components in a number of sensor, display, switch and memory devices[1-6]. Among stimulus-responsive colorimetric substances investigated thus far, conjugated polymers[7] have gained special attention owing to their unique optical properties associated with the presence of extensively delocalized π-electron networks. As localized stimulation of conjugated polymers brings about changes in absorption and emission characteristics, these materials have been effectively utilized as sensor matrices[8-17].

Polydiacetylenes (PDAs)[18-28] possess unique structural features that differ from those of other conjugated polymers. As they are prepared by polymerization of self-assembled diacetylene monomers (Fig. 1), PDAs are intrinsically supramolecular aggregates in which dense packing restricts conformational mobility of main chain backbones and enables extensive and efficient p-orbital overlap. As a result, PDAs generally have an absorption maximum at around 650 nm that corresponds to a blue colour. When the aggregated, blue-coloured supramolecular PDAs completely dissociate into individual PDAs upon exposure to an appropriate solvent, their side chains are able to rotate freely and, as a result, a blue-to-yellow colour transition occurs[29,30]. However, because of the presence of strong interchain interactions, PDAs are typically insoluble in common organic solvents. Instead, dissolution of unpolymerized residual diacetylene monomers and partially polymerized oligomers in common organic solvents creates voids in the PDA supramolecules, which leads to partial distortion of the polymer backbone and a blue-to-red (or purple) colour change[31,32].

We recently introduced conceptually new strategies, such as a protective layer approach[33] and a combinatorial method[34], for the design of solvatochromic PDA sensor systems. Our continuing efforts in this area led to development of a new PDA, which contains hygroscopic elements in its headgroup and, consequently, is colorimetrically responsive to water. The hydrochromic PDA was found to undergo a dramatic and rapid structural change in the presence of water that brings about a blue-to-red colour transition. The new solvatochromic sensor system developed in this study possesses several significant features. First, although many solvatochromic PDA sensor systems that respond to organic solvents have been described[29-34], a colorimetric PDA, which is responsive to pure water, has not yet been developed. Second, the hydrochromic PDA developed in this study is spin-coatable and inkjet-compatible, two important advantages when compared with conventional small molecule-based water-responsive sensors. Third, the colorimetric response of the sensor can be readily controlled by manipulation of the chain length of the diacetylene monomer and the hygroscopic element. This is a significantly meritorious property of PDAs that is not shared by other conjugated polymers and small molecule sensors. Fourth, the blue-to-red colorimetric transition is accompanied by the generation of red fluorescence, a feature that enables the PDA to serve as a 'Turn-On' functioning sensor. Last, the fluorescence 'Turn-On' PDA system is applicable to mapping human sweat pores. The results show that the very small amounts (<nanolitre) of water secreted from sweat pores are sufficient to promote an instantaneous blue-to-red colorimetric transition as well as a fluorescence 'Turn-On' of the polymer film, a phenomenon that enables precise mapping of active sweat pores on a human fingertip. Thus, the sensor technology developed in this study has the potential of serving as a new method for fingerprint analysis and for the clinical diagnosis of malfunctioning sweat pores. We believe that the observations made in the investigation will serve as the foundation for new applications of hydrochromic materials, beyond simply humidity sensing[35] and measurement of water contents in organic solvents[36]. The details of the investigation leading to these conclusions are described below.

Results

Discovery of the hydrochromic PDA. The investigation described below was guided by the hypothesis that introduction of hygroscopic moieties into head groups of PDAs would lead to conjugated polymers that undergo significant water-promoted electronic changes. In order to explore this proposal, PDAs containing cesium carboxylate head groups were selected as potential targets owing to the typically highly hygroscopic nature of these ion complexes and the fact that the polymers can be readily prepared from commercially available diacetylene monomers. For example, a transparent gel is generated when an equimolar mixture of 10,12-pentacosadiynoic acid (PCDA, $CH_3(CH_2)_{11}C\equiv C-C\equiv C(CH_2)_8COOH$) and cesium hydroxide in aqueous tetrahydrofuran (THF) is stirred for 1 h at an ambient temperature (Supplementary Fig. 1a). Inspection of a scanning electron microscope (SEM) image shows that the PCDA-Cs gel comprises thin-plate morphologies (Supplementary Fig. 1b) and a thin film is readily formed when the viscous gel is spin-coated on a solid substrate. UV irradiation (254 nm, 1 mW cm^{-2}, 30 s) of the coated film results in the formation of blue-coloured poly(PCDA-Cs; Fig. 2a).

Interestingly, an immediate blue-to-red colorimetric response occurs when the poly(PCDA-Cs) film is exposed to a soft blow of

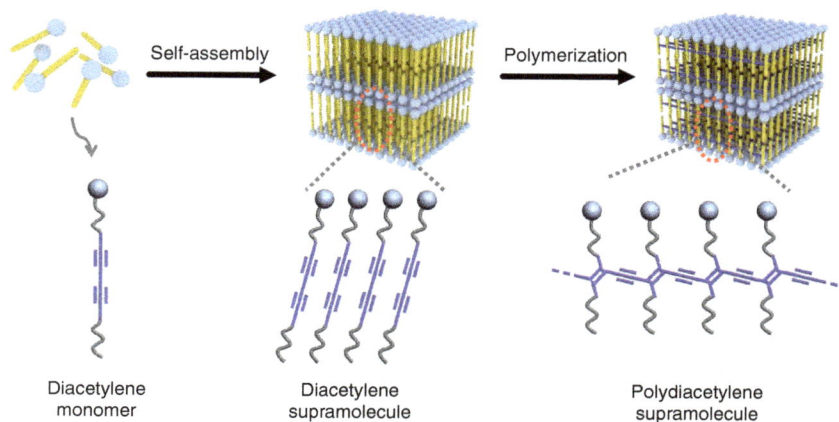

Figure 1 | Supramolecular polydiacetylene. Schematic for formation of a supramolecular PDA from a self-assembled diacetylene entity.

Diacetylene monomer — Diacetylene supramolecule — Polydiacetylene supramolecule

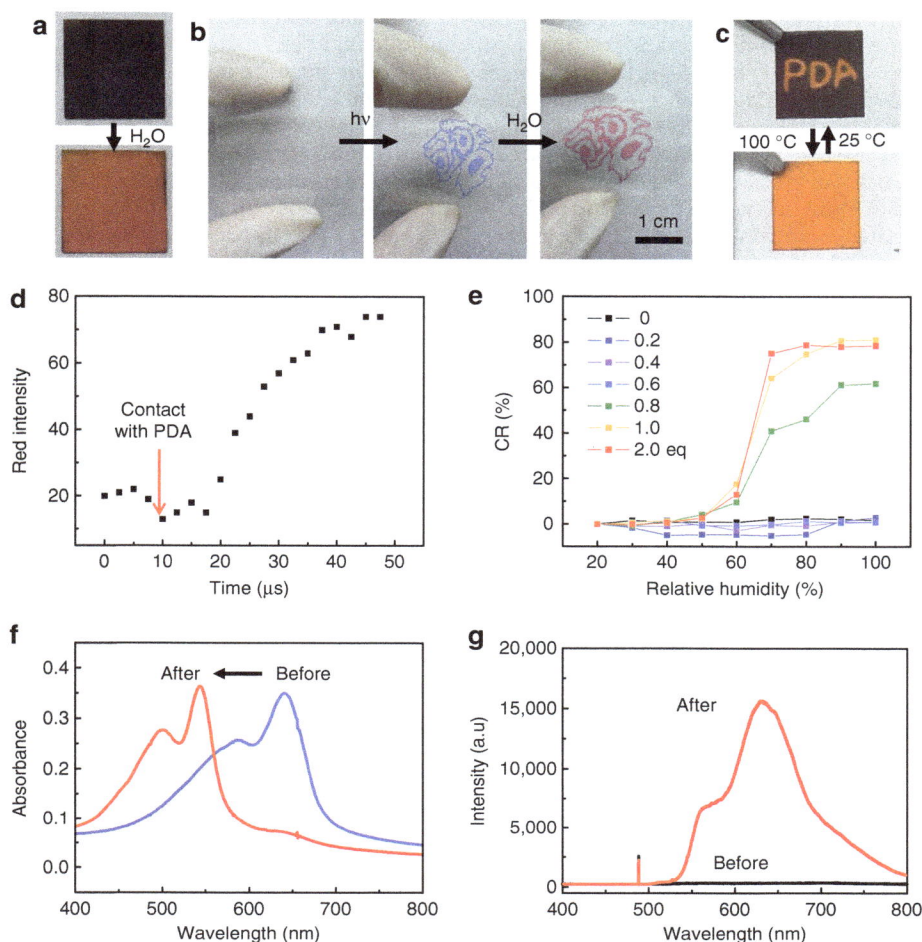

Figure 2 | Hydrochromic PDA. (**a**) Colorimetric transition of a poly(PCDA-CS) coated (thickness: *ca.* 3.8 μm) PET film upon exposure to water. (**b**) Photographs of printed image of a PET film using a PCDA-Cs ink solution immediately following printing (transparent image), after UV irradiation (254 nm, 1 mW cm^{-2}) for 30 s (blue-coloured image) and after exposure to water (red-coloured image). (**c**) A handwritten image of 'PDA' on a poly(PCDA-Cs)-coated PET film with a water-filled ballpoint pen at 25 °C (PDA image) and at 100 °C (PDA image disappears). (**d**) Time-dependent red intensity variations of a poly(PCDA-Cs) film during exposure to a water droplet. (**e**) Colorimetric response (CR) of a poly(PCDA-Cs) film as a function of molar equivalent of Cs ion to PCDA. The CR is calculated by using the well-known equation[51] CR = (PB$_O$ − PB$_f$)/PB$_O$ × 100%, where PB$_O$ and PB$_f$ are the percent blue before and after the colour transition and PB = A$_{640nm}$/[A$_{640nm}$ + A$_{550nm}$]. (**f,g**) Water induced visible absorption (**f**) and emission (**g**) (excitation at 488 nm) spectral changes of a poly(PCDA-Cs) film upon hydration.

human expiration (Fig. 2a; Supplementary Movie 1). The water-promoted colour change was found to occur even at low temperatures (Supplementary Movie 2). In addition, the poly(PCDA-Cs) can be generated in an inkjet-compatible manner (Fig. 2b, Supplementary Movie 3). To demonstrate this feature, a cartridge of a conventional thermal inkjet office printer was filled with an aqueous suspension of PCDA-Cs and then employed to print an image on a polyethylene terephthalate (PET) film (see Methods for details). Since monomeric PCDA-Cs does not absorb visible light, the printed image is not visible. However, UV irradiation with a hand-held laboratory 254 nm UV lamp for 10 s results in the formation of a blue-patterned image associated with formation of poly(PCDA-Cs). The image undergoes an instant blue-to-red colour change upon exposure to water. The versatility of the hydrochromic PDA-coated PET film was further demonstrated by utilizing it to create images with a ballpoint pen filled with water (Fig. 2c). Interestingly, the handwritten, red image disappears when the film is heated to 100 °C and reappears upon cooling the film to room temperature (Supplementary Movie 4). This colour-switching phenomenon is likely a consequence of the fact that the poly(PCDA-Cs) film displays a reversible thermochromism (Supplementary Fig. 2) and that the hydrochromism

process is irreversible. The thermally controlled, reversible on-off image process can be repeated many times (> 100) without losing the quality of the image.

Probing the nature of the hydrochromism. The rate of the water-promoted colorimetric response of the poly(PCDA-Cs) film was evaluated by employing a high-speed camera. The plot of intensity of the red colour as a function of time following exposure to a water droplet displayed in Fig. 2d demonstrates that a significant colour change occurs within 20 μs after water contacts the film and reaches a maximum after 40 μs. In order for the PDA to display effective hydrochromism, the molar equivalents of cesium ion in the PCDA complex needs to be greater than 0.8 (Fig. 2e). This observation suggests that a large portion of the PDA supramolecules needs to be in hygroscopic forms in order for an efficient colour change to occur.

The hydrochromic behaviour of the poly(PCDA-Cs)-coated PET film was also probed by using visible absorption spectroscopy (Fig. 2f). Inspection of the spectra shows that a significant shift in the wavelength maximum from 640 to 540 nm takes place when the film is treated with water. In addition, because the blue-phase poly(PCDA-Cs) is nonfluorescent while its red-phase counterpart fluoresces, the water-promoted transition was also observed using emission spectroscopy (Fig. 2g). Analysis of the emission spectra shows that a fluorescence emission band centred at 630 nm (excitation: 488 nm) arises when the blue poly(PCDA-Cs) film is exposed to water. Finally, the results of Raman spectroscopic studies (Supplementary Fig. 3) demonstrate that water causes shifts in the alkyne–alkene bands (2,079 and 1,452 cm^{-1}) associated with poly(PCDA-Cs) to higher frequencies (2,120 and 1,515 cm^{-1}), an observation that confirms that a typical[37] blue-to-red phase transition of PDA occurs in response to water.

Important observations, which contribute to an understanding of the origin of the hydrochromic behaviour of the PDAs, were made in SEM and X-ray diffraction (XRD) studies. The SEM image shows that a morphological transition from strips to isolated aggregates takes place upon exposure of the poly(PCDA-Cs) film to water (Fig. 3a). The water-promoted morphological change of the film is also reflected in the results of XRD studies (Fig. 3b). Specifically, the initial blue-coloured PDA-Cs film comprises lamellar structures with an interlamellar distance of 4.81 nm, calculated by utilizing the Bragg's law. Upon hydration, a new structure with an interlamellar distance of 6.19 nm is formed. The water-promoted increase in interchain distance should lead to partial distortion of the arrayed p-orbitals that is responsible for the blue-to-red colour change of the polymer (Fig. 3c).

Studies of complexes of PCDA that contain other metal cations indicate that films that comprises lithium, sodium and potassium salts of the polymer do not undergo the water-promoted colour transition while the film containing the poly(PCDA-Rb) complex forms a red colour upon being hydrated (Supplementary Fig. 4). The results of contact angle measurements show that the hydrochromism of the polymer film is governed by the size of the counter cation in the PDA complex (Supplementary Fig. 5).

One meritorious feature of PDAs, which is not shared by other conjugated polymers, is associated with the fact that their colorimetric responses can be readily controlled by manipulation of the chain length of the diacetylene monomer and the head groups. In general, PDAs derived from diacetylenes with shorter alkyl chains are more sensitive to environmental stimulations than are those derived from longer alkyl chain monomers[38]. We have demonstrated that this phenomenon is also observed in the hydrochromism behaviour of PDAs. For example, as pointed out above no colorimetric transition takes place when the PDA derived from PCDA-K is exposed to water (Supplementary Fig. 4). However, the PDA derived from 10,12-tricosadiynoic acid (TCDA)-K (two carbons shorter than PCDA) displays a blue-to-red colour transition initiated under 100% relative humidity (RH) conditions (Fig. 4). In addition, poly(HCDA-K) obtained from HCDA, which contains a two carbon shorter chain than does TCDA, is even more sensitive to water in that its colour change takes place under RH 90% conditions. As summarized in Fig. 4, proper combinations of diacetylene monomers and cations enable

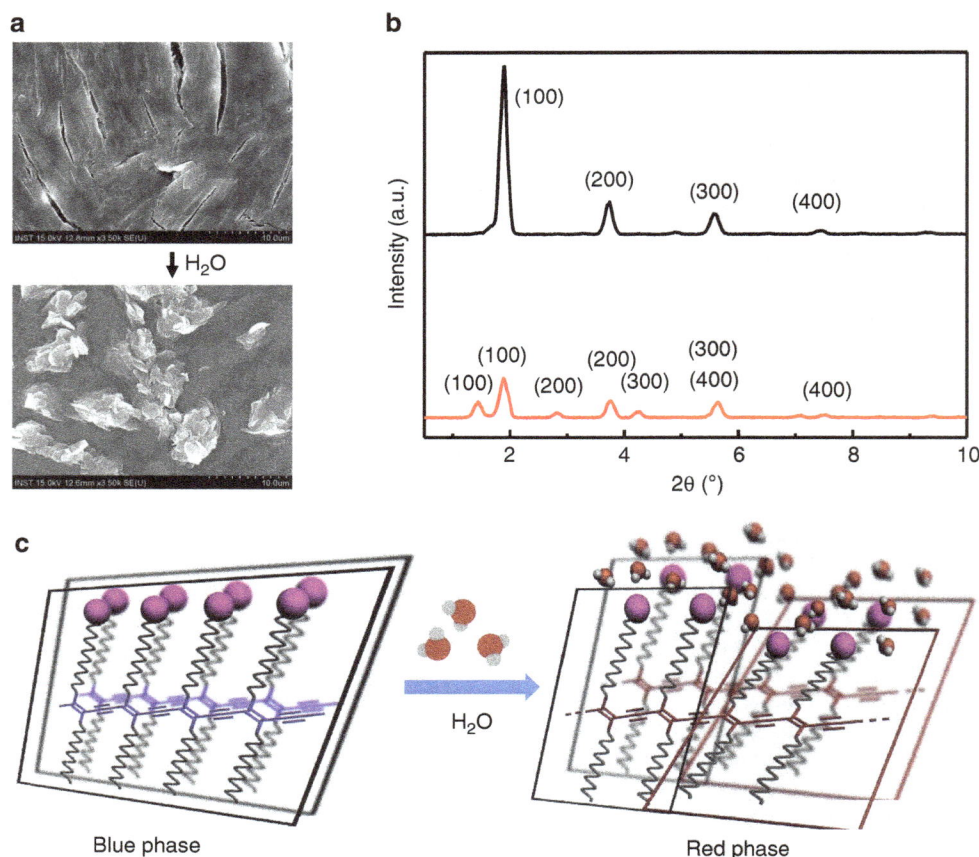

Figure 3 | Water-promoted morphological change. (a) SEM images of a poly(PCDA-Cs) film that displays a drastic morphological change upon hydration. **(b)** XRD spectra of a poly(PCDA-Cs) film before (black line) and after (red line) exposure to water. An interlamellar distance of 4.81 nm obtained before exposure to water and a new structure generated after exposure to water are indexed following the Bragg's law. **(c)** Schematic representation of the water-promoted blue-to-red phase transformation of PDA supramolecules. The purple-coloured balls represent hygroscopic elements.

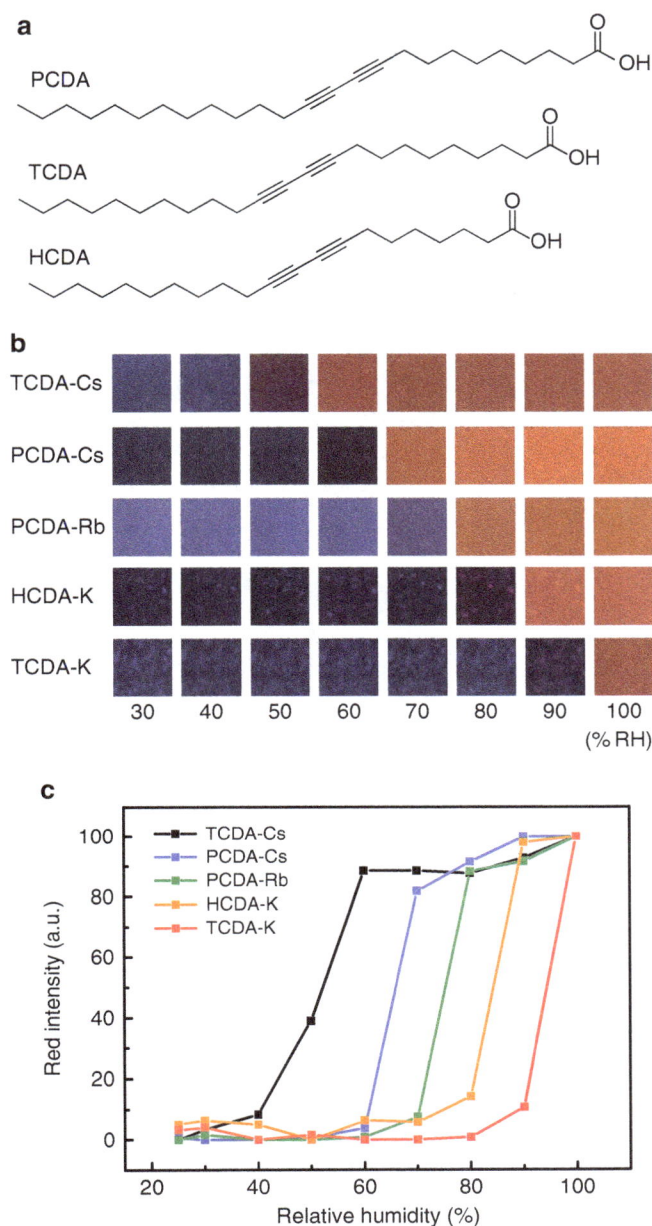

Figure 4 | Tuning of hydrochromic response. (**a**) Chemical structures of diacetylenic lipids: PCDA, TCDA, HCDA (8,10-heneicosadiynoic acid). (**b**) Colorimetric response of PDA films derived from various combinations of diacetylene monomers and ions under different humidity conditions. (**c**) Red intensity of the hydrochromic PDA films shown in **b** as a function of relative humidity. It is evident from the figures that colorimetric response of PDA films can be manipulated by choosing proper diacetylene monomers and cations.

fabrication of hydrochromic PDA films that are responsive to specific levels of humidity.

Hydrochromic mapping of human sweat pores. Every individual has discernably unique ridge patterns on their fingertips. Owing to the presence of sweat components (fatty acids, amino acids, inorganic salts and so on), these patterns are transferred to solid substrates when fingers touch the surface. Analyses of the fingerprint patterns left on solid surfaces have been unarguably one of the most convincing tools for the proof of an individual's

presence at a specific location and, as a result, a variety of techniques[39–46] have been developed for imaging and comparing these patterns.

Close inspection of fingertips reveals that they contain permanent, immutable and unique pores that are distributed on the ridges. The possible use of pore analysis for personal identification was demonstrated by Locard[47] in studies that show that 20–40 pores are sufficient to generate patterns that are required for a human's identity. When a finger touches a porous substrate, such as paper, nonvolatile sweat components from the fingertip become immobilized on the porous substrate, which because of its highly networked structure, limits free migration of the sweat components. As a result, dot images of the pores rather than complete ridge patterns that correspond to fingertip patterns become deposited on the substrate. Unfortunately, these pore dot images have been for the most part neglected owing to the fact that rapid, reliable and affordable fingertip pore mapping technologies have not been developed. Although modern high-resolution sensor systems (>1,000 d.p.i.) can be utilized to image fingertip pore patterns, they require complicated software, often contain false information owing to the presence of complicated ridge and pore structures and are incapable of distinguishing sweat-secreting active pores from those that are not functioning[48].

In light of the importance of pore mapping technologies, the water-promoted colour transition of the PDA-Cs film described above has great potential. In particular, if the blue-to-red transforming hydrochromic PDA were to be sufficiently sensitive to the small amounts of water present in sweat secreted from fingertips, it would be ideal for mapping sweat pores. In order to test this possibility, a fingerprint was deposited on the surface of a blue PDA-Cs film. Surprisingly, the polymer film became red instantaneously upon contact with the fingertip (Fig. 5a). Inspection of the contact area using an optical microscope showed that the image comprises many red-coloured dots (Fig. 5b). Fluorescence microscope analysis provided a clear image of the fingerprint in the form of emitting dots (Fig. 5c; Supplementary Movie 5). Raman spectroscopy was also employed to demonstrate that the typical structural transition of the PDA has occurred in the red microdot regions (Fig. 5d)[37].

In order to gain more information about the origin of the microdots, the fluorescence patterns corresponding to the red dots were superimposed on a digitally scanned fingertip image (Fig. 5e). Inspection of the superimposed images (Fig. 5e,f) showed that the red-coloured microdots are located on the ridges of the fingertip. In addition, black-coloured microdots seen in the images are pores on fingertip ridges, most of which match with red dots that correspond to active sweat pores. Specifically, the dots inside the blue circles in Fig. 5f correspond to pores that do not secret sweat. These observations clearly demonstrate that the red dots are generated by PDA colour transitions promoted by water secreted from sweat pores. The reproducibility of the hydrochromic method for pore mapping was demonstrated by using superimpositions of three independently deposited fingerprints (Fig. 6).

As the hydrochromic film method enables mapping of sweat-secreting pores, it should be applicable to studies of sweat pore functions. To demonstrate this feature, five independent fingerprints from the same donor were deposited on a poly(PCDA-Cs) film in the morning, afternoon and at night during a 1-week period (Fig. 7). Analysis of the resulting patterns shows that, although a majority of the sweat pores are active (dots inside the red dots in Fig. 7f), some secrete sweat in an inconsistent manner (dots inside the triangles in Fig. 7f) or not at all (dots inside the blue circles in Fig. 7f) during the observation period. Thus, the hydrochromic mapping strategy enables a clear differentiation

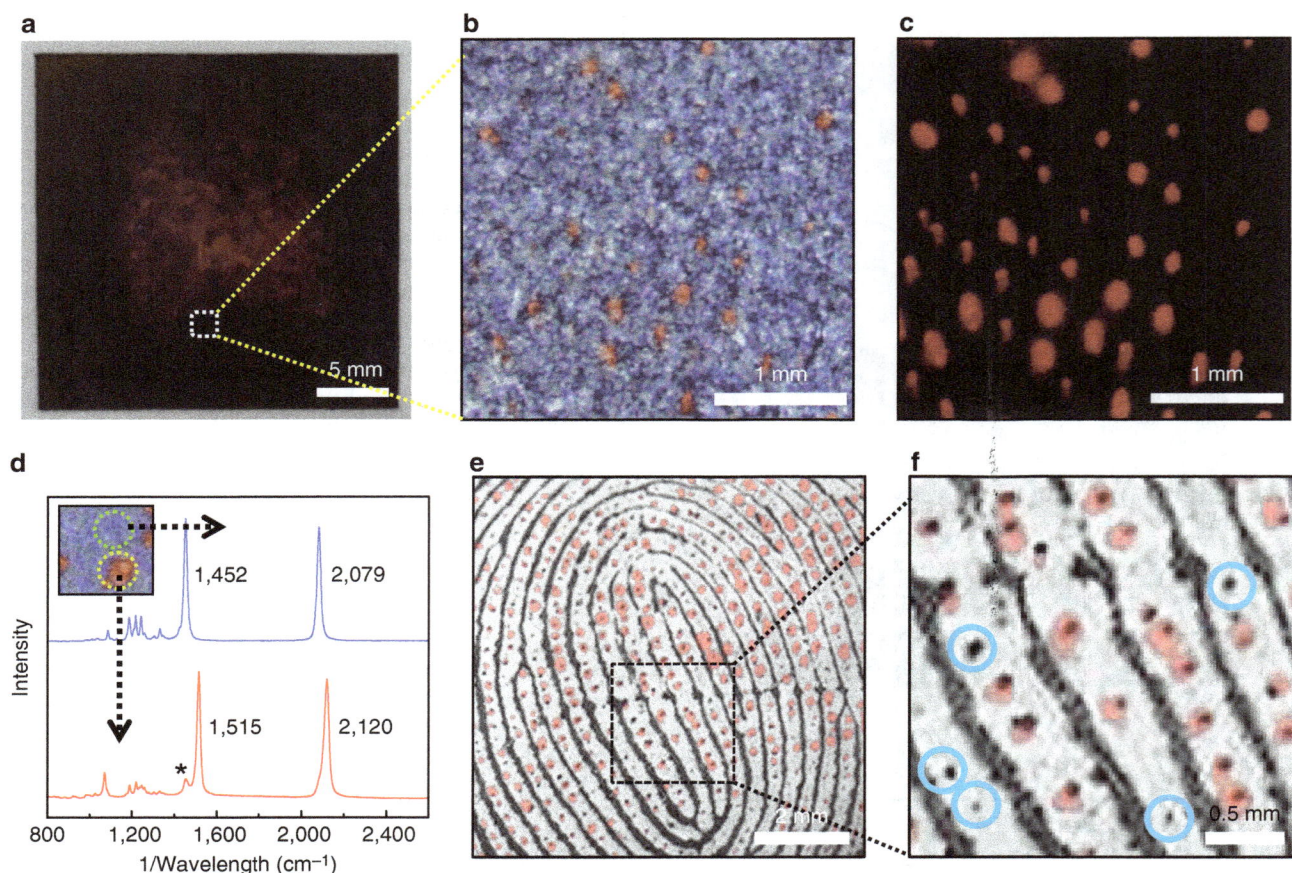

Figure 5 | Sweat pore mapping. (**a**) Photograph of a fingerprint image printed on a poly(PCDA-Cs) film (thickness: *ca.* 3.8 μm). (**b,c**) Optical (**b**) and fluorescence (**c**) microscope images of the magnified fingerprint area marked in **a**. The fluorescence image was obtained with excitation at 510–550 nm. (**d**) Raman spectra of a poly(PCDA-Cs) film after fingerprinting. The small peak marked with * corresponds to the unconverted blue-phase PDA. (**e**) Superimposed image of contrast-enhanced fluorescence dots on a scanned fingerprint image. (**f**) Magnified image of the marked area in **e**. The dots inside the blue circles are pores that do not secret sweat.

between sweat-secreting active pores and those that are inactive and, as a result, it is potentially useful for clinical diagnosis of the sweat pore function.

The feasibility of utilizing the sweat pore-based hydrochromic fingerprint detection system for latent fingerprint analysis was explored next. Java-based position tracking software (ImageJ) and a pore-pattern-matching programme implemented in MATLAB were used for the process (see Methods for details). The pore-pattern-matching programme was employed to compare a fluorescence image (Fig. 8c) and a latent fingerprint image (Fig. 8d), obtained by using ninhydrin staining (Supplementary Fig. 6). The results, displayed as blue dots in Fig. 8c,d, show that the fingerprint patterns generated by using both methods closely match one another. This finding demonstrates that the pore-pattern-matching programme is effective. To gain further support for the reliability of the pore-matching programme, a manual comparison was also carried out. Visual inspection of an image (Fig. 8g), created by superimposition of the magnified boxed areas in Fig. 8c,d (displayed in Fig. 8e,f), clearly shows that the pore-based hydrochromic sensor system is useful for fingerprint analysis (see also Supplementary Fig. 7). The meritorious feature of the new pore-based analysis system is further demonstrated by the observation that the fluorescence pore images match the ninhydrin-developed latent images including local sweat pores (Supplementary Fig. 8). The pore-pattern-matching programme enabled correct matching of 15 out of 17 partially printed latent

fingerprints with fluorescent images arising from five donors. The two incorrect matches were a consequence of the poor quality of the ninhydrin-stained latent images. Thus, if the ninhydrin-stained latent images are reasonably clear, the pore-matching programme functions with almost 100% accuracy.

Discussion

In order for it to be suitable for sweat pore mapping, a hydrochromic material must have several important characteristics. First, it should possess a balanced sensitivity to water in that it should be sufficiently responsive to display a colorimetric transition upon contact with the very low amounts of water secreted from sweat pores and not too sensitive so that it undergoes a colour change under ambient condition. Second, it is highly desirable that the water-promoted colorimeric transition be readily monitored using the naked eye and, therefore, that it takes place in the visible range. In addition, it would be more attractive if the hydrochromic phenomenon is accompanied by the generation ('Turn-On') of a colour or fluorescence upon hydration. Importantly, hydrochromic materials that display water-promoted colour or fluorescence quenching are not appropriate for sweat pore mapping because they would generate micron-sized sweat pore images that would be unrecognizable owing to the presence of dominant background signals. Finally, synthesis of the hydrochromic material and fabrication of the sensor system should be relatively simple.

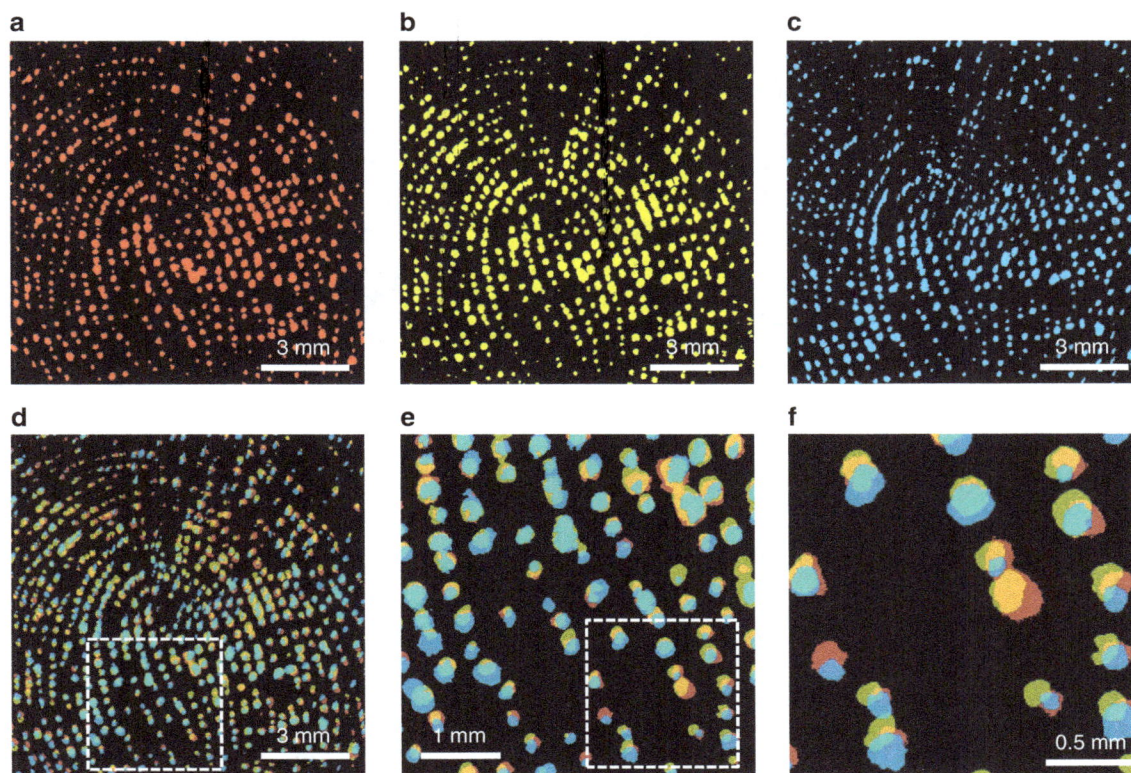

Figure 6 | Reliability test. (a–c) Three independently printed contrast-enhanced fluorescence microscope fingerprint images deposited by same donor. The yellow (b) and blue (c) coloured images are intentionally generated using a Photoshop programme for comparison purposes. (d) Superimposed images displayed in a–c. (e) Magnified image in the marked area in d. (f) Magnified image in the marked area in e.

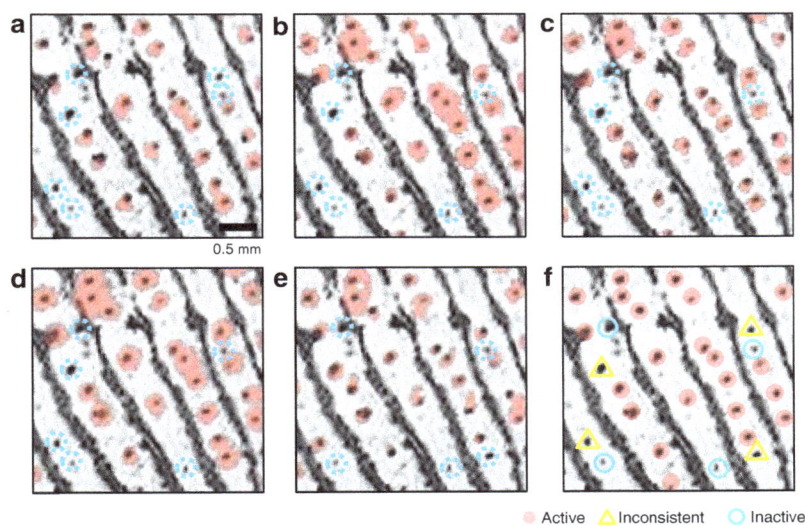

● Active   △ Inconsistent   ○ Inactive

Figure 7 | Test of active sweat pores. (a–e) Superimposed microscope images of five independently marked fingerprints on poly(PCDA-Cs) films (pale-red-coloured dots) with digitally scanned fingertips. The black-coloured microdots are pores on the ridges. The red dot images shown in a–e were obtained in the morning, afternoon and at night from the same donor for a 1-week period. (f) The dots inside the red circled areas are sweat-secreting active pores, while the microdots inside the blue circles are found to be inactive pores during the period of the experiment. The microdots inside the yellow triangles are partially active sweat pores.

Fortunately, PDA-based hydrochromic sensor systems satisfy the criteria outlined above for the efficient sweat pore mapping. The colorimetric sensitivity of hydrochromic PDAs is easy to manipulate by properly choosing the diacetylene monomers and metal cations. In addition, the blue-to-red colour change as well as fluorescence 'Turn-On' enable the PDA-based sensor of this type to be used to monitor and image sweat-secreting active pores.

Furthermore, the hydrochromic sensor film can be readily produced from commercially available diacetylene monomers. It should be noted that detection of latent fingerprint images using polymerizable diacetylenes was previously described[49]. The earlier study, however, focused on the development of latent ridge patterns.

The sweat pore-mapping strategy based on the hydrochromic PDA is highly important because conventional fingerprint

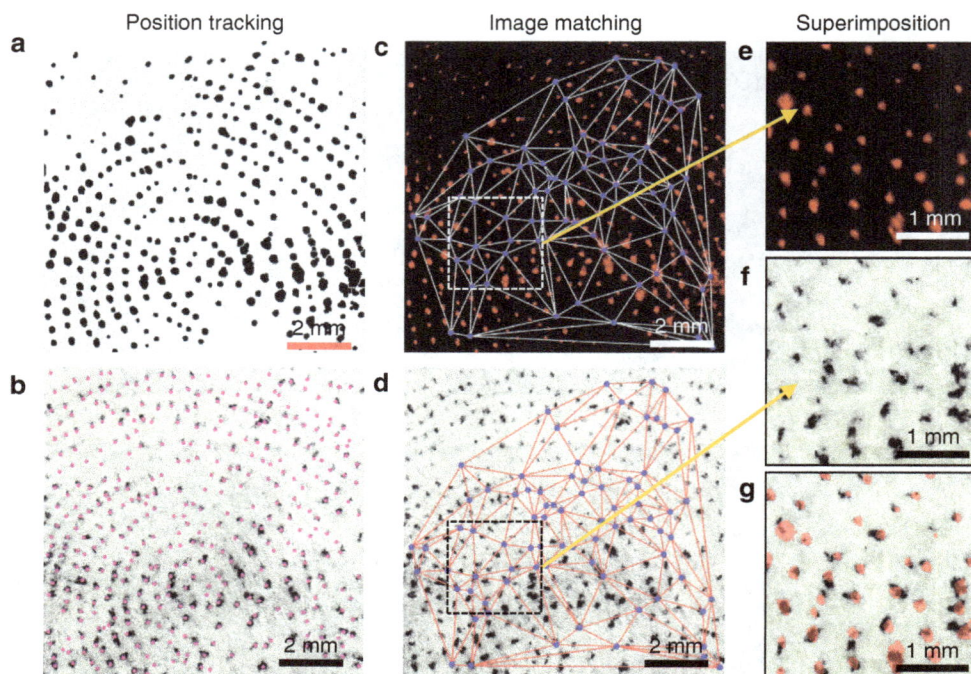

Figure 8 | Sweat pore-pattern matching. (a,b) Threshold images of fingerprint (**a**) and latent fingerprint marks (**b**) generated by using the position tracking method. The two images are extracted from the images shown in **c,d**, respectively. (**c,d**) A fluorescence microscopic fingerprint image printed on a hydrochromic poly(PCDA-Cs) film (**c**) and a latent form of the same fingerprint developed with ninhydrin (**d**). The blue-coloured dots, obtained using the image-matching process, indicate the matching points between the two images. (**e,f**) Magnified images of marked areas in **c,d**. (**g**) Superimposed image of **e,f**. The latent fingerprint image shown in **d** was obtained by pressing a fingertip on an unmodified paper followed by staining with ninhydrin. The original purple-coloured dots obtained from ninhydrin staining were transformed to black-coloured dots using a Photoshop programme because overlapping red-coloured fluorescence images (**e**) with black-coloured dots (**f**) results in a better superimposed image (**g**).

detection methods[39] depend heavily on friction ridge patterns and, consequently, they require large portions of a latent fingerprint for reliable analysis. In contrast, the new pore-based method can be employed for human identification even when a relatively small portion of a latent fingerprint exists. In addition, the newly developed technique to map active pores precisely should be highly useful in the clinical diagnosis of malfunctioning sweat pores. Although the experiments described above were carried out using poly(PCDA-Cs) films, the results of further studies demonstrate that PDA films derived from PCDA-Rb and TCDA-K are equally useful for fluorescence 'Turn-On' mapping of the active sweat pores (Supplementary Fig. 9). Notably, because the poly(TCDA-K) film displays a higher stability in humid environments (Fig. 4), it can be used to map sweat pores under a wide range (RH 30–90%) of humidity conditions (Supplementary Fig. 10). Finally, it is believed that observation made in the current effort will open new avenues for the application of hydrochromic materials, which has been limited thus far to humidity sensing and measurement of water contents in organic solvents.

Methods

General. The diacetylenic lipids, PCDA, TCDA and HCDA (8,10-heneicosadiynoic acid) were purchased from GFS Chemicals, OH, USA. Film thicknesses were measured using an alpha step instrument. Contact angle data were obtained using a PHX300, Korea. Raman spectra were obtained using laser excitation at 785 nm and a Raman microscope (Kaiser Optical Systems). SEM images were obtained with a JEOL(JSM-6330F) FE-SEM, and optical/fluorescence microscopic images were collected with a Olympus BX 51W/DP70. XRD spectra were recorded with a HR-XRD, D8 discover, Bruker. Friction ridge structures on a fingertip were obtained using a video microscope system (Camscope SV 32) with a prism. A homemade humidity-controlled chamber was used for hydrochromism tests under various humidity conditions.

Fabrication of a hydrochromic PDA film. A typical procedure for the preparation of a hydrochromic sensor film is as follows. To an aqueous solution (0.4 ml) of cesium hydroxide (750 mg, 5 mmol) was added dropwise a tetrahydrofuran solution (9.6 ml) containing PCDA 1.87 g, 5 mmol). The resulting mixture was stirred at an ambient temperature for 1 h. The formed transparent gel was spin-coated on a PET film to make a thin film (thickness: *ca.* 3.8 μm). Irradiation of the monomer film with a hand-held UV lamp (254 nm, 1 mW cm^{-2}, 10 s) resulted in the generation of a blue-coloured hydrochromic PDA.

Inkjet printing of a diacetylene suspension. Ink solution (black) from a conventional inkjet office printer (HP Deskjet D2360) cartridge was removed, and the cartridge was thoroughly washed with ethanol, water and dried with N$_2$ blowing. A suspension (0.2 ml) containing PCDA-Cs (125 mM), prepared as described above, was loaded into the cartridge. In order to avoid cartridge clogging caused by fast solvent evaporation, the content of the organic solvent was minimized (40% v/v Dioxane-H$_2$O). Printing of the diacetylene suspension on a PET film followed by UV irradiation (254 nm, 1 mW cm^{-2}, 10 s) afforded a flexible water-responsive colorimetric sensor film. It is more effective to add a small amount of dimethylsulphoxide to the PCDA-Cs ink medium (optimum condition, H$_2$O:dioxane:THF:dimethylsulphoxide = 6:4:1/20:1/40) and to filter the PCDA-Cs ink through a 0.2 μm filter to remove large aggregates. The PCDA-Cs ink prepared in this manner can be used to print more than 30 copies of clean images shown in Supplementary Movie 3 without clogging.

Sweat pore mapping. Fingerprinting was conducted by pressing a fingertip softly on a hydrochromic PDA film (see Supplementary Movie 5). Fluorescent microdots that represent the sweat-secreting active pores were analysed using a fluorescence microscope (excitation: 510–550 nm). The friction ridge patterns on a fingertip were obtained with a digital scanner equipped with a prism. An Adobe Photoshop programme was used for superimposition of the sweat pore images obtained from the hydrochromic method with the digitally scanned ridge patterns.

Latent sweat pore images on paper. A modified ninhydrin staining method was employed to obtain sweat pore images on a paper. A stock ninhydrin solution (1 g ninhydrin in 9 ml ethanol and 1 ml acetic acid) was diluted 20 times with methanol for a working staining solution. Fingerprinting was conducted by pressing a fingertip on an unmodified A4 paper. The diluted working ninhydrin solution was spray-coated to the fingerprinted area and the ninhydrin-treated paper was placed on a hot plate (80 °C) for 20 min for the generation of the latent fingerprint image.

Sweat pore-pattern-matching process. A process to match the sweat pore patterns in two images, which utilizes a Java-based image-processing tool (ImageJ) and a pore-pattern-matching programme implemented in MATLAB, was developed. The individual images are independently obtained from either a fluorescence image of a fingerprint deposited on the hydrochromic sensor or an image of a latent fingerprint obtained on paper by ninhydrin staining. To determine the sweat pore positions, the fluorescence image is thresholded to yield a binary image[50]. The coordinates of the pore positions are calculated as a centroid of each pore region in the binary image (Fig. 8a). For the latent fingerprint image (RGB), it is converted to greyscale with luminance defined as unweighted average of the RGB colours. The local maxima of the luminance under an appropriate value of noise tolerance (*ca.* 15–25) are selected as coordinates of sweat pore positions (Fig. 8b).

To compare two images, I and II, the pattern-matching programme implemented in MATLAB reads in the coordinates of sweat pores (n-positions for the image-I and m-positions for the image-II). The pore-pattern-matching algorithm is composed of two stages: (1) matching target pore positions and (2) matching background pore positions. In the first stage, the distances (d_I) between a selected pore position (i^{th}) and its 15 neighbouring pores in image-I are calculated and repeated for all other pore positions ($n-1$). The obtained distances (d_I) from the ith pore are iteratively compared with those (d_{II}) calculated for j^{th} pore ($j = 1$, $2,\ldots, m$) in image-II using the same procedure in order to find a sequence of minimum values of $s_{ij} = |d_I - d_{II}|$. Once four or five pores in the sequence of s_{ij} values satisfy the criterion, $s_{ij} < \varepsilon_1 = 0.04$ mm, the corresponding pores in both images are assigned as target pore positions. Subsequently, the relative angles (θ_I and θ_{II}) between the target pore positions are calculated for image-I and -II, respectively. The second criterion ($\Delta\theta_{ij} < \varepsilon_2 = 10°$) is then applied to the angle differences, $\Delta\theta_{ij} = |\theta_I - \theta_{II}|$, between the target positions of each image. When the second criterion is satisfied, image-II is rotated and translated so that the target positions in image-II match the corresponding positions in image-I. (If either criterion is not satisfied, the next iteration for $(i + 1)^{th}$ pore in image-I is executed.) In the second stage, background pore positions surrounding the target positions in the modified image-II are compared with those in image-I. Analysis of overlapping the two images identifies nearest pore positions in image-I with background pore positions in image-II. The displacements (λ) between the corresponding background pore positions in both images are subsequently calculated and the pore positions with the values of λ that satisfy the criterion ($\lambda < \varepsilon_1$) are selected. When the number of selected pore positions is larger than a specified percentage (ε_3) of the total m-positions in image-II, the two fingerprint images are assumed to match. Note that the range of ε_3 is from 8 to 30%, depending on the quality and the number of pores in image-II.

References

1. Morin, S. A. *et al.* Camouflage and display for soft machines. *Science* **337**, 828–832 (2012).
2. Davis, D. A. *et al.* Force-induced activation of covalent bonds in mechanoresponsive polymeric materials. *Nature* **459**, 68–72 (2009).
3. de Silva, A. P. & Uchiyama, S. Molecular logic and computing. *Nat. Nanotech.* **2**, 399–410 (2007).
4. Wenger, O. S. Vapochromism in organometallic and coordination complexes: chemical sensors for volatile organic compounds. *Chem. Rev.* **113**, 3686–3733 (2013).
5. Lim, S. H., Feng, L., Kemling, J. W., Musto, C. J. & Suslick, K. S. An optoelectronic nose for the detection of toxic gases. *Nat. Chem.* **1**, 562–567 (2009).
6. Kim, S., Yoon, S.-J. & Park, S. Y. Highly fluorescent chameleon nanoparticles and polymer films: multicomponent organic systems that combine FRET and photochromic switching. *J. Am. Chem. Soc.* **134**, 12091–12097 (2012).
7. McQuade, D. T., Pullen, A. E. & Swager, T. M. Conjugated polymer-based chemical sensors. *Chem. Rev.* **100**, 2537–2574 (2000).
8. Rochat, S. & Swager, T. M. Conjugated amplifying polymers for optical sensing applications. *ACS Appl. Mater. Interfaces* **5**, 4488–4502 (2013).
9. Rose, A., Zhu, Z., Madigan, C. F., Swager, T. M. & Bulović, V. Sensitivity gains in chemosensing by lasing action in organic polymers. *Nature* **434**, 876–879 (2005).
10. Lv, F. *et al.* Development of film sensors based on conjugated polymers for copper (II) ion detection. *Adv. Funct. Mater.* **21**, 845–850 (2011).
11. Molad, A., Goldberg, I. & Vigalok, A. Tubular conjugated polymer for chemosensory applications. *J. Am. Chem. Soc.* **134**, 7290–7292 (2012).
12. Traina, C. A., Bakus II, R. C. & Bazan, G. C. Design and synthesis of monofunctionalized, water-soluble conjugated polymers for biosensing and imaging applications. *J. Am. Chem. Soc.* **133**, 12600–12607 (2011).
13. Chen, L. *et al.* Highly sensitive biological and chemical sensors based on reversible fluorescence quenching in a conjugated polymer. *Proc. Natl Acad. Sci. USA* **96**, 12287–12292 (1999).
14. Ho, H.-A., Najari, A. & Leclerc, M. Optical detection of DNA and proteins with cationic polythiophenes. *Acc. Chem. Res.* **41**, 168–178 (2008).
15. Fan, C., Plaxco, K. W. & Heeger, A. J. High-efficiency fluorescence quenching of conjugated polymers by proteins. *J. Am. Chem. Soc.* **124**, 5642–5643 (2002).
16. Kim, I.-B. & Bunz, U. H. F. Modulating the sensory response of a conjugated polymer by proteins: an agglutination assay for mercury ions in water. *J. Am. Chem. Soc.* **128**, 2818–2819 (2006).
17. Ji, E., Wu, D. & Schanze, K. S. Intercalation-FRET biosensor with a helical conjugated polyelectrolyte. *Langmuir* **26**, 14427–14429 (2010).
18. Wegner, G. Topochemical polymerization of monomers with conjugated triple bonds. *Makromol. Chem.* **154**, 35–48 (1972).
19. Baughman, R. H. & Chance, R. R. Comments on the optical properties of fully conjugated polymers: analogy between polyenes and polydiacetylenes. *J. Polym. Sci. Polym. Phys. Ed.* **14**, 2037–2045 (1976).
20. Diegelmann, S. R. & Tovar, J. D. Polydiacetylene-peptide 1D nanomaterials. *Macromol. Rapid Commun.* **34**, 1343–1350 (2013).
21. Yarimaga, O., Jaworski, J., Yoon, B. & Kim, J.-M. Polydiacetylenes: supramolecular smart materials with a structural hierarchy for sensing, imaging and display applications. *Chem. Commun.* **48**, 2469–2485 (2012).
22. Gravel, E. *et al.* Drug delivery and imaging with polydiacetylene micelles. *Chem. Eur. J.* **18**, 400–408 (2012).
23. Chen, X., Zhou, G., Peng, X. & Yoon, J. Biosensors and chemosensors based on the optical responses of polydiacetylenes. *Chem. Soc. Rev.* **41**, 4610–4630 (2012).
24. Sun, A., Lauher, J. W. & Goroff, N. S. Preparation of poly(diiododiacetylene), an ordered conjugated polymer of carbon and iodine. *Science* **312**, 1030–1034 (2006).
25. Lu, Y. *et al.* Self-assembly of mesoscopically ordered chromatic polydiacetylene/silica nanocomposites. *Nature* **410**, 913–917 (2001).
26. Lee, J. & Kim, J. Multiphasic sensory alginate particle having polydiacetylene liposome for selective and more sensitive multitargeting detection. *Chem. Mater.* **24**, 2817–2822 (2012).
27. Peng, H. *et al.* Electrochromatic carbon nanotube/polydiacetylene nanocomposite fibres. *Nat. Nanotech.* **4**, 738–741 (2009).
28. Kolusheva, S. *et al.* Array-based disease diagnostics using lipid/polydiacetylene vesicles encapsulated in a sol-gel matrix. *Anal. Chem.* **84**, 5925–5931 (2012).
29. Park, I. S., Park, H. J. & Kim, J.-M. A soluble, low temperature thermochromic and chemically reactive polydiacetylene. *ACS Appl. Mater. Interfaces* **5**, 8805–8812 (2013).
30. Bloor, D. Dissolution and spectroscopic properties of the polydiacetylene poly(10,12-docosadiyne-1,12-diol-bisethylurethane). *Macromol. Chem. Phys.* **202**, 1410–1423 (2001).
31. Wang, X. *et al.* Colorimetric sensor based on self-assembled polydiacetylene/graphene-stacked composite film for vapor-phase volatile organic compounds. *Adv. Funct. Mater.* **23**, 6044–6050 (2013).
32. Chance, R. R. Chromism in polydiacetylene solutions and crystals. *Macromolecules* **13**, 396–398 (1980).
33. Lee, J. *et al.* A protective layer approach to solvatochromic sensors. *Nat. Commun.* **4**, 2461 (2013).
34. Yoon, J., Jung, Y.-S. & Kim, J.-M. A combinatorial approach for colorimetric differentiation of organic solvents based on conjugated polymer-embedded electrospun fibers. *Adv. Funct. Mater.* **19**, 209–214 (2009).
35. Tellis, J. C., Strulson, C. A., Myers, M. M. & Kneas, K. A. Relative humidity sensors based on an environment-sensitive fluorophore in hydrogel films. *Anal. Chem.* **83**, 928–932 (2011).
36. Ooyama, Y. *et al.* Detection of water in organic solvents by photo-induced electron transfer method. *Org. Biomol. Chem.* **9**, 1314–1316 (2011).
37. Giorgetti, E. *et al.* UV polymerization of self-assembled monolayers of a novel diacetylene on silver: a spectroscopic analysis by surface plasmon resonance and surface enhanced Raman scattering. *Langmuir* **22**, 1129–1134 (2006).
38. Phollookin, C. *et al.* Tuning down of color transition temperature of thermochromically reversible bisdiynamide polydiacetylenes. *Macromolecules* **43**, 7540–7548 (2010).
39. Ramotowski, R. S. (ed.) *Lee and Gaensslen's Advances in Fingerprint Technology* 3rd edn (CRC Press, 2012).
40. Ifa, D. R., Manicke, N. E., Dill, A. L. & Cooks, R. G. Latent fingerprint chemical imaging by mass spectrometry. *Science* **321**, 805 (2008).
41. Brown, R. M. & Hillman, A. R. Electrochromic enhancement of latent fingerprints by poly(3,4-ethylenedioxythiophene). *Phys. Chem. Chem. Phys.* **14**, 8653–8661 (2012).
42. Xu, L., Li, Y., Wu, S., Liu, X. & Su, B. Imaging latent fingerprints by electrochemiluminescence. *Angew. Chem. Int. Ed. Engl.* **51**, 8068–8072 (2012).
43. Jaber, N. *et al.* Visualization of latent fingermarks by nanotechnology: reversed development on paper-a remedy to the variation in sweat composition. *Angew. Chem. Int. Ed. Engl.* **51**, 12224–12227 (2012).
44. Kwak, G., Lee, W.-E., Kim, W.-H. & Lee, H. Fluorescence imaging of latent fingerprints on conjugated polymer films with large fractional free volume. *Chem. Commun.* **28**, 2112–2114 (2009).

45. Hazarika, P., Jickells, S. M., Wolff, K. & Russell, D. A. Imaging of latent fingerprints through the detection of drugs and metabolites. *Angew. Chem. Int. Ed. Engl.* **47**, 10167–10170 (2008).

46. Wood, M., Maynard, P., Spindler, X., Lennard, C. & Roux, C. Visualization of latent fingermarks using an aptamer-based reagent. *Angew. Chem. Int. Ed. Engl.* **51**, 12272–12274 (2012).

47. Locard, E. Les Pores et L'identification des criminels. *Biologica: Revue Scientifique de Medicine* **2**, 357–365 (1912).

48. Zhao, Q., Zhang, D., Zhang, L. & Luo, N. Adaptive fingerprint pore modeling and extraction. *Pattern. Recognit.* **43**, 2833–2844 (2010).

49. De Grazia, A. *et al.* Diacetylene copolymers for fingermark development. *Forensic Sci. Int.* **216**, 189–197 (2012).

50. Schneider, C. A., Rasband, W. S. & Eliceiri, K. W. NIH image to imageJ: 25 years of image analysis. *Nat. Methods* **9**, 671–675 (2012).

51. Okada, S., Peng, S., Spevak, W. & Charych, D. Color and chromism of polydiacetylene vesicles. *Acc. Chem. Res.* **31**, 229–239 (1998).

Acknowledgements

We thank the National Research Foundation of Korea for financial support through the Basic Science Research Program (2012R1A6A1029029, 2013004800) and the Nano-Material Technology Development Program (2012M3A7B4035286).

Author contributions

J.L., M.P., S.-h.L. and M.R. performed the experiments. J.K. and W.-Y.K analysed the data. B.J.P., C.W.L. and J.-M.K analysed the data and wrote the manuscript.

Additional information

Competing financial interests: The authors declare no competing financial interests.

Ultrasensitive and label-free molecular-level detection enabled by light phase control in magnetoplasmonic nanoantennas

Nicolò Maccaferri[1], Keith E. Gregorczyk[1], Thales V.A.G. de Oliveira[1], Mikko Kataja[2], Sebastiaan van Dijken[2], Zhaleh Pirzadeh[3], Alexandre Dmitriev[3], Johan Åkerman[4,5], Mato Knez[1,6] & Paolo Vavassori[1,6]

Systems allowing label-free molecular detection are expected to have enormous impact on biochemical sciences. Research focuses on materials and technologies based on exploiting localized surface plasmon resonances in metallic nanostructures. The reason for this focused attention is their suitability for single-molecule sensing, arising from intrinsically nanoscopic sensing volume and the high sensitivity to the local environment. Here we propose an alternative route, which enables radically improved sensitivity compared with recently reported plasmon-based sensors. Such high sensitivity is achieved by exploiting the control of the phase of light in magnetoplasmonic nanoantennas. We demonstrate a manifold improvement of refractometric sensing figure-of-merit. Most remarkably, we show a raw surface sensitivity (that is, without applying fitting procedures) of two orders of magnitude higher than the current values reported for nanoplasmonic sensors. Such sensitivity corresponds to a mass of ~ 0.8 ag per nanoantenna of polyamide-6.6 ($n = 1.51$), which is representative for a large variety of polymers, peptides and proteins.

[1] CIC nanoGUNE, 20018 Donostia-San Sebastián, Spain. [2] NanoSpin, Department of Applied Physics, Aalto University School of Science, 00076 Aalto, Finland. [3] Department of Applied Physics, Chalmers University of Technology, 41296 Gothenburg, Sweden. [4] Materials Physics, KTH Royal Institute of Technology, Electrum 229, 16440 Kista, Sweden. [5] Department of Physics, University of Gothenburg, 41296 Gothenburg, Sweden. [6] IKERBASQUE, Basque Foundation for Science, 48011 Bilbao, Spain. Correspondence and requests for materials should be addressed to P.V. (email: p.vavassori@nanogune.eu).

The most prominent routes for high-sensitivity and label-free detection in a compact device setting presently rely on localized surface plasmon resonance (LPR)-based technologies. The greatly enhanced electromagnetic fields at the surface of a resonant plasmonic nanostructure[1,2] allow for probing minute changes in the surrounding environment. Owing to the evanescent nature of the fields, the sensing volume of plasmonic nanostructures is only marginally larger than the structures themselves, making them ideal probes for localized changes in a medium. For these reasons, the utilization of LPRs for label-free molecular detection is under very active investigation for biochemical and biomedical applications[3–12]. They feature optical interrogation schemes, small footprint, high sensitivity of refractometric detection, potentially down to a single-molecule level because of the electromagnetic field enhancement at the nanoscale[3,4], and easy integration with a wide range of fluidic systems for analyte delivery.

To quantify the sensing performance of a LPR-based sensor, the bulk refractive index sensitivity $S_{RI} = \Delta\lambda^*/\Delta n$ is often considered, where $\Delta\lambda^*$ is the shift of the LPR peak position λ^* in nanometres measured in the extinction spectra over the change in the environment refractive index Δn. For detection at the molecular level, sensitivity to local variation of the index of refraction is most relevant. In this case sensor performances are quantified in terms of surface sensitivity $S_{Surf} = \Delta\lambda^*/\Delta t$ at a given refractive index n, where t is the thickness of an assembled thin layer of the material with refractive index n being sensed on top of the active surface. Since the final accuracy of the peak tracking depends both on the magnitude of the peak shift and on the resonance line-width, the most crucial performance-defining parameter is the figure-of-merit (FoM), obtained by dividing either S_{RI} or S_{Surf} by the full width at half maximum (FWHM) of the resonance. Current LPR-based sensors have a surface FoM lower or at most comparable to that of propagating surface plasmon resonance (SPR)-based sensors[13–15], which are the core and reference systems for label-free optical detection. However, SPR-based sensors are not an alternative for single-molecule level detection because of the lack of local sensitivity to the index of refraction. Therefore, tremendous research efforts aimed at improving the FoM of LPR-based sensors have been conducted[9,10,12,16–18].

Here we unveil a sensing modality that utilizes the unique optical properties of nanostructured magnetoplasmonic nano-antennas to combine and enhance all mentioned features of plasmonic sensing. Most importantly, we demonstrate that our approach delivers a manifold improvement of surface sensitivity and orders of magnitude higher surface FoM with respect to recent values reported for plasmonic-based detectors, including SPR-based sensors[14,15,18]. The proposed approach relies on magnetoplasmonic ferromagnetic (FM) nanoantennas deposited on a transparent substrate. FM nanoantennas are known to support LPRs[19] and, once activated by an external magnetic field, they acquire an intrinsic magneto-optical (MO) activity. The key point is the selection of the FM material and nanoantennas' design in order to produce exact phase compensation in the electric field components of the otherwise elliptically polarized transmitted light at a specific wavelength λ_ε. Under this condition, a vanishing ellipticity ε (ε null-point, that is, full linear polarization) is produced at λ_ε. Our underlying strategy relies on the fact that light polarization changes can be measured precisely especially near null conditions. The determination of λ_ε provides a phase-sensitive identification of the nanoantennas' LPR position and enables tracking of the spectral shift caused by local refractive index variations with unprecedented precision.

Results

Theoretical background. In ferromagnetic bulk materials and continuous films, the MO activity is governed by the spin–orbit interaction, which is a property intrinsic to a given material. Conversely, the MO response of a magnetic nanoantenna is governed also by LPRs[19–24]. As depicted in Fig. 1, light interacting with a magnetic field-activated FM nanoantenna excites two coupled LPRs, one directly driven by the electric field of the incident light and the second induced orthogonally by the inherent MO activity. Desired light polarization behaviour in transmission (and/or reflection) is achievable by tuning the relative phase of these two excited LPRs by designing the nanostructures. In detail, is the LPR polarized perpendicular to the driving electric field and light propagation direction that is governing the MO response of the nanoantenna. This fundamental physical effect can be visualized as follows: the incident electric field E_x^0 induces an electric dipole along the x axis described as $p_x = \alpha_{xx}E_x^0$, where α_{xx} is the diagonal element of the polarizability tensor. The corresponding oscillation of the conduction electrons along the x axis is spin–orbit coupled with the magnetization **M** within the nanoantenna, which is aligned along the z axis by the applied magnetic field in the experimental geometry adopted here. This coupling induces an additional oscillation motion of conduction electrons, that is, a second electric dipole p_y, along the in-plane transverse y axis direction. In general, the expression for the spin–orbit induced transverse dipole can be written conveniently in terms of the off-diagonal polarizability tensor elements $\alpha_{ij} = \varepsilon_{ij}\alpha_{ii}\alpha_{jj}/(\varepsilon_0 - \varepsilon_m)^2$, where ε_{ij} and ε_0, are the off-diagonal and diagonal elements of the dielectric tensor of the given constituent material, and ε_m is the dielectric constant of the embedding medium. This expression shows that the off-diagonal elements of the polarizability tensor are proportional to the product between the diagonal components of the tensor along the two mixed directions. In the present case, the spin–orbit transversally induced dipole is given by $p_x = \alpha_{yx}E_x^0 = [\varepsilon_{yx}\alpha_{yy}\alpha_{xx}/(\varepsilon_0 - \varepsilon_m)^2] E_x^0$ (considering the circular shape of our nanoantennas, $\alpha_{yy} = \alpha_{xx}$, that is, the two LPRs resonate at the same wavelength, and the expression above can be simplified to $p_x = \alpha_{yx}E_x^0 = [\varepsilon_{yx}\alpha^2_{xx}/(\varepsilon_0 - \varepsilon_m)^2] E_x^0$).

Notably, the transversally induced electric dipole p_y is of second order in terms of polarizability as it depends on the product of two polarizabilites, α_{xx} and α_{yy}, while the directly induced dipole p_x depends only on α_{xx}. This difference is crucial for the present application and can be understood considering that the transverse oscillation p_y is not driven directly by E_x^0 but by the induced dipole p_x with the mediation of the spin–orbit coupling. The polarization state of the far-field radiated in the z direction, either transmitted or reflected (Faraday or Kerr geometry), can be represented by the ratio of these two in-plane and mutually orthogonal oscillating electric dipoles, namely by the complex MO polarization angle $\Theta = p_y/p_x = \varepsilon_{yx}\alpha_{xx}/(\varepsilon_0 - \varepsilon_m)^2$. This expression reveals that the polarization of the transmitted and reflected light is governed by both the intrinsic properties of the constituent material $[\alpha_M = \varepsilon_{yx}/(\varepsilon_0 - \varepsilon_m)^2]$ and by the in-plane LPR in the nanoantenna (α_{yy}, which coincides with α_{xx} for circular nanoantennas). In detail, the transmitted and reflected light will have an elliptical polarization described through measurable rotation and ellipticity angles given by $\theta = \text{Re}(\Theta)$ and $\varepsilon = \text{Im}(\Theta)$, respectively. Physically, both θ and ε at any light wavelength are determined by the relative phase $\Delta\Phi$ of the two scattering electric dipoles p_x and p_y, since the complex MO angle $\Theta \propto e^{i\Delta\Phi}$, where $\Delta\Phi = \Phi[\alpha_M] + \Phi[\alpha_{yy}]$. For a given constituent material $\Phi[\alpha_M]$ is fixed. A null condition $\varepsilon = 0$ ($\Delta\Phi = 0$, π, 2π...) can be generated at a desired λ_ε simply through engineering of the size of the circular nanoantenna, which controls both the spectral position and phase $\Phi[\alpha_{yy}]$ of the LPR[21]. As a result, measuring

λ_ε provides a precise and phase-sensitive detection of any shift of the LPR position (λ_ε and $\lambda^\star$ are not bound to be identical) induced by modifications of the dielectric properties of the near-field region local environment.

Bulk refractive index sensitivity. The magnetoplasmonic nanostructures investigated here are bottom-up, short-range-ordered nickel (Ni) cylindrical nanoantennas on glass (Fig. 2a)[25]. For determining S_{RI}, we measured the extinction spectra by immersing the samples in solutions with different indices of refraction n (Fig. 2b and Supplementary Fig. 1). The observed shift of $\lambda^\star$ leads to S_{RI} of 180 and 230 nm per refraction index unit (nm RIU^{-1}) for 30-nm-thick Ni nanoantennas with diameters of 100 and 160 nm, respectively (upper panels of Fig. 2b and Supplementary Fig. 1). In parallel, we determined the null-point wavelength λ_ε by measuring the spectral dependence of the polarization ellipticity variation $\Delta\varepsilon$ of the transmitted light, induced by applying and reversing the MO activation field H (Methods and Supplementary Fig. 2). The wavelength λ_ε is more precisely visualized by plotting the quantity $1/|\Delta\varepsilon|$, which resonates at λ_ε, as shown in the top panels of Figs 1b and 1c. Plotting the data in this manner makes it much easier to visualize the enormous gain in precision (limit of detection) and sensitivity of our approach. In addition, and for the sake of direct comparison with the FoM, which is conventionally utilized for

defining the performances of a plasmonic detector, the plot of $1/|\Delta\varepsilon|$ provides the most sensible way to perform this comparison. We make it clear that comparison between sensitivity (both bulk and surface) and limit of detection performances of our approach with respect to plasmonic-based sensors are instead derived from direct measurements of $\Delta\varepsilon$ spectra and the precision (noise level) with which λ_ε is determined from such measurements.

The spectra of $1/|\Delta\varepsilon|$ for the two studied samples and for different values of n are shown in the lower panels of Fig. 2b and Supplementary Fig. 1. We observe that λ_ε undergoes a shift equal to that of $\lambda^\star$, confirming the S_{RI} values determined above. It is clear that monitoring λ_ε instead of the extinction peak $\lambda^\star$ enables tracking of the resonance shift with an exceptionally higher precision.

In principle, our approach is characterized by a virtually unlimited value of the FoM since $1/|\Delta\varepsilon|$ is diverging at the resonant wavelength λ_ε. Practically, however, we can estimate a FWHM by accounting for experimental errors. Although polarization parameters can be determined with submicro-radiant resolution (down to 10 nrad), here we assume an experimental resolution of 5 micro-radians in the determination of $\Delta\varepsilon$, which is easily achievable without the utilization of advanced tools and/or fitting procedures, and consequently truncate the $1/|\Delta\varepsilon|$ spectra at 2×10^5 rad^{-1} (see the insets of the top panels of Figs 1b and 1c). We obtain a FWHM ranging

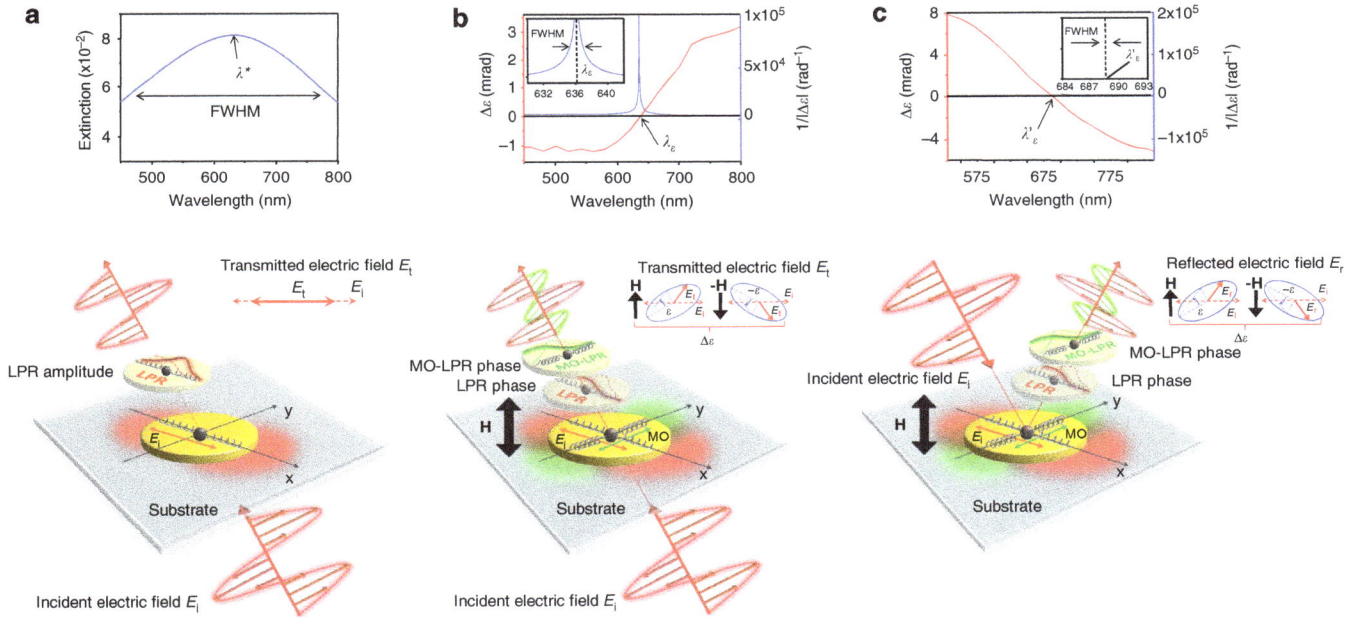

Figure 1 | LPR phase sensitivity in the transmitted light polarization. (**a**) When an incident light beam hits a ferromagnetic nanoantenna, the conduction electrons inside the nanostructure oscillate driven by the electric field E_i. These E_i-driven oscillations can be modelled as a damped spring-mass harmonic oscillator. A LPR is induced at a specific photon wavelength $\lambda^\star$, yielding a peak in the extinction spectrum ($I_0 - I_t$)/$I_0 = 1 - (E_t/E_i)^2$, displayed in the top panel. (**b**) If the nanoantenna is magnetized perpendicularly to the surface plane, a MO-activity is turned on inducing a second MO-coupled LPR (MO-LPR) orthogonal to that directly driven by E_i. In a circular nanoantenna the MO-LPR resonates at the same $\lambda^\star$. The simultaneous excitation of LPR and MO-LPR induces an elliptical polarization ε of the transmitted field E_t (refs 20,21). The null condition $\varepsilon = 0$ is generated at a desired λ_ε (in general $\lambda_\varepsilon \neq \lambda^\star$) simply through engineering of the size of the circular nanoantenna[21,34,35]. Measurement of λ_ε provides a precise phase-sensitive detection of the LPR position. The top panel displays typical $\Delta\varepsilon$ spectrum (red-line), as well as the $1/|\Delta\varepsilon|$ spectrum (blue-line) and its resonance at λ_ε. The close-up view of the $1/|\Delta\varepsilon|$ spectrum around λ_ε shown in the inset features a very narrow FWHM ($\sim$1.7 nm). (**c**) Similarly to the case described in **b**, the concerted action of the simultaneously excited LPR and MO-LPR can be exploited to actively manipulate the reflected light's polarization inducing the condition $\varepsilon = 0$ at a desired λ'_ε. In general, $\lambda'_\varepsilon \neq \lambda_\varepsilon$ since in this case also the additional phase introduced by the substrate reflectivity contributes to the polarization of the reflected field E_r. As in transmission geometry, the detection of λ'_ε provides precise phase-sensitive detection of the LPR position. The top panel displays typical $\Delta\varepsilon$ spectrum (red line), as well as the $1/|\Delta\varepsilon|$ spectrum (blue line) and its resonance at λ'_ε. The close-up view of the $1/|\Delta\varepsilon|$ spectrum around λ'_ε shown in the inset features a very narrow FWHM ($<$1.7 nm). Both in transmission and reflection, the sensitivity increases further by measuring the magnetic field-induced variation $\Delta\varepsilon$ as ε reverses its sign upon inverting **H** (see Supplementary Fig. 2).

from 1.5 to 1.8 nm for the resonance of the $1/|\Delta\varepsilon|$ spectra (Fig. 2b and Supplementary Fig. 2). Taking the average FWHM of $\sim$1.7 nm means a FoM of more than 100 RIU^{-1} and approaching 150 RIU^{-1} for Ni nanoantennas with diameters of 100 and 160 nm, respectively. Such large FoM values greatly exceed even those for SPR in the same spectral range, which are often considered as theoretical limit references[10]. Figure 2c,d shows a comparison of the sensitivities of Au and Ni cylindrical nanoantennas on glass as well as their FoM, including that of SPR-based sensors.

The outstanding sensing performance of our magnetoplasmonic nanoantennas is better appreciated if compared directly with recently developed nanoplasmonic sensors. Improvements in FoM have been achieved by boosting the S_{RI} and/or reducing the FWHM of the LPR. S_{RI} enhancement has been realized by lifting metal nanostructures above substrates with dielectric pillars in order to expose more efficiently the surrounding environment to the LPR-enhanced electric field[26]. More efforts have been devoted to a reduction of the FWHM of LPRs caused by inherent losses in metallic nanostructures. The most effective approach relies on resonant coupling of LPRs with modes that possess a smaller FWHM. Such modes could, for example, be SPRs in an optically coupled thin metallic continuous layer[16], diffractive coupling among the nanostructures arranged in periodic arrays[15], and

coherent coupling of different localized plasmon modes in nanostructures with complex shapes[9,17]. FoM values towards $\sim$60 have been reported upon resonant coupling of LPRs and SPRs[16]. More recently, a FoM value of up to 108 has been reported from periodic arrays of gold 'nano-mushrooms' that combine the two aforementioned approaches[10]. It is worth noting that all pathways towards enhancement of the FoM rely on a substantial increase in the complexity of the plasmonic nanostructures and arrangements design. In addition, the performance boost applies to a specific and narrow wavelength range where resonant coupling occurs. By exploiting the built-in phase sensitivity of the individual magnetoplasmonic nanoantennas, we circumvent the limitation due to their inherent losses, while keeping the nanostructuring process extremely simple and maintaining the outstanding performance over a broad spectral range.

Surface sensitivity. Applications to molecular detection rely on high sensitivity and FoM to the local variation of the index of refraction n. A reliable and precise assessment of the detection performance is an experimental challenge, as control of Δn on the molecular level is needed. The method of choice is the controlled deposition of extremely thin and removable films using molecular

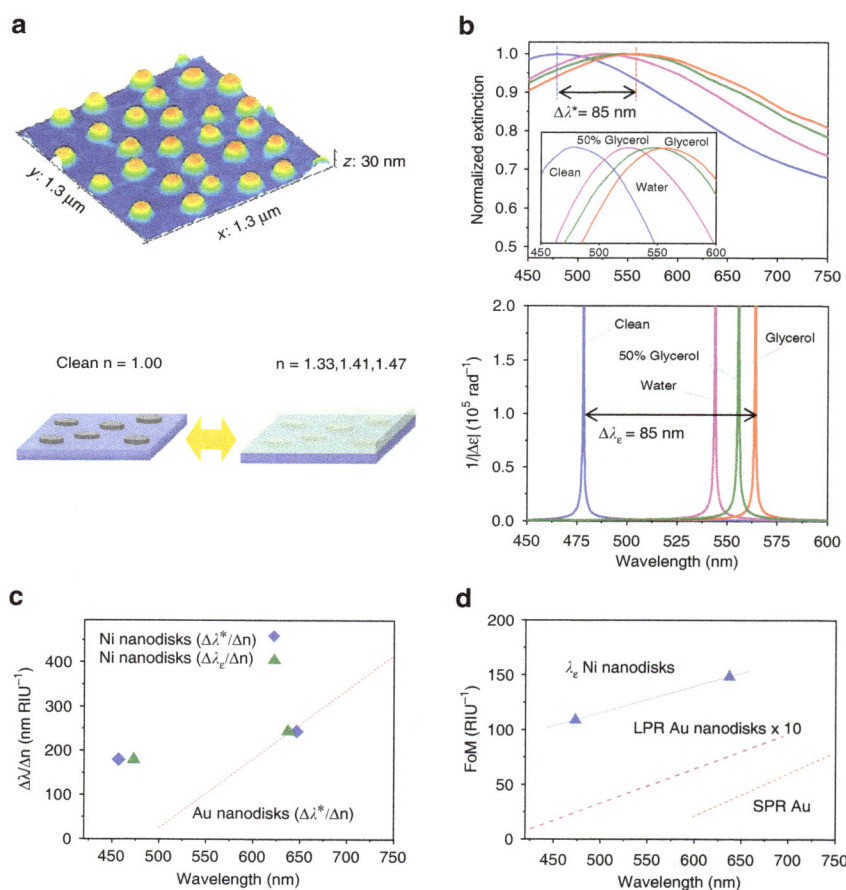

Figure 2 | Refractive index sensitivity of Ni magnetoplasmonic nanoantennas. (**a**) 3D AFM profile of Ni magnetoplasmonic nanoantennas on glass, with lateral dimensions of 103 ± 5 nm (diameter) and 30 ± 0.5 nm (thickness). (**b**; Top panel) extinction spectra of Ni cylindrical nanoantennas for different values of the embedding refractive index (clean $n = 1$, water $n = 1.33$, 50% Vol. glycerol $n = 1.41$ and glycerol $n = 1.47$; the inset shows a zoom of the resonance peaks in the spectral region 450–600 nm; bottom panel) plot of the inverse of transmitted light ellipticity variation $1/|\Delta\varepsilon|$ for the same values of the embedding refractive index as above. (**c**) Comparison between the bulk sensitivities of Au[26] ($\Delta\lambda^*/\Delta n$; red-dashed line) and Ni (both $\Delta\lambda^*/\Delta n$ and $\Delta\lambda_\varepsilon/\Delta n$; blue and green markers, respectively) cylindrical nanoantennas on glass. (**d**) Comparison between the bulk figure-of-merit of Au[26] [($\Delta\lambda^*/\Delta n$)/FWHM] (purple dashed line) and Ni [($\Delta\lambda_\varepsilon/\Delta n$)/FWHM] (blue markers are experimental data, blue dotted line is a guide for eyes) cylindrical nanoantennas, and Au surface plasmon resonance[14] (red dashed line), in the spectral range 420–750 nm.

layer deposition (MLD), an organic variant of atomic layer deposition[27,28]. The cyclic and self-terminating growth mechanisms of this method allow molecular-scale control of the polymer film growth and surface-dependent nucleation characteristics can be used for area-selective growth through matching of substrate and precursor chemistries. More importantly, MLD deposits polymer films from vapourized pure molecular fragments avoiding potentially negative impact of solvents (see also Methods).

For our experiments we used MLD of polyamide 6.6 (PA-6.6) with $n = 1.51$ (ref. 28). This polymer was chosen for convenience as the MLD process is well established and also, being a polyamide, it is representative for a large variety of polymers, peptides and proteins. In a first experiment, PA-6.6 was deposited 20 cycles at a time. The film thickness was measured by atomic force microscopy (AFM). The corresponding change of the local index of refraction was monitored by tracking λ_ε. Since the surface was modified only locally, we conducted polarization measurements also in reflection geometry as an alternative way to track λ_ε (see Fig. 1c). The experimental spectra and the resulting shift of λ_ε are shown in Fig. 3a,b. AFM images and their quantitative analysis show that PA-6.6 nucleated selectively on the Ni nanodisks and linearly grew up to ~ 100 cycles (Fig. 3c,d). Combining the two plots in Fig. 3b,d, we find S_{Surf} values of ~ 3 and ~ 5.3 in transmission and reflection geometry, respectively. The remarkable enhancement by a factor 1.7 of the S_{Surf} in reflection geometry, confirmed also by calculations, is a consequence of a larger shift of λ_ε caused by the additional phase contribution of uncovered substrate reflectivity (details in Supplementary Fig. 3). Such S_{Surf} value of ~ 5.3 is 3.5 times the

surface sensitivity reported for Au-based SPR detectors in the same spectral region for the same refractive index[13]. Even more remarkable is the surface FoM (S_{Surf}/FWHM) of our magnetoplasmonic nanoantennas approach, whose value slightly larger than three is two orders of magnitude higher than the best surface FoM values achieved with plasmonic systems[14].

In order to demonstrate that very small quantities of material can be detected, PA-6.6 deposition, polarization ellipticity measurements and AFM characterization were performed sequentially for the very first cycles of MLD. Figure 4 shows that already after two MLD cycles an average increase in t of ~ 1.6 nm is determined by AFM. The resulting $\Delta\lambda_\varepsilon$ for the two measurement geometries confirm the surface sensitivities obtained in the previous experiment. On the basis of our very conservative error bar estimate in the measurement of $\Delta\varepsilon$, we can detect λ_ε with ~ 0.5 nm precision without the application of any fitting procedure (raw limit of detection).

Given the surface sensitivity (S_{Surf}) values ~ 3 and ~ 5.3 of our approach (transmission and reflection geometry, respectively), detection of subnanometre-thick PA-6.6 coverage, namely, a discontinuous monolayer (ML) of PA-6.6 on individual cylindrical Ni nanoantennas is achievable (~ 1.7 and ~ 1.0 Å for the two measurement geometries). A ML of PA-6.6 is ~ 8.5-Å thick[29], the precision above corresponds, in the case of transmission geometry, to a coverage of less than 0.2 ML, namely a ML covering less than 20% of the nanoantenna-exposed surface. The minimum detectable coverage reduces to 0.1 ML in reflection geometry. Such sub-ML coverage values are equivalent to PA-6.6 volumes of $\sim 1,200$ and 800 nm^3, that is, a remarkable mass sensitivity of ~ 1.2 and 0.8 ag per disk corresponding to

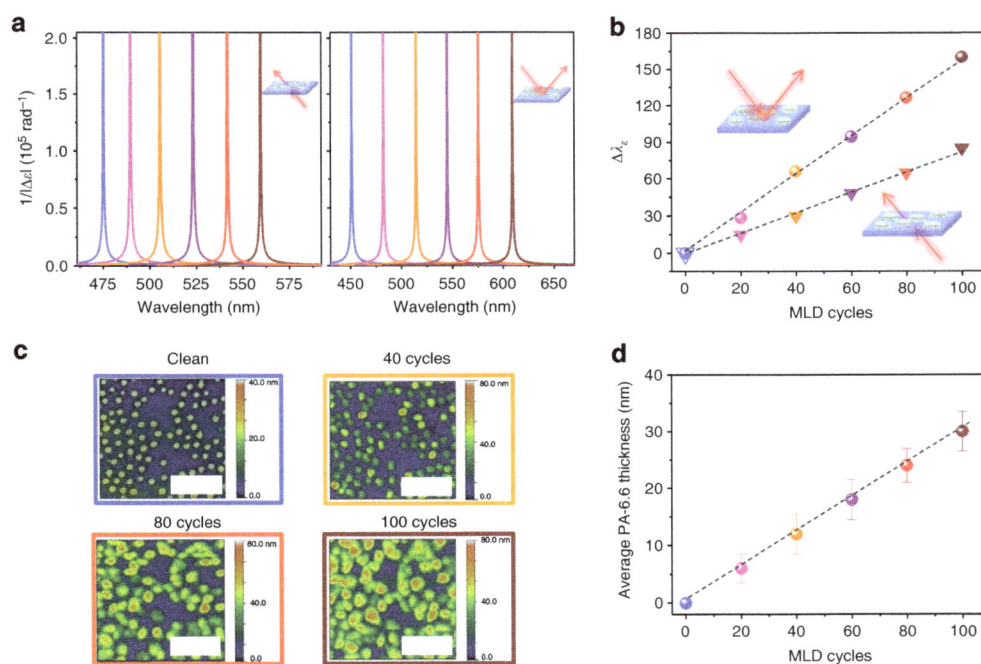

Figure 3 | Surface sensitivity assessment combining polarimetry and AFM measurements. (**a**) Plot of the inverse of the transmitted (left panel) and reflected (right panel) light ellipticity $1/|\Delta\varepsilon|$ spectra as a function of MLD cycles (at steps of 20 cycles). (**b**) $1/|\Delta\varepsilon|$ resonance wavelength λ_ε as a function of the MLD cycles for the two measurement geometries. In both cases, a linear dependence of λ_ε versus number of MLD cycles is observed (the black dashed lines are guide for eyes). The shift of λ_ε saturates for a number of MLD cycles equal to 120, corresponding to a PA-6.6 thickness of ~ 35 nm, as shown in Supplementary Fig. 7 for transmission geometry case. Such PA-6.6 thickness agrees well with the near-field spatial extension (see Supplementary Fig. 4). (**c**) AFM images taken from the same sample region (total area imaged 2.7×2.7 µm^2) after different numbers of MLD cycles. The AFM images show that PA-6.6 grows only on top of the nanoantennas. The colours of the frames refer to the corresponding coloured polarimetry (**b**) and thickness (**d**) data points. The length of the white scale bars in the images corresponds to 1µm. (**d**) PA-6.6 average thickness as function of the MLD cycles after AFM topography image analysis. Surface sensitivities (spectral variation of λ_ε divided by the average nylon thickness) of ~ 3 (transmission) and ~ 5.3 (reflection) are found combining plots (**b,d**). The error bars indicate the s.d. from the average thicknesses measured analysing the AFM images.

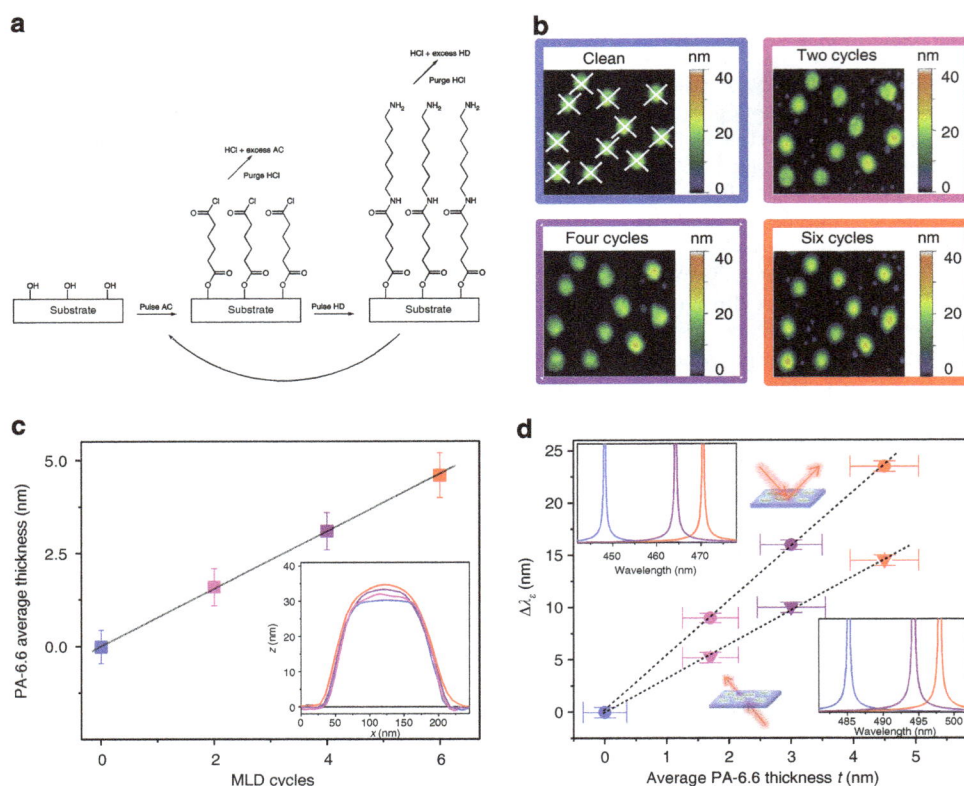

Figure 4 | Surface sensitivity in the first few cycles of PA-6.6 MLD. (**a**) Schematic of one cycle of the MLD process for PA-6.6. A substrate with –OH surface groups is exposed to a pulse of AC. The AC reacts with these –OH groups creating the by-product HCl, which is purged away along with any unreacted AC. Next, a pulse of HD is introduced to the reaction chamber and reacts with the available –Cl groups. Again, the by-product is HCl, which is purged away along with any unreacted HD. This process is repeated until the desired thickness is achieved. Nominally, the process has a growth rate of ~ 0.8 nm per cycle[29]. (**b**) AFM images taken from the same sample region (total area imaged $1.2 \times 1.2 \ \mu m^2$) before and after PA-6.6 MLD. The colours of the frames refer to the corresponding coloured thickness (**c**) and polarimetry (**d**) data points. (**c**) PA-6.6 average thickness as function of the MLD cycles after AFM topography image analysis. The error bars indicate the s.d. from the average thicknesses shown in the inset, which shows the line profiles of all the disks included in the images in **b**. The line profiles are taken along two orthogonal directions, which are shown as white dashed lines only in the AFM image of the clean sample in **b**. (**d**) Plot of the inverse of transmitted and reflected light ellipticity λ_ε as a function of MLD cycles (black dashed lines are guide for eyes). Surface sensitivities of ~ 3.1 (transmission) and ~ 5.4 (reflection) are found combining plots (**c,d**), in excellent agreement with the results presented in Fig. 3. The horizontal error bars indicate the s.d. from the average thicknesses shown in the inset in (**c**). The vertical error bars indicate the experimental error in the magneto-optical measurements. The insets show the corresponding $1/|\Delta\varepsilon|$ spectra for the two measurement geometries (reflection—top-left inset and transmission—bottom-right inset).

$\sim 3,300$ and 2,200 molecules of PA-6.6 per disk (density of amorphous PA-6.6 is $1.05 \ \mathrm{g \ cm^{-3}}$ and its molecular weight is $226 \ \mathrm{g \ mol^{-1}}$). Such minimum detectable coverage is based on extrapolation and assumes that the electric near field is uniformly distributed in the vicinity of the Ni nanoantenna surface exposed to the environment, that is, the surface sensitivity is not, or weakly, space-dependent. This condition is fulfilled to a good extent for our nanodisks with a diameter of 100 nm as we verified by the simulations of the near field produced by the excitation of a LPR performed using Lumerical (see Supplementary Fig. 4). A raw limit of detection of a few zg per nanoantenna can be achieved considering the submicro-radiant resolution of advanced polarimetry tools.

These sensing performances could be further improved by applying the same strategies as employed for noble metal nanoplasmonic systems for label-free detection, such as lifting of the nanostructures from the substrate and exploiting resonant coupling between LPRs to modes with narrow FWHM.

In addition, reducing the diameter of the cylindrical Ni nanoantennas would improve the limit of detection, since the surface sensitivity is weakly dependent on the diameter and, consequently, also the ability to detect a ML coverage of 20 and

10% of the nanoantenna-exposed surface, which for smaller nanoantennas corresponds to proportionally smaller volumes of material.

So far, the limit of detection of our approach was derived based on the signal-to-noise ratio of the measurements, without any mathematical fit of the data, and compared with those of similar 'raw' estimates of sensitivity based on absorption spectrum measurement. This comparison demonstrates the radically improved sensitivity enabled by our nanomagnetoplasmonic approach with respect to plasmon-based sensors.

Higher sensitivity and limit of detection values are reported in literature for plasmon-based sensors, which are achieved by application of fitting procedures[30,31]. We mention here that the application of fitting procedures of our data confirms the higher sensitivity of our approach as shown in the Supplementary Fig. 5. Indeed, a mass sensitivity in the sub-zg per nanoantenna, down to a few yg per nanoantenna, can be achieved through the application of fitting procedures opening a pathway to mass sensitivity corresponding to ~ 10 molecules of PA-6.6 per disk (or, equivalently, of any material having $n \sim 1.5$ and a density of $\sim 1 \ \mathrm{g \ cm^{-3}}$, which is the case of many polymers and biomolecules).

Discussion

Magnetoplasmonic nanoantennas that support magneto-optically induced localized plasmon resonances have been synthesized in order to induce a null condition of the transmitted/reflected light polarization ellipticity at desired wavelengths. The obtained null condition allows for an easily measurable and extremely precise phase-sensitive detection of localized plasmon resonances. Such magnetoplasmonic nanoantennas can be used for optical sensing of local refractive index variations with enhanced sensitivity and unrivalled values for the FoM, even exceeding the theoretically predicted upper limit for sensing based on propagating SPRs. Our approach requires extremely simple and scalable nanostructuring processes and offers a remarkably improved sensitivity performance in a large spectral range. To conclude, we would like to emphasize that the ultrasensitive sensing capabilities of our magnetoplasmonic nanoantennas can be used also in nanoplasmonic biosensing (for instance, cancer serum detection). Ni surfaces are covered by an ultrathin layer of oxidized Ni. Such surfaces can be functionalized with either silanes or even better with Histidin tags (His6)[32] on par with Au surface functionalization with thiolate chemistry, thus informing our selected Ni as the material for our nanoantennas. As an alternative, one could even deposit a thin layer of Au, which will not affect the plasmonic behaviour (even improve it, according to literature), and use conventional thiolate chemistry. In addition to biosensing, there are also many other potential civil and/or military applications that do not require surface functionalization and would enormously benefit from our approach like chemical sensing of toxic materials, explosives and ultra-precise thickness-monitoring applications.

Methods

Optical and MO measurements. The extinction spectra $(I_0 - I_t)/I_0$, where I_0 and I_t are the intensities of the incident and transmitted light, respectively, were taken in the wavelength range 420–900 nm. The intensity of the light passing through the substrate without nanostructures on top was taken as the reference I_0 signal.

The wavelength dependence of the magneto-optically induced ellipticity change $\Delta\varepsilon$ of the transmitted or reflected light was measured using MO Faraday (incidence angle 0°) and Kerr (incidence angle 2.5°) effect spectrometers working in polar geometry in the wavelength range 420–950 nm. The incident light beam was linearly polarized with either p- or s-polarization. $\Delta\varepsilon$ was measured by switching the polarity of a magnetic field $H = 4$ kOe applied normal to the sample plane to activate the MO coupling in the nanoantennas' constituent material. As shown schematically in Supplementary Fig. 2, in the polar-Kerr geometry $\Delta\varepsilon$ was measured at each wavelength with the transmitted beam passing through a photoelastic phase modulator and a polarizer before detection. A lock-in amplifier was used to filter the signal at the modulation frequency in order to retrieve $\Delta\varepsilon$ (ref. 33). Our experimental set-up allows measuring polarization parameters, like ε and $\Delta\varepsilon$, with submicro-radiant precision, although in our analysis of the measured spectra we assumed a 5 micro-radians error bar, for the sake of demonstrating that our approach guarantees a high sensitivity even using extremely simple measurement set-ups as that shown in the Supplementary Fig. 6. It is worth mentioning that $\Delta\varepsilon$ can be equivalently measured in reflection geometry with the light impinging on either magnetoplasmonic sample surface side (Fig. 1c). Measurement of $\Delta\varepsilon$ with the light beam impinging from the glass bottom surface can be particularly appealing for implementation of our approach in practical devices (Supplementary Fig. 6).

Sensing experiments. Bulk sensitivity experiments were performed in a microfluidic cell where different liquids, with different refractive index, were injected every time. We used liquids and mixtures whose refractive indexes are well known, namely water ($n = 1.33$), glycerol ($n = 1.47$) and 50% glycerol volume in water ($n = 1.41$).

For surface sensitivity assessment, MLD of polyamide 6.6 (PA-6.6) was carried out in a commercially available Cambridge Nanotech/Ultratech Savannah system using alternating pulses of adipoyl chloride (AC; Sigma Aldrich) and 1,6-hexamethylenediamine (HD; Sigma Aldrich) with the following pulse/purge/pulse/purge parameters; 0.5 s/30 s/0.5 s/30 s. The AC and HD were heated to 70 and 80 °C, respectively. The reaction chamber was maintained at 85 °C throughout the reaction. Before deposition the samples were treated using 5-min cycles of acetone and isopropanol ultrasonication, and then exposed to Ar plasma at 90 W for 60 s to eliminate residues of organic materials. Coating and cleaning procedures were repeated several times to ensure the reproducibility of the experiment (both polarization and thickness of PA-6.6 deposits). The thickness of the PA-6.6 deposits was monitored via AFM. The AFM measurements were carried out in the same region after each MLD cycles and MOKE measurement step, using an Agilent 5500 AFM microscope with a Si tip operated in tapping mode. The images were analysed using the Gwyddion software package in order to extract the average thickness values of PA-6.6 deposits.

Both bulk and surface sensitivity experiments were repeated five times each, giving the same results within the experimental errors reported above.

Sample fabrication. Bottom-up hole mask colloidal lithography was used to pattern the Ni nanodisks[25]. The following process steps were applied in all presented examples. The substrates used are $10 \times 10 \times 1$ mm³ pieces of microscope slide glass (VWR International). The glass substrates were first cleaned through 5-min cycles of acetone, isopropanol and water ultrasonication. A poly(methyl methacrylate) (PMMA) film (2 wt % PMMA diluted in anisole, MW = 950,000) was spin-coated on a clean surface and followed bysoft baking (170 °C, 10 min on a hot plate). Reactive oxygen plasma treatment (50 W, 5 s, 250 mTorr, Plasma Therm Batchtop RIE 95 m) was applied in order to decrease the polymer film hydrophobicity and avoid spontaneous de-wetting of the surface during subsequent polyelectrolyte and particle deposition steps, which would introduce inhomogenities in the particle distribution. Providing a net charge to the PMMA surface by pipetting a solution containing a positively charged polyelectrolyte on the film (polydiallyldimethylammonium MW 200,000–350,000, Sigma Aldrich, 0.2 wt % in Milli-Q water, Millipore), followed by careful rinsing with de-ionized water in order to remove excess polydiallyldimethylammonium and blow-drying with a N_2 stream. Deposition of a water suspension containing negatively charged polystyrene particles (sulfate latex, Invitrogen, 0.2 wt % in Mili-Q water) and N_2 drying in a similar manner as described above, leaving the PMMA surface covered with uniformly distributed PS-spheres. Evaporation of an oxygen plasma-resistant thin film of Au. Removing the PS spheres using tape stripping (SWT-10 tape, Nitto Scandinavia AB), resulting in a mask with holes arranged in a pattern determined by the self-assembled colloidal particles. Transfer of the hole-mask pattern into the sacrificial layer via an oxygen plasma treatment (50 W, 250 mTorr, Plasma Therm Batchtop RIE 95 m), which effectively removes all PMMA situated underneath the holes in the film, leaving the surface covered with a thin-film mask supported on a perforated, undercut polymer film. Ni deposition is carried out by e-beam-assisted evaporation (AVACHVC600). Lift off was carried out using acetone at room temperature or 50 °C for 5–10 min. The filling factor (surface covered by Ni nanoantennas normalized to area of the sample surface) can be varied between 5 and 30% in a controlled manner. The filling factor of the samples utilized in our experiments can be estimated to be around 15%.

References

1. Sandtke, M. & Kuipers, L. Slow guided surface plasmons at telecom frequencies. *Nat. Photon.* **1**, 573–576 (2007).
2. Ferry, V. E., Sweatlock, L. A., Pacifici, D. & Atwater, H. A. Plasmonic nanostructure design for efficient light coupling into solar cells. *Nano Lett.* **8**, 4391–4397 (2008).
3. Cui, Y., Wei, Q. Q., Park, H. K. & Lieber, C. M. Nanowire nanosensors for highly sensitive and selective detection of biological and chemical species. *Science* **293**, 1289–1292 (2001).
4. Anker, J. N. *et al.* Biosensing with plasmonic nanosensors. *Nat. Mater.* **7**, 442–453 (2008).
5. Stewart, M. E. *et al.* Nanostructured plasmonic sensors. *Chem. Rev.* **108**, 494–521 (2008).
6. Larsson, E. M., Syrenova, S. & Langhammer, C. Nanoplasmonic sensing for nanomaterials science. *Nanophotonics* **1**, 249–266 (2012).
7. Mayer, K. M. & Hafner, J. H. Localized surface plasmon resonance sensors. *Chem. Rev.* **111**, 3828–3857 (2011).
8. Chung, T., Lee, S. Y., Song, E. Y., Chun, H. & Lee, B. Plasmonic nanostructures for nano-scale bio-sensing. *Sensors* **11**, 10907–10929 (2011).
9. Kravets, V. G. *et al.* Singular phase nano-optics in plasmonic metamaterials for label-free single-molecule detection. *Nat. Mater.* **12**, 304–309 (2013).
10. Shen, Y. *et al.* Plasmonic gold mushroom arrays with refractive index sensing figures of merit approaching the theoretical limit. *Nat. Commun.* **4**, 2381 (2013).
11. Aćimović, S. S. *et al.* LSPR chip for parallel, rapid, and sensitive detection of cancer markers in Serum. *Nano Lett.* **14**, 2636–2641 (2014).
12. Kabashin, A. V. *et al.* Plasmonic nanorod metamaterials for biosensing. *Nat. Mater.* **8**, 867–871 (2009).
13. Svedendahl, M., Chen, S., Dmitriev, A. & Käll, M. Refractometric sensing using propagating versus localized surface plasmons: a direct comparison. *Nano Lett.* **9**, 4428–4433 (2009).
14. Otte, M. A. *et al.* Identification of the optimal spectral region for plasmonic and nanoplasmonic sensing. *ACS Nano* **4**, 349–357 (2010).
15. Offermans, P. *et al.* Universal scaling of the figure of merit of plasmonic sensors. *ACS Nano* **5**, 5151–5157 (2011).
16. Lodewijks, K. *et al.* Tuning the Fano resonance between localized and propagating surface plasmon resonances for refractive index sensing applications. *Plasmonics* **8**, 1379–1385 (2013).

17. Verellen, N. *et al.* Plasmon line shaping using nanocrosses for high sensitivity localized surface Plasmon resonance sensing. *Nano Lett.* **11**, 391–397 (2011).

18. Lodewijks, K., Van Roy, W., Borghs, G., Lagae, L. & Van Dorpe, P. Boosting the Figure-Of-Merit of LSPR-based refractive index sensing by phase-sensitive measurements. *Nano Lett.* **12**, 1655–1659 (2012).

19. Chen, J. *et al.* Plasmonic nickel nanoantennas. *Small.* **7**, 2341–2347 (2011).

20. Bonanni, V. *et al.* Designer magnetoplasmonics with nickel nanoferromagnets. *Nano Lett.* **11**, 5333–5338 (2011).

21. Maccaferri, N. *et al.* Tuning the magneto-optical response of nanosize ferromagnetic Ni disks using the phase of localized plasmons. *Phys. Rev. Lett.* **111**, 167401 (2013).

22. Belotelov, V. I. *et al.* Enhanced magneto-optical effects in magnetoplasmonic crystals. *Nat. Nanotech.* **6**, 370–376 (2011).

23. Armelles, G., Cebollada, A., García-Martín, A. & González, M. U. Magnetoplasmonics: combining magnetic and plasmonic functionalities *Adv. Opt. Mater.* **1**, 10–35 (2013).

24. Lodewijks, K. *et al.* Magnetoplasmonic design rules for active magneto-optics. *Nano Lett.* **14**, 7207–7214 (2014).

25. Fredriksson, H. *et al.* Hole-Mask Colloidal Lithography. *Adv. Mater.* **19**, 4297–4302 (2007).

26. Dmitriev, A. *et al.* Enhanced nanoplasmonic optical sensors with reduced substrate effect. *Nano Lett.* **8**, 3893–3898 (2008).

27. Knez, M., Nielsch, K. & Niinistö, L. Synthesis and surface engineering of complex nanostructures by atomic layer deposition. *Adv. Mater.* **19**, 3425–3438 (2007).

28. George, S. M. Atomic layer deposition: An overview. *Chem. Rev.* **110**, 111–131 (2010).

29. Du, Y. & George, S. M. Molecular layer deposition of nylon 66 films examined using in situ FTIR spectroscopy. *J. Phys. Chem. C* **111**, 8509–8517 (2007).

30. Dahlin, A. B., Tegenfeldt, J. O. & Höök, F. Improving the instrumental resolution of sensors based on localized surface plasmon resonance. *Anal. Chem.* **78**, 4416–4423 (2006).

31. Chen, S., Svedendahl, M., Käll, M., Gunnarsson, L. & Dmitriev, A. Ultrahigh sensitivity made simple: nanoplasmonic label-free biosensing with an extremely low limit-of-detection for bacterial and cancer diagnostics. *Nanotechnology* **20**, 434015 (2009).

32. Hochuli, E., Bannwarth, W., Döbeli, H., Gentz, R. & Stüber, D. Genetic Approach to Facilitate Purification of Recombinant Proteins with a Novel Metal Chelate Adsorbent. *Nat. Biotechnol.* **6**, 1321–1325 (1988).

33. Vavassori, P. Polarization modulation technique for magneto-optical quantitative vector magnetometry. *Appl. Phys. Lett.* **77**, 1605–1607 (2000).

34. Maccaferri, N. *et al.* Polarizability and magnetoplasmonic properties of magnetic general nanoellipsoids. *Opt. Express* **21**, 9875–9889 (2013).

35. Maccaferri, N. *et al.* Effects of a non-absorbing substrate on the magneto-optical Kerr response of plasmonic ferromagnetic nanodisks. *Phys. Status Solidi A* **211**, 1067–1075 (2014).

Acknowledgements

P.V., N.M., K.E.G. and M.Kn. acknowledge financial support from the Basque Government (Programs No. PI2012-47 and PI2013-56) and the Spanish Ministry of Economy and Competitiveness (Projects No. MAT2012-36844 and MAT2012-38161). N.M. acknowledges support from the Doctoral Programme of the Department of Education, Linguistic Policy and Culture of the Basque Government (Grant No. PRE_2013_1_975). P.V. and N.M. also thank Paolo Biagioni for support on simulations using Lumerical FDTD Solutions Software. T.V.A.G.d.O. acknowledges support from the European Research Council (Grant No. 257654-SPINTROS). Z.P., A.D. and J.Å. acknowledge the Swedish Research Council (VR) and the Swedish Foundation for Strategic Research, SSF (Z.P. and A.D.: Framework programme Functional Electro-magnetic Metamaterials, Project No. RMA08; J.Å.: Successful Research Leader Program). M.K. and S.v.D. acknowledge support from the National Doctoral Programme in Nanoscience and the Academy of Finland (Grant No. 263510).

Author contributions

P.V. conceived the experiments. N.M. performed optical and magneto-optical measurements and FDTD simulations. K.E.G. and M.Kn. performed molecular layer deposition. T.V.A.G.d.O. performed AFM measurements and developed non-destructive sample cleaning procedures. M.K. and S.v.D. contributed to the measurements and data analysis. Z.P. and A.D. fabricated the samples. N.M. and P.V. analysed the data and all authors wrote the manuscript.

Additional information

Structural analysis and mapping of individual protein complexes by infrared nanospectroscopy

Iban Amenabar[1], Simon Poly[1], Wiwat Nuansing[1], Elmar H. Hubrich[2], Alexander A. Govyadinov[1], Florian Huth[1,3], Roman Krutokhvostov[1], Lianbing Zhang[1], Mato Knez[1,4], Joachim Heberle[2], Alexander M. Bittner[1,4] & Rainer Hillenbrand[1,4]

Mid-infrared spectroscopy is a widely used tool for material identification and secondary structure analysis in chemistry, biology and biochemistry. However, the diffraction limit prevents nanoscale protein studies. Here we introduce mapping of protein structure with 30 nm lateral resolution and sensitivity to individual protein complexes by Fourier transform infrared nanospectroscopy (nano-FTIR). We present local broadband spectra of one virus, ferritin complexes, purple membranes and insulin aggregates, which can be interpreted in terms of their α-helical and/or β-sheet structure. Applying nano-FTIR for studying insulin fibrils—a model system widely used in neurodegenerative disease research—we find clear evidence that 3-nm-thin amyloid-like fibrils contain a large amount of α-helical structure. This reveals the surprisingly high level of protein organization in the fibril's periphery, which might explain why fibrils associate. We envision a wide application potential of nano-FTIR, including cellular receptor *in vitro* mapping and analysis of proteins within quaternary structures.

[1] CIC nanoGUNE Consolider, 20018 Donostia—San Sebastián, Spain. [2] Experimental Molecular Biophysics, Department of Physics, Freie Universität Berlin, 14195 Berlin, Germany. [3] Neaspec GmbH, 82152 Martinsried, Germany. [4] IKERBASQUE, Basque Foundation for Science, 48011 Bilbao, Spain. Correspondence and requests for materials should be addressed to R.H. (email: r.hillenbrand@nanogune.eu).

Fourier transform infrared spectroscopy (FTIR) allows for the analysis of chemical bonds, protein morphology or secondary structure[1-4]. Applications include the tracing of protein modifications that may originate from interactions with other biomaterials (for example, cellular receptors and enzymatic cofactor)[5] or inorganic matter (for example, heavy metal ions and nanoparticles)[6]. Such interactions are of high relevance in various physiological cellular processes (for example, cellular attachment and genomic regulation), pathological processes of a large variety of diseases (for example, viral infection and neurodegenerative diseases)[7,8], as well as in biomedical diagnostics and equipment (for example, support for human cell culture)[9]. The analysis of amide vibrations provides valuable insights into the secondary structures of proteins, revealing information regarding conformation and folding[3,10-13]. However, because of the diffraction-limited resolution of FTIR, infrared spectroscopy applications on the nanometre scale or even single-molecule level have so far proved elusive.

The diffraction-limited resolution of FTIR is circumvented by Fourier transform infrared nanospectroscopy (nano-FTIR)[14-19]. This technique is based on scattering-type scanning near-field optical microscopy (s-SNOM)[20] where infrared images with nanoscale spatial resolution are obtained by recording the infrared light scattered at a scanning probe tip. The probe is typically a metalized atomic force microscope (AFM) tip. Acting as an antenna, it concentrates the incident field at the tip apex[15,20] for local probing of molecular vibrations, similar to tip-enhanced Raman spectroscopy[21-24]. When the AFM tip is illuminated by the broadband infrared radiation, Fourier transform spectroscopy of the scattered light yields infrared spectra with a spatial resolution down to 20 nm (refs 16–19). However, broadband nano-FTIR spectra of protein complexes have neither been observed so far nor shown to be suitable for the analysis of their structure. Biological objects such as protein fibrils and membranes have been already studied by s-SNOM imaging at selected wavelengths[25-29] but it has not been demonstrated that the most important secondary structures—α-helical and β-sheet structures—can be identified on the level of individual protein complexes.

To demonstrate that nano-FTIR can probe protein secondary structure on the nanometre scale and with sensitivity to individual protein complexes, we have chosen to study the amide bands of individual tobacco mosaic viruses (TMVs) and ferritin complexes[30] for their well-defined, robust and dominantly α-helical protein structure[31], insulin aggregates for exhibiting β-sheet structure[32] and purple membranes (PMs) because of the well-defined orientation of the transmembrane α-helices[33]. Broadband infrared spectra of a ferritin complex indicate that nano-FTIR is sensitive to about 4,000 amino acids, corresponding to about 5,000 C=O and N-H bonds, respectively. Most importantly, we find that the amide I band in the nano-FTIR spectra can be analysed and interpreted in the framework of standard infrared spectroscopy. Our studies establish a solid foundation for infrared nanobiospectroscopy, which is a prerequisite for nano-FTIR applications in biochemical and biomedical research. We demonstrate the potential for real-life applications of nano-FTIR in these fields by exploring the still not fully clarified protein conformation in individual insulin fibrils.

Results

s-SNOM and nano-FTIR set-up.
The s-SNOM and nano-FTIR set-up (Neaspec GmbH, Germany) used in this work is based on an AFM where the tip is oscillating vertically at the mechanical resonance frequency Ω of the cantilever. Infrared near-field imaging and spectroscopy are performed by interferometric

detection of the light scattered from a gold-coated AFM tip, which is illuminated by the radiation from a tunable quantum cascade laser (QCL) or a broadband mid-infrared laser beam, respectively (Fig. 1a,b).

For s-SNOM imaging, the tip is illuminated with radiation from the wavelength-tunable QCL at individual wavelengths. The tip-scattered light is detected by operating the interferometer in pseudoheterodyne mode[34]. Demodulating the detector signal at a higher harmonic $n\Omega$ of the tip oscillation frequency yields infrared amplitude and phase images, s_n and φ_n (see Methods section).

In nano-FTIR spectroscopy, broadband infrared amplitude and phase spectra are obtained by employing the interferometer for Fourier transform spectroscopy of the scattered light[14-19]. In this case, the tip is illuminated with a broadband laser of a total power of about 100 µW (refs 17,35). As the tip and the sample are located in one of the interferometer arms (in contrast to standard FTIR), we can measure both local amplitude and phase spectra, $s_n(\omega)$ and $\varphi_n(\omega)$. These spectra are normalized to the references, $s_{ref,n}(\omega)$ and $\varphi_{ref,n}(\omega)$, obtained on a clean area on the sample support (marked R in the topography images presented in the work). We thus obtain normalized amplitude and phase spectra form which we can calculate the nano-FTIR absorption spectrum $a_n(\omega) = s_n/s_{ref,n} \sin(\varphi_n - \varphi_{ref,n})$. As demonstrated in refs 17,19, the nano-FTIR spectrum $a_n(\omega)$ reveals the local infrared absorption in the sample.

Setting the foundations for protein analysis with nano-FTIR. We first verify the ability of nano-FTIR to measure nanoscale

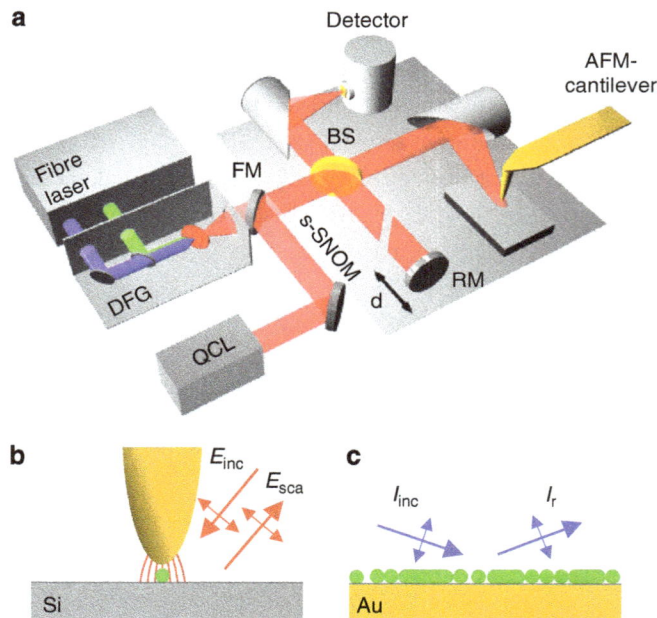

Figure 1 | Infrared spectroscopy of protein nanostructures. (a) s-SNOM and nano-FTIR set-up employing a tunable single line laser (quantum cascade laser, QCL) or a mid-infrared continuum source (fibre laser plus difference frequency generator) for tip illumination, respectively. The illumination can be chosen with a flip mirror (FM). The light backscattered from the tip is analysed with a Michelson interferometer that is operated in either pseudoheterodyne mode (for s-SNOM imaging) or as a Fourier transform spectrometer (for nano-FTIR spectroscopy). It comprises a beam splitter (BS, uncoated ZnSe), a reference mirror (RM) and a detector. **(b)** Schematics of s-SNOM and nano-FTIR spectroscopy. **(c)** Schematics of far-field GI-FTIR spectroscopy of protein nanostructure ensembles.

broadband infrared spectra of the amide I and II bands over the range from 1,400 to 1,800 cm^{-1}. To that end, we studied TMVs (Fig. 2a). These well-defined protein complexes have a diameter of 18 nm and a length of 300 nm, and consist of 2,140 identical proteins assembled helically around an RNA strand[31]. Figure 2c shows near-field infrared phase φ_3 images taken at 1,660 and 1,720 cm^{-1}, that is, on and off resonance with the amide I vibration. At 1,660 cm^{-1}, the phase image exhibits strong contrast owing to the amide I absorption[25]. As expected, the phase contrast vanishes when the illumination is tuned to 1,720 cm^{-1} where the protein does not absorb. From nano-FTIR amplitude $s_3(\omega)$ and phase $\varphi_3(\omega)$ spectra recorded on top of the virus (position marked by the red dot in the topography image) and normalized to those taken on a clean silicon area, we obtained the local infrared absorption spectrum a_3 (red spectrum in Fig. 2d). For comparison, we recorded a FTIR spectrum of a large TMV ensemble (blue spectrum in Fig. 2d) by employing p-polarized-grazing incidence (GI-FTIR, schematics see Fig. 1c). We find a good agreement between the GI-FTIR and nano-FTIR spectra. Both reveal the amide I and amide II bands associated with combinations of C=O and C-N stretching with C-N-H-bending vibrations, respectively. We note that the agreement can vary among different amide bands, since the exact peak position and shape of an absorption band depend on the band strength and the employed FTIR technique[36] (see also Supplementary Fig. S1).

In Fig. 2e–h, we perform nano-FTIR studies of ferritin, a globular protein complex of 12 nm in diameter. Ferritin comprises 24 subunits that form a cage around a ferrihydrite nanoparticle (Fig. 2e). Each ferritin subunit is composed of six α-helixes and one β-sheet[30]. The topographical image in Fig. 2g shows two particles of about 10 and 8 nm, and one of about 6 nm height, which appear much broader because of the convolution with tip apex that has a radius of about 35 nm (see Fig. 2f and Supplementary Fig. S2). The infrared phase image φ_3 at 1,660 cm^{-1} reveals a strong absorption for the 10- and 8-nm particles, indicating that these are ferritin complexes. No infrared contrast is seen for the 6-nm particle, which we interpret as ferrihydrite core that does not absorb at 1,660 cm^{-1}. Figure 2h shows the nano-FTIR spectrum (red curve) of the ferritin marked by the red dot in Fig. 2g in comparison with a GI-FTIR spectrum of a large ferritin ensemble (blue curve). As demonstrated with the TMV (*vide supra*), the nano-FTIR spectrum reveals the amide I and II bands in excellent agreement with GI-FTIR. Thus, we conclude that nano-FTIR allows for measuring infrared spectra of individual protein complexes, which can be interpreted by comparison with standard GI-FTIR absorbance spectra. The analysis provided in Supplementary Fig. S2 and Supplementary Note 1 indicates that the marked object is one ferritin complex.

To explore whether and how nano-FTIR absorption spectra depend on the protein orientation, we studied the PMs of *Halobacterium salinarum* (Fig. 3), which is composed of a double layer of polar and neutral lipids, and the integral membrane protein bacteriorhodopsin. The secondary structure of bacteriorhodopsin comprises seven transmembrane α-helices and an extracellular β-sheet[33]. As the helices are predominantly oriented

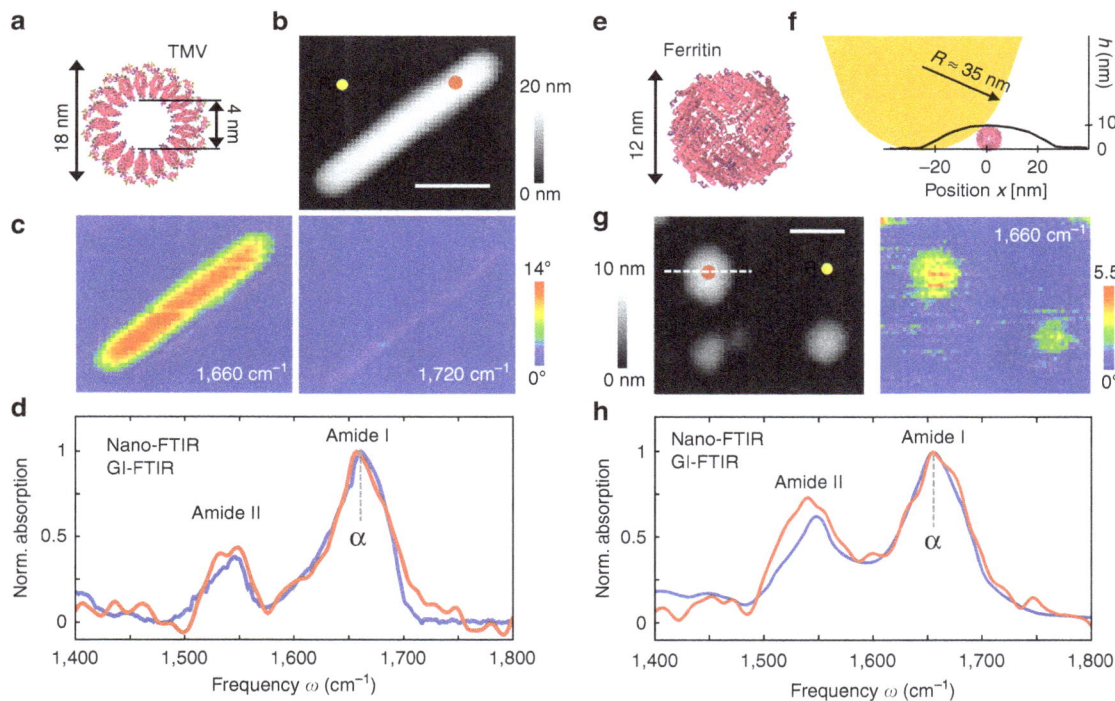

Figure 2 | Infrared nanospectroscopy of TMV and individual ferritin protein complexes. (**a**) Illustration of the protein structure of TMV. (**b**) Topography of TMV on a silicon substrate. Scale bar, 100 nm. (**c**) Infrared near-field phase images φ_3 at two different frequencies, 1,660 cm^{-1} (left) and 1,720 cm^{-1} (right; 2.5 min total image acquisition time). (**d**) Nano-FTIR spectrum of TMV taken at the position marked by the red dot in the topography image (red, average of 40 interferograms; 22 min total acquisition time; 8 cm^{-1} resolution; ×4 zero filling) and a GI-FTIR spectra of a large TMV ensemble on a gold substrate (blue). (**e**) Illustration of the protein structure of ferritin. (**f**) Illustration of the tip probing a ferritin particle. The graph shows the measured height profile h when the tip is scanned along the dashed line in Fig. 2g. (**g**) Topography and infrared near-field image φ_3 of two ferritin molecules and a ferrihydrite core on a silicon substrate. Scale bar, 50 nm (1.5 min total image acquisition time). (**h**) Nano-FTIR (red) and GI-FTIR spectra of a large ferritin ensemble on a gold substrate (blue). The red dot in the topography image marks the position where the nano-FTIR spectra (average of 60 interferograms; 23 min total acquisition time; 8 cm^{-1} resolution; ×4 zero filling) was recorded. The position marked by the yellow dot and R in **b** and **g** indicates where the reference spectrum was recorded.

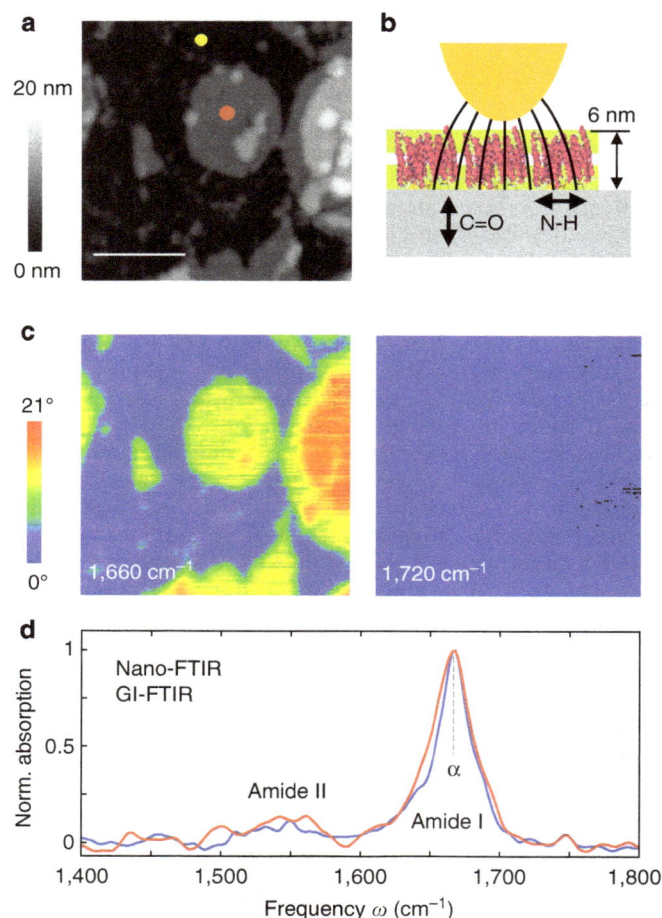

Figure 3 | Nano-FTIR sensitivity to protein orientation demonstrated with PM. (**a**) Topography of PMs on a silicon substrate. Scale bar, 400 nm. (**b**) Illustration of near-field probing of PM. (**c**) Infrared near-field phase images φ_3 at two different frequencies, 1,660 cm^{-1} (left) and 1,720 cm^{-1} (right; 19 min total image acquisition time). (**d**) Nano-FTIR (red) and GI-FTIR spectra of an ensemble of horizontally absorbed PMs on a gold substrate (blue). The red dot in the topography image marks the position where the nano-FTIR spectra (average of 60 interferograms; 23 min total acquisition time; 8 cm^{-1} resolution; $\times$ 4 zero filling) was recorded. The position marked by the yellow dot and R indicates where the reference spectrum was recorded.

perpendicular to the membrane plane, the infrared amide I vibration is normal to the membrane surface, whereas the amide II vibration is parallel to the membrane surface[37] (see illustration in Fig. 3b). The well-defined directions of the amide vibrations render PM an excellent protein structure for studying the sensitivity of nano-FTIR to protein orientation. In the topography image (Fig. 3a), the membranes appear as 6 nm high flat layers. As before with the TMV, the infrared phase contrast (Fig. 3c) at 1,660 cm^{-1} exhibits the typical amide I absorption, which vanishes at 1,720 cm^{-1} because the archaeobacterial PM is devoid of any ester lipids that usually give rise to carbonyl stretching vibrations in lipid membranes of eubacteria and eukarya. A nano-FTIR spectrum a_3 (Fig. 3d, red curve) is obtained by placing the tip atop the PM (marked by red dot in Fig. 3a), analogous to the TMV experiment. The GI-FTIR spectrum (p-polarized incident field) of horizontally adsorbed PMs was recorded for comparison (Fig. 3d, blue curve; for schematics see Fig. 1c). Interestingly, the amide II band does not appear in both spectra. In the GI-FTIR experiment, the electric

field is perpendicular to the PM surface and thus perpendicular to the direction of the amide II vibrations. For that reason, amide II vibrations are not excited[37]. The absence of the amide II band in the nano-FTIR spectrum is explained by the field distribution at the tip apex, which essentially is that of a vertically oriented point dipole located in the tip apex (Fig. 3b)[20]. The strongest field, located directly below the tip apex, is vertically oriented, and efficiently couples the protein vibrations normal to the PM surface (amide I) but not the protein vibrations parallel to the PM surface (amide II). This orientation-dependent effect is well known from surface-enhanced IR absorption spectroscopy[38]. Thus, we conclude that nano-FTIR primarily probes molecular vibrations that oscillate perpendicular to the sample surface.

In Fig. 4, we evaluate the spatial resolution and reproducibility of nano-FTIR protein spectroscopy by recording 20 spectra, while the tip is scanning in steps of 10 nm across the PM. The spectral line scan (Fig. 4b) along the line of red and blue dots in the topography image (Fig. 4a) reveals the amide I peak on top of the membrane until the edge is reached (at position $x \approx 80$ nm). Plotting the nano-FTIR absorption a_3 at 1,660 cm^{-1} as a function of the position x (Fig. 4c) shows that absorption vanishes within three scan steps. This demonstrates a spatial resolution of about 30 nm, which is an improvement by more than 100 compared with micro-FTIR mapping[39]. We further note that the eight local infrared spectra on the PM (P1–P8, between $x = 0$ and 70 nm) exhibit a stable peak position and peak height, which illustrates the high reproducibility of nano-FTIR (Fig. 4d). On the other hand, these results show that the infrared absorption of PM does not exhibit significant spatial variations, confirming the homogeneity of the protein structure within the PM.

Nanoscale mapping of structural protein heterogeneity. To establish nanoscale identification of α-helical and β-sheet structure in proteins, and to demonstrate heterogeneity in a sample at the nanometre scale that is only resolvable by nano-FTIR, we studied insulin aggregates deliberately contaminated with a low amount of TMV (Fig. 5). While α-helices dominate in TMV, the insulin aggregates are composed of essentially two β-sheets[32]. The TMV can be thus considered a diluted α-helical structure within a β-sheet sample. Note that for sample preparation, we used a 2-year old insulin solution (see Methods section) because we found that such a solution yields insulin aggregates rather than well-defined fibrils. For such a sample, we expect that the orientation of the infrared-active β-sheet dipole of the amide I band is random, thus yielding a signal in nano-FTIR. Note that only vertically oriented dipoles yield a signal in nano-FTIR (see Fig. 3).

In Fig. 5a, we show attenuated total reflectance (ATR-)FTIR spectra of pure insulin aggregates (black curve) and of the insulin/TMV mixture (blue curve). Both pure insulin and insulin/TMV samples were deposited on a silicon support. The ATR-FTIR spectrum of the insulin/TMV mixture closely matches the ATR-FTIR spectrum of pure insulin and does not reveal the presence of TMV. The topographical image (Fig. 5b) of the mixture shows rod-like structures and aggregates of up to 32 nm in height. Still, TMV and insulin cannot be clearly discriminated. Figure 5c shows the near-field infrared phase images at 1,634 and 1,660 cm^{-1}, corresponding to the centre frequencies of the amide I band for β-sheet and α-helical structure, respectively[11]. In both infrared images, all structures show a distinct absorption contrast. As expected, the phase contrast to the substrate (that is, absorption) increases with particle height (that is, volume). Comparing the two infrared images, we find a significant increase of the phase contrast from 1,634 to 1,660 cm^{-1} for the smooth rod in the lower left corner of the image, while for all other

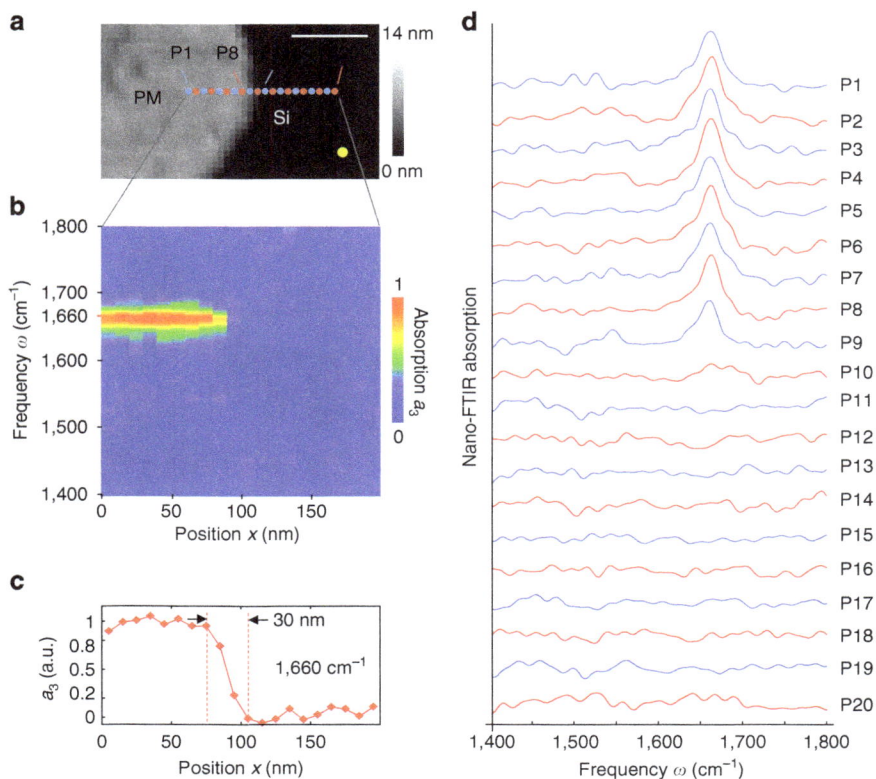

Figure 4 | Nano-FTIR mapping of PM. (**a**) Topography image of PM on a silicon substrate. Scale bar, 100 nm. (**b**) Infrared spectroscopic line scan recorded while the tip was scanned in 10 nm steps along the line of red and blue dots in **a**. At the position of each dot, a spectrum was taken (each spectrum is an average of 17 interferograms; 2.3 min acquisition time; 16 cm^{-1} resolution; × 6 zero filling). (**c**) Absorption signal a_3 at 1,660 cm^{-1} as a function of position x. (**d**) Twenty nano-FTIR spectra recorded at the positions P1–P20 marked by red and blue dots in **a**. The spectra are the same as those presented in **b**. For better visibility, each spectrum is vertically offset. All spectra have been normalized to the reference spectrum recorded at the position marked by the yellow dot and R in **a**.

structures the phase contrast did not change with respect to the substrate. From this observation, we conclude that α-helices form this rod-like structure, which is thus most probably a TMV. For identification of the secondary structure, we recorded nano-FTIR spectra (Fig. 5d) on the smooth rod at the position marked by the red dot in the topography image (red spectrum), and on the aggregate marked by the green dot (green spectrum). The red spectrum exhibits the absorption maximum at 1,660 cm^{-1}, which corresponds to the amide I resonance frequency of α-helices. Further, the shape fits well to the infrared spectrum of TMV (Fig. 2d). These results lead us to identify this rod-like structure as a TMV particle. In contrast, the green spectrum significantly differs from the red, exhibiting two peaks at 1,660 and 1,634 cm^{-1}, which are attributed to the presence of α-helices and β-sheets, respectively. Thus, we can identify this particle as an insulin aggregate, as its nano-FTIR spectrum agrees with the ATR-FTIR spectrum of the pure insulin sample (Fig. 5a, black curve). Note that we compare the nano-FTIR spectrum with an ATR-FTIR spectrum, in contrast to Fig. 2, as the infrared signal was too weak to obtain a GI-FTIR spectrum. As ATR-FTIR spectra are typically red shifted relative to GI-FTIR spectra[36] (see Supplementary Fig. S1), the ATR-FTIR spectrum is slightly red shifted compared with the nano-FTIR spectrum.

To visualize the nanoscale distribution of TMV and insulin aggregates, that is, the nanoscale structural heterogeneity of the sample, we calculated the ratio between the infrared phase images at 1,660 and 1,634 cm^{-1}. From the spectra displayed in Fig. 5d, we know that the absorption of TMV is stronger at 1,660 than

at 1,634 cm^{-1}. For insulin, the absorption is stronger at 1,634 cm^{-1}. Every image pixel where we obtain a ratio that is significantly larger than 1 thus reveals TMV. Pixel where we obtain a ratio that is smaller than 1 reveals insulin. Figure 5e shows a map where pixel with ratio larger than 1.5 are depicted in purple, and pixel with ratio smaller than 1 in yellow. The map clearly reveals the TMV (purple) and shows that all other protein structures (yellow) have a strong signal at 1,634 cm^{-1}, which can be attributed to the presence of β-sheets. This allows us to identify them as insulin aggregates.

Nano-FTIR studies of individual insulin fibrils. Having demonstrated the capability of mapping secondary structure on the nanometre scale, nano-FTIR is well prepared for applications in biochemical and biomedical research. We explore as a first application example the protein conformation in individual insulin fibrils (Fig. 6). Insulin can form amyloid-like fibrils and fibres composed of filament-shaped protein aggregates[11,32,40], which renders it an excellent and widely used model system for neurodegenerative disease research (that is, Alzheimer and Parkinson). Studies[32,41] show that the filaments have a core composed of β-sheets (Fig. 6c). It is assumed that this core is surrounded by randomly oriented secondary structures, including α-helices, β-turns and unordered structures[11]. However, the exact structure of the shell is still an open question of high biological relevance[40]. Recent studies by tip-enhanced Raman spectroscopy indicate the presence of α-helices/unordered structures at the surface of the fibrils[42,43]. In the following, we apply nano-FTIR to study the protein conformation in insulin fibrils.

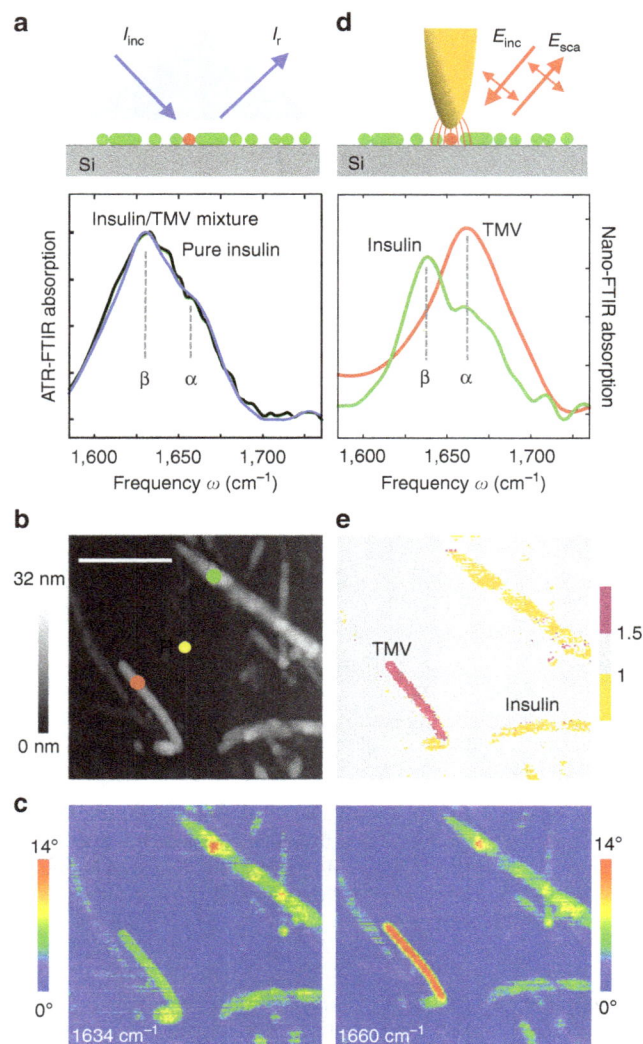

Figure 5 | Nanoscale mapping of α-helical and β-sheet secondary structure. (**a**) ATR-FTIR spectrum of a pure insulin aggregate sample (black) and a mixture of insulin aggregates and TMV (blue) on a silicon support. (**b**) Topography of a mixture of TMV and insulin aggregates on silicon. Scale bar, 500 nm. (**c**) Infrared near-field phase images φ_3 at two different frequencies, 1,634 cm^{-1} (left) and 1,660 cm^{-1} (right; 19 min total image acquisition time). (**d**) Nano-FTIR spectra of TMV (red, average of 70 interferograms; 19 min total acquisition time; 16 cm^{-1} resolution; × 4 zero filling) and insulin aggregate (green, average of 40 interferograms; 22 min total acquisition time; 8 cm^{-1} resolution × 4 zero filling). Red and green dots in the topography image mark the positions where the nano-FTIR spectra were taken. The position marked by the yellow dot and R indicates where the reference spectrum was recorded. (**e**) Map of TMV (purple) and insulin (yellow) aggregates.

Figure 6 | Infrared nanospectroscopy and nanoimaging of secondary structure in individual insulin fibrils. (**a**) Topography of insulin fibrils on a silicon substrate. Scale bar, 200 nm. The arrows indicate a type I fibril (I) and a 9-nm-thick fibril composed of several protofilaments (X), respectively. (**b**) Nano-FTIR spectrum of a 9-nm-thick insulin fibril (red, average of 154 interferograms; 8 cm^{-1} resolution; × 4 zero filling) recorded at the position marked by the red symbol in **a**. The position marked by the yellow dot and R indicates where the reference spectrum was recorded. The dashed blue line shows for comparison a GI-FTIR spectrum of insulin monomers on a gold substrate. (**c**) Illustration of the structure of an amyloid-like insulin protofibril (a type I fibril consists of two protofibrils). (**d**) Band decomposition of the nano-FTIR spectrum (red curve) based on five absorption bands (thin black curves). The dashed black curve shows the resulting fit. (**e**) s-SNOM phase images of the fibrils shown in **a**. Scale bar, 300 nm. (**f**) Local infrared absorption spectra (symbols) depicting the normalized imaginary part of the near-field signal at the positions marked in **a**. The data points were extracted from 12 near-field amplitude and phase images and were normalized to the imaginary part of the near-field signal on the silicon substrate. Four of the phase images are shown in **e**. For comparison, the nano-FTIR spectrum of **b** is depicted by the red thick curve. All spectra are normalized to their maximum value.

In contrast to Fig. 5, insulin fibrils were grown by incubating insulin protein at 60 °C in a pH 2 buffer for 30 h, thus representing a biologically relevant model system. In Fig. 6a, the topography of a sample area is shown where we found type I (3 nm high, consisting of two protofibrils, respectively four protofilaments, marked I) and thicker insulin fibrils (with increasing thickness the number of protofilaments increases) with heights ranging from 5 to 10 nm[11,44,45]. In the nano-FTIR spectrum of a 9-nm-thick fibril (Fig. 6b, red curve, taken at the position marked by the red dot in the topographical image), we find the strongest peak at 1,669 cm^{-1}. By comparison with a GI-FTIR spectrum of monomeric insulin

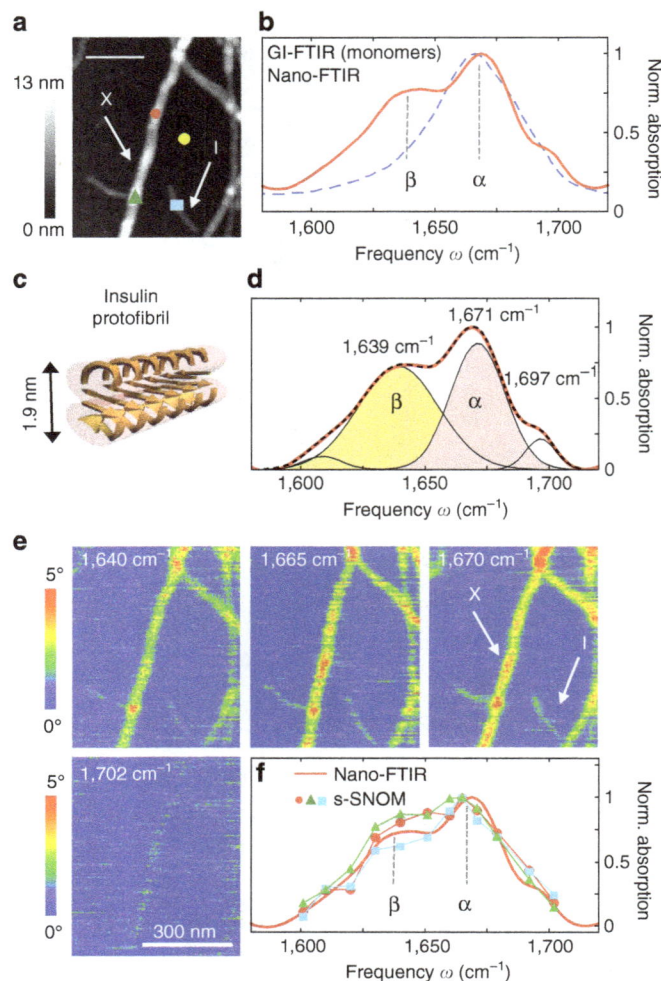

(predominantly α-helical structure, blue dashed curve), we can assign this peak to the presence of α-helices. The peak at 1,638 cm^{-1} is assigned to β-sheets. However, it is relatively weak compared with the nano-FTIR spectrum of the insulin aggregates (Fig. 5). We explain this observation by the well-aligned β-sheet structure forming the core of the fibrils seen in Fig. 6, where the dipole orientation of the amide I band is mostly parallel to the filament axes (that is, to the substrate surface)[13,46]. As seen before in Fig. 3 with PM, protein vibrations parallel to the substrate couple only weakly to the near field at the tip apex and thus are suppressed in the nano-FTIR spectra.

Standard band decomposition of the nano-FTIR spectrum (Fig. 6d) reveals two major bands at 1,639 and at 1,671 cm^{-1}, confirming that β-sheet and α-helical structures are the predominant contributions. The band at 1,697 cm^{-1} could be caused by β-turns or even antiparallel β-sheet structure and the weak band at 1,609 cm^{-1} might indicate side chains. Most importantly, no band is observed in between the β- and α-peaks where typically disordered structure is located[10]. Thus, we conclude that disordered structures are almost absent in the fibrils. Note that the dipole of the amide I band is isotropically oriented in disordered structures and thus cannot be the cause for this observation.

To explore the presence of α-helices in type I insulin fibrils (which are only 3 nm thick), we performed infrared s-SNOM imaging at different frequencies provided by our QCL. For nano-FTIR of such thin fibrils, the broadband laser source does not yet provide enough infrared power. Figure 6e shows infrared near-field phase images at four different wavelengths, exhibiting clear contrast for both type I and 9 nm thick fibrils. From altogether 12 images, we extracted local infrared spectra at the positions marked by the red, green and blue symbols in Fig. 6a. The spectra are plotted in Fig. 6f using the corresponding symbols. For both the 9-nm-thick fibril (red and green symbols) and type I (blue symbols) fibril, we find the same spectral signature as observed in the nano-FTIR spectrum (thick red curve in Fig. 6f). The spectrum acquired at the position marked by the blue symbol in Fig. 6a thus provides experimental evidence that α-helices are also present in type I fibrils. The current resolution of about 30 nm does not allow for concluding whether the α-helices are inside the core or forming a shell. Assuming that the core is formed of purely β-sheets (according to current models[41]), our findings suggest that the shell is highly structured (mainly α-helical structure) and not randomly organized. The presence of α-helices in the shell could explain the tendency of fibrils to associate.

Discussion

Nano-FTIR enables reliable probing and mapping of protein secondary structure with nanoscale resolution, and close to single-protein sensitivity. Our results indicate that infrared spectra of one ferritin complex can be obtained, which demonstrates extraordinary sensitivity of nano-FTIR to ultra-small amounts of material, in case of one ferritin complex 0.8×10^{-18} g (about 1 attogram) of protein, respectively, 5,000 C=O bonds. By further sharpening the tips and optimizing their antenna performance[47], we envision single-protein spectroscopy in the future, paving the way to a new era in infrared biospectroscopy. We foresee manifold applications, such as studies of conformational changes in amyloid structures on the molecular level, the mapping of nanoscale protein modifications in biomedical tissue or the label-free mapping of membrane proteins. To operate under physiological conditions, for example, to study the membrane of living cells, s-SNOM and nano-FTIR have to operate in liquids. Indeed, scanning near-field optical optical microscopy in liquids and under physiological conditions has been already demonstrated[48,49]. To perform nano-FTIR, the water could be exchanged by D_2O, which is typically done in conventional FTIR to eliminate infrared absorption by H_2O in the region of the amide I band[50].

Methods

s-SNOM. The AFM is operated in dynamic mode, where a standard Au-coated tip (PPP-NCSTAu, Nanosensors) is vertically vibrating with an amplitude of about 40 nm at the mechanical resonance frequency Ω of the cantilever, in this work at about $\Omega \sim 135$ kHz. For infrared imaging, we use a QCL (Daylight Solutions, USA), tunable between 1,560 and 1,750 cm^{-1}. A Michelson interferometer is used to analyse the tip-scattered light. Unavoidable background signals can be efficiently suppressed by demodulating the detector signal at a higher harmonic of the tapping frequency, $n\Omega$ (ref. 34). Employing a pseudoheterodyne detection scheme[34] (where the reference mirror oscillates at a frequency M) enables the simultaneous detection of both amplitude and phase signals. In all presented images, the demodulation order was $n = 3$. The images were recorded with an integration time of 25 ms per pixel, resulting in a total image acquisition time of several minutes, depending on the number of pixel.

nano-FTIR. For broadband spectroscopy, the AFM tip is illuminated with a coherent mid-infrared beam of an average output power of about 100 µW (refs 16,17,35). The broadband mid-infrared laser beam is generated by a difference frequency generator (Lasnix, Germany) where two near-infrared, 100-fs pulse trains from a fibre-laser system (FemtoFiber pro IR and SCIR, Toptica Germany) are superimposed in a GaSe crystal. This mid-infrared source emits a continuous spectrum with a usable width up to 700 cm^{-1}, which can be tuned within the limits 700–2,500 cm^{-1} dependent on difference frequency generator settings (for example, the crystal orientation). FTIR spectroscopy of the light scattered by the tip is performed by recording the demodulated detector signal I as a function of the reference mirror position d. Because of the asymmetry of the interferometer (that is, the sample is located in one interferometer arm), Fourier transformation of the interferogram $I(d)$ subsequently yields the near-field amplitude and phase spectra. In all presented nano-FTIR spectra, the demodulation order was $n = 3$ except for the nano-FTIR spectra shown in Fig. 3 where it was $n = 2$.

As the sample stage is subject to unavoidable mechanical drift in the range of several nanometres per minute, we recorded individual interferograms of the object (TMV, ferritin, insulin aggregates and fibrils) with an acquisition time of 23 s. To improve the signal-to-noise ratio, the interferograms were recorded multiple times (the numbers of interferograms are provided in the figure captions) and an average was calculated. Between each of the nano-FTIR measurements, the tip was relocated to the original point on the object. Relocation was done with the help of topography images, which were recorded in between each interferogram acquisition. Subsequent Fourier transformation of the averaged interferograms yields the complex valued near-field spectra of the object, $E_n(\omega) = s_n(\omega)e^{i\varphi_n(\omega)}$.

For normalizing the spectra of the objects, the tip was positioned on a clean area of the silicon substrate close to the object and reference interferograms were recorded. We recorded a similar number of interferograms as on the object, each one in 23 s. Fourier transformation of the averaged interferograms yields the near-field reference spectrum $E_{ref,n}(\omega) = s_{ref,n}(\omega)e^{i\varphi_{ref,n}(\omega)}$, that is, the near-field spectrum of the clean silicon surface. By complex-valued division, $E_n(\omega)/E_{ref,n}(\omega)$, we obtained the normalized amplitude $s_n(\omega)/s_{ref,n}(\omega)$ and phase $\varphi_n(\omega) - \varphi_{ref,n}(\omega)$ spectra of the object, yielding the nano-FTIR absorption spectra of the object, $a_n(\omega) = s_n/s_{ref,n} \sin(\varphi_n - \varphi_{ref,n})$.

FTIR. We used a Bruker Hyperion 2000 microscope coupled to a Vertex 70 FTIR spectrometer equipped with GI and ATR modules and a liquid-nitrogen-cooled mid-band mercury cadmium telluride detector. All FTIR spectra show the absorbance.

The GI module ($\times$ 15 GI reflection objective, Bruker) comprises a plane mirror that reflects the beam onto the sample surface at grazing angle incidence. The reflected beam is backreflected and refocused onto the same sample area and then collected and detected. The spectra were measured with a spectral resolution of 4 cm^{-1} and present an average over 1,000 scans with a total acquisition time of 860 s. All samples were prepared on a 150-nm-thick gold film on silicon. The reference spectra were recorded on a cleaned 150 nm thick gold film on a silicon wafer.

The ATR module ($\times$ 20 ATR objective, Bruker, single internal reflection) comprises a germanium crystal with a diameter of about 100 µm at the point of contact with the sample. The spectra were measured with a spectral resolution of 4 cm^{-1} and present an average over 1,000 scans with a total acquisition time of 860 s. The samples were prepared on a silicon wafer or on a gold-coated silicon wafer. The reference spectra were recorded on a cleaned silicon wafer or gold-coated silicon wafer, respectively.

Decomposition of the amide I band. The band decomposition was performed with the OPUS software package (version 4.2) supplied by Bruker. As a starting point for the curve-fitting procedure, five individual absorption bands were proposed at 1,638, 1,655, 1,667, 1,685 and 1,705 cm^{-1}, defining parallel β-sheets, unordered, α-helices, β-turn and antiparallel β-sheet structures, respectively. The curve fitting was successfully performed based on the damped least-squares optimization algorithm developed by Levenberg–Marquardt and assuming Gaussian band envelopes. The obtained residual root mean squared error was 0.0048.

Comparing the nano-FTIR spectrum of the insulin fibrils with a GI-FTIR spectrum of insulin monomers (peak maximum at 1,666 cm^{-1}), we can assign the peak at 1,671 cm^{-1} to α-helical structure. The insulin monomers are dominantly α-helical structure, as we confirmed by circular dichroism measurements shown in

Supplementary Fig. S3. The frequency difference between 1,666 and 1,671 cm^{-1} might be caused by the structural change from monomer to fibril structure or owing to the different sample support, which in nano-FTIR is silicon and in GI-FTIR is gold.

We note that a peak frequency of 1,666 cm^{-1} for α-helical structure seems to be unusually high. However, it can be readily explained by the grazing-incidence FTIR modality. It is well known that GI-FTIR spectra can be significantly blue shifted compared with standard transmission FTIR spectra[36]. We demonstrate this phenomenon in the Supplementary Fig. S1 where we compare a standard transmission FTIR spectrum with the GI-FTIR spectrum of the insulin monomer sample. The peak in the transmission FTIR spectrum occurs at the much lower frequency 1,656 cm^{-1}, which matches literature data (ranging from 1,654 to 1,658 cm^{-1} (refs 50–52)).

Sample preparation. All water was deionized ultrapure (18 MΩcm). Substrate surfaces were used as provided, but in some cases additionally cleaned by ultrasonication in acetone, isopropanol and water (each for 10 min), and blow dried with nitrogen. For all s-SNOM and nano-FTIR measurements, all samples were prepared on silicon wafers. For GI-FTIR measurements, gold substrates were used (150 nm of thermally evaporated gold deposited on a silicon wafer), while for ATR-FTIR measurements silicon or gold-coated silicon wafers were used.

TMV samples. TMV suspension: A suspension of TMV (strain *vulgare*), purified from systemically infected tobacco plants (*Nicotiana tabacum cv Samsun nn*) was kindly provided by Professor C. Wege (University of Stuttgart, Germany). TMV was dialyzed against water in a 10-kDa Slide-A-Lyzer dialysis unit (Thermo Fisher Scientific), and diluted with water.

TMV for s-SNOM: Ten microlitres of 1 µg ml^{-1} suspension was spin coated (spin coater SCI-20, Schaefer Technologie) on a cleaned silicon wafer and dried in air.

TMV for GI-FTIR: Ten microlitres of 1 µg ml^{-1} suspension was spin coated on a gold wafer, and dried in air.

Ferritin samples. Ferritin solution: Horse spleen ferritin solution (44 mg ml^{-1}, Sigma-Aldrich) was diluted 500-fold with water.

Ferritin for s-SNOM: A droplet of Ferritin solution was deposited on a silicon wafer and incubated for 2 min. Excess of solution was removed with filter paper (Sigma) and the sample was dried in air.

Ferritin for GI-FTIR: Ferritin solution was first purified with a spin desalting column (7 K MWCO, Thermo Scientific) and followed with the procedure as above, but on a gold substrate.

PM samples. Halobacteria (*H. salinarum* strain S9) were cultivated in high-salt pepton medium (10 g l^{-1} pepton, 4.3 M NaCl, 80 mM $MgSO_4$, 27 mM KCl, 10 mM sodium citrate, pH 6.5) for 6 days. The collected cells were lysed by osmotic shock with distilled water. PM fragments were purified by fractionated centrifugation[53].

PM for s-SNOM: A 30-µl droplet of a buffer pH 7.8, containing 150 mM of KCl and 20 mM of Tris, was deposited on a silicon substrate and 0.2 µl of purified PM (6 mg ml^{-1}) was added into the droplet. Fifteen microlitres of imaging buffer was subsequently added into the droplet to ensure maximum dispersion of previously added PMs. The resulting droplet was incubated on the silicon substrate for 25 min and then rinsed several times with water and dried in air.

PM for GI-FTIR: Same sample preparation procedure as for s-SNOM, but a gold substrate was used and the PM was incubated for 40 min.

Insulin aggregate samples. Insulin aggregate solution: Dispersions were prepared solubilizing 20 mg dehydrated human insulin (Sigma-Aldrich) in 1 ml 1 mM HCl solution. The solution was then re-diluted in 10 mM HCl/KCl at 2 mg ml^{-1}, and incubated at 60 °C for 24 h in a thermomixer (Eppendorf), followed by 2 years storage at 4 °C.

Insulin aggregates (+ TMV) for s-SNOM: Ten microlitres of insulin aggregate solution was spin coated (as for TMV) on a silicon substrate, and rinsed with water. Ten microlitres of 1 µg ml^{-1} TMV suspension was then spin coated on the dried insulin aggregates.

Insulin aggregates for ATR-FTIR: Twenty microlitres of insulin aggregate solution was deposited on a silicon wafer, dried and rinsed with water.

Insulin fibril samples. Insulin fibril solution: Dispersions were prepared as above, but incubated at 60 °C for 30 h.

Insulin fibrils for s-SNOM: Ten microlitres of insulin fibril solution was spin coated on a silicon wafer and rinsed with water.

Insulin monomer samples. Insulin monomer solution: Dispersions were prepared as above, but without any incubation.

Insulin monomer for GI-FTIR: Twenty microlitres of insulin monomer solution was spin coated on a gold substrate and rinsed with water.

References

1. Venyaminov, S. Y. & Kalnin, N. N. Quantitative IR spectrophotometry of peptide compounds in water (H_2O) solutions. I. Spectral parameters of amino-acid residue absorption-bands. *Biopolymers* **30**, 1243–1257 (1990).
2. Jackson, M. & Mantsch, H. H. The use and misuse of FTIR spectroscopy in the determination of protein-structure. *Crit. Rev. Biochem. Mol. Biol.* **30**, 95–120 (1995).
3. Garczarek, F. & Gerwert, K. Functional waters in intraprotein proton transfer monitored by FTIR difference spectroscopy. *Nature* **439**, 109–112 (2006).
4. Ataka, K., Kottke, T. & Heberle, J. Thinner, smaller, faster: IR techniques to probe the functionality of biological and biomimetic systems. *Angew. Chem Int. Ed.* **49**, 5416–5424 (2010).
5. Gupta, S. C. et al. Multitargeting by curcumin as revealed by molecular interaction studies. *Natural Product Reports* **28**, 1937–1955 (2011).
6. Mahmoudi, M. et al. Protein-nanoparticle interactions: opportunities and challenges. *Chem. Rev.* **111**, 5610–5637 (2011).
7. Lasch, P., Haensch, W., Naumann, D. & Diem, M. Imaging of colorectal adenocarcinoma using FT-IR microspectroscopy and cluster analysis. *Biochim. Biophys. Acta* **1688**, 176–186 (2004).
8. Caine, S., Heraud, P., Tobin, M. J., McNaughton, D. & Bernard, C. C. A. The application of Fourier transform infrared microspectroscopy for the study of diseased central nervous system tissue. *Neuroimage* **59**, 3624–3640 (2012).
9. Xia, Z., Yu, X. & Wei, M. Biomimetic collagen/apatite coating formation on Ti6Al4V substrates. *J Biomed. Mater. Res.B Appl. Biomater.* **100B**, 871–881 (2012).
10. Byler, D. M. & Susi, H. Examination of the secondary structure of proteins by deconvolved FTIR spectra. *Biopolymers* **25**, 469–487 (1986).
11. Bouchard, M., Zurdo, J., Nettleton, E. J., Dobson, C. M. & Robinson, C. V. Formation of insulin amyloid fibrils followed by FTIR simultaneously with CD and electron microscopy. *Prot. Sci.* **9**, 1960–1967 (2000).
12. Ritter, C. et al. Correlation of structural elements and infectivity of the HET-s prion. *Nature* **435**, 844–848 (2005).
13. Rodriguez-Perez, J. C., Hamley, I. W., Grasb, S. L. & Squires, A. M. Local orientational disorder in peptide fibrils probed by a combination of residue-specific C-13-O-18 labelling, polarised infrared spectroscopy and molecular combing. *Chem. Comm.* **48**, 11835–11837 (2012).
14. Amarie, S. & Keilmann, F. Broadband-infrared assessment of phonon resonance in scattering-type near-field microscopy. *Phys. Rev. B* **83**, 045404 (2011).
15. Huth, F., Schnell, M., Wittborn, J., Ocelic, N. & Hillenbrand, R. Infrared-spectroscopic nanoimaging with a thermal source. *Nat. Mater.* **10**, 352–356 (2011).
16. Amarie, S. et al. Nano-FTIR chemical mapping of minerals in biological materials. *Beilstein J. Nanotechnol.* **3**, 312–323 (2012).
17. Huth, F. et al. Nano-FTIR absorption spectroscopy of molecular fingerprints at 20 nm spatial resolution. *Nano Lett.* **12**, 3973–3978 (2012).
18. Xu, X. J. G., Rang, M., Craig, I. M. & Raschke, M. B. Pushing the sample-size limit of infrared vibrational nanospectroscopy: from monolayer toward single molecule sensitivity. *J. Phys. Chem. Lett.* **3**, 1836–1841 (2012).
19. Govyadinov, A. A., Amenabar, I., Huth, F., Carney, P. S. & Hillenbrand, R. Quantitative measurement of local infrared absorption and dielectric function with tip-enhanced near-field microscopy. *J. Phys. Chem. Lett.* **4**, 1526–1531 (2013).
20. Keilmann, F. & Hillenbrand, R. *Nano-Optics and Near-Field Optical Microscopy* (Artech House, 2008).
21. Stöckle, R. M., Suh, Y. D., Deckert, V. & Zenobi, R. Nanoscale chemical analysis by tip-enhanced Raman spectroscopy. *Chem. Phys. Lett.* **318**, 131–136 (2000).
22. Hartschuh, A., Sanchez, E. J., Xie, X. S. & Novotny, L. High-resolution near-field Raman microscopy of single-walled carbon nanotubes. *Phys. Rev. Lett.* **90**, 095503 (2003).
23. Ichimura, T., Hayazawa, N., Hashimoto, M., Inouye, Y. & Kawata, S. Tip-enhanced coherent anti-Stokes Raman scattering for vibrational nanoimaging. *Phys. Rev. Lett.* **92**, 220801 (2004).
24. Verma, P., Ichimura, T., Yano, T.-a., Saito, Y. & Kawata, S. Nano-imaging through tip-enhanced Raman spectroscopy: stepping beyond the classical limits. *Laser Photonics Rev.* **4**, 548–561 (2010).
25. Brehm, M., Taubner, T., Hillenbrand, R. & Keilmann, F. Infrared spectroscopic mapping of single nanoparticles and viruses at nanoscale resolution. *Nano Lett.* **6**, 1307–1310 (2006).
26. Wollny, G., Bruendermann, E., Arsov, Z., Quaroni, L. & Havenith, M Nanoscale depth resolution in scanning near-field infrared microscopy. *Opt. Express* **16**, 7453–7459 (2008).
27. Ballout, F. et al. Scanning near-field IR microscopy of proteins in lipid bilayers. *Phys. Chem. Chem. Phys.* **13**, 21432–21436 (2011).
28. Paulite, M. et al. Imaging secondary structure of individual amyloid fibrils of a β2-microglobulin fragment using near-field infrared spectroscopy. *J. Am. Chem. Soc.* **133**, 7376–7383 (2011).

29. Li, J. J. & Yip, C. M. Super-resolved FT-IR spectroscopy: strategies, challenges, and opportunities for membrane biophysics. *Biochim. Biophys. Acta* **1828**, 2272–2282 (2013).

30. Theil, E. C., Behera, R. K. & Tosha, T. Ferritins for chemistry and for life. *Coord. Chem. Rev.* **257**, 579–586 (2013).

31. Ge, P. & Zhou, Z. H. Hydrogen-bonding networks and RNA bases revealed by cryo electron microscopy suggest a triggering mechanism for calcium switches. *Proc. Natl Acad. Sci. USA* **108**, 9637–9642 (2011).

32. Ivanova, M. I., Sievers, S. A., Sawaya, M. R., Wall, J. S. & Eisenberg, D. Molecular basis for insulin fibril assembly. *Proc. Natl Acad. Sci. USA* **106**, 18990–18995 (2009).

33. Henderson, R. & Unwin, P. N. T. Three-dimensional model of purple membrane obtained by electron-microscopy. *Nature* **257**, 28–32 (1975).

34. Ocelic, N., Huber, A. & Hillenbrand, R. Pseudoheterodyne detection for background-free near-field spectroscopy. *Appl. Phys. Lett.* **89**, 101124 (2006).

35. Keilmann, F. & Amarie, S. Mid-infrared frequency comb spanning an octave based on an er fiber laser and difference-frequency generation. *J. Infrared Millimeter Terahertz Waves* **33**, 479–484 (2012).

36. Griffiths, P. R. & de Haseth, J. A. *Fourier Transform Infrared Spectrometry* (Wiley, 2007).

37. Rothschild, K. J. & Clark, N. A. Polarized infrared spectroscopy of oriented purple membrane. *Biophys. J.* **25**, 473–487 (1979).

38. Jiang, X. *et al.* Resolving voltage-dependent structural changes of a membrane photoreceptor by surface-enhanced IR difference spectroscopy. *Proc. Natl Acad. Sci. USA* **105**, 12113–12117 (2008).

39. Salzer, R. & Siesler, H. W. *Infrared and Raman Spectroscopic Imaging* (Wiley. com, 2009).

40. Selivanova, O. M. & Galzitskaya, O. V. Structural polymorphism and possible pathways of amyloid fibril formation on the example of insulin protein. *Biochemistry* **77**, 1237–1247 (2012).

41. Jimenez, J. L. *et al.* The protofilament structure of insulin amyloid fibrils. *Proc. Natl Acad. Sci. USA* **99**, 9196–9201 (2002).

42. Deckert-Gaudig, T. & Deckert, V. Tip-enhanced Raman scattering (TERS) and high-resolution bio nano-analysis - a comparison. *Phys. Chem. Chem. Phys.* **12**, 12040–12049 (2010).

43. Kurouski, D., Deckert-Gaudig, T., Deckert, V. & Lednev, I. K. Structure and composition of insulin fibril surfaces probed by TERS. *J. Am. Chem. Soc.* **134**, 13323–13329 (2012).

44. Khurana, R. *et al.* A general model for amyloid fibril assembly based on morphological studies using atomic force microscopy. *Biophys. J.* **85**, 1135–1144 (2003).

45. Guo, S. & Akhremitchev, B. B. Packing density and structural heterogeneity of insulin amyloid fibrils measured by AFM nanoindentation. *Biomacromolecules* **7**, 1630–1636 (2006).

46. Cheatum, C. M., Tokmakoff, A. & Knoester, J. Signatures of beta-sheet secondary structures in linear and two-dimensional infrared spectroscopy. *J. Chem. Phys.* **120**, 8201–8215 (2004).

47. Huth, F. *et al.* Resonant antenna probes for tip-enhanced infrared near-field microscopy. *Nano Lett.* **13**, 1065–1072 (2013).

48. Fragola, A., Aigouy, L., Mignotte, P. Y., Formanek, F. & De Wilde, Y. Apertureless scanning near-field fluorescence microscopy in liquids. *Ultramicroscopy* **101**, 47–54 (2004).

49. Hoppener, C., Siebrasse, J. P., Peters, R., Kubitscheck, U. & Naber, A. High-resolution near-field optical imaging of single nuclear pore complexes under physiological conditions. *Biophys. J.* **88**, 3681–3688 (2005).

50. Dzwolak, W., Loksztejn, A. & Smirnovas, V. New insights into the self-assembly of insulin amyloid fibrils: an HD exchange FT-IR study. *Biochemistry* **45**, 8143–8151 (2006).

51. Nielsen, L., Frokjaer, S., Carpenter, J. F. & Brange, J. Studies of the structure of insulin fibrils by Fourier transform infrared (FTIR) spectroscopy and electron microscopy. *J. Pharm. Sci.* **90**, 29–37 (2001).

52. Webster, G. T., Dusting, J., Balabani, S. & Blanch, E. W. Detecting the early onset of shear-induced fibril formation of insulin in situ. *J. Phys. Chem. B* **115**, 2617–2626 (2011).

53. Oesterhelt, D. & Stoeckenius, W. Isolation of the cell membrane of Halobacterium halobium and its fractionation into red and purple membrane. *Methods Enzymol.* **31**, 667–678 (1974).

Acknowledgements

We thank Professor Christina Wege (Universität Stuttgart) for providing the TMV. This work was financially supported by an ERC Starting Grant (TERATOMO), the Spanish Ministerio de Economía y Competitividad (Projects MAT2012-36580 and MAT2010-16184) and the Basque Government (Project number PI2010-7 and nanoIKER Etortek-2011).

Author contributions

All authors were involved in designing the research, performing the research and writing the manuscript.

Additional information

Intracellular temperature mapping with a fluorescent polymeric thermometer and fluorescence lifetime imaging microscopy

Kohki Okabe[1], Noriko Inada[2], Chie Gota[1], Yoshie Harada[3], Takashi Funatsu[1] & Seiichi Uchiyama[1]

Cellular functions are fundamentally regulated by intracellular temperature, which influences biochemical reactions inside a cell. Despite the important contributions to biological and medical applications that it would offer, intracellular temperature mapping has not been achieved. Here we demonstrate the first intracellular temperature mapping based on a fluorescent polymeric thermometer and fluorescence lifetime imaging microscopy. The spatial and temperature resolutions of our thermometry were at the diffraction limited level (200 nm) and 0.18–0.58 °C. The intracellular temperature distribution we observed indicated that the nucleus and centrosome of a COS7 cell, both showed a significantly higher temperature than the cytoplasm and that the temperature gap between the nucleus and the cytoplasm differed depending on the cell cycle. The heat production from mitochondria was also observed as a proximal local temperature increase. These results showed that our new intracellular thermometry could determine an intrinsic relationship between the temperature and organelle function.

[1] Graduate School of Pharmaceutical Sciences, The University of Tokyo, 7-3-1 Hongo Bunkyo-ku, Tokyo 113-0033, Japan. [2] The Graduate School of Biological Sciences, Nara Institute of Science and Technology, 8916-5 Takayama-Cho Ikoma-shi, Nara 630-0101, Japan. [3] Institute for Integrated Cell-Material Sciences (WPI-iCeMS), Kyoto University, Yoshida-Honmachi Sakyo-ku, Kyoto 606-8501, Japan. Correspondence and requests for materials should be addressed to S.U. (email: seiichi@mol.f.u-tokyo.ac.jp).

Temperature is a fundamental physical quantity that governs every biological reaction within living cells[1–4]. The biological reactions responsible for cellular functions occur either exothermically or endothermically at particular locations within a cell, such as inside organelles. Thus, temperature distributions inside a living cell reflect the thermodynamics and functions of cellular components[5]. In medical studies, the cellular pathogenesis of diseases (for example, cancer) is characterized by extraordinary heat production[6]. Therefore, intracellular temperature mapping of living cells should promote better understanding of cellular events and the establishment of novel diagnoses and therapies[7].

In spite of the considerable demand and expectation in various fields of life science, intracellular temperature distributions within living cells have not yet been observed, because no thermometry was capable of intracellular temperature mapping. First of all, conventional thermometers (that is, thermography and thermocouples) have low spatial resolution (up to $10\,\mu m$) and are unable to function within cells. In contrast to conventional thermometers, fluorescent molecular thermometers[8–12] are promising tools for intracellular thermometry, as they function at the molecular level. Nevertheless, when considered as a tool to observe intracellular temperature distribution, the fluorescent molecular thermometers should simultaneously satisfy multiple requirements, that is, high temperature resolution, high spatial resolution, functional independency of environmental changes in pH and ionic strength, as well as functional independency of surrounding biomacromolecules, and concentration-independent output. The unavailability of these features has been an obstacle to intracellular temperature mapping. For example, we created a fluorescent nanogel thermometer (FNT) by combining a thermo-responsive polymer with a water-sensitive fluorophore in addition to intensive developments of the fluorescent molecular thermometer over the past few years[13–21]. As we reported in a previous paper, FNT functioned even in living cells and showed high sensitivity to changes in temperature, thus enabling intracellular thermometry with high temperature resolution (better than $0.5\,^{\circ}C$ in the range between 27 and $33\,^{\circ}C$)[22]. However, FNT could measure only the average temperature of the whole cell. It could not reveal any intracellular temperature distribution, because the relatively large size (62 nm or more in hydrodynamic diameter) and low hydrophilicity (aggregation was induced at temperatures higher than $27\,^{\circ}C$) of FNT hindered the dispersion of this temperature probe throughout the cell.

In this study, we developed a novel fluorescent polymeric thermometer (FPT) that could diffuse throughout the cell and applied it to intracellular temperature mapping, where the fluorescence lifetime of FPT was adopted as a temperature-dependent variable. In intracellular temperature mapping, the fluorescence lifetime permits an accurate measurement of temperature, because it is independent of fluctuations under experimental conditions, such as the concentration of FPT at a given location within a cell or the strength of the excitation source. Specifically, time-correlated single photon counting (TCSPC) system-based fluorescence lifetime imaging microcopy (FLIM)[23] was used in our intracellular thermometry to measure the temperature-dependent fluorescence lifetime of FPT with the highest accuracy[24]. Our new thermometric methodology using FPT and TCSPC-FLIM had a high spatial and temperature resolution. This method was sufficient to perform temperature mapping in living cells in which novel biological insights, such as intracellular temperature gradients and organelle-specific thermogenesis, were unequivocally revealed.

Results

Fluorescent polymeric thermometer.
To create a fluorescent molecular thermometer that would diffuse throughout a living cell, we prepared an upgraded FPT (Fig. 1a, $M_n = 19,300$, $M_w/M_n = 2.1$) of a reduced size (8.9 nm in hydrodynamic diameter in

the globular state, Supplementary Fig. S1) and with sufficiently hydrophilic residues. FPT works using an established temperature-sensing mechanism (Fig. 1b)[25–27]. At low temperatures, a thermo-responsive poly-N-n-propylacrylamide (NNPAM) sequence in FPT assumes an extended structure with hydration of amide linkages, in which a water-sensitive N-{2-[(7-N,N-dimethylaminosulfonyl)-2,1,3-benzoxadiazol-4-yl](methyl)amino}ethyl-N-methylacrylamide (DBD-AA) unit can be quenched by neighbouring water molecules. At higher temperatures, which weaken hydration, the polyNNPAM sequence shrinks because of the hydrophobic interaction among the NNPAM units, resulting in the release of water molecules and strong fluorescence from the DBD-AA unit. Additionally, an ionic potassium 3-sulfopropyl acrylate (SPA) unit enriches the hydrophilicity of FPT to prevent interpolymeric aggregation within a cell.

The fluorescence properties of FPT were examined using a COS7 cell extract. Complementary experiments indicated that the COS7 cell extract had an environment that was nearly identical to the interior of living COS7 cells (Supplementary Fig. S2). As shown in Fig. 1c, d, FPT showed stronger fluorescence in a COS7 cell extract in accordance with temperature increases. This fluorescence response was independent of ionic strength (in the range of 0.1–0.2) and environmental pH (6–10; Supplementary Fig. S3) at values that include the ranges of physiological variations in the cytosol[28,29].

A control copolymer without thermosensitivity (Fig. 1e, f, $M_n = 15,100$, $M_w/M_n = 2.6$) was also prepared. The only difference in chemical structure between FPT and the control copolymer is the methylene chain length of the major unit (that is, 3 and 2, respectively). The diminished hydrophobicity of the control copolymer renders itself extended even at high temperatures in an aqueous solution. Aside from the response to temperature, its structural similarity to FPT ensured comparable behaviour in living cells.

To examine the intracellular distribution of FPT, it was microinjected into the cytoplasm of COS7 cells, and the resulting fluorescence images were recorded using a confocal fluorescence microscope. The obtained fluorescence images of the COS7 cells (Fig. 1g) showed that regardless of temperature, FPT could diffuse throughout the cells (both within the nucleus and in the cytoplasm). This distribution of FPT throughout the entire cell suggested the feasibility of intracellular mapping with high spatial resolution.

Fluorescence lifetime as a temperature-dependent parameter. As shown in Fig. 1g, we found that the fluorescence intensity of FPT was higher in the nucleus than in the surrounding cytoplasm; this difference in intensity was significant when the temperature of the medium was $36\,^{\circ}C$. The observed difference in intensity between the nucleus and cytoplasm could reflect the distribution of intracellular temperatures, but a difference in probe concentration between the two compartments would also considerably influence the fluorescence intensity. Therefore, the fluorescence lifetime (an index of the relaxation time from the excited state), which is independent of concentration[30], was adopted as a temperature-dependent variable of FPT for the imaging of intracellular temperature. The fluorescence lifetime measurements of FPT in a COS7 cell extract indicated that the fluorescence decay curves consisted of two exponential components (Fig. 2a). From the previous photophysical study on a copolymer labelled with DBD-AA units[30], it can be affirmed that fast and slow fluorescence decays of FPT originated from the hydrogen-bonded and the hydrogen bond-free DBD-AA units, respectively. Thus, the fluorescence lifetime of FPT was calculated with the measured coefficients of the double-exponential function (see Methods). In a COS7 cell extract, the fluorescence lifetime of FPT was prolonged as the solution temperature increased (Fig. 2b, closed circle), and was shortened as the solution temperature decreased (Supplementary Fig. S4). Moreover, this fluorescence lifetime response of FPT was not sensitive to either the variation in probe concentration

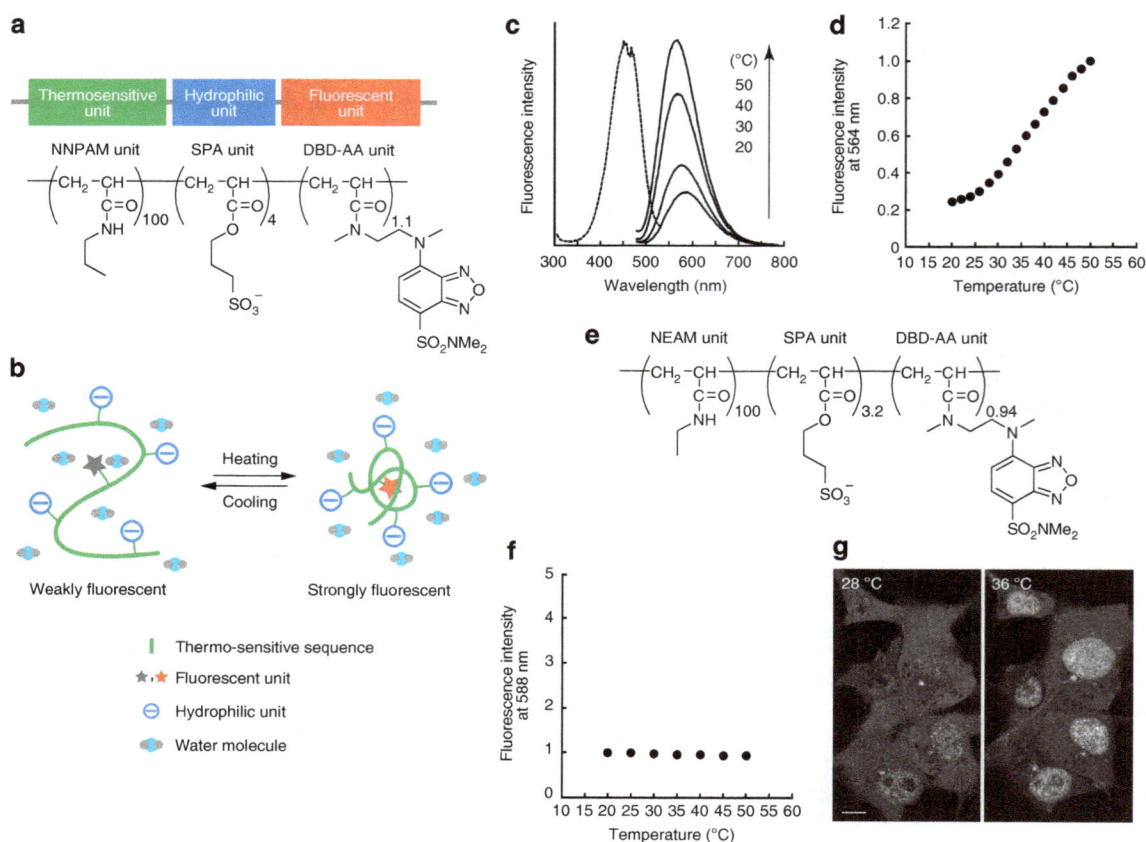

Figure 1 | FPT for intracellular temperature mapping. (**a**) Chemical structure. The original name of each unit is described in the main text. (**b**) Functional diagram in an aqueous medium. (**c**) Fluorescence spectra in a COS7 cell extract. An excitation spectrum (broken, at 30 °C) was obtained from emissions at 565 nm and normalized. Emission spectra (solid) were obtained with an excitation at 456 nm. (**d**) Fluorescence intensity response to the temperature variation in a COS7 cell extract. The fluorescence quantum yield was 0.25 at 50 °C. (**e**) Chemical structure of the control copolymer. (**f**) Relationship between the fluorescence intensity of the control copolymer in a COS7 cell extract (excited at 456 nm) and the temperature. (**g**) Confocal fluorescence images of FPT in living COS7 cells. Scale bar represents 10 μm.

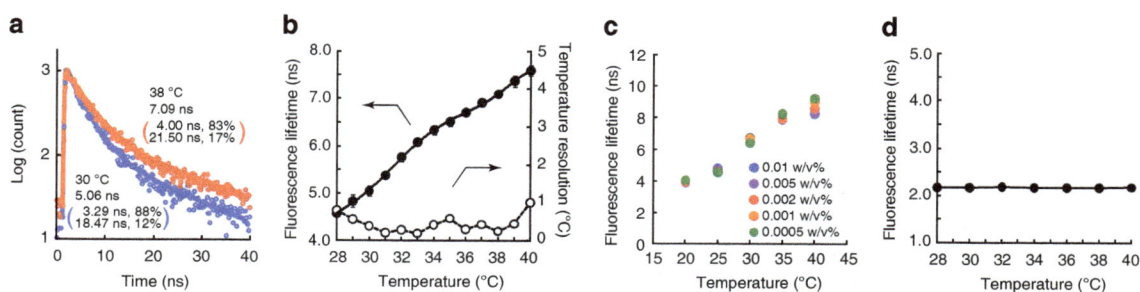

Figure 2 | Temperature-dependent fluorescence lifetime of FPT. (**a**) Representative fluorescence decay curves of FPT in a COS7 cell extract. The fluorescence lifetime and its two components are also indicated. (**b**) Calibration curve (closed, left axis) and temperature resolution (open, right axis) of FPT. (**c**) Concentration independence of FPT in COS7 cell extracts. (**d**) Relationship between the fluorescence lifetime of the control copolymer in a COS7 cell extract and temperature.

(Fig. 2c), the surrounding proteins (Supplementary Fig. S5) or environmental viscosity (Supplementary Fig. S6). The relationship between the fluorescence lifetime of FPT in COS7 cell extracts and temperature (Fig. 2b, closed circle) was used as a calibration curve for temperature imaging. The temperature resolution in the thermometry of the COS7 cells was evaluated to be 0.18–0.58 °C within the range of 29–39 °C (Fig. 2b, open circle). In contrast to FPT, the fluorescence lifetime of the control copolymer showed a complete lack of sensitivity to temperature variation (Fig. 2d).

In performing TCSPC-FLIM on living COS7 cells, the fluorescence decay curve of FPT at each pixel was obtained with high repeatability unless the experimental conditions (for example, sample and temperature) were changed, and was a double-exponential function at every pixel within the cells. Then, the fluorescence lifetime at each pixel was calculated from the corresponding double-exponential function (Supplementary Fig. S7), which allowed the construction of fluorescence lifetime images. The fluorescence lifetime of each pixel within the COS7 cells was prolonged with increased

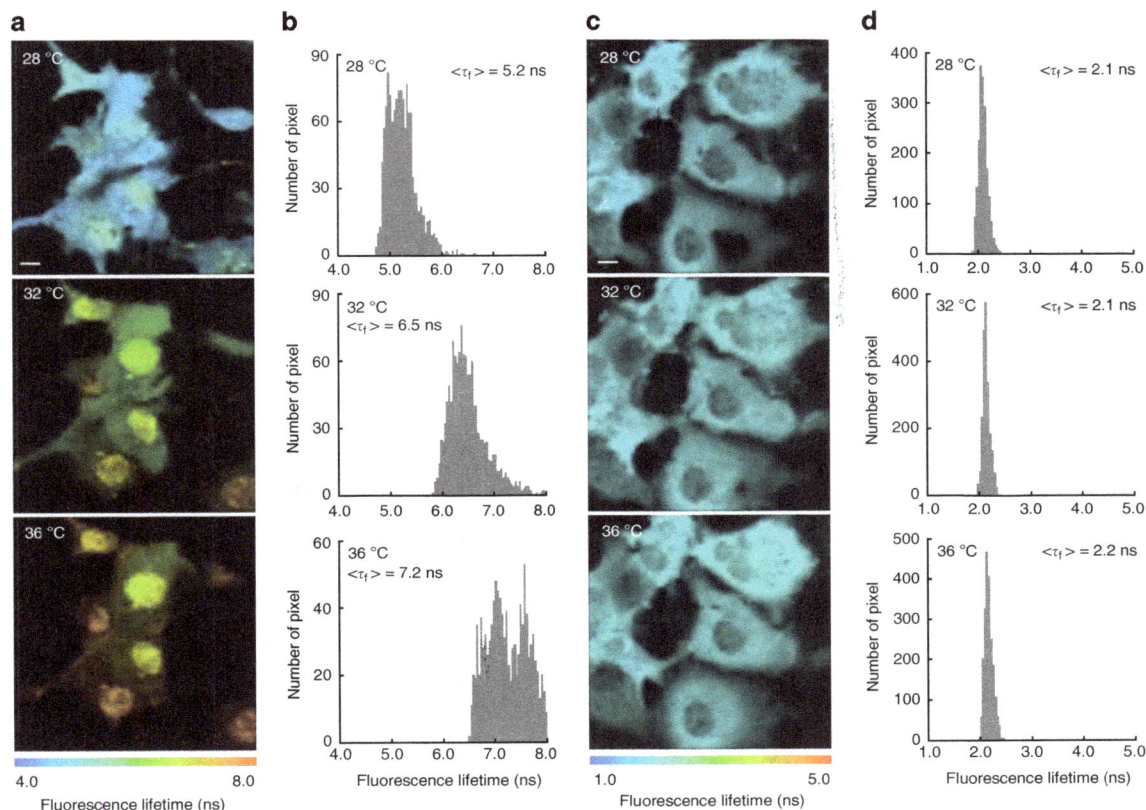

Figure 3 | Temperature imaging of living COS7 cells by TCSPC-FLIM with FPT. (**a, b**) Fluorescence lifetime images of FPT (**a**) and histograms of fluorescence lifetime in whole living cells (**b**). (**c,d**) Fluorescence lifetime images of the control copolymer (**c**) and histograms of fluorescence lifetime in whole living cells (**d**). The temperatures indicated in the images were of the culture medium. $<\tau_f>$ represents an average of the histogram. Scale bar represents 10 μm.

culture medium temperature (Fig. 3a, b). It should be noted that we have confirmed that the fluorescence signal in fluorescence lifetime images was derived from FPT, not from auto-fluorescence of living cells (Supplementary Fig. S8). Moreover, the temperature-dependent change in fluorescence lifetime was not observed when the control copolymer was used (Fig. 3c, d). The fluorescence lifetime images of living COS7 cells with FPT (Fig. 3a) had larger deviations in fluorescence lifetime (5.3–5.7%) than those of a COS7 cell extract containing FPT (Supplementary Fig. S9) (1.9–2.1%), indicating that intracellular temperature distribution takes place within the living COS7 cells. Related fluorescence lifetime images of a dividing COS7 cell were also successfully obtained with FPT (and the control copolymer; Supplementary Fig. S10). From these results, it can be concluded that the fluorescence lifetime is a suitable variable in intracellular temperature mapping using the novel sensor FPT.

Imaging of the intracellular temperature distributions. Next, the temperature distributions in living COS7 cells were investigated with a particular focus on organelles. It is noteworthy that the energy of glucose metabolism in cells is equivalent to a temperature increase of at least 2 °C within an entire cell, because of the following reasons: the total free energy released by the oxidation of glucose (glucose $+6O_2 \rightarrow 6CO_2 + 6H_2O$) is 2870 kJ mol^{-1}; the specific heat capacity of the cell can be estimated to be 4.184 J (gK)$^{-1}$ (similar to water); the intracellular glucose concentration is kept to at least 3–5 mM, when the culture medium includes 25 mM glucose (that is, in the present condition)[31]. This encouraged us to utilize FPT (with a temperature resolution of 0.18–0.58 °C) for intracellular temperature imaging. We set the temperature of the culture medium to ca. 30 °C, the temperature at which a specimen can readily and

consistently be maintained due to only a small difference from the ambient temperature (ca. 24–27 °C). Biological activities such as cell proliferation were confirmed at this temperature. Culture medium of higher temperature (for example, 37 °C) resulted in a larger error of fluorescence lifetime of FPT within the imaging field, most likely due to the considerable temperature gap between the optical set-up in the microscope and the sample, which unignorably reduced the accuracy of intracellular thermometry (cf. maximum errors in determining temperature within the imaging field at 30 and 37 °C are 0.35 and 1.3 °C, respectively). By performing TCSPC-FLIM on living COS7 cells containing FPT (Fig. 4a), we attributed the heterogeneous fluorescence intensities of COS7 cells (as discussed in the previous section) to the temperature differences inside the cell.

The most notable feature of the resultant temperature distributions is the difference in temperature between the nucleus and the cytoplasm, as observed in numerous cells. In a representative cell, the temperature of the nucleus, evaluated from the fluorescence lifetime at each pixel in the image, was significantly higher than that of the cytoplasm (Fig. 4b). Analysis of many cell samples ($n = 62$) revealed that the average temperature difference between the nucleus and the cytoplasm was 0.96 °C (Fig. 4c). This thermal pool in the nucleus may originate from its activities, such as DNA replication, transcription and RNA processing, as well as its structural separation by the nuclear membrane[32]. Strikingly, we found that this temperature gap between the nucleus and cytoplasm was dependent on the cell cycle: the nucleus was warmer (by 0.70 °C on average) in cells synchronized to the G1 phase (Fig. 4d), whereas there was little temperature gap in S/G2-phased cells (Fig. 4e). It is likely that this temperature homogeneity at the S/G2 phase was the result of a temperature increase in the cytoplasm rather than a

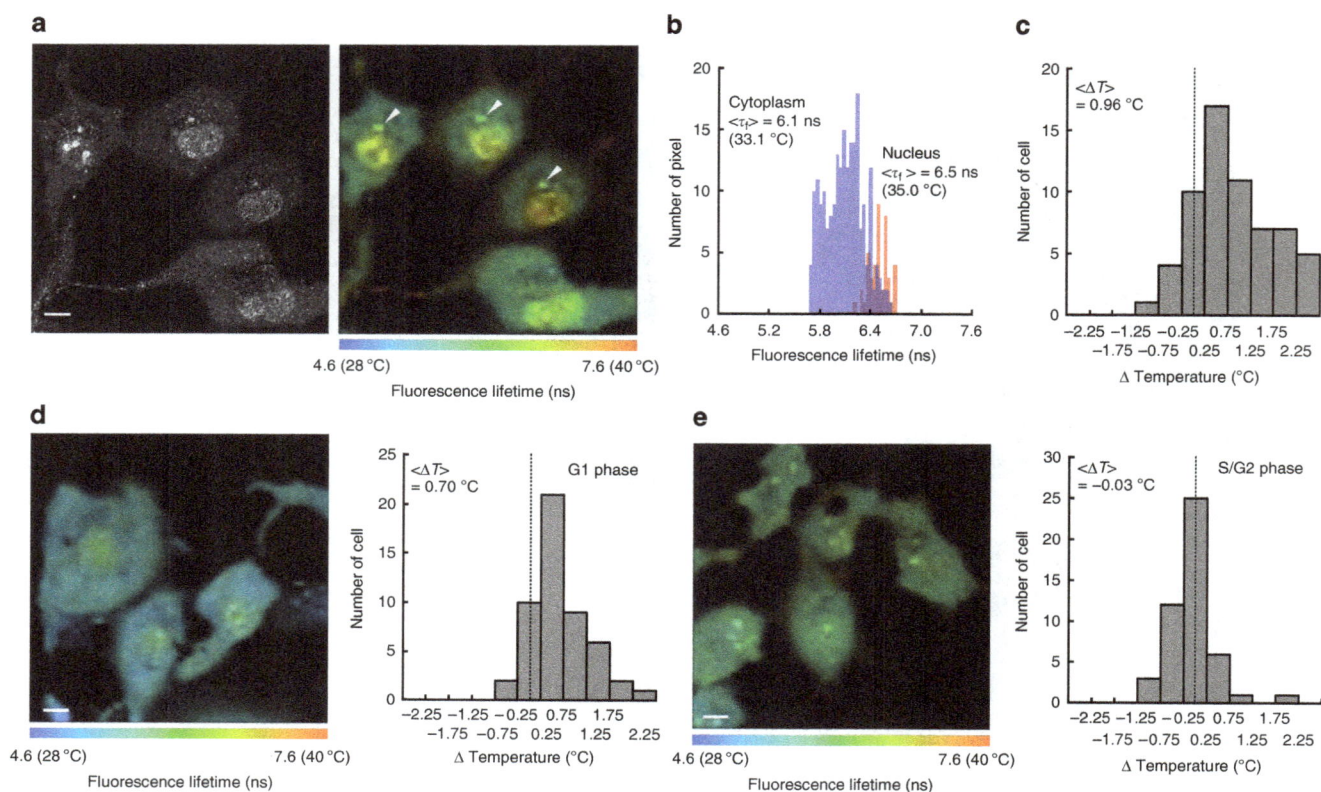

Figure 4 | Temperature mapping in living COS7 cells. (**a**) Confocal fluorescence image (left) and fluorescence lifetime image (right) of FPT. (**b**) Higher temperature in the nucleus. Histograms of the fluorescence lifetime in the nucleus and in the cytoplasm in a representative cell (the leftmost cell in **a**). The s.e. in determining a temperature was 0.38 °C. (**c**) Histogram of the temperature differences between the nucleus and the cytoplasm ($n = 62$). ΔTemperature was calculated by subtracting the average temperature of the cytoplasm from that of the nucleus. $<\Delta T>$ represents an average of the histogram. (**d,e**) Cell-cycle-dependent thermogenesis in the nucleus. Representative fluorescence lifetime images and histograms of ΔTemperature in living cells synchronized to G1 phase ($n = 51$) (**d**) and S/G2 phase ($n = 48$) (**e**). The potential maximum error in determining ΔTemperature value was 0.35 °C. In (**a–e**), the temperature of the medium was maintained at 30 °C. Scale bar represents 10 μm.

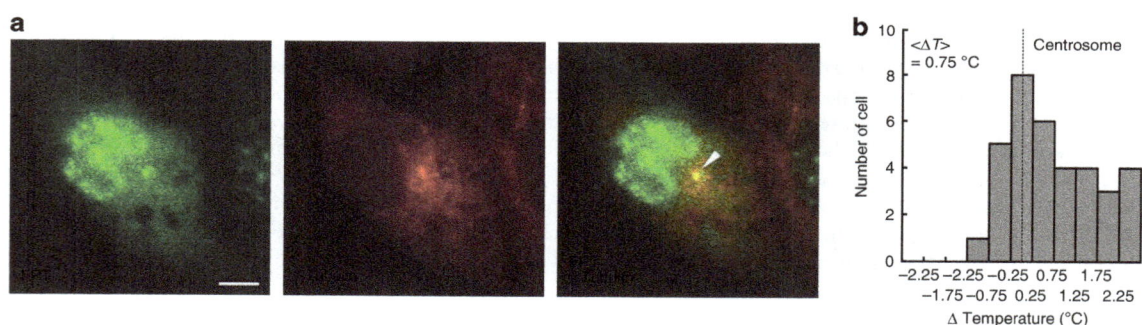

Figure 5 | Heat production from the centrosome in living COS7 cells. (**a**) Heat production near the centrosome (arrowhead). Epifluorescence images of FPT (left), the centrosome visualized with γ-tubulin antibody (middle) and their merged image (right). (**b**) Histogram of temperature differences between the centrosome and the cytoplasm ($n = 35$). ΔTemperature was calculated by subtracting the average temperature of the cytoplasm from that of the centrosome with the potential maximum error of 0.35 °C. The temperature of the medium was maintained at 30 °C. Scale bar represents 10 μm.

temperature decrease in the nucleus. Consistent with our results, this cell-cycle-dependent heat production was previously observed from a mass of cells in a study using a microcalorimeter[33].

Another remarkable feature of the temperature imaging (Fig. 4a) was a single spot in the perinuclear region (arrowheads). This spot, observed in 56% of the cells imaged, was identified as a centrosome by co-localization with γ-tubulin (Fig. 5a). Temperature imaging with FPT by TCSPC-FLIM revealed that the centrosome is also significantly warmer (by 0.75 °C on average) than the cytoplasm, with considerable cell-to-cell variation (Fig. 5b). This centrosome-specific thermogenesis might be associated with its diverse functions, such

as mitosis and the organization of microtubules[34]. The hydrolysis of tubulin GTP, ATP-driven motion of motor proteins (for example, dynein and kinesin), and phosphorylation/dephosphorylation of centrosomal proteins by kinase/phosphatase are all possible heat source at the centrosome[35].

Imaging of local thermogenesis at mitochondria. Finally, we studied the local heat production by mitochondria. Mitochondria release surplus energy in the form of heat through respiration. To image this local thermogenesis, the temperature distribution and locations of the mitochondria were co-visualized by FPT and

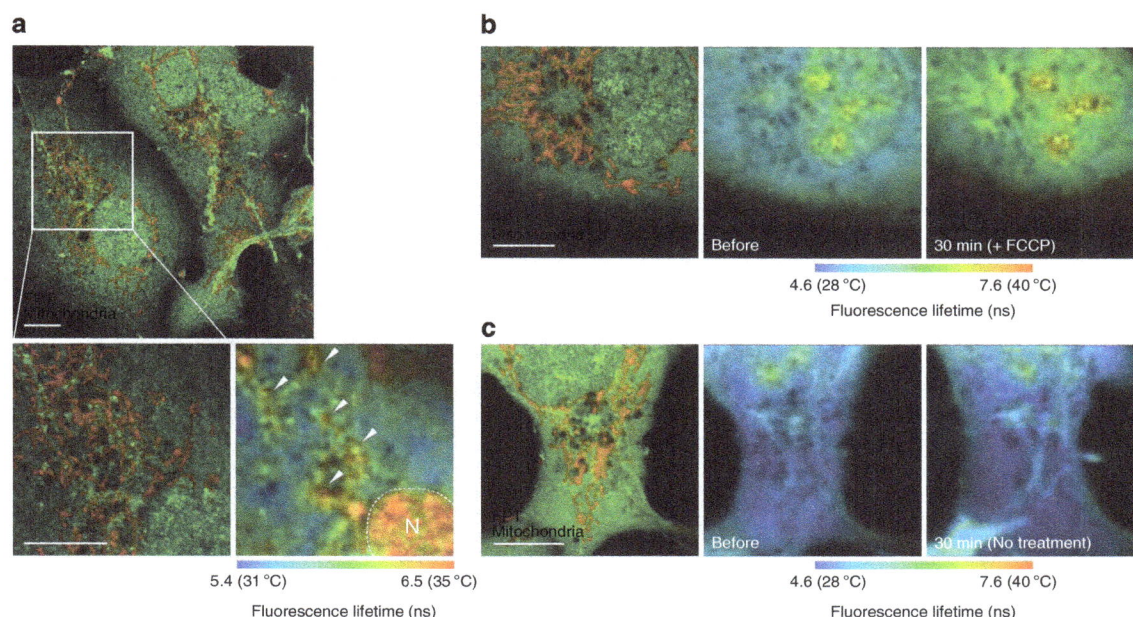

Figure 6 | Local heat production near the mitochondria in living COS7 cells. (**a**) Confocal fluorescence images of FPT (green) and MitoTracker Deep Red FM (red; upper and left lower) and fluorescence lifetime image of FPT (right lower). The region of interest, as shown in the square in the upper panel, is enlarged in the lower panels. Arrowheads point to local heat production. In the right lower panel, N indicates the nucleus. (**b**) Temperatures increase near the mitochondria after the inhibition of ATP synthesis by the uncoupler FCCP. Confocal fluorescence images of FPT (green) and MitoTracker Deep Red FM (red; left) and fluorescence lifetime images of FPT (middle and right). (**c**) Control experiment for (**b**) No significant change in the fluorescence lifetime of FPT was detected without a chemical stimulus. In (**a–c**), the temperature of the medium was maintained at 30 °C. Scale bar represents 10 μm.

a mitochondria indicator, respectively (Fig. 6a). The enlarged fluorescence lifetime image clearly showed localized thermogenesis near the mitochondria (Fig. 6a, arrowheads). It is notable that this mitochondria-mediated heat production was supported by the intracellular distribution of our previous nanogel thermometer FNT (Supplementary Figs S11 and S12). As described in our previous paper[22], FNT inside living COS7 cells moderately aggregates at culture medium temperatures greater than 27 °C (Supplementary Fig. S11a). Considering that the aggregation of FNT is triggered by heat, it probably began at the warmest place within the cytosol. To determine the location of aggregated FNT, representative organelles (that is, mitochondria, lysosomes, endoplasmic reticulum (ER) and the Golgi apparatus) were co-visualized. As a result, aggregated FNT was found to reside near mitochondria (Supplementary Fig. S11b, c), but not near other organelles, including lysosomes, ER and the Golgi apparatus (Supplementary Fig. S12). Thus, it could be concluded that the local temperature near the mitochondria is higher than the temperature of the rest of the space in the cytosol (aside from the centrosome).

Furthermore, this local heat release from the mitochondria is accelerated when ATP synthesis is stalled by an uncoupling reagent[36]. Thus, a temperature increase due to local heat production from the mitochondria following the addition of an uncoupling reagent, 4-(trifluoromethoxy)phenylhydrazone (FCCP)[37], was imaged in FPT-filled COS7 cells (Fig. 6b, c). As evaluated from fluorescence lifetime images of whole COS7 cells, this heat production by FCCP resulted in an average temperature increase of 1.02 ± 0.17 °C (avg.± s.d., $n = 7$) inside COS7 cells 30 min after the chemical stimuli.

Temperature mapping in a different cell line. With the method established in this study, intracellular temperature imaging was also performed in living HeLa cells. The functionality of FPT was confirmed in a HeLa cell extract (Supplementary Fig. S13) and in living HeLa cells (Supplementary Fig. S14). Intracellular temperature imaging in HeLa cells indicated the higher temperature of the

nucleus (Supplementary Fig. S15), the centrosome (Supplementary Fig. S16) and mitochondria (Supplementary Fig. S17), as seen in the imaging of COS7 cells. The results demonstrating the temperature gradient in both living HeLa and COS7 cells display the universal characteristics of our findings among mammalian cells.

Discussion

In this study, we have established a method for imaging the temperature distribution inside living cells by performing TCSPC-FLIM of our novel FPT. By exploiting the high spatial resolution of FPT, intracellular temperature imaging revealed the thermal profiles of living cells, that is, the higher temperatures of the nucleus, centrosome and the areas near the mitochondria.

Recently, numerous temperature sensors have been developed to meet the urgent demand for measuring temperature in a tiny intracellular space. Among these emerging thermometers, which are based on lanthanide nanoparticles[13,16,17,19,20], dye-coated nanoparticles[14], quantum dots[15], semiconducting polymer dots[18] and temperature-responsive polymers[21,22], FPT in this study is the only thermometer that allows the mapping of intracellular temperatures due to its ability to diffuse throughout living cells and its sensitivity to temperature. In terms of sensitivity, the intracellular temperature variations unveiled in this study indicated that fluorescent molecular thermometers for intracellular temperature mapping must have the ability to distinguish a temperature difference of less than 1 °C. Related to this requirement, an appropriate evaluation of temperature resolution should be carefully considered for reliable intracellular thermometry (cf. Fig. 2b and Methods). It should also be noted that a fluorescence response of FPT to temperature change was observed in the range of 20–50 °C (Fig. 1c, d), which was much wider than that of our previous thermometer, FNT (27–33 °C). This is a bonus property of FPT. Consequently, the intracellular temperature of a wider range of subjects can be measured using our present thermometry.

In addition to our novel FPT, the TCSPC-FLIM technique also contributed to the realization of intracellular temperature mapping with high spatial and temperature resolution. In measuring local intracellular temperature that differed among sub-cellular locations, the fluorescence intensity of a fluorescent molecular thermometer was not an appropriate variable for calculating temperature, because it also strongly correlates with the concentration of the fluorescent molecular thermometer. As temperature-dependent but concentration-independent variables, fluorescence lifetime[38,39], maximum emission wavelength[15] and a ratio of fluorescence intensities at two different wavelengths[10,13,17,18] are all useful. In our methodology, the fluorescence lifetime was chosen because of its high sensitivity and accuracy when the robust TCSPC-FLIM technique was utilized[24,40]. Another advantage of the fluorescence lifetime in bioimaging is multidimensionality[41]. As for FPT, the number of components in the fluorescence lifetime represents the number of states that the fluorophore assumes. TCSPC-FLIM showed that the fluorescence lifetime of FPT consisted of two components in both a cell extract and a living cell (Fig. 2a and Supplementary Fig. S7), indicating that the DBD-AA units in FPT assumed two defined states (that is, hydrogen-bonded and hydrogen bond-free states)[30] in both aqueous media, and that the interaction between these media and FPT was analogous. Thus, the calibration curve (Fig. 2b) obtained using the cell extract was effective for intracellular temperature mapping in living cells. In contrast, single-dimensional variables (that is, maximum emission wavelength and fluorescence intensity ratio) would not allow us to compare the interactions between different environments (that is, in vitro and in vivo in our study) and the fluorophore; hence, the accuracy of thermometry would be compromised.

The fluorescence lifetime of FPT in living cells is potentially influenced by changes in environmental viscosity. Here, we indicate that the heterogeneous fluorescence lifetime of FPT in the fluorescence lifetime images of living cells (for example, Figs 3, 4 and 6) did not originate from variations in viscosity. The fluorescence image of FPT dispersed in living COS7 cells (Fig. 1g) and the recovery of fluorescence signal in the FRAP (fluorescence recovery after photobleaching) experiments using FPT and living COS7 cells (Supplementary Fig. S18) showed that FPT microinjected into cells could freely move in the cytoplasm and the nucleus without any localization in organelles. Therefore, the variation in viscosity, which may have an effect on FPT in living cells, can be estimated in the range from 0.9 to 1.1 cP (in the cytoplasm) to 1.4 cP (in the nucleus)[42–44]. The fluorescence lifetime measurements of FPT in Ficoll solutions with different concentration (that is, viscosity; Supplementary Fig. S6) indicated that the fluorescence lifetime of FPT was almost identical in all concentrations. Because of this functional independency of FPT from the variations in environmental viscosity, the accuracy of our intracellular thermometry was assured.

The intracellular temperature map we drew in this study demonstrates the existence of a temperature gradient inside a living cell. Furthermore, this inhomogeneous temperature distribution is intrinsically related to fundamental cellular processes, such as the cell cycle and the stimulation of the mitochondria. Our results are in a good accordance with the study by Zohar et al.[11], in which temperature imaging of cell surfaces was performed by locating a fluorescent molecular thermometer on the cell membrane, in that both results indicate the spontaneous and significant thermogenesis upon the stimulation of steady-state live cells. In comparison, our methodology of imaging intracellular temperature is of distinct superiority with respect to the sensitivity, the spatial resolution, functional independency and ability to function inside living cells. These findings have a significant impact on the comprehension of cell function in different aspects of molecular and cellular structure, and provide insights into the regulatory mechanisms of intracellular signalling. Therefore, in future, intracellular temperature imaging, relating specifically to various cell species (that is, animal and plant cells in both normal and pathological conditions) and cellular

activities (that is, cell division, differentiation, or stress responses) will assist in the study of intracellular thermodynamics and thereby aid future discoveries of the profound mechanisms of life.

Methods

Preparation of FPT and the control copolymer. NNPAM[45](2.4375 mmol), SPA(62.5 μmol, Aldrich), DBD-AA(25 μmol)[26] and α,α′-azobisisobutyronitrile (25 μmol) were dissolved in N,N-dimethylformamide (5 ml), and the solution was bubbled with dry nitrogen for 30 min to remove dissolved oxygen. The solution was heated to 60 °C for 12 h and then cooled to room temperature. The reaction mixture was then poured into diethyl ether (200 ml). The obtained FPT was purified by reprecipitation using methanol (3 ml)-diethyl ether (200 ml) and dialysis (yield: 35%). For the preparation of the control copolymer, N-ethylacrylamide (NEAM, 2.4375 mmol, Monomer-Polymer & Dajac Labs) was used instead of NNPAM (yield: 16%).

The contents of the NNPAM, NEAM and SPA units in the copolymers were determined from [1]H-nuclear magnetic resonance spectra (Bruker AVANCE400). The proportions of the DBD-AA units in the copolymers were determined from their absorbance in methanol compared with the model fluorophore N,2-dimethyl-N-(2-{methyl[7-(dimethylsulfamoyl)-2,1,3-benzoxadiazol-4-yl]amino}ethyl) propanamide (DBD-IA)[30] ($\varepsilon = 10{,}800\,M^{-1}\,cm^{-1}$ at 444 nm). The molecular weights of the copolymers were determined using gel permeation chromatography. A calibration curve was obtained using a polystyrene standard, and 1-methyl-2-pyrrolidinone containing LiBr (5 mM) was used as the eluent. The hydrodynamic diameter of FPT (0.001 w/v% in water) was estimated by a dynamic light-scattering measurement at 40 °C with a Zetasizer Nano ZS (Malvern).

Fluorescence lifetime measurements in cell extracts. The fluorescence lifetime of FPT (or the control copolymer) was measured using a FluoroCube 3000U (Horiba Jobin Yvon) (Fig. 2c, Supplementary Figs S4–S6) or a TCSPC system SPC830 (ver. 3.0, Becker & Hickl) with a TCS SP5 confocal laser-scanning microscope (Leica; Fig. 2a,b,d, Supplementary Figs S9 and S13).

For the measurements with the FluoroCube, the sample was excited with a pulsed diode laser (NanoLED-405L, Horiba, 405 nm) at a repetition rate of 1 MHz, and the emission from 554 to 566 nm was collected. The temperature of the sample was controlled by an ETC-273T controller (JASCO). The obtained fluorescence decay curve was fitted by a double exponential function:

$$I(t) = A_1 \exp(-t/\tau_1) + A_2 \exp(-t/\tau_2) \tag{1}$$

Then, the fluorescence lifetime (τ_f) was calculated using the following equation:

$$\tau_f = (A_1\tau_1^2 + A_2\tau_2^2)/(A_1\tau_1 + A_2\tau_2) \tag{2}$$

For the measurements with the SPC830, 200 μl of the sample (0.02 w/v%) was excited in a 35-mm glass base dish (No. 3911–035, ASAHI Techno GLASS) with a pulsed diode laser (LDH-P-C-405B, PicoQuant, 405 nm) at a repetition rate of 20 MHz, and the emission from 500 to 700 nm was collected through an HCX PL APO Ibd.BL 63×1.4 NA oil objective (Leica Microsystems) in a 64×64 pixel format. The sample temperature was controlled using a GSI-H1R stage with an INUB-F1 controller (Tokai Hit), and monitored with a thermocouple (TSU-0125 thermometer with a TSU-7225 probe, Tokai Hit). The fluorescence lifetime data were collected for 60 s and analysed using the SPCImage software (Becker & Hickl) with the binning procedure (factor: 5 or 6). The fluorescence lifetime for each pixel was calculated using equation (2) and averaged within the whole image.

The calibration curve for the temperature imaging of COS7 cells with FPT was obtained by approximating the relationship between the averaged fluorescence lifetime of FPT in a COS7 cell extract (in triplicate) and the temperature to the sixth-degree polynomial (correlation coefficient $r = 0.998$)

$$\begin{aligned} \tau_f(T) = {} & -1.868 \times 10^{-5} T^6 + 3.767 \times 10^{-3} T^5 \\ & -3.152 \times 10^{-1} T^4 + 1.400 \times 10^{1} T^3 \\ & -3.482 \times 10^{2} T^2 + 4.596 \times 10^{3} T - 2.516 \times 10^{4} \end{aligned} \tag{3}$$

where T and $\tau_f(T)$ represent the temperature (°C) and the fluorescence lifetime (ns) at T°C, respectively.

The temperature resolution (δT) of the fluorescent thermometer was evaluated by the following equation:

$$\delta T = \left(\frac{\partial T}{\partial \tau_f}\right)\delta\tau_f \tag{4}$$

where $\partial T/\partial\tau_f$ and $\delta\tau_f$ represent the inverse of the slope in the fluorescence lifetime temperature diagram and the s.d. of the averaged fluorescent lifetime, respectively.

Preparation of cell extract. The cell pellets (0.5 ml) were collected and resuspended in hypertonic buffer (2.5 ml, containing 0.42 M KCl, 50 mM HEPES (4-(2-hydroxyethyl)-1-piperazineethanesulfonic acid)-KOH, 5 mM $MgCl_2$, 0.1 mM EDTA, 20% glycerol, pH 7.8), followed by lysis using a 25-G needle with a syringe and centrifugation (11,000 r.p.m., 15 min, 4 °C). The supernatant was diluted by 40% with water to adjust its KCl concentration to 0.15 M. The adequacy of using the cell extract was assessed by comparing the fluorescence response of FPT in a COS7 cell extract (obtained by fluorescence spectroscopy) with that in living COS7 cells (obtained by epifluorescence microscopy; Supplementary Fig. S2).

Fluorescence imaging of the cells. Microinjection of FPT, the control copolymer and FNT into living cells was performed with Femtojet (Eppendorf), controlled by a micromanipulator (Eppendorf). The fluorescent thermometers were dissolved in an aqueous solution (1 w/v%) containing 80 mM KCl, 10 mM K_2HPO_4 and 4 mM NaCl. The solution was filtered using an Ultrafree-MC filter (Millipore) and microinjected into the cytoplasm with a glass capillary needle (Femtotips II, Eppendorf). The volume of the injected solution was estimated to be 2 fl.

Epifluorescence imaging was performed on an IX70 inverted microscope (Olympus) equipped with an objective lens (×60, UplanApo NA 1.40, Olympus). The temperature of the culture medium was controlled with a stage and a microscope objective lens heater with a controller (Olympus MI-IBC) and monitored using a thermocouple (TSU-0125 thermometer with a TSU-7225 probe, Tokai Hit). A cooled CCD camera (ORCA-ER, Hamamatsu) was used to acquire cell images. The fluorescence images were taken using a sapphire laser (Model 488–30 CDRH, Coherent), a DM505 dichroic mirror (Olympus) and a BA515–550 emission filter (Olympus). The obtained images were analysed using AQUA-Lite ver. 10 (Hamamatsu).

Mitochondria, lysosomes, ER and the Golgi apparatus were stained with MitoTracker Deep Red FM (Invitrogen), CellLight BacMam 2.0 for lysosome C10597 (Invitrogen), ER-tracker Red (Invitrogen) and BODIPY TR C5 ceramide complexed to BSA (Invitrogen), respectively, before microinjection of FNT. The organelles and FNT were then imaged by epifluorescence microscopy (at 28–37 °C). Mitochondria and the Golgi apparatus were visualized using a He-Ne laser (GLS5360, 633 nm; Showa Optronics), a 660LP dichroic mirror (Chroma Technology) and a 610/75M emission filter (Chroma Technology). Lysosomes and ER were imaged using a solid-state laser (Compass 315M-100, 532 nm; Coherent), a 565LP dichroic mirror (Chroma Technology) and a 610/75M emission filter.

Confocal fluorescence imaging and fluorescence ifetime imaging were performed on a TCS SP5 confocal laser-scanning microscope equipped with a TCSPC module SPC-830 (as described above, except that the emission was collected from 500 to 620 nm in co-visualization of FPT and mitochondria). The fluorescence was captured through an HCX PL APO lbd.BL 63×1.4 NA oil objective (Leica Microsystems) with 1~9 zoom factors in a 1024×1024 pixel format for confocal fluorescence imaging and in a 64×64 pixel format for fluorescence lifetime imaging (for 60 s). The obtained fluorescence decay curve in each pixel was fitted with a double exponential function using SPCImage software (Becker & Hickl) after the binning procedure (factor: 2 or 3). In evaluating the intracellular temperature by FLIM, the cell periphery regions were omitted from the temperature averaging within a cell, because they gave less than the required 1,000 photon count at the peak of the fluorescence decay curve, which is below that required to obtain a reliable fluorescence lifetime[46] (see Supplementary Fig. S19). The s.e. in determining temperature from fluorescence lifetime was evaluated from a deviation of averaged fluorescence lifetime in fluorescence lifetime images of different samples (that is, equal to the temperature resolution). The potential maximum error in determining ΔTemperature in living cells (in Figs 4, 5, Supplementary Figs S15 and S16) was evaluated from a deviation of fluorescence lifetime within a fluorescence lifetime image, which was examined using a cell extract containing FPT.

Mitochondria were labelled with MitoTracker Deep Red FM (Invitrogen), and the fluorescence from 645 to 730 nm was collected using confocal microscopy.

The chemical stimulus was conducted by quickly adding 100 μl of 2 mM FCCP (Sigma) in phenol red-free culture medium to the studied cells in 1.9 ml of phenol red-free culture medium in a glass base dish. The temperature of the culture medium was maintained at 30.0 °C.

Immunofluorescence imaging of γ-tubulin. COS7 cells were fixed with 4% paraformaldehyde in PBS (137 mM NaCl, 2.68 mM KCl, 8.10 mM Na_2HPO_4, 1.47 mM KH_2PO_4, pH 7.7) for 30 min at room temperature, permeabilized with 0.1% Triton X-100 (Wako) in PBS for 30 min at room temperature and blocked with 1% BSA (Sigma-Aldrich) and 0.1% Tween 20 (Pharmacia Biotech) in PBS. Subsequently, the cells were incubated with a mouse monoclonal anti-γ-tubulin antibody (Abcam). The primary antibodies were detected with secondary anti-mouse IgG1 antibodies labelled with Cy5 (Beckman Coulter). After being washed several times with PBS, the cells were imaged with the epifluorescence microscope using GLS5360, a 660LP dichroic mirror and a 700/75 M emission filter (Chroma technology).

References

1. Bahat, A. *et al.* Thermotaxis of mammalian sperm cells: a potential navigation mechanism in the female genital tract. *Nature Med.* **9**, 149–150 (2003).
2. Warner, D. A. & Shine, R. The adaptive significance of temperature-dependent sex determination in a reptile. *Nature* **451**, 566–569 (2008).
3. Seymour, R. S. Biophysics and physiology of temperature regulation in thermogenic flowers. *Biosci. Rep.* **21**, 223–236 (2001).
4. Patel, D. & Franklin, K. A. Temperature-regulation of plant architecture. *Plant Signal. Behav.* **4**, 577–579 (2009).
5. Lowell, B. B. & Spiegelman, B. M. Towards a molecular understanding of adaptive thermogenesis. *Nature* **404**, 652–660 (2000).
6. Monti, M., Brandt, L., Ikomi-Kumm, J. & Olsson, H. Microcalorimetric investigation of cell metabolism in tumour cells from patients with non-Hodgkin lymphoma (NHL). *Scand. J. Haematol.* **36**, 353–357 (1986).
7. McCabe, K. M. & Hernandez, M. Molecular thermometry. *Pediatr. Res.* **67**, 469–475 (2010).
8. Uchiyama, S., de Silva, A. P. & Iwai, K. Luminescent molecular thermometers. *J. Chem. Educ.* **83**, 720–727 (2006).
9. Lee, J. & Kotov, N. A. Thermometer design at the nanoscale. *Nano Today* **2**, 48–51 (2007).
10. Chapman, C. F., Liu, Y., Sonek, G. J. & Tromberg, B. J. The use of exogenous fluorescent probes for temperature measurements in single living cells. *Photochem. Photobiol.* **62**, 416–425 (1995).
11. Zohar, O. *et al.* Thermal imaging of receptor-activated heat production in single cells. *Biophys. J.* **74**, 82–89 (1998).
12. Tseeb, V., Suzuki, M., Oyama, K., Iwai, K. & Ishiwata, S. Highly thermosensitive Ca^{2+} dynamics in a HeLa cell through IP_3 receptors. *HFSP J.* **3**, 117–123 (2009).
13. Vetrone, F. *et al.* Temperature sensing using fluorescent nanothermometers. *ACS Nano* **4**, 3254–3258 (2010).
14. Huang, H., Delikanli, S., Zeng, H., Ferkey, D. M. & Pralle, A. Remote control of ion channels and neurons through magnetic-field heating of nanoparticles. *Nat. Nanotechnol.* **5**, 602–606 (2010).
15. Yang, J.-M., Yang, H. & Lin, L. Thermogenesis detection of single living cells via quantum dots. *IEEE 23rd Int. Conf. Microelectromech. S. (MEMS)* 963–966 (2010).
16. Peng, H. *et al.* Luminescent europium(III) nanoparticles for sensing and imaging of temperature in the physiological range. *Adv. Mater.* **22**, 716–719 (2010).
17. Brites, C. D.S. *et al.* A luminescent molecular thermometer for long-term absolute temperature measurements at the nanoscale. *Adv. Mater.* **22**, 4499–4504 (2010).
18. Ye, F. *et al.* Ratiometric temperature sensing with semiconducting polymer dots. *J. Am. Chem. Soc.* **133**, 8146–8149 (2011).
19. Fischer, L. H., Harms, G. S. & Wolfbeis, O. S. Upconverting nanoparticles for nanoscale thermometry. *Angew. Chem. Int. Ed.* **50**, 4546–4551 (2011).
20. Brites, C. D.S. *et al.* Lanthanide-based luminescent molecular thermometers. *New J. Chem.* **35**, 1177–1183 (2011).
21. Chen, C.-Y. & Chen, C.-T. A PNIPAM-based fluorescent nanothermometer with ratiometric readout. *Chem. Commun.* **47**, 994–996 (2011).
22. Gota, C., Okabe, K., Funatsu, T., Harada, Y. & Uchiyama, S. Hydrophilic fluorescent nanogel thermometer for intracellular thermometry. *J. Am. Chem. Soc.* **131**, 2766–2767 (2009).
23. Berezin, M. Y. & Achilefu, S. Fluorescence lifetime measurements and biological imaging. *Chem. Rev.* **110**, 2641–2684 (2010).
24. Becker, W. *Advanced Time-Correlated Single Photon Counting Techniques* (Springer: Berlin, 2005).
25. Uchiyama, S., Matsumura, Y., de Silva, A. P. & Iwai, K. Fluorescent molecular thermometers based on polymers showing temperature-induced phase transitions and labeled with polarity-responsive benzofurazans. *Anal. Chem.* **75**, 5926–5935 (2003).
26. Gota, C., Uchiyama, S. & Ohwada, T. Accurate fluorescent polymeric thermometers containing an ionic component. *Analyst* **132**, 121–126 (2007).
27. Pietsch, C., Schubert, U. S. & Hoogenboom, R. Aqueous polymeric sensors based on temperature-induced polymer phase transitions and solvatochromic dyes. *Chem. Commun.* **47**, 8750–8765 (2011).
28. Bright, G. R., Fisher, G. W., Rogowska, J. & Taylor, D. L. Fluorescence ratio imaging microscopy: temporal and spatial measurements of cytoplasmic pH. *J. Cell Biol.* **104**, 1019–1033 (1987).
29. Lodish, H. *et al. Molecular Cell Biology* 6th edn, 448–449 (W.H. Freeman: New York, 2007).
30. Gota, C., Uchiyama, S., Yoshihara, T., Tobita, S. & Ohwada, T. Temperature-dependent fluorescence lifetime of a fluorescent polymeric thermometer, poly (N-isopropylacrylamide), labeled by polarity and hydrogen bonding sensitive 4-sulfamoyl-7-aminobenzofurazan. *J. Phys. Chem. B* **112**, 2829–2836 (2008).
31. Foley, J. E., Cushman, S. W. & Salans, L. B. Intracellular glucose concentration in small and large rat adipose cells. *Am. J. Physiol. Endocrinol. Metab.* **238**, E180–E185 (1980).
32. Lamond, A. I. & Earnshaw, W. C. Structure and function in the nucleus. *Science* **280**, 547–553 (1998).
33. Yamamura, M., Hayatsu, H. & Miyamae, T. Heat production as a cell cycle monitoring parameter. *Biochem. Biophys. Res. Commun.* **140**, 414–418 (1986).

34. Doxsey, S. Re-evaluating centrosome function. *Nat. Rev. Mol. Cell Biol.* **2,** 688–698 (2001).

35. Andersen, J. S. *et al.* Proteomic characterization of the human centrosome by protein correlation profiling. *Nature* **426,** 570–574 (2003).

36. Nakamura, T. & Matsuoka, I. Calorimetric studies of heat of respiration of mitochondria. *J. Biochem.* **84,** 39–46 (1978).

37. Heytler, P. G. & Prichard, W. W. A new class of uncoupling agents–carbonyl cyanide phenylhydrazones. *Biochem. Biophys. Res. Commun.* **7,** 272–275 (1962).

38. Benninger, R. K. P. *et al.* Quantitative 3D mapping of fluidic temperatures within microchannel networks using fluorescence lifetime imaging. *Anal. Chem.* **78,** 2272–2278 (2006).

39. Graham, E. M. *et al.* Quantitative mapping of aqueous microfluidic temperature with sub-degree resolution using fluorescence lifetime imaging microscopy. *Lab Chip* **10,** 1267–1273 (2010).

40. Valeur, B. *Molecular Fluorescence* 155–199 (Wiley-VCH: Weinheim, 2002).

41. Lakowicz, J. R. *Principles of Fluorescence Spectroscopy* 3rd edn, 97–155 (Springer: New York, 2006).

42. Fushimi, K. & Verkman, A. S. Low viscosity in the aqueous domain of cell cytoplasm measured by picosecond polarization microfluorimetry. *J. Cell. Biol.* **112,** 719–725 (1991).

43. Luby-Phelps, K. *et al.* A novel fluorescence ratiometric method confirms the low solvent viscosity of the cytoplasm. *Biophys. J.* **65,** 236–242 (1993).

44. Liang, L., Wang, X., Xing, D., Chen, T. & Chen, W. R. Noninvasive determination of cell nucleoplasmic viscosity by fluorescence correlation spectroscopy. *J. Biomed. Opt.* **14,** 024013 (2009).

45. Maeda, Y., , Nakamura, T. & Ikeda, I. Changes in the hydration state of poly(*N*-alkylacrylamide)s during their phase transition in water observed by FTIR spectroscopy. *Macromolecules* **34,** 1391–1399 (2001).

46. SPCImage ver. 3.1 Manual, p. 10 (Becker & Hickl GmbH: Berlin, 2010).

Acknowledgements

We are grateful for support from Professor A. P. de Silva, Mr. S. P. McKelvey, the Development of Advanced Measurement and Analysis System by JST, Grant-in-Aid for Young Scientists (A) by JSPS, Grants-in-Aid for Scientific Research for Plant Graduate Students from NAIST by MEXT, and the Mitsubishi Chemical Corporation Fund.

Author contributions

K. O., N. I., and S. U. designed this work. C. G. and S. U. prepared FPT. S. U. performed the spectroscopic experiments. K. O. performed the cell preparation and all imaging experiments. K. O., N. I., Y. H. and T. F. constructed the fluorescence microscope. K. O. and N. I. set up the FLIM system. K. O., N. I. and S. U. analysed the data. K. O. and S. U. wrote the paper. S. U. organized this work. All authors discussed the results and commented on the manuscript.

Additional information

Direct mechanochemical cleavage of functional groups from graphene

Jonathan R. Felts[1], Andrew J. Oyer[2], Sandra C. Hernández[3], Keith E. Whitener Jr[2], Jeremy T. Robinson[4], Scott G. Walton[3] & Paul E. Sheehan[5]

Mechanical stress can drive chemical reactions and is unique in that the reaction product can depend on both the magnitude and the direction of the applied force. Indeed, this directionality can drive chemical reactions impossible through conventional means. However, unlike heat- or pressure-driven reactions, mechanical stress is rarely applied isometrically, obscuring how mechanical inputs relate to the force applied to the bond. Here we report an atomic force microscope technique that can measure mechanically induced bond scission on graphene in real time with sensitivity to atomic-scale interactions. Quantitative measurements of the stress-driven reaction dynamics show that the reaction rate depends both on the bond being broken and on the tip material. Oxygen cleaves from graphene more readily than fluorine, which in turn cleaves more readily than hydrogen. The technique may be extended to study the mechanochemistry of any arbitrary combination of tip material, chemical group and substrate.

[1] Mechanical Engineering Department, Texas A&M University, 3123 TAMU, College Station, Texas 77843, USA. [2] National Research Council, US Naval Research Laboratory, 4555 Overlook Avenue SW, Washington, District Of Columbia 20375, USA. [3] Plasma Physics Division, US Naval Research Laboratory, 4555 Overlook Avenue SW, Washington, District Of Columbia 20375, USA. [4] Electronics Science and Technology Division, US Naval Research Laboratory, 4555 Overlook Avenue SW, Washington, District Of Columbia 20375, USA. [5] Chemistry Division, US Naval Research Laboratory, Washington, District Of Columbia 20375, USA. Correspondence and requests for materials should be addressed to J.R.F. (email: jonathan.felts@tamu.edu) or to P.E.S. (email: paul.sheehan@nrl.navy.mil).

pplying heat, pressure or light are common approaches to accelerate a chemical reaction. A less common but highly efficient alternative drives the reaction through direct mechanical stress, often yielding reaction products that are otherwise difficult to synthesize[1,2]. Much about mechanochemical reactions remains poorly understood. Macroscale mechanochemical processes such as ball milling and ultrasonic cavitation are often used to drive such reactions[3,4]; however, in situ characterization is difficult and the delivery of precise stresses is not possible. Pulling on individual molecules with atomic force microscope (AFM)[5] tips can rupture bonds with pN resolution, but the technique is limited to molecules long enough to detect rupture and systems where the bonds between tip and sample are stronger than the bond under investigation[6–9]. There is a need for an experimental system capable of measuring mechanochemical bond scission that can be applied universally to a wide variety of material systems and that produces quantitative information about the chemical reaction occurring.

While mechanically induced chemistry is a well-established phenomenon, it has historically been difficult to generate well-defined experiments. For instance, mechanochemical studies with relatively precise control over stresses exist for hard lubricating films such as diamond-like carbon using pin-on-disk tribometers where a millimetre diameter ball continuously rubs against a film[10]. Decades of careful research have shown that the performance of such carbon films depends on multiple factors including the relative amount of sp^2 versus sp^3 carbon centres, surface roughness, the presence of other chemical functional groups, the substrate and the environment. However, despite extensive study, these wear studies have produced only empirical relationships between mechanical load and film chemistry, due in part to poor chemical definition of the starting material, film inhomogeneity and environmental contaminants[10]. Consequently, there has been a growing interest in using nanometre scale, single asperity sliding contacts to probe mechanochemical bond scission at the tip–substrate interface, and recent work has shown volumetric resolution approaching a few atoms[11,12]. It remains unclear, however, what role surface topography plays, which bond or bonds are being broken and which properties—length, energy, polarization and so on—of the bond are important.

Although the experimental apparatus has been highly developed, what has been lacking is a more defined material system that enables unambiguous identification of the bonds broken, removes the complication of corrugation and allows bonds to be broken singly. While this is difficult to achieve in three-dimensional (3D) crystalline materials, the 2D carbon lattice found in graphene is ideal for mechanochemistry since it is atomically flat, can be produced with exceptional crystalline quality, can be functionalized with a wide range of chemical groups and is easily characterized by conventional surface science techniques[13,14]. Chemically modifying graphene markedly alters its optical[15], electronic[16] and lubricating[17,18] properties. It should be noted that while prior work has shown that scanning probes can remove functional groups by locally applying heat[16,19] or electronic potential[20]; mechanochemical cleavage by a scanning probe has not been addressed.

Here we introduce an AFM technique to study mechanochemical bond cleavage of arbitrary organic bonds within zeptolitre volumes on chemically modified graphene (CMG) sheets. Monitoring the friction between the tip and graphene provides an in situ method to measure the kinetics of functional group removal and the relative bond strength of the functional groups. The chemical flexibility of graphene enabled us to mechanically cleave several different functional groups including hydrogenated graphene (HG), plasma oxygenated graphene (OG)

and fluorinated graphene (FG), and so directly determine how the character of the bond impacts its scission. Moreover, the approach clearly demonstrates that the chemistry of the tip material itself can alter the reaction by altering the required stress for the reaction. Indeed, this robust experimental arrangement enables all the mechanochemical reactants—the tip (Si, Si_3N_4 and diamond), the functional groups on the film (O, F and H) and the substrate (SiO_2 and Cu)—to be interchanged to measure the reaction kinetics in detail. This technique could be extended to study mechanochemical bond scission of any arbitrary organic bond using any tip–molecule–substrate system.

Results

Relating chemical composition to lateral force. Mechanochemical cleavage occurred for all systems examined and under a wide range of conditions. Figure 1a illustrates the process for removing oxygen-rich functional groups from an isolated square region on an OG sheet. Applying loads between 10 and 500 nN with an AFM tip controllably cleaved functional groups from OG, and hundreds of lithographic features ranging in width from 18 nm to over 10 μm were patterned at speeds up to 500 μm s^{-1}. For the starting OG material, X-ray photoelectron spectroscopy (XPS) showed that most of the functional groups were carbonyls (C = O, 48%) followed by other oxygen bonds (C–OH or C–O- or C–O–C, 30%; O–C = O, 22%; see Methods section for details)[21]. Since the areas for mechanochemical cleavage were carefully chosen away from tears and edges, the bonds cleaved were predominately epoxide and hydroxide groups that reside on the basal plane and not the carbonyl and carboxyl groups, which reside at the graphene edges[22]. Because graphene has lower friction than all the CMGs used, monitoring the friction provided a direct, in situ measure of the local chemistry, as discussed below. Longer scans and higher loads resulted in greater removal of the functional groups, providing a tool to tailor the local surface chemistry exactly. This was achieved by intermittently measuring the friction force during mechanochemical bond cleavage and stopping the reaction at the desired friction, and hence surface concentration, by decreasing the load. For example, Fig. 1b shows a friction force image of four patches where the desired levels of friction were programmed to be 20, 40, 60 and 80% of the original friction value, as indicated by the friction force linescan shown in Fig. 1c. Such cleavage was not observed in previous work on CMG friction measurements likely due to low tip loading, short scan times and surface contamination; although, there is insufficient information present to make contact stress comparisons[23].

The measured change in friction directly relates to the mass removed from the surface in the form of covalently bound chemical groups. Previous studies of friction on CMG have established that the added chemical groups interact more strongly with the tip than does the graphene, thus raising the friction[18,23,24]. Consequently, the relative friction difference between functionalized and pristine graphene is a direct measure of the degree of chemical functionalization, where lower-friction force values correspond to fewer chemical groups attached to the basal plane. This point may be reinforced by noting that the change in friction force is linearly proportional to the volume of functional groups cleaved from the surface and that the ultimate height decrease of 3.7 ± 1.3 Å corresponds roughly to the length of an –OH group (Supplementary Fig. 1 and Supplementary Discussion, section 'Friction versus removed mass'). Thus, friction force provides a robust in situ means to track the areal concentration of functional groups bound to the graphene surface. Two additional experiments confirm bond cleavage: Raman spectroscopy shows functional group removal after mechanical processing, and conductivity measurements

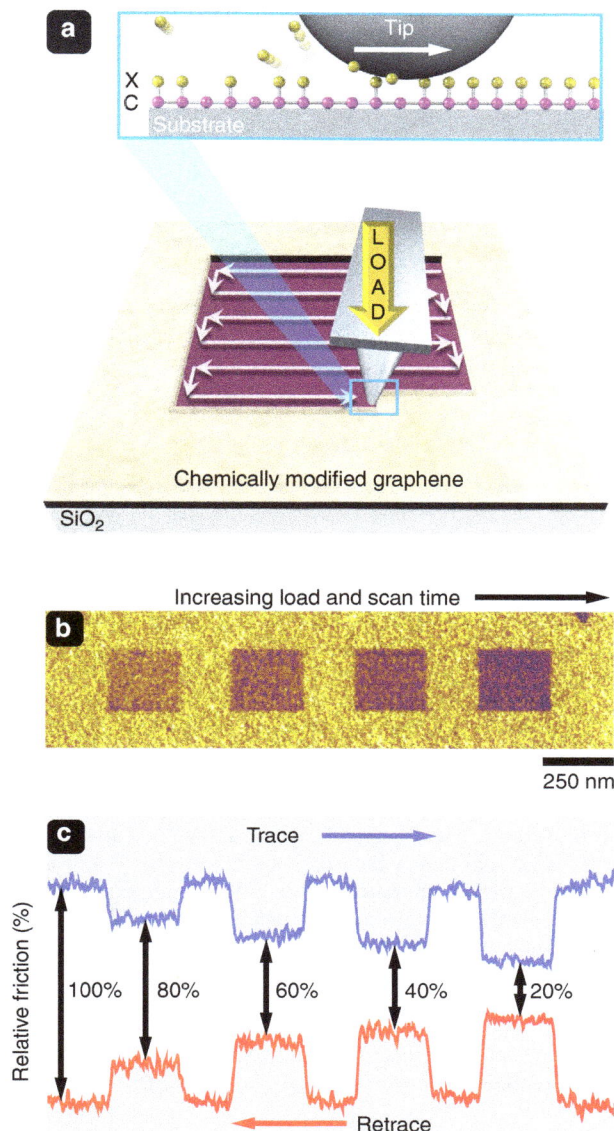

Figure 1 | Measuring and controlling chemical bond scission on graphene (**a**) Schematic of the AFM mechanochemical process where a scanning AFM tip with a known normal load removes chemical groups from a graphene sheet and measures the removal with lateral friction measurements. (**b**) A lateral friction scan on plasma OG with four areas reduced with increasing local contact stress and scan time. The desired friction level was set before mechanochemical processing. (**c**) A linescan of the four squares showing up to a 5× reduction in friction due to the removal of oxygen groups. Load and scan time were tuned during the removal process to quickly reach the desired friction reduction, where higher loads and longer scan times led to faster and more complete removal of functional groups.

taken during the removal show increased electronic conductivity as the insulating CMG is converted to more conductive graphene (Supplementary Fig. 2 and Supplementary Discussion, section 'Film characterization'). Finally, the basal plane remains intact throughout the experiment, since rupturing the basal plane leads to easily detectible film failure (Supplementary Fig. 3).

Bond scission dynamics. The removal rate of oxygen-rich groups from OG shown in Fig. 2a was determined by repeatedly scanning

a single square and monitoring the monotonic decrease in friction force as a function of cumulative tip dwell time and tip normal load. Tip dwell time depends critically on the contact radius between the tip and surface, so contact radius was routinely measured (Supplementary Fig. 4 and Supplementary Discussion Calculating tip dwell time and contact stress). For each point in a single reaction curve at one load (Fig. 2b), a 250×250 nm^2 square region ($N = 65,536$) was scanned (Supplementary Movie 1). Critically, increasing the applied load significantly increases the mechanochemical reaction rate. For a given load, the relative decrease in friction force with tip dwell time was well fit ($R^2 = 0.95$) with an exponential:

$$\Delta f(t) = A_1 e^{-\lambda_1 t} + y_0 \qquad (1)$$

representing a first-order process governed by reaction rate λ_1. Friction force measurements spanning days after removal of oxygen groups showed a small increase in friction with an exponential rise time of ≈ 4 h (Fig. 2b inset), suggesting that the initial rapid decay not captured by the exponential fit may be attributed to adsorbed environmental contaminants that slowly resorb onto the surface after removal, as seen before in carbon systems by Erdemir and colleagues[25]. Additional measurements of friction change behaviour for graphene oxygenated as grown on a copper foil further showed that the change in friction is not due to contaminants from the graphene transfer process (Supplementary Fig. 5). Despite the reversible friction contribution from contaminants, the friction force reduction represented by this term persists indefinitely as a result of permanent removal of functional groups.

Fitting equation (1) to the data in Fig. 2a yielded values for λ_1 as a function of applied contact stress (Fig. 2c), showing an exponential relationship (Supplementary Figs 6 and 7, and Supplementary Discussion, section 'Calculating tip dwell time and contact stress). Recent work has shown that atomic-scale stress-assisted mechanochemical processes can be modelled with a modified version of the Arrhenius thermal activation model

$$\lambda = \lambda_0 \exp\left(-\frac{\Delta U_{act}}{k_B T}\right) \exp\left(\frac{\Delta V_{act} \sigma}{k_B T}\right) \qquad (2)$$

where ΔU_{act} is the energy activation barrier, σ is the applied stress, ΔV_{act} is the activation volume acted on by the stress, k_B is Boltzmann's constant, T is absolute temperature and λ_0 is an effective attempt frequency pre-factor (where reasonable attempt frequencies are based on atomic vibrations in the range 10^{13}–10^{15} s^{-1}; refs 11,26). Fitting the reaction rate data as a function of applied stress from Fig. 2c at a known temperature (30 °C) resulted in an activation volume of 10.6 ± 1.6 Å^3 and an activation energy of 0.73 ± 0.06 eV. Since we are likely removing hydroxyls and epoxides from the graphene basal plane, the calculated activation volume and activation energy should correspond to removal of these structures. The activation volume is generally held to be the volume over which the stress must dissipate for the reaction to occur. The measured activation volume 10.6 ± 1.6 Å^3 compares favourably with literature values for the volumes of epoxides (13.1 Å^3) and for hydroxyls (14.8 Å^3), confirming that bonds should be removed one by one as expected[27]. Second, the activation energy of 0.73 ± 0.06 eV is comparable to previous density functional theory calculations, which found a binding energy of 0.67–0.70 eV for hydroxyls on graphene and 1.9–2.7 eV for epoxides on graphene[28,29]. Note that because we did not vary the temperature, the value of the activation energy depends on the assumed attempt frequency. Thus, the values for both the activation volume and activation energy are reasonable for the current system, suggesting removal of hydroxyl groups and epoxides from the graphene lattice.

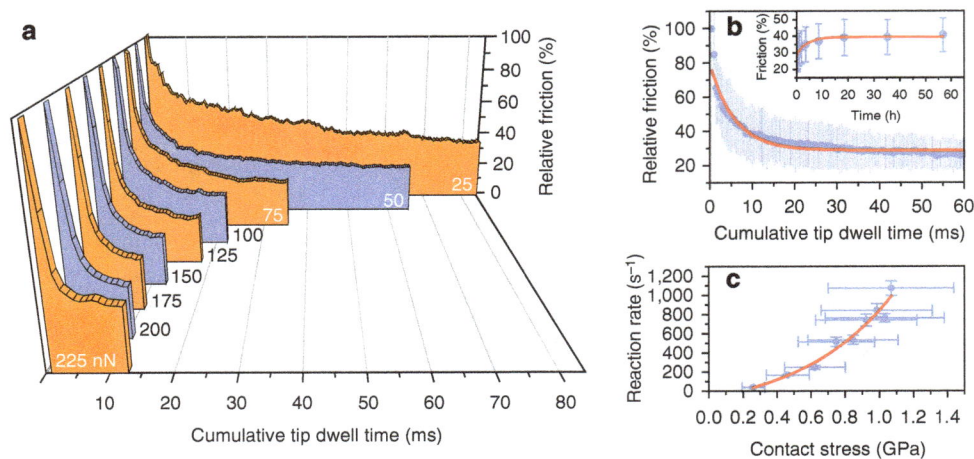

Figure 2 | *In situ* measurement of bond scission dynamics (**a**) Relative friction force measurements as a function of cumulative tip dwell time for normal loads 25–225 nN, showing an increasing exponential decay rate with increasing normal load. (**b**) Reaction curve for 50 nN normal load showing the kinetics of bond scission follows an exponential decay typical of a first-order chemical reaction. At low loads, a contamination layer contributes an additional decay term. This contamination layer slowly resorbs over many hours (inset). The error bars are the s.d. of the friction force values for each AFM scan. (**c**) Reaction rate as a function of measured contact stress, showing an exponential dependence consistent with an atomic-by-atom stress-assisted reaction mechanism. The x axis error bars were calculated by propagating uncertainty values from the linear fit of tip contact radius measurements to the calculation of contact stress. The y axis error comes from the calculated uncertainty of the exponential fit to the data in **a**.

Relative bond strength. If the bond rupture is purely mechanical, then higher bond strengths should require larger forces to break. We studied the relationship between bond strength and mechanochemical removal by monitoring the friction force on different CMGs while linearly ramping the normal load on the tip between each scan. This is similar to the conventional method for taking friction-load plots. Figure 3a shows the normalized friction force as a function of tip load for pristine graphene such as OG, FG and HG. Pristine graphene gave the expected response, with friction increasing monotonically and approximately linearly with normal load. In contrast, the friction on all the CMG sheets showed an unexpected nonlinear behaviour—at low contact stresses the friction forces increased linearly but then decreased as the functional groups were removed to reveal lower-friction graphene. Once the groups were fully removed, the friction force again increased linearly with normal load. Although this is another example of a 'negative friction coefficient', the mechanism here differs significantly from that recently reported by Cannara and colleagues, wherein the wetting of the AFM tip by OG causes sheet delamination and thus higher dissipation at lower loads[17]. Rather, we observed a persistent, repeatable chemical change in the CMG (Supplementary Fig. 8 and Supplementary Discussion, section 'Repeatability').

Monitoring the relative mass loss from the graphene film as a function of applied stress provides a measure of the strength of the ruptured bonds. After each scan at linearly increasing loads, the friction values for the square and the adjacent nonreduced film were measured at low loads (≈ 10 nN). Figure 3b uses the ratio of these values to remove any potential effect of tip shape or intermittent contamination. The derivative of this curve (Fig. 3c) gives the rate of loss with contact stress and enables the determination of the contact stress required to remove different functional groups. These contact stresses were as follows: 0.37 ± 0.18 GPa for OG, 0.72 ± 0.39 GPa for FG and 0.89 ± 0.21 GPa for HG. The position of these values may be understood by recognizing that the force required to rupture a bond, including thermal energy, must exceed the slope (dE/dx) of the potential energy curve under extension. Higher bond energies clearly should require greater forces; and it has been suggested that longer bond lengths should require lower forces[3]. Density

functional theory calculations of bond energy and bond length exist for each system studied here[28,30] and are summarized in Table 1. The force required to remove each bond can be estimated from the ratios of the bond energy E_{bond} to the bond length l_{bond} given in Table 1, and the trend in the forces calculated from the density functional theory data corresponds to the observed trend in contact stresses measured here for CMG. The table also demonstrates the importance of bond length since both –H and $>O_{epoxy}$ have comparable energies, but the much shorter bond length of –H increases the required pressure for removal. Clearly, direct measurements of these bond lengths and energies would improve our understanding of these processes, as would refined models to relate the macroscopic quantity of contact stress to atomic interactions between the tip and 2D films.

Effect of tip material. A final benefit of this experimental approach is that the tip can be treated as a reactant in the mechanochemical reaction. Typically, the scanning probe is contaminated by a transfer film; however, as explained above, the presence of transfer films can be quickly detected in this experimental configuration. Consequently, we explored how the chemical nature of the tip, in addition to the stress applied, impacts the reaction. Figure 4a shows relative friction reduction as a function of contact stress for silicon, silicon nitride and diamond (Advanced Diamond Technologies, Inc.) tips. As above, Fig. 4b shows the derivative of the friction, revealing the contact stresses for the fastest removal of oxygen groups. Notably, both the Si and Si_3N_4 tips remove functional groups with the same dependence on contact stress—0.37 ± 0.18 GPa for Si and 0.40 ± 0.23 GPa for Si_3N_4. In contrast, the diamond tip drives the reaction at a much lower stress of 0.08 ± 0.02 GPa (Supplementary Fig. 8). This lower reaction barrier suggests that the removal of functional groups from the graphene depends on the tip chemistry, either through direct reaction of the graphene oxygen groups with tip material or through a catalytic process mediated by the surface chemistry of the tip. In the latter case, the silicon, silicon nitride and diamond tips all present hydroxyl groups at their surfaces. However, the acidity of tertiary carbon hydroxyls is quite different from that of silanol hydroxyls, and the possibility exists for acid- or base-

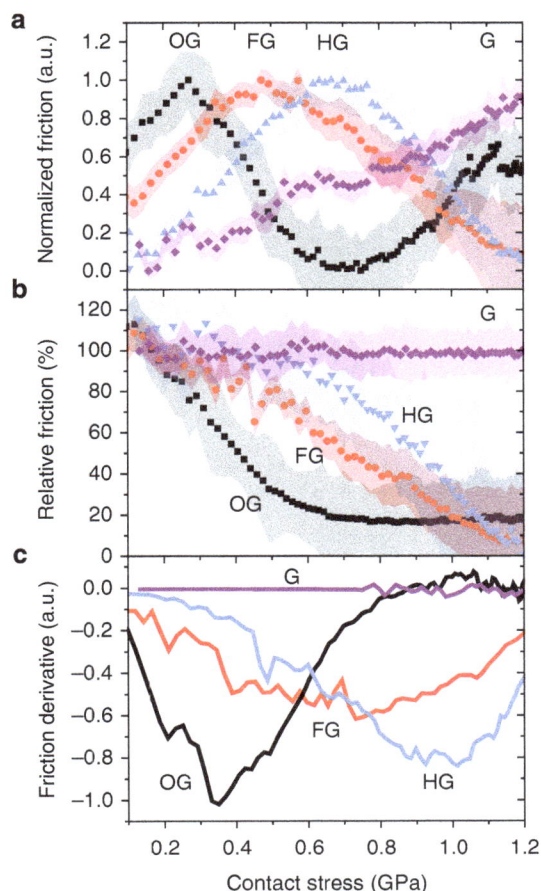

Figure 3 | Chemical composition changes during a contact stress ramp
(**a**) Normalized friction force as a function of applied contact stress for plasma OG, FG, HG and an unmodified graphene sheet. The error bars were determined from the s.d. of each lateral force scan. (**b**) Relative friction compared with the CMG sheet outside of the scan area, showing the degree of functional group removal. The data were captured by a tip scanning at a set low load (10 nN) after a scan at the load indicated on the abscissa. The error bars were calculated as the s.d. of lateral force divided by the average lateral force for each data point. (**c**) Derivative of friction force, where the minima indicate the maximum reduction rate in friction. The data was smoothed using the Savitzky–Golay smoothing algorithm to clearly show the trends.

Table 1 | Density functional theory calculated bond lengths and binding energies.

Functional groups	l_{bond} (Å)	E_{bond} (eV)	E_{bond}/l_{bond}(nN)
− OH	1.57	0.67	0.68
> O$_{(epoxy)}$	1.50	2.69	2.87
− F	1.37	2.86	3.34
− H	1.10	2.48	3.60

Bond lengths and binding energies calculated in the literature for − OH, > O$_{(epoxy)}$, − F and − H bonded to a graphene sheet. The estimated force required to break the bond, calculated as the ratio of the bond energy to the bond length, matches the trend for the calculated contact stresses.

mediated hydroxyl removal from the graphene sheet. Alternatively, high stress may promote wear of the diamond tip and activate the hydroxyls on the graphene and tip surfaces, facilitating a redox disproportionation reaction suggestive of the well-known Boudouard reaction[31]. In contrast to previous AFM pulling force

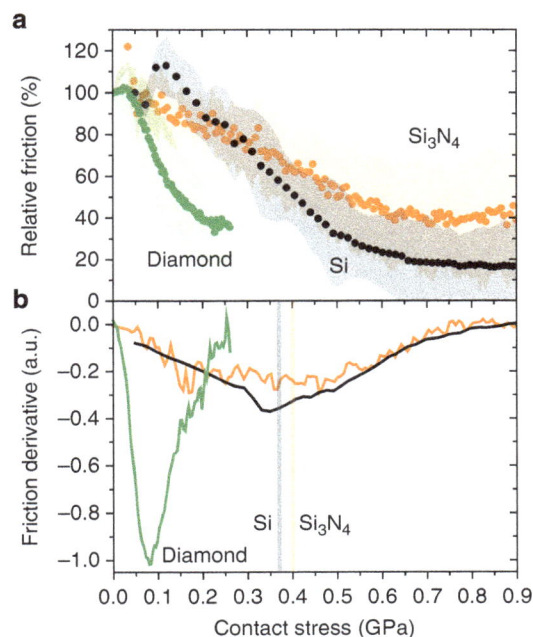

Figure 4 | The effect of tip material on the local chemical reaction (**a**) Relative friction drop on OG as a function of contact stress for a diamond (green), silicon (black) and silicon nitride (orange) tip. For reference, the data for Si is the same as for Fig. 3. The error bars were calculated as the s.d. of lateral force divided by the average lateral force for each data point. (**b**) Derivative of the friction decrease for all three tips, showing that the kinetics of oxygen group removal is the same for SiN and Si tips, while the diamond tip removes the oxygen groups with ∼4× lower contact stress. The data was smoothed using the Savitzky–Golay smoothing algorithm to clearly show the trends.

measurements on long molecule chains, where the AFM tip is distant from the ruptured bond, here we show that the tip material actively contributes to bond scission. This behaviour helps confirm that a mechanochemical reaction between the tip material and the exposed covalent bond on the graphene is occurring, and provides a route to understanding and quantifying these reactions by simultaneously monitoring wear of both the surface and the tip.

Discussion

In conclusion, mechanochemical AFM on CMG sheets provides a flexible and powerful new tool to study stress-driven chemistry. Quantitative determinations of the activation energy and activation volumes are directly obtained. The known orientations of the chemical bond relative to the tip and substrate simplify interpretation of the results and would expedite efforts to model the system via density functional theory. More importantly, the tip, the functionalization and the substrate can all be varied, enabling a refined approach in studying arbitrary C–X bonds. Judicious choice of tip material, substrate, chemical functional groups and scanning parameters will provide insight into the reaction pathway occurring under the tip. That is, whether the reaction proceeds under compressive or tensile stress, the effect of substrate[32] and if the tip material acts as a reaction catalyst. Finally, although the work presented here focuses on graphene, this methodology can be applied to any covalently functionalized 2D material, many of which are currently actively studied[33–35].

Methods

Material preparation. Graphene was grown using chemical vapour deposition on Cu foil substrates[36] in an 'enclosure' geometry[37]. The growth was carried out at

1,030 °C for ~1 h with flowing methane ($P_{CH4} \approx$ 30–50 mTorr) and hydrogen ($P_{H2} \approx$ 5 mTorr) and subsequently quenched. The CVD graphene was transferred using conventional wet etching techniques, using a poly(methyl methacrylate) PMMA protective coating and Transene Cu etchant. Graphene films were transferred to SiO$_2$ (100 nm)/Si substrates, spun dry and baked at 150 °C for 15 min before soaking in acetone for 1 min to remove the PMMA coating.

After the transfer process, the graphene sheets were functionalized with either oxygen or fluorine groups using a plasma process, or with hydrogen using the Birch reduction. Oxygen or fluorine moieties were introduced using pulsed electron-beam-generated plasmas. The electron beam is produced by applying a − 2-kV pulse to a linear hollow cathode. The beam emerges from the hollow cathode and passes through a slotted anode, and terminates at a second grounded anode located further downstream. The resulting electron beam is magnetically confined, to minimize spreading, producing a sheet-like plasma in background gasses of O$_2$/Ar or SF$_6$/Ar mixtures to produce the desired functionalities. The system base pressure is maintained at ~1 × 10^{-6} Torr before processing by a turbo molecular pump. Reactive gases are introduced at 5% of the total flow rate (180 sccm) with argon providing the balance to achieve an operating pressure of 90 mTorr. For this work, the pulse width was 2 ms and the duty factor was maintained at 10%, O$_2$ or SF$_6$ were used as the reactive gas, plasma processing time of 1 min and system pressure of 90 mTorr. All processing was performed at room temperature. Graphene samples were placed on a processing stage adjacent to the plasma at a distance of 2.5 cm from the electron-beam axis.

Graphene was hydrogenated via the Birch reduction as described in detail elsewhere[38]. Briefly, substrate-supported graphene was placed in a nitrogen-purged vessel cooled to − 78 °C in a dry ice bath. Approximately 10 ml of anhydrous ammonia was distilled into the vessel and 50 mg of lithium metal was added to the liquid in small pieces. Removing the vessel from the cold bath and gently swirling turned the mixture a homogeneous dark blue colour. After 2 min, the reaction was quenched with dropwise addition of 10 ml of ethanol. The graphene was removed from the reaction vessel, rinsed with additional ethanol and dried under nitrogen.

Following plasma functionalization, *ex-situ* XPS measurements using a monochromatic X-ray photoelectron spectrometer (K-Alpha XPS System) with a spot size of 400 μm were performed to assess starting chemical composition (see Supplementary Fig. 9 and Supplementary Methods XPS analysis for more details). All experiments on functionalized graphene sheets were performed within their known lifetimes. Functional groups persist for weeks or months on plasma OG[39] and HG[38], while the lifetime of FG is on the order of days[40]. To mitigate degradation effects, comparative experiments with fluorinated films were performed within hours of functionalizing the graphene sheets with fluorine. All samples were stored in dry nitrogen when not in use.

Experimental setup. AFM experiments were performed using an Asylum Research Cypher with an automatic temperature control unit modified to circulate dry nitrogen through the AFM chamber. The ambient temperature was maintained at 29 °C, and O$_2$ content was measured at <3% within the chamber for all experiments. All experiments were performed with either NSC35 rectangular AFM tips from MikroMasch, with a nominal spring constant of 5.4 N m^{-1}, AC240TS AFM tips from Olympus with a spring constant of 1 N m^{-1}, RC800PSA silicon nitride tips from Olympus with a nominal spring constant of 0.76 N m^{-1} or ultrananocrystalline diamond tips from Advanced Diamond Technologies, Inc. with a spring constant of 0.46 N m^{-1}. A relationship between cantilever force and vertical deflection was obtained using force–distance curves and thermomechanical noise measurements[41]. These relationships are used to calculate the deflection voltage necessary to apply a specified force. The raw data collected from the experiments were pull-off force, lateral deflection and relative friction as a function of AFM scans and tip load (Supplementary Fig. 10).

Bond scission dynamics measurements. Friction force was monitored as a function of AFM scan passes within a specified area to measure the chemical dynamics of bond scission during scanning (Supplementary Fig. 10). The AFM tip scans perpendicular relative to the length of the cantilever and measures the relative friction force by taking the difference between the lateral trace and retrace signals. After the first scan, we measure the relative decrease in friction by scanning an area encompassing our original scan area and the unscanned surroundings at low force. The ratio of the friction values between the scanned area and its surroundings provides the calibration between friction signal and relative friction decrease, and is used to calculate the relative friction at all subsequent scans. Another comparison between scanned area and surroundings at the end of the experiment confirms that the ratio is valid throughout the whole experiment. Tip pull-off force measurements after each AFM scan provides one measure of tip wear during the experiment. The adhesion force measured during each force–distance curve is subtracted from the setpoint calculation to prevent underestimating the applied force on the tip.

References

1. Hickenbooth, C. R. *et al.* Biasing reaction pathways with mechanical force. *Nature* **446**, 423–427 (2007).
2. Gilman, J. J. Mechanochemistry. *Science* **274**, 65–65 (1996).
3. Beyer, M. K. & Clausen-Schaumann, H. Mechanochemistry: the mechanical activation of covalent bonds. *Chem. Rev.* **105**, 2921–2948 (2005).
4. Brantley, J. N., Wiggins, K. M. & Bielawski, C. W. Unclicking the click: mechanically facilitated 1,3-dipolar cycloreversions. *Science* **333**, 1606–1609 (2011).
5. Binnig, G., Quate, C. F. & Gerber, C. Atomic force microscope. *Phys. Rev. Lett.* **56**, 930–933 (1986).
6. Duwez, A. S. *et al.* Mechanochemistry: targeted delivery of single molecules. *Nat. Nanotechnol.* **1**, 122–125 (2006).
7. Rief, M., Gautel, M., Oesterhelt, F., Fernandez, J. M. & Gaub, H. E. Reversible unfolding of individual titin immunoglobulin domains by AFM. *Science* **276**, 1109–1112 (1997).
8. Merkel, R., Nassoy, P., Leung, A., Ritchie, K. & Evans, E. Energy landscapes of receptor-ligand bonds explored with dynamic force spectroscopy. *Nature* **397**, 50–53 (1999).
9. Rico, F., Gonzalez, L., Casuso, I., Puig-Vidal, M. & Scheuring, S. High-speed force spectroscopy unfolds titin at the velocity of molecular dynamics simulations. *Science* **342**, 741–743 (2013).
10. Erdemir, A. & Donnet, C. Tribology of diamond-like carbon films: recent progress and future prospects. *J. Phys. D Appl. Phys.* **39**, R311–R327 (2006).
11. Jacobs, T. D. B. & Carpick, R. W. Nanoscale wear as a stress-assisted chemical reaction. *Nat. Nanotechnol.* **8**, 108–112 (2013).
12. Vahdat, V. *et al.* Atomic-scale wear of amorphous hydrogenated carbon during intermittent contact: a combined study using experiment, simulation, and theory. *ACS Nano* **8**, 7027–7040 (2014).
13. Georgakilas, V. *et al.* Functionalization of graphene: covalent and non-covalent approaches, derivatives and applications. *Chem. Rev.* **112**, 6156–6214 (2012).
14. Kim, S. *et al.* Room-temperature metastability of multilayer graphene oxide films. *Nat. Mater.* **11**, 544–549 (2012).
15. Sokolov, D. A. *et al.* Direct observation of single layer graphene oxide reduction through spatially resolved, single sheet absorption/emission microscopy. *Nano Lett.* **14**, 3172–3179 (2014).
16. Wei, Z. Q. *et al.* Nanoscale tunable reduction of graphene oxide for graphene electronics. *Science* **328**, 1373–1376 (2010).
17. Deng, Z., Smolyanitsky, A., Li, Q. Y., Feng, X. Q. & Cannara, R. J. Adhesion-dependent negative friction coefficient on chemically modified graphite at the nanoscale. *Nat. Mater.* **11**, 1032–1037 (2012).
18. Kwon, S., Ko, J.-H., Jeon, K.-J., Kim, Y.-H. & Park, J. Y. Enhanced nanoscale friction on fluorinated graphene. *Nano Lett.* **12**, 6043–6048 (2012).
19. Lee, W. K. *et al.* Nanoscale reduction of graphene fluoride via thermochemical nanolithography. *ACS Nano* **7**, 6219–6224 (2013).
20. Byun, I. S. *et al.* Nanoscale lithography on mono layer graphene using hydrogenation and oxidation. *ACS Nano* **5**, 6417–6424 (2011).
21. Hernández, S. C. *et al.* Chemical gradients on graphene to drive droplet motion. *ACS Nano* **7**, 4746–4755 (2013).
22. De Jesus, L. R. *et al.* Inside and outside: X-ray absorption spectroscopy mapping of chemical domains in graphene oxide. *J. Phys. Chem. Lett.* **4**, 3144–3151 (2013).
23. Jae-Hyeon, K. *et al.* Nanotribological properties of fluorinated, hydrogenated, and oxidized graphenes. *Tribol. Lett.* **50**, 137–144 (2013).
24. Li, Q. *et al.* Fluorination of graphene enhances friction due to increased corrugation. *Nano Lett.* **14**, 5212–5217 (2014).
25. Dickrell, P. L. *et al.* A gas-surface interaction model for spatial and time-dependent friction coefficient in reciprocating contacts: applications to near-frictionless carbon. *J. Tribol.* **127**, 82–88 (2005).
26. Zhurkov, S. N. Kinetic concept of the strength of solids. *Int. J. Fract. Mech.* **1**, 311–322 (1965).
27. Ammon, H. L. New atom/functional group volume additivity data bases for the calculation of the crystal densities of C-, H-, N-, O-, F-, S-, P-, Cl-, and Br-containing compounds. *Struct. Chem.* **12**, 205–212 (2001).
28. Kim, M. C., Hwang, G. S. & Ruoff, R. S. Epoxide reduction with hydrazine on graphene: a first principles study. *J. Chem. Phys.* **131**, 064704 (2009).
29. Lahaye, R. J. W. E., Jeong, H. K., Park, C. Y. & Lee, Y. H. Density functional theory study of graphite oxide for different oxidation levels. *Phys. Rev. B* **79**, 125435 (2009).
30. Leenaerts, O., Peelaers, H., Hernandez-Nieves, A. D., Partoens, B. & Peeters, F. M. First-principles investigation of graphene fluoride and graphane. *Phys. Rev. B* **82**, 195436 (2010).
31. Walker, P. L., Rusinko, F. & Austin, L. G. Gas reactions of carbon. *Adv. Catal.* **11**, 133–221 (1959).
32. Wang, Q. H. *et al.* Understanding and controlling the substrate effect on graphene electron-transfer chemistry via reactivity imprint lithography. *Nat. Chem.* **4**, 724–732 (2012).
33. Liu, H. *et al.* Phosphorene: an unexplored 2D semiconductor with a high hole mobility. *ACS Nano* **8**, 4033–4041 (2014).
34. Radisavljevic, B., Whitwick, M. B. & Kis, A. Integrated circuits and logic operations based on single-layer MoS$_2$. *ACS Nano* **5**, 9934–9938 (2011).
35. Houssa, M. *et al.* Electronic properties of hydrogenated silicene and germanene. *Appl. Phys. Lett.* **98**, 223107 (2011).

36. Li, X. S. *et al.* Large-area synthesis of high-quality and uniform graphene films on copper foils. *Science* **324**, 1312–1314 (2009).
37. Li, X. S. *et al.* Large-area graphene single crystals grown by low-pressure chemical vapor deposition of methane on copper. *J. Am. Chem. Soc.* **133**, 2816–2819 (2011).
38. Whitener, K. E., Lee, W. K., Campbell, P. M., Robinson, J. T. & Sheehan, P. E. Chemical hydrogenation of single-layer graphene enables completely reversible removal of electrical conductivity. *Carbon* **72**, 348–353 (2014).
39. Hernandez, S. C. *et al.* Plasma-based chemical modification of epitaxial graphene with oxygen functionalities. *Surf. Coat. Technol.* **241**, 8–12 (2014).
40. Stine, R., Lee, W. K., Whitener, K. E., Robinson, J. T. & Sheehan, P. E. Chemical stability of graphene fluoride produced by exposure to XeF_2. *Nano Lett.* **13**, 4311–4316 (2013).
41. Levy, R. & Maaloum, M. Measuring the spring constant of atomic force microscope cantilevers: thermal fluctuations and other methods. *Nanotechnology* **13**, 33–37 (2002).

Acknowledgements

This work has been supported by the Naval Research Laboratory Nanoscale Science Institute, the Naval Research Laboratory Base Program, and by the Office of Naval Research (N0001412WX21684). J.R.F., A.J.O. and K.E.W. were supported by a National Research Council fellowship. We thank Kathy Wahl for insight into the observed contamination layer.

Author contributions

J.R.F., A.J.O. and P.E.S. conceived and designed the experiments; J.R.F., A.J.O., S.C.H., S.G.W., K.E.W. and J.T.R. performed experiments; J.R.F., S.C.H. and P.E.S. analysed the data; J.R.F., A.J.O., S.C.H., K.E.W., J.T.R. and S.G.W. contributed materials and analysis tools; and J.R.F., A.J.O., S.C.H., K.E.W. and P.E.S. wrote the manuscript.

Additional information

Competing financial interests: The authors declare no competing financial interests.

13

Enantioselective recognition at mesoporous chiral metal surfaces

Chularat Wattanakit[1,2], Yémima Bon Saint Côme[1], Veronique Lapeyre[1], Philippe A. Bopp[3], Matthias Heim[1], Sudarat Yadnum[1,2], Somkiat Nokbin[2], Chompunuch Warakulwit[2], Jumras Limtrakul[2,4] & Alexander Kuhn[1]

Chirality is widespread in natural systems, and artificial reproduction of chiral recognition is a major scientific challenge, especially owing to various potential applications ranging from catalysis to sensing and separation science. In this context, molecular imprinting is a well-known approach for generating materials with enantioselective properties, and it has been successfully employed using polymers. However, it is particularly difficult to synthesize chiral metal matrices by this method. Here we report the fabrication of a chirally imprinted mesoporous metal, obtained by the electrochemical reduction of platinum salts in the presence of a liquid crystal phase and chiral template molecules. The porous platinum retains a chiral character after removal of the template molecules. A matrix obtained in this way exhibits a large active surface area due to its mesoporosity, and also shows a significant discrimination between two enantiomers, when they are probed using such materials as electrodes.

[1] Univ. de Bordeaux, CNRS, ISM, UMR 5255, ENSCBP, 16 Avenue Pey Berland, Pessac FR-33607, France. [2] Department of Chemistry and NANOTEC Center for Nanoscale Materials Design for Green Nanotechnology, Kasetsart University, Bangkok 10900, Thailand. [3] Univ. de Bordeaux, CNRS, ISM, UMR 5255, 351 cours de la Libération, Talence FR-33405, France. [4] PTT Group Frontier Research Center, PTT Public Company Limited, 555 Vibhavadi Rangsit Road, Chatuchak, Bangkok 10900, Thailand. Correspondence and requests for materials should be addressed to A.K. (email: kuhn@enscbp.fr).

The development of surfaces and materials with chiral features is a major scientific challenge due to the large number of potential applications that can be addressed. Many biological molecules and pharmaceutical compounds are chiral, one enantiomer exhibiting the desired effect while the other one is, in the best case, inactive or can even be toxic. There is thus an increasing demand to either separate enantiomers or to selectively generate them[1]. This results in a continuous growth of the number of methods for asymmetric synthesis[2–6], chiral separation[7,8] and selective chiral detection[9–12]. Although surfaces with chiral characteristics have been successfully obtained by many different approaches, including at the very small scale of metal nanoclusters[13,14], the most popular one is based on molecular imprinting with chiral molecules as templates[10,15–17]. This allows designing chiral materials with specific recognition properties, corresponding to the template used[18,19]. The approach has been very successfully employed to elaborate molecularly imprinted polymers (MIPs)[20]. However, it sometimes also suffers from disadvantages, such as difficult template removal, poor mass transfer, low binding constants and slow binding kinetics[21].

An alternative way is to generate chirality on metallic surfaces. Chiral metal surfaces have been studied over the past decade and were mainly obtained by one of the following three approaches[22]: (i) cutting a bulk metal along a low symmetry plane, liberating in this way a surface that lacks mirror symmetry[23,24]; (ii) adsorption of chiral molecules at metal surfaces[25,26]; (iii) adsorption of species that play the role of chiral templates on metal surfaces[27]. The limitation of the last two approaches is that the adsorbed chiral molecules can desorb, resulting in weak or no surface chirality[28]. The imprinting approach is expected to lead to intrinsically chiral metal surfaces, retaining their enantio-selectivity even after removal of the chiral template. Such surfaces have been successfully produced by electrodeposition of copper oxide films in the presence of chiral tartrate ions. They exhibit moderate enantioselective recognition properties when used for the oxidation of tartrate enantiomers[29]. To date, the synthesis of such intrinsically chiral metal structures by the imprinting approach is still in an early stage of development[30,31]. Engineering of chiral Pd, Au, Pt and Ag has been reported[30–32]; however, the imprinted chirality could not be used for enantioselective recognition after removal of the template.

One reason for the low selectivity might be the relatively small surface area, and therefore the small number of imprinted recognition sites, that are available when using flat metal surfaces. Therefore, a promising strategy to improve the amplitude of chiral recognition consists in using porous metals.

Mesoporous materials have played an important role in a wide range of applications, such as catalysis[33], electronic devices[34], chemical detection[35] and drug delivery[36]. This is because of their attractive features such as high surface area, high stability, well-defined and tunable pore size as well as a predefined organization[37]. Mesoporous metals can greatly improve the accessibility of the metallic framework for guest molecules, and the mesoporous structure can be easily controlled by adapting the lyotropic liquid crystalline phase, which acts as porogen[38].

In the present study, we imprint the structure of chiral molecules at the internal surface of mesoporous platinum. This metal has been chosen since it has a higher structural stability compared with other, much softer metals such as gold. It should therefore retain more efficiently the structure of the template molecule. In contrast to the above-mentioned MIPs, the chiral information is in this case encoded only via the geometry of the cavity and not due to its complementary chemical functionality. This means that, in principle, this approach might be more general and would not need too many adjustments in terms of reagents, when changing the target molecule. Chiral recognition in such porous systems has not yet been demonstrated, even though this would have the twofold advantage of an active surface area that can be two to three orders of magnitude higher compared with flat surfaces, and easy access of the chiral target molecules to the recognition sites. We show that when using such chiral imprinted mesoporous metal surfaces (CIMMS) as electrodes, a very significant discrimination between two enantiomers can be achieved.

Results

Engineering mesoporous chiral platinum surfaces. In this study, chiral imprinted mesoporous platinum films have been prepared by electrodeposition in the simultaneous presence of a lyotropic liquid crystal phase, which serves as a template to generate the mesopores, and, of the chiral molecule, which acts as the template to obtain chiral cavities in the walls of the mesopores (see Fig. 1 and Methods section for more details of the pre-paration process). The morphology of the lyotropic liquid crystal chosen for this study is the (H_1) phase, leading to an alignment of the final mesopores in a hexagonal lattice[39]. The electrochemical reduction of the metal salt occurs around the lyotropic phase and the chiral molecules. This results in the formation of mesoporous channels with chiral cavities at their inner surface after removal of both types of templates.

L-DOPA (3,4-dihydroxy-L-phenylalanine) has been chosen as the chiral template (Fig. 2a, **I**), since its electroactivity is compatible with the potential window where platinum is subject to neither oxidation nor hydrogen adsorption or evolution. Therefore, when it is used as a probe molecule after the imprinting, its reaction on the platinum electrode can be easily monitored without altering the metal structure.

L-DOPA is known to play an important role in pharmaceutics and neurochemistry[40]. DOPA and other catechol compounds have also been employed as reactants to modify inorganic and biological materials[41–43]. Therefore, the transfer of the chiral features from a DOPA enantiomer to the internal pore walls of mesoporous platinum is an interesting choice for this first proof-of-principle study. As the mixture of surfactant and DOPA contains also $PtCl_6^{2-}$ as platinum precursor, the two hydroxyl groups that are normally located on the aromatic ring of DOPA will be in their quinoic form due to the strong oxidizing character of the platinum salt (Fig. 2a, **II**). DOPA molecules are also pH-sensitive, and the electrodeposition is carried out in a mixture with pH = 2, meaning that the protonated form of quinoic DOPA is the molecule present in the mixture.

Different interactions of DOPA with the aggregated Brij56 molecules are possible. On the one hand, the two oxygen atoms (indicated as O1 and O2, Fig. 2a) connected to the ring can be directed towards the hydrophilic surface of the Brij56 columns. On the other hand, the carboxyl group (hydrogen (H) and oxygen (O3) atoms) of the chiral molecule can also interact with the surface of the surfactant phase. Considering the possible establishment of hydrogen bonds between the terminal OH groups of Brij56 and the quinoic part of DOPA as well as the carboxyl group, it is reasonable to assume that the chiral template molecules stay at the outer surface of the surfactant columns and do not penetrate into the hydrophobic core of the columns (Fig. 2b). As the platinum salt is dissolved in the aqueous fraction present around the surfactant columns, the metal deposition is expected to occur around the chiral DOPA molecules present at the surface of these template columns, resulting in a transfer of chirality to the walls of the so-generated metallic mesopores.

Characterization of mesoporous chiral platinum films. To verify the chirality transfer experimentally, platinum-based

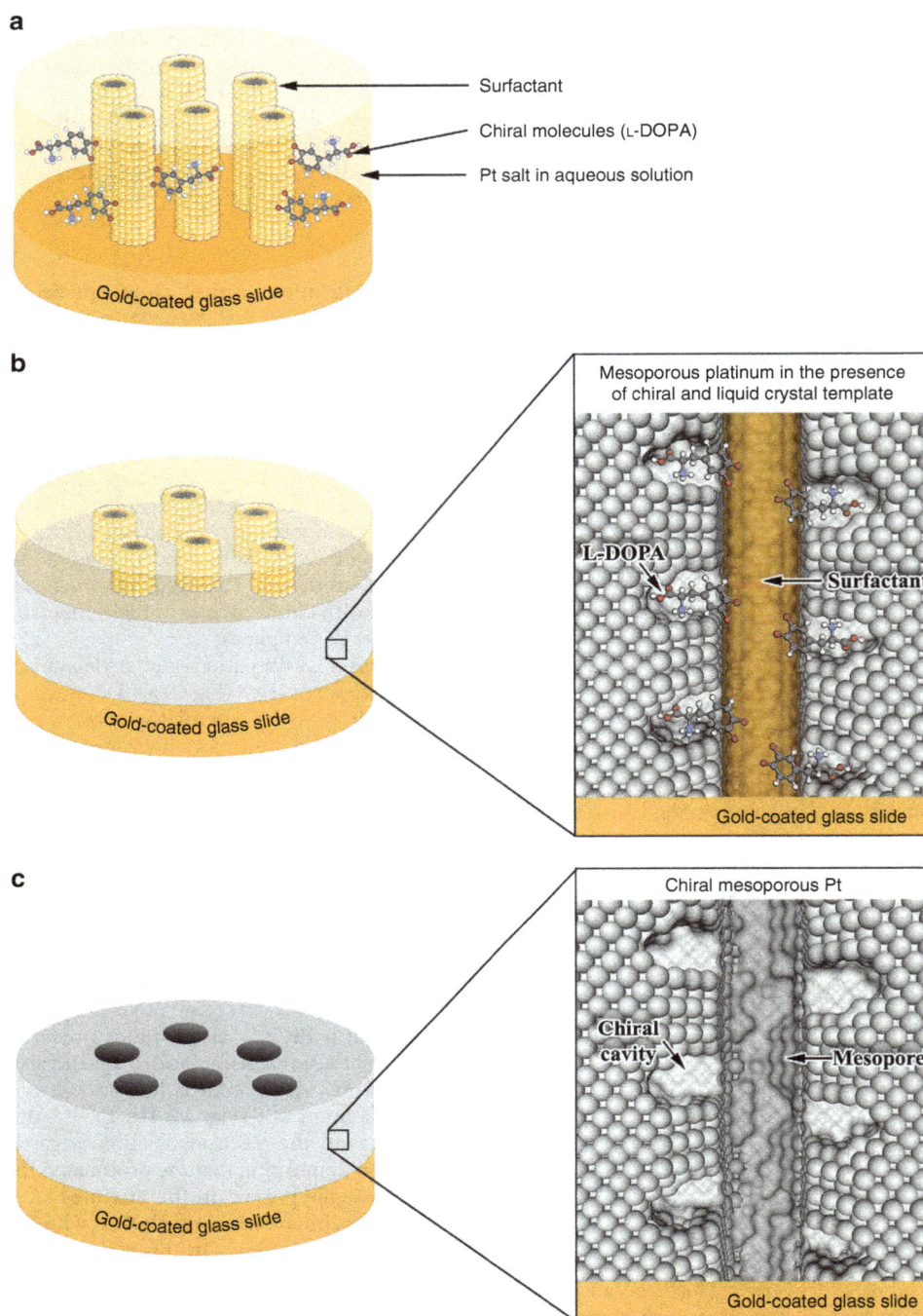

Figure 1 | Fabrication of chiral-imprinted mesoporous platinum films. (**a**) Interaction of the liquid crystal phase with the chiral template molecules, (**b**) Electrodeposition of platinum around the self-assembled structure and (**c**) Structure after template dissolution.

CIMMS have been generated on gold electrodes. The thickness of the metal film strongly depends on the injected charge density during the electrodeposition. Typically, the thickness of the platinum film varies between 0.3 and 2 μm when charge densities between 2 and 12 C cm^{-2} are used. Scanning electron microscopy studies show that the thickness of these films is very uniform over the entire area (Fig. 3a,b). Transmission electron microscopy reveals a mesoporous structure with pores of about 5 nm diameter (Fig. 3c). The active surface area of the porous film is measured by cyclic voltammetry[44]. The voltammograms exhibit the characteristic features of polycrystalline platinum (Fig. 3d) with the adsorption and desorption of hydrogen (H_a and H_d) and the oxidation and reduction of Pt and PtO_n, respectively. The effective active surface area is calculated from the charge associated with the hydrogen adsorption[45]. Compared with a polished flat platinum electrode, the CIMMS show a strong increase in surface area. The calculated surface area is used to estimate a roughness factor, which is defined as the ratio between the active surface area and the geometric surface area of the electrode. Compared with polished platinum, the roughness is increased by a factor of about 80 in the present case. This result confirms that a mesoporous network has been successfully generated.

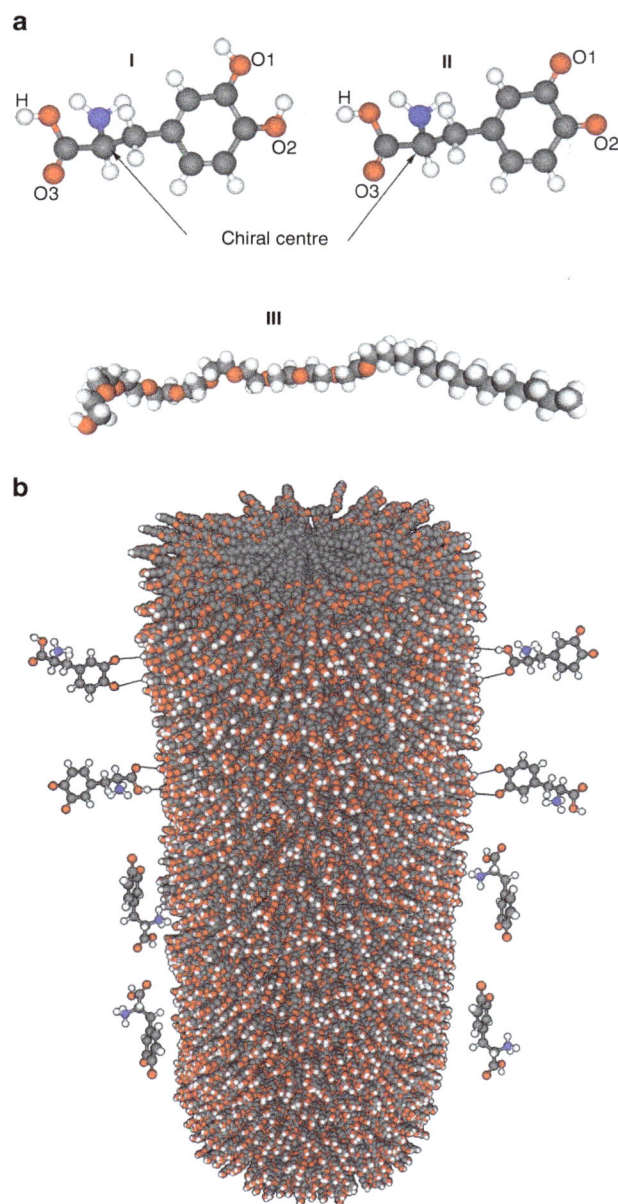

Figure 2 | Interaction between the chiral molecules and the surfactant phase. (**a**) L-DOPA (**I**), the oxidized (quinoic) form of L-DOPA (**II**), and the Brij56 surfactant [(C$_{16}$H$_{33}$(OCH$_2$CH$_2$)$_{10}$OH] (**III**). (**b**) Schematic view of a column of aggregated Brij56 molecules, present in the lyotropic liquid crystal phase, in interaction with the oxidized DOPA molecules.

Electrochemical enantioselectivity. To illustrate the enantioselective properties of such CIMMS, the electrodes were studied by differential pulse voltammetry (DPV). As shown in Fig. 3d, different electrochemical processes, including hydrogen sorption and the oxidation of platinum, can occur during a scan of the potential. To avoid interference of these reactions with the enantioselective recognition, the electrochemical activity of the probe molecule should be located in a potential range where no such competing faradaic processes occur (typically in the range between 0.2 and 0.7 V). It is therefore important to note that DOPA undergoes electro-oxidation in a potential window between 0.35 and 0.60 V. From an electrochemical point of view, the behaviour of L-DOPA and D-DOPA on non-imprinted flat platinum electrodes is, as expected, identical (Fig. 4a). In addition,

the DPV responses of the two enantiomers are also identical on non-imprinted mesoporous platinum (Supplementary Fig. 1). From these observations it is clear that the two enantiomers, as expected, cannot be discriminated on both non-imprinted flat and mesoporous electrodes. However, the current density of DOPA electro-oxidation on non-imprinted mesoporous platinum is strongly increased compared with a flat polished platinum electrode (shown in Supplementary Fig. 1b).

In strong contrast to this, the chiral imprinted mesoporous platinum shows significant differences in the electro-oxidation current densities between L-DOPA and D-DOPA. We found that chiral mesoporous platinum electrodes imprinted by L-DOPA are much more active with respect to the electrooxidation of L-DOPA compared with that of D-DOPA (Fig. 4b). In contrast, those electrodes imprinted in control experiments with D-DOPA are more active for D-DOPA oxidation (Fig. 4c). This difference in activity can be attributed to the chiral 'footprints' that have been generated in the pore walls during the imprinting process, thus facilitating the oxidation of the DOPA molecules that have the right configuration. To quantify the enantioselectivity in both cases 4b and 4c, one needs to compare the relative current changes. If these values are calculated by taking the absolute current differences and dividing them by the peak current of the imprinted species with respect to the baseline, ~ 0.4 is obtained for both sets of experiments. This means that there is a 40% decrease of current when comparing the signal of the imprinted versus non-imprinted species. Small variations between two sets of experiments can be attributed to the fact that different parameters, like for example temperature, can have an impact on the efficiency of the imprinting process.

To exclude that the observed enantioselectivity is generated by a layer of DOPA molecules remaining inside the pores, even after extensive washing of the electrodes after the imprinting process, we carried out a series of control experiments. Indeed DOPA is known to adsorb on platinum[46,47], and this could eventually lead to supramolecular diastereomeric interactions with DOPA molecules in solution, leading *in fine* to chiral discrimination. Such a mechanism could be excluded as shown in Supplementary Figs 2–7).

To verify that no other artifacts, such as a change in active surface area, lead to these observations, we carried out another control experiment. As stated above, and also illustrated in Fig. 3d, for potentials more positive than 0.6 V the platinum surface starts to become oxidized. When performing such an oxidation, followed by the back-reduction of the formed platinum oxide, the platinum atoms at the inner surface of the pore walls, that initially encode the chiral information, will reorganize. This should erase the chiral surface imprints. To verify this effect, the electrooxidation of L-DOPA and D-DOPA was reinvestigated once the electrodes had been oxidized and reduced during a full potential cycle in sulfuric acid from -0.25 to 1.25 V and back. In this case, no selectivity between the two enantiomers is found (Fig. 4d), confirming that oxidation of the chiral-imprinted mesoporous material destroyed its chirality. The active surface areas of all chiral electrodes are obtained by cyclic voltammetry experiments (Supplementary Fig. 8), subsequent to the chiral recognition DPV experiment. This allows the renormalization of the current densities to compare all electrodes with each other, and Fig. 4d confirms that the electrodeposition process itself is very reproducible.

The enantioselective recognition properties should depend on the amount of imprinted recognition sites. We therefore carried out several experiments with varying amounts of DOPA in the electroplating mixture. The enantioselectivity, calculated from the electro-oxidation current density ratio of L-DOPA to D-DOPA ($J_{\text{L-dopa}}/J_{\text{D-dopa}}$), significantly rises when the content of chiral template is increased. For instance, the current density ratio of L-DOPA to D-DOPA increases from 1.2 to 2.4 for a chiral-imprinted

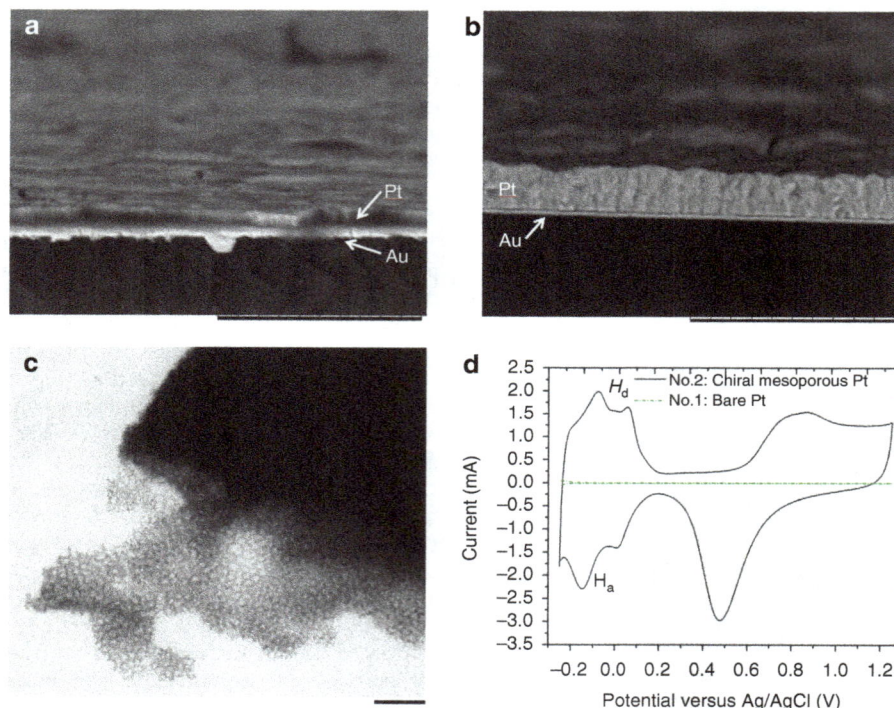

Figure 3 | Electrode modified with a chiral-imprinted mesoporous platinum film. (a,b) Scanning electron microscopy images of typical cross sections of metal films obtained for injected charge densities of 2 and 12 C cm^{-2}, respectively. Scale bar, 10 μm. **(c)** Transmission electron microscope image of a chiral mesoporous Pt film. Scale bar, 50 nm. **(d)** Cyclic voltammograms of flat-polished platinum (No. 1) and a chiral-imprinted mesoporous platinum film obtained by injecting a charge density of 2 C cm^{-2} (No. 2), recorded in 0.5 M H$_2$SO$_4$ at 100 mVs^{-1}. H$_d$ (desorption of hydrogen) and H$_a$ (adsorption of hydrogen).

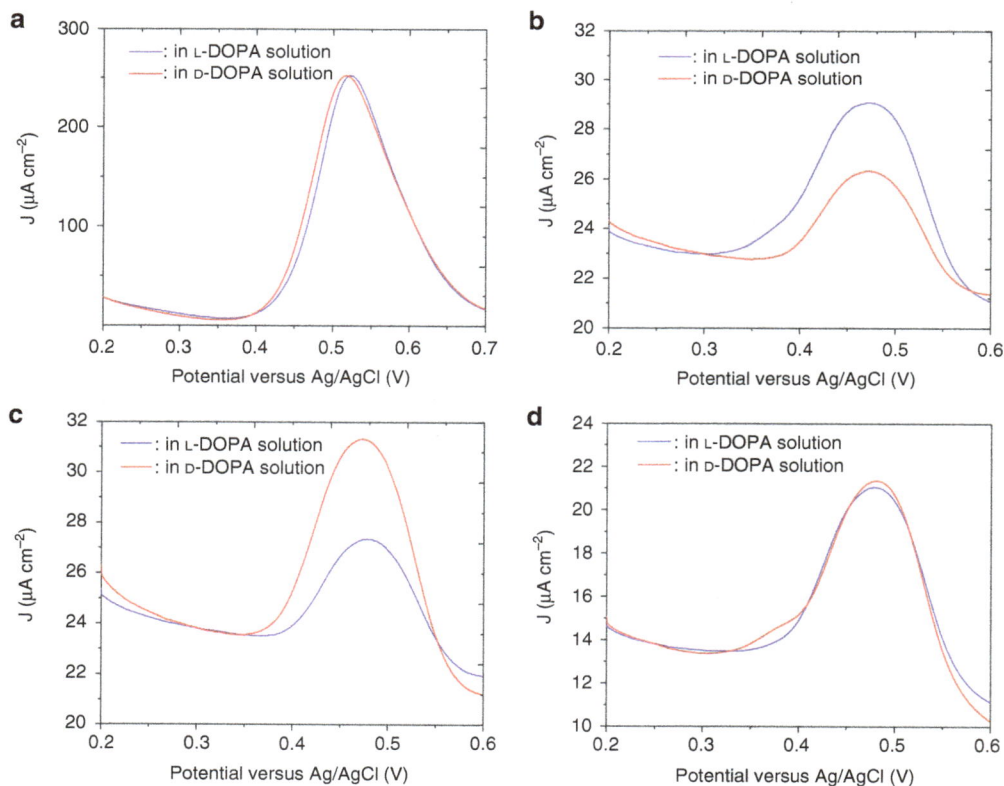

Figure 4 | Characterization of the enantioselectivity by differential pulse voltammetry. DPV in 4 mM L-DOPA (blue) and D-DOPA (red) using 50 mM HCl as supporting electrolyte with **(a)** a flat platinum electrode, **(b)** a chiral mesoporous platinum electrode imprinted with L-DOPA using a L-DOPA/PtCl$_6^{2-}$ ratio of 1/25, **(c)** a chiral mesoporous platinum electrode imprinted with D-DOPA using a D-DOPA/PtCl$_6^{2-}$ ratio of 1/25, **(d)** chiral mesoporous platinum electrodes initially imprinted with L-DOPA, but measured after deliberately destroying the chiral information by scanning the potential between -0.2 V and +1.25 V in 0.5 M H$_2$SO$_4$.

mesoporous electrode obtained using L-DOPA/PtCl$_6^{2-}$ molar ratios of 1/50 and 1/17, respectively. The reason for this increase in enantioselectivity relates to the fact that the increase of the chiral template content is directly influencing the number of created chiral cavities in the mesoporous structure. In contrast, for even higher amounts of chiral template (molar ratio of L-DOPA/PtCl$_6^{2-}$ > 1/12), the efficiency of the enantioselective recognition starts to decrease. This is most likely due to the fact that a large concentration of chiral template destabilizes the lyotropic liquid crystal phase of the surfactant used as mesopore directing agent. An indication for this is the observation that in the presence of large concentrations of DOPA template, the final metal layer is much more brittle and tends to detach from the electrode surface.

To illustrate the beneficial effect of mesoporosity on the enantioselectivity, the chiral recognition of the two enantiomers of L- and D-DOPA was also investigated on chiral imprinted non-mesoporous platinum deposits. In this case, the electrode was simply prepared by an electrochemical reduction of the platinum salt in the presence of the chiral molecules (L-DOPA/Pt = 1/25), but without surfactant. Interestingly, only a slight difference in the electro-oxidation current is observed between the two enantiomers (Supplementary Fig. 9). This makes it clear that the mesoporous structure is a crucial feature for the enantioselectivity because almost no chiral cavities seem to be generated in the absence of surfactant.

This is consistent with the view that the chiral cavities in the walls of the mesopores are generated due to the interaction between DOPA and the surfactant molecules, trapping DOPA on the surface of the pillars of the lyotropic liquid crystal phase. The chiral templates seem to remain in this position during the reduction of the platinum salt, whereas in the absence of surfactant only a few chiral recognition sites are imprinted on the outermost surface of the platinum layer. As a result, a significant discrimination between the two enantiomers of DOPA is observed for the mesoporous structure, whereas the non-mesoporous platinum deposit shows only a very modest selectivity.

It was also found that the mechanical stability of chiral imprinted non-mesoporous platinum films is low, they are easily destroyed during oxidation by cyclic voltammetry in sulfuric acid, whereas the films of chiral mesoporous platinum are stable, even when the electrode is oxidized and reduced several times.

Enantioselective adsorption. It has been reported[24,26] that enantioselective properties at metal surfaces can be successfully generated by introducing defects, for example, steps, kinks and vacancies on the metal, resulting in slightly different adsorption energies for the two enantiomers at such surfaces. It is therefore reasonable to assume that in the case of the imprinting approach it should also be possible to generate sites with different adsorption enthalpies for the two enantiomers. Macroscopically, this should lead to a difference in partition coefficient between the outer solution phase and the inside of the porous structure when comparing the two enantiomers.

To confirm this hypothesis, adsorption experiments were performed by exposing chiral mesoporous platinum films to racemic mixtures of L-DOPA and D-DOPA and monitoring the composition of the supernatant solution by HPLC with a chiral stationnary phase. In Table 1, the D/L DOPA separation factors ($\alpha_{D/L}$) were calculated as the quotient of the ratios of D-DOPA to L-DOPA concentrations in the solution, after and before being in contact with the platinum films for a certain time (1, 3 and 5 h). The $\alpha_{D/L}$ of a non-imprinted mesoporous platinum film (NIM-Pt 4C cm^{-2}) is, as expected, close to one, within experimental errors, even after an adsorption time of 5 h (Entry 1), implying no enantioselective adsorption on such materials. In

Table 1 | Enantioselective adsorption of a racemic DOPA solution on chiral mesoporous platinum films.

Entry	Sample name	Temperature (°C)	Time (h)	$\alpha_{D/L}$*
1	NIM-Pt 4C cm^{-2}†	22	1	0.99 ± 0.05
			5	0.98 ± 0.05
2	CIM-Pt-L 4C cm^{-2}‡	22	1	1.02 ± 0.05
			5	1.29 ± 0.05
3	CIM-Pt-L 8C cm^{-2}§	22	1	0.96 ± 0.05
			5	1.32 ± 0.05
4	CIM-Pt-L 8C cm^{-2}§	2	1	1.18 ± 0.05
			3	1.52 ± 0.05
5	CIM-Pt-D 4C cm^{-2}‖	2	1	0.98 ± 0.05
			5	0.93 ± 0.05
6	CIM-Pt-D 8C cm^{-2}¶	2	1	0.91 ± 0.05
			3	0.85 ± 0.05

*The D/L DOPA separation factor ($\alpha_{D/L}$) was calculated as the quotient of the ratios of D-DOPA to L-DOPA concentrations in the solution, after and before being in contact with the platinum films for a certain time (1, 3 and 5 h).
†Non-imprinted mesoporous platinum, deposition charge density of 4 C cm^{-2}.
‡Chiral-imprinted mesoporous platinum obtained using L-DOPA as template, deposition charge density of 4 C cm^{-2}.
§Chiral-imprinted mesoporous platinum obtained using L-DOPA as template, deposition charge density of 8 C cm^{-2}.
‖Chiral-imprinted mesoporous platinum obtained using D-DOPA as template, deposition charge density of 4 C cm^{-2}.
¶Chiral-imprinted mesoporous platinum obtained using D-DOPA as template, deposition charge density of 8 C cm^{-2}.

contrast, the chiral mesoporous platinum films obtained with L-DOPA as chiral template (CIM-Pt-L 4C cm^{-2}) show a significant increase in enantioselectivity, especially after longer exposure times (Entry 2). It is noteworthy that this characteristic time is much longer than the duration of the electrochemical experiments. This may be explained by the fact that dry platinum films were used for the adsorption experiments, while the electrochemical experiments were carried out with already wet electrodes. The wetting and the diffusion of DOPA into the mesoporous channels takes more time in the case of an initially dry electrode.

The $\alpha_{D/L}$ values were compared for different experimental conditions. When increasing the thickness of the platinum film by using a higher injected charge density during the electrodeposition, the enantioselective adsorption does not change very much at room temperature (22 °C) (Entries 2 and 3). However, the enantioselective adsorption is significantly increased at low temperature (Entry 4). The $\alpha_{D/L}$ at 22 and 2 °C of chiral mesoporous platinum, imprinted by L-DOPA for an injected charge density of 8 C cm^{-2}, is 1.32 and 1.52, respectively. This indicates that L-DOPA is preferentially adsorbed in the metal matrix, leaving an excess of D-DOPA in the supernatant phase. The opposite effect is observed for chiral platinum films that have been imprinted with D-DOPA. The $\alpha_{D/L}$ becomes smaller than 1, suggesting that the adsorption of D-DOPA is favoured in these materials (Entries 5 and 6). Interestingly, an increase in thickness leads to a more pronounced change in enantioselectivity when measured at low temperatures (Entries 5 and 6), compared with what is observed at higher temperatures (Entry 2 and 3). Although the difference is not very large, this is in agreement with an energetic stabilization of the preferred adsorption state in such imprinted materials[48,49].

Discussion

The obtained results indicate that applying the general philosophy of molecular imprinting to organized porous metals, having high surface areas, results in CIMMS that show a very significant selectivity for one or the other enantiomer. Control experiments could exclude that this selectivity is due to the presence of chiral molecules left over in the cavities after imprinting. The observed enantioselectivity strongly depends on several parameters, like the thickness of the imprinted metal layer, the type and amount of chiral template present during the imprinting, as well as the temperature. Playing with these parameters allows the rational design of enantioselective surfaces, but, depending on the surface mobility of the metal atoms, not all metals might be equally appropriate for maintaining the chiral information.

In conclusion, our results demonstrate that mesoporous platinum with a pronounced chiral character has been successfully obtained by electrodeposition of platinum in the simultaneous presence of a surfactant, forming a lyotropic crystal phase and a chiral template molecule. Because of interactions between the surfactant and the chiral template, it has been possible to generate chiral footprints in the mesopores, allowing the discrimination between two enantiomers in electrochemical and adsorption experiments. The effect of this chirality transfer from a molecular species to a metal phase is amplified by the high active surface area of the mesoporous matrix. This first example of a chiral imprinted metal, which retains its chirality even after removal of the template, is complementary to the family of molecular imprinted polymers, and it opens up interesting perspectives for the development of new materials for a wide range of applications from chiral synthesis and enantioselective sensing to chiral separation and purification.

Methods

Chemicals. Hexachloroplatinic acid hydrate ($H_2PtCl_6 \cdot xH_2O$), polyoxyethylene (10) cetyl ether (Brij 56), 3,4-dihydroxy-D-phenylalanine (D-DOPA), 3,4-dihydroxy-L-phenylalanine (L-DOPA) and the racemic mixture of 3,4-dihydroxy-DL-phenylalanine (DL-DOPA) were purchased from Sigma-Aldrich (Missouri, USA). 1 M hydrochloric acid (HCl) was purchased from J.T. Baker (Deventer, Holland). All chemicals were used without further purification. All solutions were prepared with milliQ water.

Synthesis of chiral-imprinted mesoporous platinum electrodes. Liquid crystal plating mixtures were prepared as a quaternary system composed of 42 wt% of nonionic surfactant (Brij 56), 29 wt% of chloroplatinic acid, 29 wt% of DI water and the desired amount of L-DOPA or D-DOPA. Electrochemical reduction of platinum salts in the presence of surfactant and DOPA enantiomer was carried out at $-0.1\,V$ on gold-coated glass slides (0.25 cm²). After the electrodeposition process, the prepared samples were rinsed for several hours with a large amount of water to remove the surfactant and the chiral template. To ensure the complete removal of the chiral template during the washing, all electrodes were checked by differential pulse voltammetry (DPV) in 50 mM HCl for the electrochemical signal of eventually remaining DOPA. Only completely DOPA-free electrodes were used for the subsequent experiments.

For non-imprinted mesoporous platinum, the plating mixture was a ternary system composed of 42 wt% of Brij 56, 29 wt% of chloroplatinic acid and 29 wt% of DI water. For generating chiral-imprinted non-mesoporous platinum electrodes, the electrodes were prepared by electrodeposition at $-0.1\,V$ in an aqueous mixture of 60 mM chloroplatinic acid and L-DOPA (L-DOPA/Pt = 1/25) without adding surfactant.

Characterization and chiral recognition studies. SEM and TEM experiments were carried out on a Hitachi TM-1000 tabletop microscope and a JEOL JEM-2010 TEM, respectively. For TEM measurements, chiral-imprinted mesoporous platinum films on Au-coated glass slides were exposed to an aqueous solution of 4 wt% KI and 1 wt% I_2 for 20 min to dissolve the underlying Au layer. The Pt film could then be easily removed from the electrode and floated on the water surface after slow immersion of the samples into DI water. The freestanding films were then transferred onto TEM grids.

All electrochemical experiments were performed with a μ-Autolab Type III using Ag/AgCl (sat. KCl), a Pt mesh and the prepared mesoporous electrodes as reference, counter and working electrodes, respectively. The chiral recognition studies were based on two independent techniques. The first one is differential pulse voltammetry (DPV). The parameters of the DPV used here were a pulse

modulation of $+50\,mV$ in amplitude, a pulse duration of 50 ms and an interval time of 0.1 s.

The second technique is based on adsorption experiments, exposing a 100 μM racemic mixture of DL-DOPA to the chiral-imprinted mesoporous platinum layers (0.25 cm²), obtained by injecting various charge densities during the electroplating. After a certain time, the supernatant solution was collected and analyzed by HPLC. HPLC was performed on a Merck L-6200A instrument with detection at 230 nm using a UV detector (L-4000A Model). The HPLC analytical assays were carried out on a 150 mm × 3 mm ID CHIRALPAK ZWIX (+) column (Chiral Technologies Europe). All analyses were performed at a flow rate of 0.5 ml min^{-1}, using as mobile phase a mixture of 50/50 (v/v) methanol/acetonitrile and 50 mM formic acid as well as 25 mM diethylamine. Before analysis of the samples by HPLC, the remaining supernatant solution was evaporated to remove all solvent, and redissolved in the mobile phase.

References

1. Stinson, S. C. Chiral drugs. *Chem. Eng. News* **70**, 46–79 (1992).
2. Feringa, B. L. & Van Delden, R. A. Absolute asymmetric synthesis: the origin, control, and amplification of chirality. *Angew. Chem. Int. Ed.* **38**, 3418–3438 (1999).
3. Noyori, R. Asymmetric catalysis: science and opportunities (Nobel Lecture 2001). *Adv. Synth. Catal.* **345**, 15–32 (2003).
4. Dalko, P. I. & Moisan, L. In the golden age of organocatalysis. *Angew. Chem. Int. Ed.* **43**, 5138–5175 (2004).
5. Mohr, J. T., Krout, M. R. & Stoltz, B. M. Natural products as inspiration for the development of asymmetric catalysis. *Nature* **455**, 323–332 (2008).
6. Kuhn, A. & Fischer, P. Absolute asymmetric reduction based on the relative orientation of achiral reactants. *Angew. Chem. Int. Ed.* **48**, 6857–6860 (2009).
7. Seo, J. S. *et al.* A homochiral metal-organic porous material for enantioselective separation and catalysis. *Nature* **404**, 982–986 (2000).
8. Maier, N. M., Franco, P. & Lindner, W. Separation of enantiomers: needs, challenges, perspectives. *J. Chromatogr. A* **906**, 3–33 (2001).
9. Nonokawa, R. & Yashima, E. Detection and amplification of a small enantiomeric imbalance in α-amino acids by a helical poly(phenylacetylene) with crown ether pendants. *J. Am. Chem. Soc.* **125**, 1278–1283 (2003).
10. Pernites, R. B., Venkata, S. K., Tiu, B. D. B., Yago, A. C. C. & Advincula, R. C. Nanostructured, molecularly imprinted, and template-patterned polythiophenes for chiral sensing and differentiation. *Small* **8**, 1669–1674 (2012).
11. Ariga, K., Richards, G. J., Ishihara, S., Izawa, H. & Hill, J. P. Intelligent chiral sensing based on supramolecular and interfacial concepts. *Sensors* **10**, 6796–6820 (2010).
12. Torsi, L. *et al.* A sensitivity-enhanced field-effect chiral sensor. *Nat. Mater.* **7**, 412–417 (2008).
13. Jadzinsky, P. D., Calero, G., Ackerson, C. J., Bushnell, D. A. & Kornberg, R. D. Structure of a thiol monolayer-protected gold nanoparticle at 1.1 Å resolution. *Science* **318**, 430–433 (2007).
14. Dolamic, I., Knoppe, S., Dass, A. & Bürgi, T. First enantioseparation and circular dichroism spectra of Au$_{38}$ clusters protected by achiral ligands. *Nat. Commun.* **3**, 1–6 (2012).
15. Vlatakis, G., Andersson, L. I., Muller, R. & Mosbach, K. Drug assay using antibody mimics made by molecular imprinting. *Nature* **361**, 645–647 (1993).
16. Ramström, O. & Ansell, R. J. Molecular imprinting technology: challenges and prospects for the future. *Chirality* **10**, 195–209 (1998).
17. Tada, M. & Iwasawa, Y. Design of molecular-imprinting metal-complex catalysts. *J. Mol. Catal. A Chem.* **199**, 115–137 (2003).
18. Wang, H. F., Zhu, Y. Z., Yan, X. P., Gao, R. Y. & Zheng, J. Y. A room temperature ionic liquid (rtil)-mediated, non-hydrolytic sol-gel methodology to prepare molecularly imprinted, silica-based hybrid monoliths for chiral separation. *Adv. Mater.* **18**, 3266–3270 (2006).
19. Qu, P., Lei, J., Ouyang, R. & Ju, H. Enantioseparation and amperometric detection of chiral compounds by in situ molecular imprinting on the microchannel wall. *Anal. Chem.* **81**, 9651–9656 (2009).
20. Haupt, K., Linares, A. V., Bompart, M. & Bui, B. T. S. Molecularly imprinted polymers. *Top. Curr. Chem.* **325**, 1–28 (2012).
21. Chen, L., Xu, S. & Li, J. Recent advances in molecular imprinting technology: current status, challenges and highlighted applications. *Chem. Soc. Rev.* **40**, 2922–2942 (2011).
22. Gellman, A. J. Chiral surfaces: accomplishments and challenges. *ACS Nano* **4**, 5–10 (2010).
23. McFadden, C. F., Cremer, P. S. & Gellman, A. J. Adsorption of chiral alcohols on "chiral" metal surfaces. *Langmuir* **12**, 2483–2487 (1996).
24. Attard, G. A. Electrochemical studies of enantioselectivity at chiral metal surfaces. *J. Phys. Chem. B* **105**, 3158–3167 (2001).
25. Kühnle, A., Linderoth, T. R., Hammer, B. & Besenbacher, F. Chiral recognition in dimerization of adsorbed cysteine observed by scanning tunnelling microscopy. *Nature* **415**, 891–893 (2002).
26. Baddeley, C. J. & Richardson, N. V. *Chirality at Metal Surfaces* (Wiley-VCH Verlag GmbH & Co. KGaA, 2009).

27. Bombis, C. *et al.* Steering organizational and conformational surface chirality by controlling molecular chemical functionality. *ACS Nano* **4**, 297–311 (2010).

28. LeBlond, C., Wang, J., Liu, J., Andrews, A. T. & Sun, Y. K. Highly enantioselective heterogeneously catalyzed hydrogenation of α- ketoesters under mild conditions. *J. Am. Chem. Soc.* **121**, 4920–4921 (1999).

29. Switzer, J. A., Kothari, H. M., Poizot, P., Nakanishi, S. & Bohannan, E. W. Enantiospecific electrodeposition of a chiral catalyst. *Nature* **425**, 490–493 (2003).

30. Durán Pachón, L. *et al.* Chiral imprinting of palladium with cinchona alkaloids. *Nat. Chem.* **1**, 160–164 (2009).

31. Attard, G., Casadesús, M., Macaskie, L. E. & Deplanche, K. Biosynthesis of platinum nanoparticles by Escherichia coli MC4100: can such nanoparticles exhibit intrinsic surface enantioselectivity? *Langmuir* **28**, 5267–5274 (2012).

32. Behar-Levy, H., Neumann, O., Naaman, R. & Avnir, D. Chirality induction in bulk gold and silver. *Adv. Mater.* **19**, 1207–1211 (2007).

33. Corma, A. From microporous to mesoporous molecular sieve materials and their use in catalysis. *Chem. Rev.* **97**, 2373–2419 (1997).

34. Stein, A. Advances in microporous and mesoporous solids-highlights of recent progress. *Adv. Mater.* **15**, 763–775 (2003).

35. Wagner, T., Haffer, S., Weinberger, C., Klaus, D. & Tiemann, M. Mesoporous materials as gas sensors. *Chem. Soc. Rev.* **42**, 4036–4053 (2013).

36. Botella, P., Corma, A. & Quesada, M. Synthesis of ordered mesoporous silica templated with biocompatible surfactants and applications in controlled release of drugs. *J. Mater. Chem.* **22**, 6394–6401 (2012).

37. Kresge, C. T., Leonowicz, M. E., Roth, W. J., Vartuli, J. C. & Beck, J. S. Ordered mesoporous molecular sieves synthesized by a liquid-crystal template mechanism. *Nature* **359**, 710–712 (1992).

38. Attard, G. S., Göltner, C. G., Corker, J. M., Henke, S. & Templer, R. H. Liquid-Crystal Templates for Nanostructured Metals. *Angew. Chem. Int. Ed.* **36**, 1315–1317 (1997).

39. Attard, G. S. *et al.* Mesoporous platinum films from lyotropic liquid crystalline phases. *Science* **278**, 838–840 (1997).

40. Sweetman, S. C. *Martindale: The Complete Drug Reference* (Pharmaceutical Press, London, UK, 2007).

41. Lee, H., Dellatore, S. M., Miller, W. M. & Messersmith, P. B. Mussel-inspired surface chemistry for multifunctional coatings. *Science* **318**, 426–430 (2007).

42. Lee, H., Lee, B. P. & Messersmith, P. B. A reversible wet/dry adhesive inspired by mussels and geckos. *Nature* **448**, 338–341 (2007).

43. Lee, G. *et al.* Nanomechanical characterization of chemical interaction between gold nanoparticles and chemical functional groups. *Nanoscale Res. Lett.* **7**, 2–11 (2012).

44. Doña Rodríguez, J. M., Melián, J. A. H. & Peña, J. P. Determination of the real surface area of Pt electrodes by hydrogen adsorption using cyclic voltammetry. *J. Chem. Educ.* **77**, 1195–1197 (2000).

45. Trasatti, S. & Petrii, O. A. Real surface area measurements in electrochemistry. *J. Electroanal. Chem.* **327**, 353–376 (1992).

46. Stern, D. A. *et al.* Studies of L-DOPA and related compounds adsorbed from aqueous solutions at platinum(100) and platinum(111): electron energy-loss spectroscopy, Auger spectroscopy, and electrochemistry. *Langmuir* **4**, 711–722 (1988).

47. Hazzazi, O. A., Attard, G. A. & Wells, P. B. Molecular recognition in adsorption and electro-oxidation at chiral platinum surfaces. *J. Mol. Catal. A Chem.* **216**, 247–255 (2004).

48. Lin, J. M., Nakagama, T., Uchiyama, K. & Hobo, T. Temperature effect on chiral recognition of some amino acids with molecularly imprinted polymer filled capillary electrochromatography. *Biomed. Chromatogr.* **11**, 298–302 (1997).

49. Kida, T., Iwamoto, T., Asahara, H., Hinoue, T. & Akashi, M. Chiral recognition and kinetic resolution of aromatic amines via supramolecular chiral nanocapsules in nonpolar solvents. *J. Am. Chem. Soc.* **135**, 3371–3374 (2013).

Acknowledgements

This work was supported by the French Ministry of Research, CNRS, and ENSCBP. A.K. thanks the Institut Universitaire de France for financial support. Chu.W. and J.L. thank the Thailand Research Fund (TRF) for a Royal Golden Jubilee Ph.D. Fellowship, the French Government for its contribution to the Royal Golden Jubilee-Ph.D. Program (3.C.KU/50/A.2) and the Crown Property Bureau for the NSTDA Chair Professor as well as the NANOTEC Center funded by the National Nanotechnology Center.

Author contributions

Chu.W. and A.K. conceived and designed the experiments; Chu.W., V.L., S.Y., M.H. and Y.B.S.C performed the experiments; Chu.W., S.N., Cho.W., P.B. and J.L. provided model calculations. Chu.W., Cho.W., S.N. P.B., J.L. and A.K. co-produced the manuscript.

Additional information

Solution-based circuits enable rapid and multiplexed pathogen detection

Brian Lam[1], Jagotamoy Das[2], Richard D. Holmes[3], Ludovic Live[2], Andrew Sage[2], Edward H. Sargent[4] & Shana O. Kelley[1,2,3,5]

Electronic readout of markers of disease provides compelling simplicity, sensitivity and specificity in the detection of small panels of biomarkers in clinical samples; however, the most important emerging tests for disease, such as infectious disease speciation and antibiotic-resistance profiling, will need to interrogate samples for many dozens of biomarkers. Electronic readout of large panels of markers has been hampered by the difficulty of addressing large arrays of electrode-based sensors on inexpensive platforms. Here we report a new concept—solution-based circuits formed on chip—that makes highly multiplexed electrochemical sensing feasible on passive chips. The solution-based circuits switch the information-carrying signal readout channels and eliminate all measurable crosstalk from adjacent, biomolecule-specific microsensors. We build chips that feature this advance and prove that they analyse unpurified samples successfully, and accurately classify pathogens at clinically relevant concentrations. We also show that signature molecules can be accurately read 2 minutes after sample introduction.

[1] Department of Chemistry, Faculty of Arts and Sciences, University of Toronto, Toronto, Ontario, Canada M5S 3M2. [2] Department of Pharmaceutical Sciences, Leslie Dan Faculty of Pharmacy, University of Toronto, Toronto, Ontario, Canada M5S 3M2. [3] Institute for Biomaterials and Biomedical Engineering, University of Toronto, Toronto, Ontario, Canada M5S 3M2. [4] Department of Electrical and Computer Engineering, Faculty of Engineering, University of Toronto, Toronto, Ontario, Canada M5S 3M2. [5] Department of Biochemistry, Faculty of Medicine, University of Toronto, Toronto, Ontario, Canada M5S 3M2. Correspondence and requests for materials should be addressed to E.H.S. (email: ted.sargent@utoronto.ca) or to S.O.K. (email: shana.kelley@utoronto.ca).

Electronic readout of the presence of specific biological molecules in solution represents a powerful means to detect disease-related markers[1-4]. In particular, highly sensitive and specific methods have been developed to detect nucleic acids[5-13], proteins[14-18] and small molecules[19-22], using electrochemical readout, and it has been shown that the robustness of electrochemistry allows accurate detection to be done in the presence of heterogeneous, unpurified samples[20,23-25]. Numerous studies documenting the application of electrochemical sensing to cancer[17,23,26,27] and infectious disease[25,28] markers have illustrated the promise of this strategy for future clinical diagnostics, and the cost-effectiveness of the simple instrumentation needed for analysis further enhances the appeal of electrochemical detection for further development[1].

Arrays of serially addressed biosensors can be fabricated to work in conjunction with electrochemical reporter systems, enabling multiplexing and detection of several analytes simultaneously[29]. However, the need for independently addressed electrical contacts corresponding to each sensor, as well as reference and counter electrodes, requires that highly multiplexed arrays employ an active multiplexing strategy. The additional complexity of integrated active electronics explains why, before this work, electrochemical biodetection reports describe studies that employ very low levels of multiplexing[5-29].

Here we create distinct columns of connected biosensors, perpendicular to which we array distinct rows of electrochemical solution. Using this two-dimensional array of electrodes, we programme transient, solution-based circuits that permit individual analysis of sensors using shared contacts. This approach allows a much higher level of multiplexing than was attainable previously using a small set of contacts. We apply this advance towards the detection of panels of pathogenic bacteria and antibiotic-resistance markers.

Results

Overview of approach.
The solution circuit chip (SCC) is depicted in Fig. 1. It consisted of 100 working electrodes (WEs) with 30 off-chip contacts. It included 20 common WEs and 5 counter/reference (CE/RE) electrode pairs, with 25 probe wells to facilitate manual probe deposition and 5 separate liquid channels (Fig. 1a,b). We defined templates for the sensing regions of the WEs by opening 5 μm apertures (Fig. 1c,d) in the top passivation layer. By restricting growth within these apertures, we then grew micron-sized tree-like electrodes via electrodeposition, resulting in nanostructured microelectrodes that protrude from the surface and reach into solution. It was previously shown that the micron-sized scale of the protruding electrodes increases the cross-section for interaction with analyte molecules[30], whereas the nanostructuring maximizes sensitivity by enhancing hybridization efficiency between tethered probe and the analyte in solution[31]. Here these structures are being used to enable the creation of layered, insulated electrode arrays (Fig. 1d).

By patterning channels on the SCC, we created separate liquid compartments (Fig. 1a). WEs are multiplexed on common leads such that they are physically isolated because of the air/water interface in liquid channels separating them. Reference and counter electrodes are routed along the liquid channels. The electrical isolation of the reference and counter layers from the WEs within the SCC, and the ability to bring the electrodes into contact at specific positions using the liquid channels, are the essential elements that allow solution-based circuits to be formed. The contacts made through the conductive solution allow transient circuits to be formed and allow individual contacts to address many sensors in series.

The patterned microsensors are functionalized with peptide nucleic acid (PNA) probes specific to regions of targeted pathogens (Fig. 1f). Electrodes are exposed to samples of interest and binding occurs if a target nucleic acid is present. To detect positive target binding we use an electrocatalytic reporter pair[32] comprising $Ru(NH_3)_6^{3+}$ and $Fe(CN)_6^{3-}$. $Ru(NH_3)_6^{3+}$ is electrostatically attracted to the phosphate backbone of nucleic acids bound near the surface of electrodes by probe molecules and is reduced to $Ru(NH_3)_6^{2+}$ when the electrode is biased at the reduction potential. The $Fe(CN)_6^{3-}$ present in solution auto-oxidizes $Ru(NH_3)_6^{2+}$ back to $Ru(NH_3)_6^{3+}$, which allows for multiple turnovers of $Ru(NH_3)_6^{3+}$ and generates an electrocatalytic current. The difference between pre-hybridization and post-hybridization currents are used as a metric to determine positive target binding.

Characterization of the SCC.
The SCC was fabricated using a series of simple lithographic steps followed by the electrodeposition of three-dimensional microsensors. The electrodeposition process produces sensors with somewhat variable nanoscopic morphologies, but as it is programmed to deposit the same number of gold atoms in each structure, it produces sensors with surface areas that vary by less than 10% (Fig. 1e).

We employed electrochemical analysis to determine whether SCCs provided the necessary level of electrochemical isolation. We also devised a spectroscopic approach to confirm the isolation analysis. Cyclic voltammetry of ferrocyanide was used to evaluate electrochemical characteristics of SCCs versus standard serially connected chips (Fig. 2a). We observed that SCCs have nearly identical electrochemical signals to serially wired chips. It is noteworthy that grounding of all other unbiased counter, reference and working electrodes is required to eliminate crosstalk, as evidenced by the differential pulse voltammetry scans of ferrocyanide with and without the grounding of these electrodes (Fig. 2b). To further investigate whether signals from adjacent liquid channels were picked up by individual sensors, we added ferrocyanide adjacent to channels and monitored the effect on the signals obtained (Fig. 2c). It was observed that when ferrocyanide is added to an adjacent channel, there is minimal perturbation of the signal of ferrocyanide within the channel of interest. The effects of repeated scanning were also explored and were found to be minimal (data not shown).

A non-electrochemical strategy was also pursued to verify independently that the solution circuits were formed as desired, and to study any evolving leakage over longer time periods. To investigate electrochemical isolation using a spectroscopic approach we utilized methylene blue (MB), which can be electrochemically reduced to a colourless form (Fig. 2d). MB was loaded into all liquid channels, and WEs within the middle channel of the chip were held at the reduction potential of MB for 1 h. Visual evidence of the reduction of MB was observed exclusively in the selected channel, providing a qualitative measure of electrical isolation. Measurements of the absorbance (Fig. 2e) of each channel quantitatively confirmed that the significant loss in MB absorbance is detected in the middle lane only.

To investigate whether signals from adjacent channels would interfere with our electrochemical nucleic acid assay, we investigated different orientations of sensors that would yield positive and negative electrochemical responses in the presence of a target DNA sequence. A small area of sensors was functionalized with a PNA probe sequence that would not bind a specific target DNA sequence, and this area was surrounded by sensors functionalized with a PNA probe that would bind the DNA target (Fig. 2f).

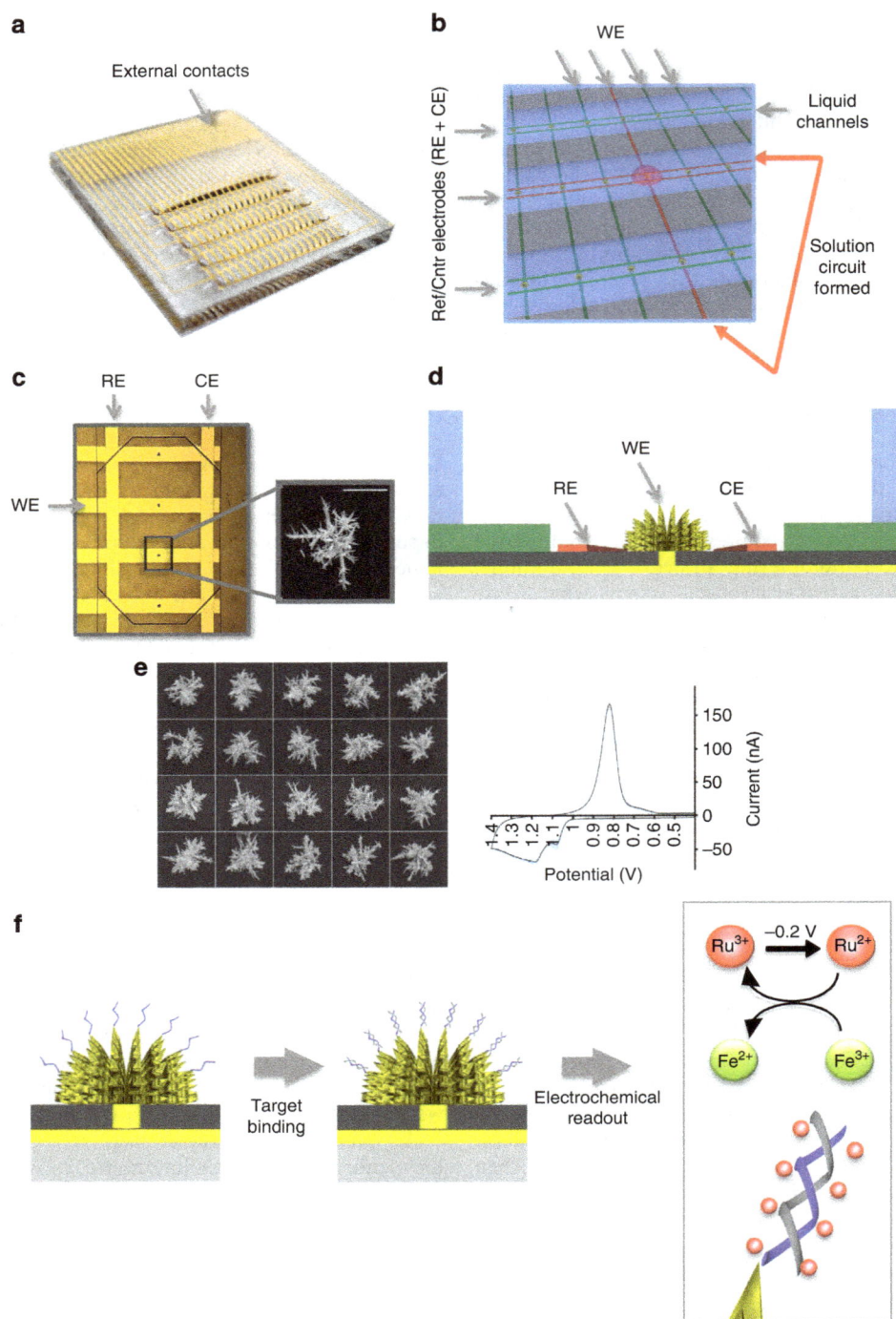

Figure 1 | The solution circuit chip. (a) An SCC featuring 5 liquid channels containing 20 sensors each. **(b)** An SCC featuring common WEs, and counter and reference electrode pairs (CE and RE) that can be activated in sets to form solution-based circuits (red). **(c)** Optical image of single probe well with 4 WEs. Inset: a scanning electron microscopy(SEM) image of a nanostructured microelectrode. Scale bar, 50 μm. **(d)** Cross-section looking down liquid channel of a sensor on a SCC: glass substrate (light grey), common WE (yellow), SU-8 passivation/aperture layer (dark grey), CE/RE (red), SU-8 probe wells (green) and SU-8 liquid channel barrier (blue). **(e)** Sensor-to-sensor comparison of SEM images and acid stripping scans for 20 sensors. Although morphological differences may exist, the surface areas of the sensors are highly consistent, as evidenced by the low levels of s.e. (blue error bars) observed for scans of the 20 sensors conducted in acidic solution. Gold oxide is formed at 1.05 V–1.3 V and reduced at 0.85 V. **(f)** Electrochemical nucleic acid assay scheme. PNA probes are immobilized on microsensors, and in the presence of a complementary target the electrostatic charge on the sensors is increased. This charge change is read out in the presence of $Ru(NH_3)_6^{3+}$ and $Fe(CN)_6^{3-}$, a mixture that yields currents that report on the amount of nucleic acid bound to the sensor.

On this chip, a large positive signal change was obtained with the positive sensors, and a small negative signal change was observed on the much smaller number of negative sensors. This indicates that with the nanoampere levels of current generated during sequence analysis, crosstalk between sensors does not influence the results obtained. When trials were conducted that reversed

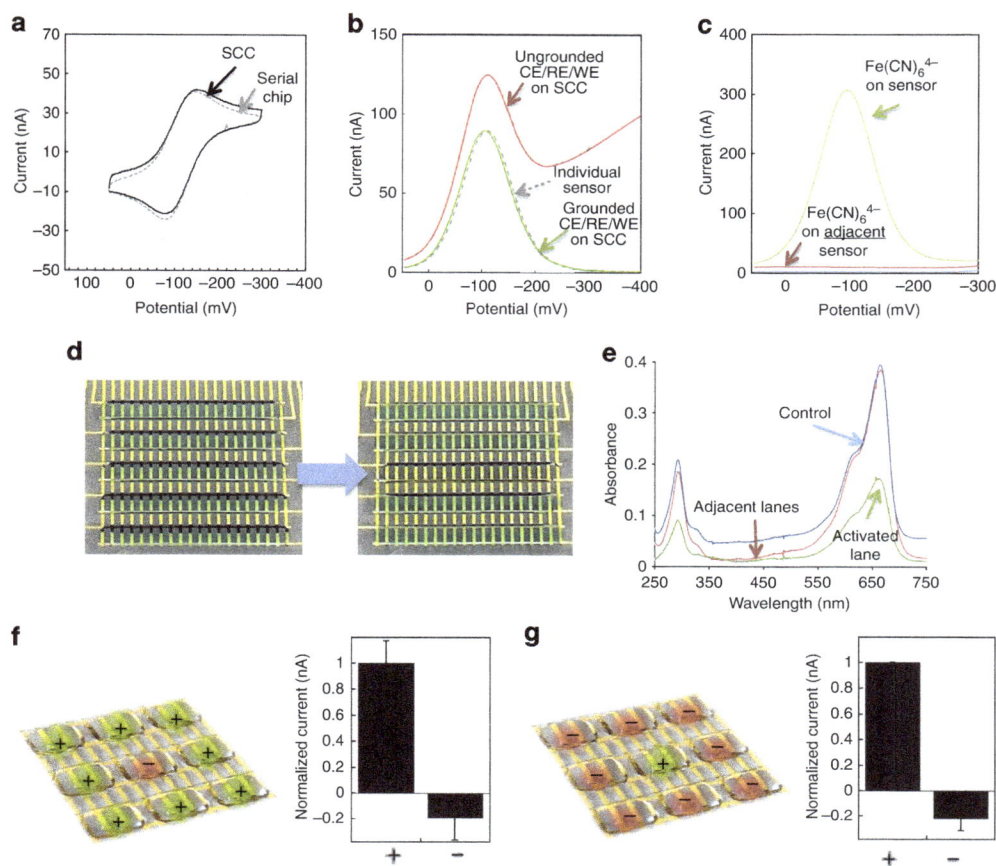

Figure 2 | Electrochemical validation of the SCC. (**a**) Cyclic voltammograms collected with standard chips with individually addressable electrodes and SCC electrodes in 2 mM ferrocyanide. (**b**) Differential pulse voltammetry of biased WE and CE + RE solution circuit in 2 mM ferrocyanide when unbiased CE, RE and WEs are ungrounded (red) and grounded (green) versus standard serially connected WE (dotted gray). (**c**) Differential pulse voltammetry of 2 mM ferrocyanide within liquid channel (green) and on adjacent liquid channels (red) of interest. (**d**) Evaluation of cross-talk by monitoring MB electrochemical bleaching. Middle channel was activated. (**e**) MB absorbance measurements for middle and adjacent lanes of SCC. (**f**) Analysis of signals obtained when sensors that are positive or (**g**) negative for a target sequence are surrounded by sensors yielding the opposite result.

the positions of the positive and negative sensors, the expected reversed results were obtained (Fig. 2g). These results indicate that the solution-based circuits are indeed suitable for multiplexed sequence detection.

Detection of urinary tract infection pathogens. A compelling application for electrochemical analysis is the identification and classification of pathogens. Infectious disease can be difficult to diagnose accurately, because symptoms caused by different pathogens can be quite similar. However, it is important that a causative organism is identified correctly before a treatment is selected. We therefore elected to adapt the multiplexing capabilities of our chips to look for many different types of pathogens, and to also probe for antibiotic resistance. A set of PNA probes were designed, synthesized and tested, and validated for sensitivity and specificity (Fig. 3). It is these probes that make individual sensors specific for pathogens, as they are targeted to unique sequences within the genomes of the organisms. The probe set developed covered 90% of the common urinary tract pathogens[33] and many of the major types of drug resistance that are encountered in the clinic[34].

We used our multiplexed chips to achieve the parallelized assessment of 30 probes for pathogens and antibiotic-resistance markers (Fig. 3a,b). The pathogen probes were targeted against either the RNA polymerase β mRNA (*rpoβ*), or a ribosomal RNA,

and the antibiotic-resistance probes were targeted against known sequences correlated with drug deactivation. To screen sequences for specificity, we compared the response obtained from a solution containing a 1-nM concentration of a complementary target with the response from a solution containing a 100-nM concentration of a non-complementary target. The background-subtracted current generated was then analysed, and for the large majority of the probes that were tested the current obtained was greater than baseline by three standard deviations. However, it is noteworthy that the amount of current generated with each probe varies for the different probes. Nonetheless, the magnitude of current generated for each probe was highly reproducible and hence, despite this effect, the approach can be used for specific pathogen detection.

The sensitivity of sensors modified with these probes was then tested against a panel of pathogens amenable to culture (Fig. 3c). Unpurified bacterial lysates containing 1 cfu μl^{-1} and 100 cfu μl^{-1} were incubated with the sensors, and electro-chemical signals were compared with those obtained when the same sensor type was exposed to 100 cfu μl^{-1} of non-target bacteria (*Escherichia coli* for each trial, except for those testing the sensitivity of the *E. coli* probes, where *Staphylococcus saprophyticus* was used). The signals obtained for each pathogen differed, which likely reflects idiosyncratic nucleic acid structures, but in each case excellent sensitivity and specificity were obtained using samples that have undergone minimal processing. In each case,

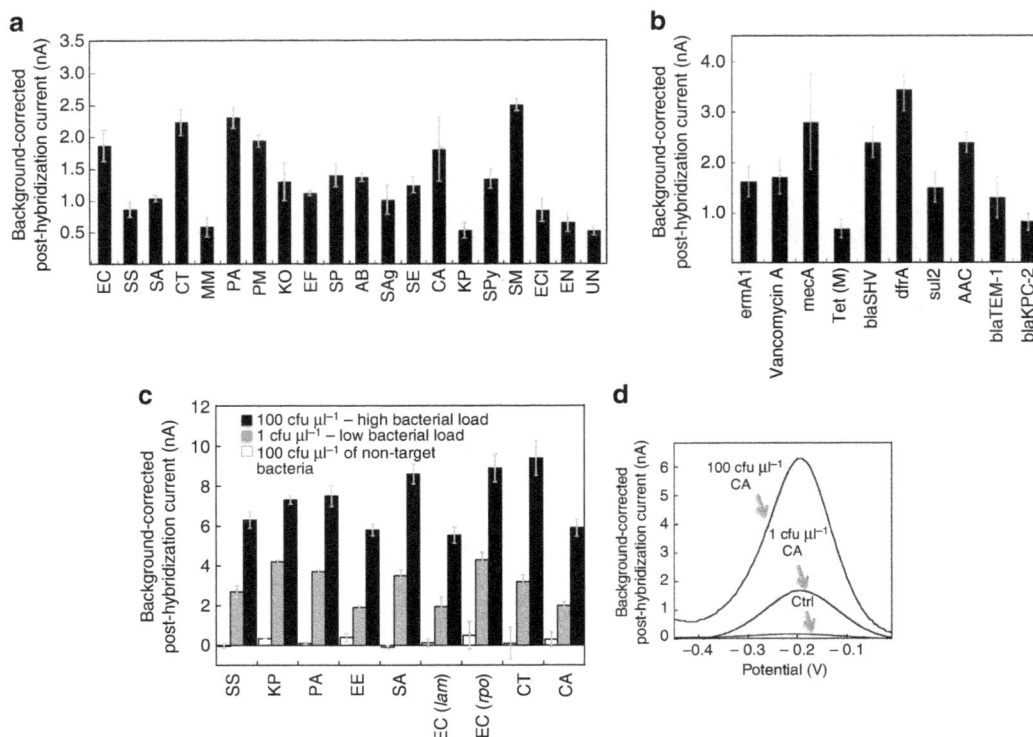

Figure 3 | Validation of pathogen and antibiotic-resistance probes. (a) Background-corrected post-current values for pathogen probe set. Sensors were challenged with 1 nM of a synthetic DNA complement and the signal generated was compared with that obtained with a 100-nM solution of a non-complementary DNA sequence to obtain the background-corrected value. **(b)** Background-corrected post-current values for antibiotic-resistance probe set. Sensors were challenged with 1 nM of a synthetic DNA complement, and these signals were compared with those obtained with a 100-nM solution of a non-complementary DNA sequence to obtain the background-corrected value. **(c)** Pathogen probes validated versus bacterial lysates (EC, *E. coli*; SS, *S. saprophyticus*; SA, *S. aureus*;. CT, *C. trachomatis*; MM, *M. morganii*; PA, *P. aeruginosa*; PM, *P. mirabilis*; KO, *K. oxytoca*; EF, *E. faecalis*; SP, *S. pneumonia*; AB, *A. baumannii*; Sag, *S. agalactiae*; SE, *S. epidermis*; CA, *C. albicans*; KP, *K. pneumonia*; SP, *S. pyogenes*; Spy, *S. pyogenes*; SM, *S. marescens*; ECl, *E. cloacae*; EN, *enterobacter* genus; UN, universal bacteria probe). Error bars reflect s.d. collected from data sets obtained with >3 independent chips. **(d)** Representative differential pulse voltammograms (DPVs) obtained for a control sample (Ctrl), a lysate containing 1 cfu µl^{-1}, and a lysate containing 100 cfu µl^{-1}.

the signals obtained with solutions containing 1 cfu µl^{-1} were greater by a factor of at least three standard deviations relative to background signals, indicating that the limits of detection reside at or below this concentration.

The detection limit obtained here of 1 cfu µl^{-1} is clinically significant. Many applications in infectious disease testing—including testing of swab samples for health-care-associated infections or sexually transmitted infections, or the testing of urine for infectious pathogens—yield samples that contain concentrations higher than 1 cfu µl^{-1}. This detection and speed combination also approaches the diffusional limits for large molecules in static solutions[30].

Multiplexed detection of urinary tract infection pathogens. The SCC was then put to the ultimate test: the analysis of samples containing clinically relevant concentrations of pathogens for panels of markers. SCCs were functionalized simultaneously using nine different probes (Fig. 4a) and were challenged with bacterial lysates at 100 cfu µl^{-1}. Chips were first challenged with lysates of *E. coli*, the most common urinary tract infection-causing pathogen (Fig. 4b). The response of sensors modified with the *E. coli* probe targeted against the RNA polymerase gene (*rpoβ*) was significant, whereas no other probes showed significant response to the lysate. We also challenged chips prepared with the same sensors and probes with a form of antibiotic-

resistant *E. coli* that contains the β-lactamase (*β-lac*) gene (Fig. 4c). With this sample, only EC and β-lactamase sensors exhibited a significant response, indicating that the SCC can classify pathogens and detect antibiotic resistance simultaneously. SCCs were also challenged with lysates of *Staphylococcus aureus* to confirm successful detection of Gram-positive pathogens (Fig. 4d). Only electrodes functionalized with SA probe showed a significant electrochemical response. In addition, we challenged chips with a mixture of *S. aureus* and antibiotic-resistant *E. coli* (+ *β-lac*) to evaluate the performance of chips brought into contact with several analytes producing a positive response (Fig. 4e). Only electrodes functionalized with matching probes exhibited significant electrochemical responses to the mixed lysed sample. These results illustrate that the multiplexing provided by solution-based circuits enables the parallelized detection of multiple analytes at clinically relevant levels.

As a final test of the utility of the SCC, we investigated whether very short hybridization times could be used to enable very rapid bacterial detection (Fig. 4f). We investigated the evolution of signals that could be detected on chip with a 2- and 5-min hybridization, and determined that a pathogen-specific response could be obtained even with the shortest time studied. These results indicate that SCCs can be used for highly multiplexed pathogen detection, and can also be used to deliver the very rapid results needed to make diagnostic information clinically actionable.

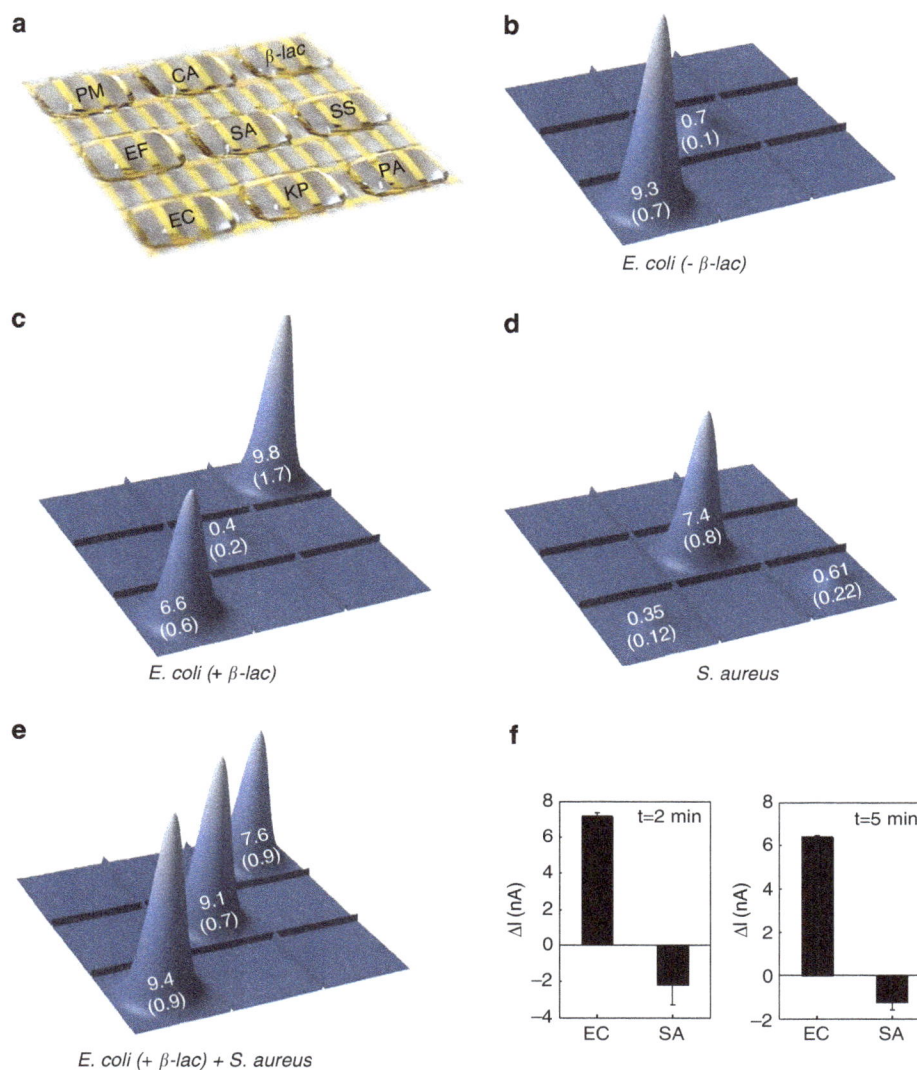

Figure 4 | Multiplexed pathogen and antibiotic-resistance testing on an SCC. (**a**) Arrangement of sensors on a 100-plexed SCC. (**b**) Response of SCC challenged with 100 cfu µl⁻¹ *E. coli.* (− *β-lac*). (**c**) Response of SCC challenged with 100 cfu µl⁻¹ *E. coli.* (+ *β-lac*). (**d**) Response of SCC challenged with 100 cfu µl⁻¹ *S. aureus.* (**e**) Response of SCC challenged with 100 cfu µl⁻¹ *S. aureus* + *E. coli* (+ *β-lac*) bacterial lysate. (**f**) Response of SCC at 2 min and 5 min challenged with *E. coli* lysates at 100 cfu µl⁻¹.

Discussion

Using the simple idea that solution-based circuits could be harnessed on chip to enhance the level of multiplexing that could be achieved on a passive chip, we created a new method for electrochemical multiplexing. Compared with complicated and expensive active electronics on the surface of silicon, this approach has significant advantages for the development of low-cost diagnostic tools. A simple six-step fabrication approach, where most of the steps were performed using transparency masks and low-grade glass was used as a substrate, is advantageous over more involved protocols used to create active silicon electronics. The approach reported herein is scalable in that it could be deployed to produce chips with higher levels of multiplexing than shown here. Given the current capabilities of photolithography and technologies that can be used to deliver probe molecules to high-density arrays, thousands of sensors could be addressed using the solution circuit strategy. The development of strategies to maintain the large number of liquid channels would represent the greatest challenge in realizing this level of multiplexing.

The sensor technology used here, which relies on the electrodeposition of microscale, three-dimensional sensors, is a powerful tool in the analysis of complex samples. It has been shown to perform well in the presence of blood[23] and urine[25], and can also discriminate single base changes in sequence[35]. As demonstrated here, it is compatible with unpurified bacterial lysates and, therefore, is easily integrated for sample-to-answer testing. These features provide a significant advantage over gold standard molecular testing methods, such as the PCR, which typically requires sample clean-up and the use of costly enzymatic reagents. It is also compatible with samples ranging from microlitres to millilitres, which broadens the potential clinical utility. Here we have demonstrated that in addition to possessing these features, it can also be used with a highly multiplexed format if the solution circuit approach is used to contact the sensors. The solution-based circuit chip, as it yields rapid and accurate information on the identity and antibiotic resistance of pathogens, represents a significant advance in the field of biomolecular sensing.

Methods

Multiplexed chip fabrication. Multiplexed chips were fabricated in house on plain glass substrates obtained from Telic Company (Valencia, CA). The substrates are precoated with 5 nm Cr followed by 50 nm Au and positive photoresist (AZ1600). The common WEs were imaged using standard contact lithography and etched using Au and Cr wet etchants, followed by removal of the positive photoresist. A negative photoresist SU-8 2002 was spin coated at 5,000 r.p.m. for 30 s and the aperture layer was imaged into the surface. A positive photoresist S1811 was spin coated on the surface and imaged to act as a lift-off layer. The AUX/REF electrodes (5 nm Cr–50 nm Au) were fabricated onto the surface using standard e-beam evaporation followed by lift-off by sonication in acetone. The probe well layer was created using SU-8 2002 spin coated at 3,000 r.p.m. for 30 s. A thicker liquid channel layer was fabricated with SU-8 3025 at 2,000 r.p.m. for 30 s. A plasma etch mask layer with openings defined for the probe wells was patterned on the surface using SPR 7.0 at 2,000 r.p.m. for 30 s. Chips were diced in house using a standard straight edge scriber. Before electroplating, chips were O_2 plasma etched in a Samco-RIE-1C reactive ion etcher at 15 W for 15 s to create a hydrophilic probe well area. The core structure of the nanostructured microelectrodes was fabricated by electroplating chips in a solution of 50 mM $HAuCl_4$ and 0.5 M HCl at 0 mV for 100 s. The nanostructured surface of the nanostructured microelectrodes was generated by electrodeposition of a Pd coating in a solution of 5 mM $PdCl_2$ and 0.5 $HClO_4$ at -250 mV for 10 s.

Electrochemical chip characterization. Ferricyanide was used to evaluate the electrochemical characteristics of the multiplexed chip. Standard cyclic voltammetry in 2 mM $Fe(CN)_6^{3-}$ in 0.1 × PBS scanned from 50 mV to -300 mV at 100 mV s^{-1} versus on-chip Au AUX/REF was performed on the surface of chips of similar layout that have WEs that are individually addressed versus the multiplexed chips with solution circuits. In addition, ferricyanide crosstalk was evaluated by filling a single channel with 2 mM $Fe(CN)_6^{3-}$ in 0.1 × PBS and adjacent channels with 0.1 × PBS. Differential pulse voltammetry from 50 mV to -300 mV was performed in the ferricyanide-containing channel and adjacent channels without ferricyanide.

Spectroscopic chip characterization. The MB crosstalk experiment was performed on the surface of SCCs by first loading each liquid channel with a solution of 100 μM MB and 50 mM NaCl. Chips were placed in a humidity chamber where the middle lane WEs and AUX/REF were held under bias for 1 h at -650 mV, which is a sufficient potential for reduction of MB. The MB in each channel was then diluted by a factor of 10 and the corresponding absorbance was measured with a ultraviolet–visible spectrophotometer.

Electrochemical assay crosstalk evaluation. Probe wells on the surface of multiplexed chips were used to functionalize with target-specific PNA probes. Each well was filled with 1.5 μl of complementary or non-complementary probe solution initially heated to 60 °C for 5 min, containing 100 nM probe and 900 nM mercaptohexanol, and was incubated directly on chip for 30 min at room temperature. Chips were washed with 0.1 × PBS twice for 5 min after probe deposition, and an initial DPV background scan in electrocatalytic solution (10 μM $Ru(NH_3)_6^{3+}$ and 4 mM $Fe(CN)_6^{3-}$ in a 0.1 × PBS) was performed. Chips were then hybridized with 10 nM complementary target in 1 × PBS for 30 min at 37 °C. Chips were washed with 0.1 × PBS twice for 5 min after target hybridization and differential pulse voltammetry hybridization scan in electrocatalytic solution was performed.

Identification of pathogen-specific probes. A list of potential probes based on the rpoβ gene sequences of the bacteria under study was first generated. All sequences were obtained from the NCBI Nucleotide Database. For a given bacterial sequence, a BLAST search was performed to identify the rpoβ sequence of the most similar bacterial species that could potentially cross-hybridize. This sequence was retrieved and aligned with the targeted bacterial sequence using CLUSTALW2. A computer script was used to identify regions of greatest variability, as they can be used to best differentiate the target species from non-target species. Potential rpoβ probes that could cross-hybridize with non-target molecules in patient samples were eliminated first. Other probe characteristics, such as secondary structure melting temperature, were also analysed to ensure optimal specificity. If suitable probes could not be identified targeting the rpoβ mRNA, then the 16S rRNA (Morganella morganii, Pseudomonas aeruginosa, Enterobacteria family and universal probe) or the 28S rRNA (Candida albicans) was used as a target instead.

Synthesis of probes. Synthesis of PNA probes was performed in house using a Protein Technologies Prelude peptide synthesizer. The following pathogen probe sequences (NH₂-Cys-Gly-Asp SEQUENCE Asp-CONH₂) that are specific to mRNA targets were synthesized: E. coli, 5'-ATC-TGC-TCT-GTG-GTG-TAG-TT-3'; Proteus mirabilis, 5'-AAG-CGA-GCT-AAC-ACA-TCT-AA-3'; S. saprophyticus, 5'-AAG-TAA-GAC-ATT-GAT-GCA-AT-3'; S. aureus, 5'-CCA-CAC-ATC-TTA-TCA-CCA-AC-3'; Klebsiella pneumoniae, 5'-GTT-TAG-CCA-CGG-CAG-TAA-CA-3'; M. morganii, 5'-CGC-TTT-GGT-CCG-AAG-ACA-TTA-T-3'; P. aeruginosa, 5'-CCC-GGG-GAT-TTA-CAA-TCC-AAC-TT-3'; K. oxytoca, 5'-CCA-GTA-GAT-TCG-TCA-ACA-TA-3'; Serratia marescens, 5'-TGC-GAG-TAA-CGT-CAA-TTG-ATG-A-3'; Enterococcus faecalis, 5'-CGA-CAC-CCG-AAA-GCG-CCT-

TT-3'; Acinetobacter baumannii, 5'-CGT-CAA-GTC-AGC-ACG-TAA-TG-3'; Streptococcus pyogenes, 5'-TCT-TGA-CGA-CGG-ATT-TCC-AC-3'; Streptococcus agalactiae, 5'-GTT-CAG-TAA-CTA-CAG-CAT-AA-3'; Staphylococcus epidermidis, 5'-AAA-TAA-CTC-ATT-GAG-GCA-AC-3'; Enterobacter cloacae, 5'-TCA-ACG-TAA-TCT-TTC-GCG-GC-3'; Streptococcus pneumoniae, 5'-GTT-ACG-ACG-CGA-TCT-GGA-TC-3'; Providencia stuartii, 5'-GCC-AAG-TGC-CAA-TTC-ACC-TAG-3'; C. albicans, 5'-GCT-ATA-ACA-CAC-AGC-AGA-AG-3'; Chlamydia trachomatis, 5'-TGC-ATT-TGC-CGT-CAA-CTG-3'; Enterobacteriaceae, 5'-ACT-TTA-TGA-GGT-CCG-CTT-GCT-CT-3'; and Universal bacteria probe, 5'-GGT-TAC-CTT-GTT-ACG-ACT-T-3'. After synthesis, all probes were stringently purified using reverse-phase HPLC. Excitation coefficients were calculated from http://www.panagene.com and concentrations of probe molecules were determined by measuring absorbance at 260 nm with a NanoDrop spectrophotometer.

Preparation of bacterial samples and lysis. The following bacterial strains were used in this study: K. pneumoniae ATCC 27799, E. coli K12 ATCC 33876, E. coli Invitrogen 18265-017, S. saprophyticus ATCC 15305, P. aeruginosa PAO1, E. faecalis ATCC 29212, S. aureus, C. trachomatis and C. albicans. All bacteria were grown in the appropriate growth media and conditions. Approximate quantification of bacteria was performed by measuring the optical density at 600 nm with an Agilent 8453 ultraviolet–visible spectrometer. After the desired population was reached, the growth media was replaced with 1 × PBS. Lysis of bacteria was performed utilizing the Claremont BioSolutions OmniLyse rapid cell lysis kit.

Probe validation. Probe solutions were initially heated to 60 °C for 5 min before deposition, and contained 100 nM probe and 900 nM mercaptohexanol. Deposition was carried out for 30 min at room temperature. Chips were washed with 0.1 × PBS twice for 5 min after probe deposition and chips were then challenged with a non-complementary oligomer at 100 nM in 1 × PBS for 30 min. Chips were washed with 0.1 × PBS twice for 5 min, and then an initial non-complementary DPV background scan was performed in electrocatalytic solution (10 $Ru(NH_3)_6^{3+}$ and 4 mM $Fe(CN)_6^{3-}$ in a 0.1 × PBS). Chips were then hybridized with 1 nM complementary target in 1 × PBS for 30 min at 37 °C. Chips were washed with 0.1 × PBS twice for 5 min after target hybridization and a DPV in electrocatalytic solution was then performed. A similar protocol was followed for lysate testing. Chips were hybridized with lysates in 1 × PBS for 30 min at 37 °C. Chips were washed with 0.1 × PBS twice for 5 min after target hybridization and DPV hybridization scan in electrocatalytic solution was performed.

Multiplexed bacterial lysate experiments. Multiplexed chips for bacterial lysate testing were functionalized with nine probes using nine probe wells. Probe solutions were initially heated to 60 °C for 5 min, and contained 100 nM probe and 900 nM mercaptohexanol that was incubated directly on chip for 30 min at room temperature. Chips were washed with 0.1 × PBS twice for 5 min after probe deposition and an initial DPV background scan was collected in electrocatalytic solution. Multiplexed chips were then hybridized with individual and mixed lysates at 100 cells per μl in 1 × PBS for 30 min at 37 °C in a 200-μl volume.

References

1. Ronkainen, N. J., Halsall, H. B. & Heineman, W. R. Electrochemical biosensors. Chem. Soc. Rev. 39, 1747–1763 (2010).
2. Drummond, T. G., Hill, M. G. & Barton, J. K. Electrochemical DNA sensors. Nat. Biotechnol. 21, 1192–1199 (2003).
3. Song, S. et al. Functional nanoprobes for ultrasensitive detection of biomolecules. Chem. Soc. Rev. 39, 4234–4243 (2010).
4. Lin, P. & Yan, F. Organic thin-film transistors for chemical and biological sensing. Adv. Mater. 24, 34–51 (2012).
5. Soleymani, L., Fang, Z., Sargent, E. H. & Kelley, S. O. Programming the detection limits of biosensors through controlled nanostructuring. Nat. Nanotechnol. 4, 844–848 (2009).
6. Alligrant, T. M., Nettleton, E. G. & Crooks, R. M. Electrochemical detection of individual DNA hybridization events. Lab. Chip. 13, 349–354 (2013).
7. Sorgenfrei, S. et al. Label-free single-molecule detection of DNA-hybridization kinetics with a carbon nanotube field-effect transistor. Nat. Nanotechnol. 6, 126–132 (2011).
8. Slinker, J. D., Muren, N. B., Gorodetsky, A. A. & Barton, J. K. Multiplexed DNA-modified electrodes. J. Am. Chem. Soc. 132, 2769–2774 (2010).
9. Fan, C., Plaxco, K. W. & Heeger, A. J. Electrochemical interrogation of conformational changes as a reagentless method for the sequence-specific detection of DNA. Proc. Natl Acad. Sci. USA 100, 9134–9137 (2003).
10. Patolsky, F., Lichtenstein, A. & Willner, I. Detection of single-base DNA mutations by enzyme-amplified electronic transduction. Nat. Biotechnol. 19, 253–257 (2001).
11. Pheeney, C. G., Guerra, L. F. & Barton, J. K. DNA sensing by electrocatalysis with hemoglobin. Proc. Natl Acad. Sci. 109, 11528–11533 (2012).
12. Wanunu, M. et al. Rapid electronic detection of probe-specific microRNAs using thin nanopore sensors. Nat. Nano 5, 807–814 (2010).

13. Khan, H. U. *et al.* In situ, label-free DNA detection using organic transistor sensors. *Adv. Mater.* **22,** 4452–4456 (2010).

14. Das, J. & Kelley, S. O. Protein detection using arrayed microsensor chips: tuning sensor footprint to achieve ultrasensitive readout of CA-125 in serum and whole blood. *Anal. Chem.* **83,** 1167–1172 (2011).

15. Duan, X. *et al.* Quantification of the affinities and kinetics of protein interactions using silicon nanowire biosensors. *Nat. Nano* **7,** 401–407 (2012).

16. Tang, D., Yuan, R. & Chai, Y. Ultrasensitive electrochemical immunosensor for clinical immunoassay using thionine-doped magnetic gold nanospheres as labels and horseradish peroxidase as enhancer. *Anal. Chem.* **80,** 1582–1588 (2008).

17. Malhotra, R., Patel, V., Vaqué, J. P., Gutkind, J. S. & Rusling, J. F. Ultrasensitive electrochemical immunosensor for oral cancer biomarker IL-6 using carbon nanotube forest electrodes and multilabel amplification. *Anal. Chem.* **82,** 3118–3123 (2010).

18. Xiang, Y., Xie, M., Bash, R., Chen, J. J. L. & Wang, J. Ultrasensitive label-free aptamer-based electronic detection. *Angew. Chem. Int. Ed.* **46,** 9054–9056 (2007).

19. Das, J. *et al.* An ultrasensitive universal detector based on neutralizer displacement. *Nat. Chem.* **4,** 642–648 (2012).

20. Zuo, X., Xiao, Y. & Plaxco, K. W. High specificity, electrochemical sandwich assays based on single aptamer sequences and suitable for the direct detection of small-molecule targets in blood and other complex matrices. *J. Am. Chem. Soc.* **131,** 6944–6945 (2009).

21. Liu, H., Xiang, Y., Lu, Y. & Crooks, R. M. Aptamer-based origami paper analytical device for electrochemical detection of adenosine. *Angew. Chem. Int. Ed.* **51,** 6925–6928 (2012).

22. Kuang, Z., Kim, S. N., Crookes-Goodson, W. J., Farmer, B. L. & Naik, R. R. Biomimetic chemosensor: designing peptide recognition elements for surface functionalization of carbon nanotube field effect transistors. *ACS Nano* **4,** 452–458 (2010).

23. Vasilyeva, E. *et al.* Direct genetic analysis of ten cancer cells: tuning sensor structure and molecular probe design for efficient mRNA capture. *Angew. Chem. Int. Ed.* **50,** 4137–4141 (2011).

24. Swensen, J. S. *et al.* Continuous, real-time monitoring of cocaine in undiluted blood serum via a microfluidic, electrochemical aptamer-based sensor. *J. Am. Chem. Soc.* **131,** 4262–4266 (2009).

25. Lam, B., Fang, Z., Sargent, E. H. & Kelley, S. O. Polymerase chain reaction-free, sample-to-answer bacterial detection in 30 min with integrated cell lysis. *Anal. Chem.* **84,** 21–25 (2012).

26. Mani, V., Chikkaveeraiah, B. V., Patel, V., Gutkind, J. S. & Rusling, J. F. Ultrasensitive immunosensor for cancer multienzyme-particle amplification. *ACS Nano* **3,** 585–594 (2009).

27. Lerner, M. B. *et al.* Hybrids of a genetically engineered antibody and a carbon nanotube transistor for detection of prostate cancer biomarkers. *ACS Nano* **6,** 5143–5149 (2012).

28. Mannoor, M. S., Zhang, S., Link, A. J. & McAlpine, M. C. Electrical detection of pathogenic bacteria via immobilized antimicrobial peptides. *Proc. Natl Acad. Sci. USA* **107,** 19207–19212 (2010).

29. Zheng, G., Patolsky, F., Cui, Y., Wang, W. U. & Lieber, C. M. Multiplexed electrical detection of cancer markers with nanowire sensor arrays. *Nat. Biotechnol.* **23,** 1294–1301 (2005).

30. Soleymani, L. *et al.* Hierarchical nanotextured microelectrodes overcome the molecular transport barrier to achieve rapid, direct bacterial detection. *ACS Nano* **5,** 3360–3366 (2011).

31. Bin, X., Sargent, E. H. & Kelley, S. O. Nanostructuring of sensors determines the efficiency of biomolecular capture. *Anal. Chem.* **82,** 5928–5931 (2010).

32. Lapierre, M. A., O'Keefe, M., Taft, B. J. & Kelley, S. O. Electrocatalytic detection of pathogenic DNA sequences and antibiotic resistance markers. *Anal. Chem.* **75,** 6327–6333 (2003).

33. Foxman, B. The epidemiology of urinary tract infection. *Nat. Rev. Urol.* **7,** 653–660 (2010).

34. Levy, S. B. & Marshall, B. Antibacterial resistance worldwide: causes, challenges and responses. *Nat. Med.* **10,** S122–S129 (2004).

35. Yang, H. *et al.* Direct, electronic microRNA detection for the rapid determination of differential expression profiles. *Angew. Chem. Int. Ed.* **48,** 8461–8464 (2009).

Acknowledgements

This research was sponsored by the Defense Advanced Research Projects Agency through the Autonomous Diagnostics to Enable Prevention and Therapeutics: Diagnostics on Demand—Point-of-Care (ADEPT:DxOD—POC) programme.

Author contributions

B.L., J.D., R.D.H., L.L., A.S., E.H.S. and S.O.K. developed the concepts described and designed the experiments; B.L., J.D., R.D.H., L.L. and A.S. performed experiments; B.L., J.D., E.H.S. and S.O.K. wrote the manuscript with contributions from all of the other authors.

Additional information

Coherent anti-Stokes Raman scattering with single-molecule sensitivity using a plasmonic Fano resonance

Yu Zhang[1,2], Yu-Rong Zhen[1,2], Oara Neumann[2,3], Jared K. Day[2,3], Peter Nordlander[1,2,3] & Naomi J. Halas[1,2,3]

Plasmonic nanostructures are of particular interest as substrates for the spectroscopic detection and identification of individual molecules. Single-molecule sensitivity Raman detection has been achieved by combining resonant molecular excitation with large electromagnetic field enhancements experienced by a molecule associated with an inter-particle junction. Detection of molecules with extremely small Raman cross-sections ($\sim 10^{-30}\,cm^2\,sr^{-1}$), however, has remained elusive. Here we show that coherent anti-Stokes Raman spectroscopy (CARS), a nonlinear spectroscopy of great utility and potential for molecular sensing, can be used to obtain single-molecule detection sensitivity, by exploiting the unique light harvesting properties of plasmonic Fano resonances. The CARS signal is enhanced by ~ 11 orders of magnitude relative to spontaneous Raman scattering, enabling the detection of single molecules, which is verified using a statistically rigorous bi-analyte method. This approach combines unprecedented single-molecule spectral sensitivity with plasmonic substrates that can be fabricated using top-down lithographic strategies.

[1] Department of Physics and Astronomy, Rice University, Houston, Texas 77005, USA. [2] Laboratory for Nanophotonics, Rice University, Houston, Texas 77005, USA. [3] Department of Electrical and Computer Engineering, Rice University, Houston, Texas 77005, USA. Correspondence and requests for materials should be addressed to N.J.H. (email: halas@rice.edu).

Raman spectroscopy measures the vibrational modes of a molecule, providing information about its molecular structure, conformation and temperature. Surface-enhanced Raman scattering (SERS)[1-3] has advanced to the detection limit of individual resonant molecules when the molecule is closely associated with the nanoscale junction between chemically fabricated nanoparticles[4-12]. Here two enhancement mechanisms are at play: the enhanced Raman cross-section of a molecule when the excitation laser is resonant with an electronic excitation of the molecule; and the extremely large local electromagnetic (EM) field enhancement generated by optically exciting the collective plasmon mode of the two closely adjacent nanoparticles. Due to the magnitude of the EM enhancement achievable in nanoparticle junctions, it appears doubtful whether this same approach could be used to generally detect random 'unknown' molecules at the single-molecule level.

One could, however, improve molecular detection sensitivity further by combining plasmonic (SERS) enhancements with coherence. Coherent anti-Stokes Raman scattering (CARS)[13-17] is a third-order nonlinear optical process (a specific type of four-wave mixing (FWM) spectroscopy) that employs molecular coherence. In CARS, the pump (ω_P) and Stokes (ω_S) fields interact coherently through the third-order polarizability of the dipole-forbidden vibronic modes of a molecule, generating an anti-Stokes signal $\omega_{AS} = 2\omega_P - \omega_S$ with the intensity dependence of a third-order nonlinear optical process:

$$I_{CARS} \propto |\chi^{(3)}|^2 I_P^2 I_S. \tag{1}$$

When the frequency difference $\omega_P - \omega_S$ coincides with a Raman (dipole-forbidden vibrational) band of the molecule, the CARS signal is resonantly enhanced as a scattering peak. This two-beam, coherent process is usually orders of magnitude stronger than spontaneous Raman scattering[18]. Coherent and spontaneous Raman scattering processes are diagrammed in Fig. 1a.

The sensitivity of CARS can be increased even further by designing an appropriately resonant plasmonic substrate. If the input (ω_P, ω_S) or output (ω_{AS}) frequencies in CARS are in resonance with the collective modes of the plasmonic nano-structure, the surface-enhanced CARS (SECARS)[18-21] signal from molecules adsorbed onto the nanostructure will be further enhanced by the local fields of the excited plasmon modes. The EM enhancement factor in SECARS is given by

$$G_{SECARS} = |E(\omega_P)/E_0(\omega_P)|^4 \\ \times |E(\omega_S)/E_0(\omega_S)|^2 |E(\omega_{AS})/E_0(\omega_{AS})|^2, \tag{2}$$

the product of the enhanced local EM fields at the characteristic frequencies of the spectroscopy: an appropriate plasmonic substrate for CARS would provide EM enhancements at these three characteristic frequencies.

In this Article, we show that plasmonic nanostructures supporting Fano resonances can be designed to enhance the CARS with single-molecule sensitivity. Plasmonic Fano resonances by their nature as coupled subradiant and superradiant modes can provide much larger field enhancements than bright modes[22-26]. The substrate we use is a quadrumer, that is, four coupled nanodisks (Fig. 1b) that possess a strong Fano resonance[27]. The local fields give rise to a highly localized SECARS enhancement in a single junction at the center of the quadrumer structure, a field distribution particularly well suited for achieving single-molecule detection sensitivity. We verify SECARS of single molecules with Raman cross-sections as low as 10^{-30} cm^2 sr^{-1}, using a bi-analyte method.

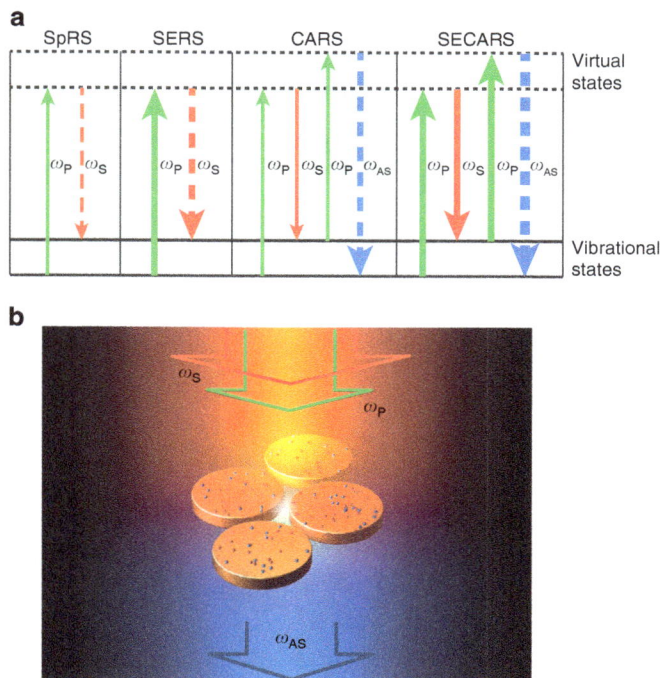

Figure 1 | Different types of Raman processes and the SECARS configuration. (**a**) Energy level diagram shows four types of Raman processes of a molecule. The arrow thickness indicates the transition strength. (**b**) SECARS configuration of two diluted molecules on a nanoquadrumer. A single-wavelength pump laser (ω_P) and a supercontinuum Stokes laser (ω_S) generate an enhanced anti-Stokes scattering (ω_S) of a molecule in the quadrumer central gap.

Results

SECARS substrates and experimental setup. The plasmonic quadrumer[27] (Fig. 2a, inset) is composed of four Au disks of diameter 158.9 ± 2.2 nm and height 51.1 ± 2.5 nm with an interparticle spacing of 15.4 ± 1.1 nm, evaporated onto a fused silica substrate (see Methods). The optical spectrum of a plasmonic quadrumer cluster clearly shows a strong Fano resonance due to its coupled subradiant and superradiant plasmon modes (Fig. 2a). Its linear (Rayleigh) scattering spectrum (Fig. 2a) reveals a narrow minimum centered at ~ 772 nm on top of a far broader resonance, which red-shifts to ~ 800 nm after functionalization of a monolayer of *para*-mercaptoaniline (*p*-MA) molecules. To exploit these spectral properties for SECARS, it is most advantageous to minimize losses at the pump wavelength while maximizing far-field coupling for the wavelength range of the anti-Stokes emission. The reduction of scattered light at frequencies corresponding to the Fano dip maximizes energy coupling into the structure at those frequencies (Fig. 2a, wavelength marked in green). For the anti-Stokes output light, the superradiant shoulder of the optical spectrum acts as an antenna, maximizing the propagation of the anti-Stokes light to the far field (Fig. 2a, blue wavelength band).

The nature and origin of the Fano resonance can be better understood by examining the surface charge densities at different wavelengths associated with this resonance (Fig. 2b). The Fano resonance is a result of interference between a subradiant mode and a superradiant mode. For the resonant frequency of the Fano dip, dipoles of the left and right nanodisks oscillate out of phase with the top and bottom disks in this symmetry-breaking quadrumer (Fig. 2b, green), indicating a small net dipole moment

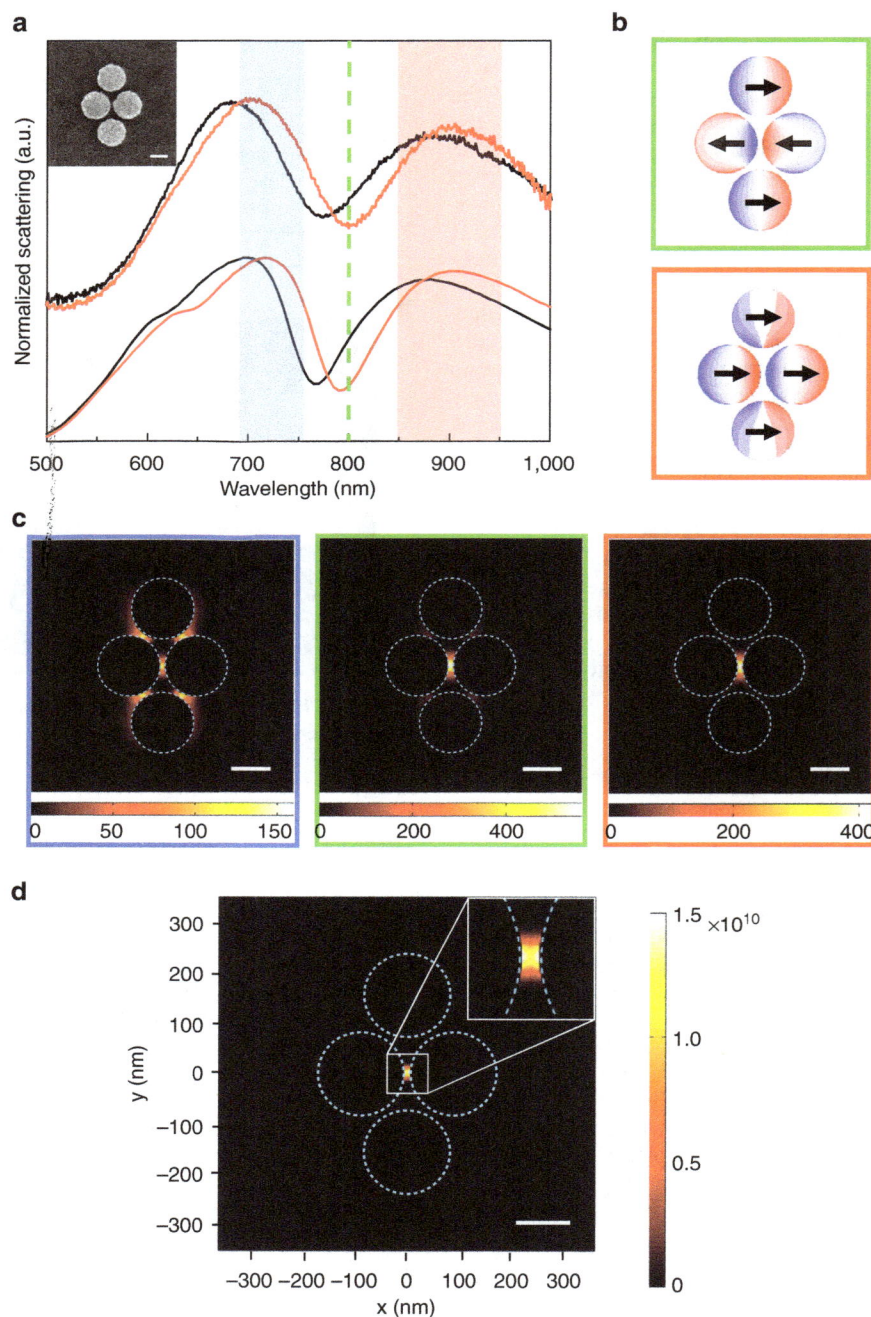

Figure 2 | Experimental and FDTD simulated properties of the quadrumer with *p*-MA molecule monolayer coating. (a) Experimental (top) and calculated (bottom) linear scattering spectra of a single quadrumer before (black) and after (red) the *p*-MA absorption, obtained with horizontal polarization. The calculated spectra for both cases are consistent with their corresponding experimental results. Green dashed line: the pump beam (800 nm); red zone: the Stokes scattering region (848–952 nm); blue zone: the anti-Stokes scattering region (756–690 nm). The Stokes and anti-Stokes regions correspond to Raman shifts of 700–2,000 cm^{-1}. The inset shows an s.e.m. image of a gold quadrumer. **(b)** Charge densities on the top surface of the quadrumer excited at 800 nm pump (top) and 900 nm Stokes (bottom), corresponding to the subradiant and superradiant modes, respectively. **(c)** Field enhancement intensity (g^2) distribution at the anti-Stokes (left), pump (middle) and Stokes (right) frequencies for the *p*-MA 1,070 cm^{-1} mode evaluated at mid-height of the quadrumer. **(d)** SECARS enhancement ($G_{SECARS} = g_P^4 g_S^2 g_{AS}^2$) map for the mode in **c**. The maximum enhancement factor is $\sim 1.5 \times 10^{10}$ in the central gap, and significantly lower ($\sim 2.5 \times 10^6$) in the four peripheral gaps. Scale bar, 100 nm. FDTD, finite-difference-time-domain.

and subradiant behaviour. For the broad plasmon peak, the dipoles of all four disks oscillate in phase, indicating superradiant behaviour (Fig. 2b, red). The pronounced Fano dip in the otherwise superradiant scattering continuum indicates resonant energy storage at the frequencies corresponding to this feature[27].

By optimizing the nanodisk diameters and gap sizes, we can tune the plasmonic quadrumer geometry so that its Fano resonance corresponds to the frequency of the pump laser; then the superradiant shoulders to the red and blue of the Fano minimum correspond to the Stokes and anti-Stokes scanning regions of the SECARS spectrum, respectively.

These simultaneous resonances in the same spatial location result in a 'mixed frequency coherent mode' as previously described on more complex plasmonic clusters[22].

The field intensity distributions for the plasmonic quadrumer at the pump, Stokes and anti-Stokes frequencies, for the $1,070\,\mathrm{cm}^{-1}$ Raman mode of the adsorbate molecule p-MA, are shown in Fig. 2c. The strong spatial localization at all three frequencies leads to a tightly confined 'SECARS hot spot' in the center of the cluster with an $\sim 250\,\mathrm{nm}^2$ size (only $\sim 0.31\%$ of the surface area of a quadrumer) (Fig. 2d). With horizontally polarized excitation, the four peripheral gaps have a SECARS enhancement $\sim 6,000$ times weaker than the central gap, thus a large CARS polarization could be effectively induced only by the molecules located in the region of this central gap. In this region, we calculated that the SECARS over CARS enhancement is $\sim 1.5 \times 10^{10}$ for this mode, and is 0.7–2.3×10^{10} for Raman modes spanning the range of 700–$2,000\,\mathrm{cm}^{-1}$ (Supplementary Fig. 1). In comparison, SERS on this same structure and pump laser frequency was calculated to enhance only $\sim 2.5 \times 10^5$ over spontaneous Raman scattering. It is also useful to compare the SECARS enhancement (over CARS) for this Fano-resonant cluster with other plasmonic but non-Fano clusters: SECARS by a broadly resonant (non-Fano) plasmonic quadrumer is enhanced by $\sim 7.5 \times 10^8$ over unenhanced CARS (Supplementary Fig. 2); SECARS by a nanovoid gold surface[18] is enhanced $\sim 10^5$ over CARS when all frequencies coincide with its plasmon resonances. Taking into consideration that CARS is usually orders of magnitude stronger than spontaneous Raman[18], SECARS due to this Fano-resonant quadrumer should be anticipated to be enhanced more than 10 orders of magnitude over the spontaneous Raman process.

In the SECARS experiments, one laser beam from a single 76 MHz pulsed Ti:sapphire oscillator was split into two beams, consisting of a pump beam and a continuum Stokes beam, the latter generated by propagating a fraction of the beam through a nonlinear photonic crystal fibre. These two horizontally polarized, collinear and coherent pulse trains were focused onto the sample, which was mounted on a microscope with a high-precision positioning stage to place the desired quadrumer at the laser focus (Methods and Supplementary Fig. 3). The focused laser spot was $\sim 1\,\mu\mathrm{m}$ in diameter, where both incident beams behave like normal-incident plane waves on the quadrumer surface[22]. The CARS emission was collected by an objective in transmission and analysed by a spectrometer. When the CARS polarizations are generated in a volume much smaller than its propagation wavelength, and/or in the focused laser fields, the phase matching requirements for nonlinear optics can be fulfilled automatically ($|\Delta k \times z \ll \pi|$ when $z \ll \lambda$)[16,28].

Ensemble-molecule detection. We first performed SECARS with this quadrumer structure on ensemble molecules (see Methods). The input (Fig. 3a) consisted of a pump beam centered at $\omega_\mathrm{P} = 800\,\mathrm{nm}$, and a Stokes continuum $\omega_\mathrm{S} = 830$–$1,050\,\mathrm{nm}$ spanning several hundred nanometers to the long wavelength side of the pump wavelength. Several types of molecules nonresonant at these laser excitation wavelengths have been investigated (Supplementary Fig. 4). The resulting spectrum (Fig. 3b) from p-MA molecules ($\sim 10^3$ molecules bound to the Au surfaces of the probe region of the quadrumer structure) consists of three sharp anti-Stokes features on top of a broadband background. The background has two main contributions: a dominant two-photon luminescence signal[29,30], as well as a much weaker FWM signal[22], both due to the Au nanostructure (Supplementary Fig. 5 and Supplementary Note 1). These background signals were measured before the SECARS experiments on a pristine quadrumer sample. The applied laser power was too low to

affect the stability of the molecule, thus the output spectra were stable during the ~ 10-min acquisition time. Interestingly, the CARS spectrum of p-MA showed quadrumer-to-quadrumer variations, characteristic of a depolarization behaviour with random fluctuations, likely due to variable orientations of molecules within the gap[31,32].

Figure 3c shows the reconstructed SECARS spectrum of p-MA, obtained by averaging three output spectra at different pump-Stokes delays (Supplementary Fig. 6 and Supplementary Note 2), correcting (subtracting) for background noise (Supplementary Fig. 5a), and calibrated (divided) by the quadrumer SECARS enhancement[33] (Supplementary Fig. 1b) and instrument efficiency spectrum at each wavelength. The obtained spectra in wavelengths (λ) were then converted into wavenumbers (k) by the following equation: $\Delta k = 10^7 \times (1/800 - 1/\lambda)$. The broad Raman peaks ($\sim 100\,\mathrm{cm}^{-1}$) are mostly due to the spectral bandwidth of the pump ($\sim 5\,\mathrm{nm}$). Two prominent p-MA modes due to the EM field enhancement[23], a peak at approximately $-1,800\,\mathrm{cm}^{-1}$ due to the overtone of the ring mode[34], and a weak mode at $\sim 1,290\,\mathrm{cm}^{-1}$ possibly due to the combined contribution of CT (charge-transfer between Au surface and adsorbed molecules) and EM mechanisms can be observed. See Supplementary Fig. 7 for the control experiments of normal Raman and SERS, Supplementary Tables 1–3 for complete Raman-band assignments and Supplementary Note 3 for the explanation of the CT contribution. The intensity of the $-1,070\,\mathrm{cm}^{-1}$ p-MA mode (Fig. 3d) was observed to depend cubically on the total input power P_tot, quadratically on the pump power P_P, and linearly with Stokes power P_S. These results follow the third-order nonlinearity scaling of equation (1), further confirming the CARS origin of the spectrum, and demonstrate that the intensities in our experiment have not reached the coherent-Raman saturation threshold[35].

Using this approach we were able to obtain SECARS spectra from a variety of other small molecules. The SECARS spectrum of adenine, one of the constituent bases of DNA and RNA, is dominated by the $-740\,\mathrm{cm}^{-1}$ ring-breathing mode[36] (Fig. 3e). Benzocaine, another polyheterocyclic aromatic molecule, has a SECARS spectrum (Fig. 3f) characterized by five primary modes[37]. The clearly resolvable features of these SECARS spectra indicate the potential for vibrational spectroscopy of these and similar molecular species in ultrasmall quantities.

Single-molecule detection. We then examined whether SECARS obtained using this approach could achieve single-molecule sensitivity for small molecules, which would require[38] an enhancement factor of $\sim 10^{10}$–10^{11}. A polarization-sensitive CARS geometry[39] (with a polarization analyser oriented to transmit y-polarized light placed before the detector) was used to suppress $\sim 99\%$ of the essentially x-polarized[22] background FWM signal, with the quadrumer dimensions slightly modified to reduce the unpolarized background two-photon luminescence (Supplementary Fig. 8 and Supplementary Note 4). We used the bi-analyte method[7,10], where p-MA and adenine (both Raman cross-sections $\sim 10^{-30}\,\mathrm{cm}^2\,\mathrm{sr}^{-1}$ in air for NIR excitation[38,40]) have clearly distinguishable Raman spectra allowing for the unequivocal identification of each molecule. A mixed solution of equal concentrations (100 nM) of the two molecules was drop-casted onto multiple (49) quadrumers for SECARS. This concentration was chosen to ensure that ~ 1.5 molecules fall into each probe region ($250\,\mathrm{nm}^2$ area; $6,250\,\mathrm{nm}^3$ volume) onto the substrate[7,10], based on our estimation (see Methods). Each molecule type should have similar probability of deposition by the drop-casting method at this sufficiently low coverage, as two molecules will not compete for binding on the Au surface

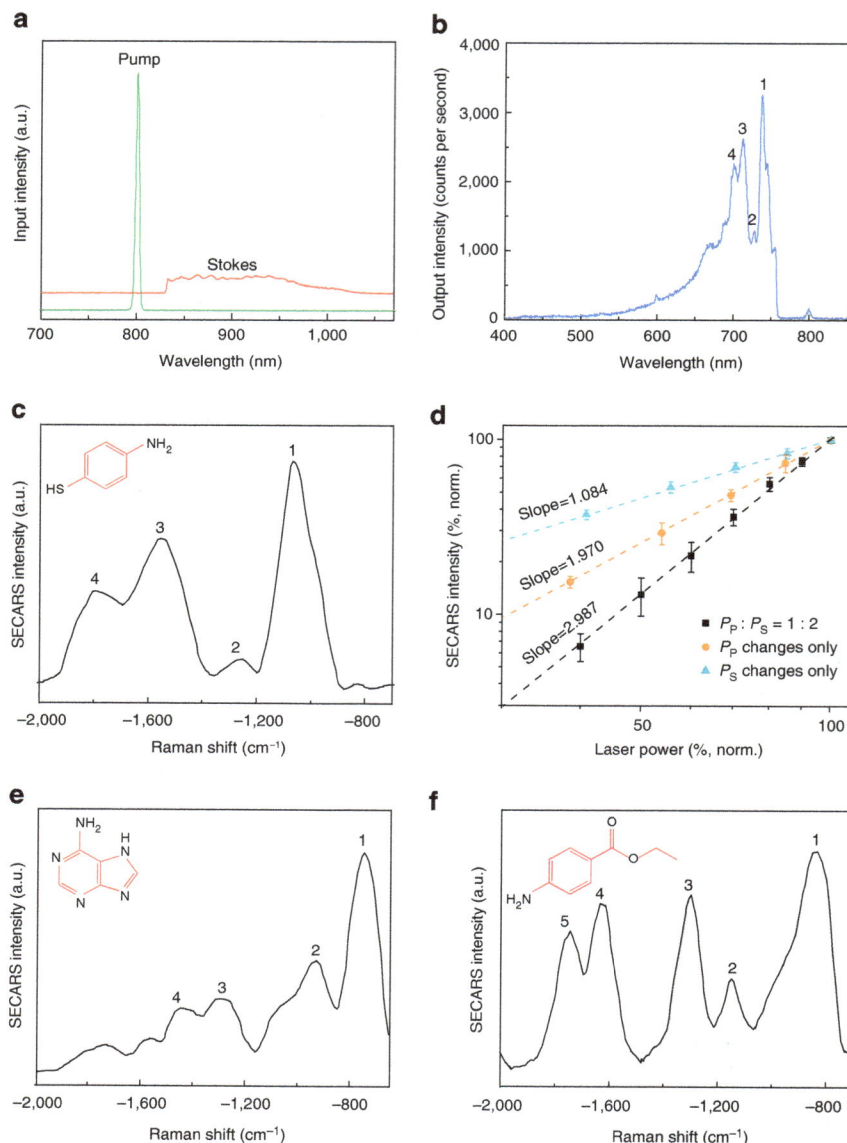

Figure 3 | SECARS experiments of ensemble molecules on the quadrumer. (**a**) Input spectra of the pump (green) and Stokes (red) beams. (**b**) Example output signal of p-MA obtained with $P_P = 33\,\mu W$ and $P_S = 67\,\mu W$. The numbers of photons were recorded with pixel size ~0.4 nm. (**c**) Reconstructed SECARS spectrum versus Raman shift of p-MA shows two prominent peaks at $-1,070\,cm^{-1}$ (1) and $-1,580\,cm^{-1}$ (3), a weak charge-transfer mode at approximately $-1,290\,cm^{-1}$ (2) and an overtone of ring mode at approximately $-1,800\,cm^{-1}$ (4), all corresponding to peaks with same numbers in **b**. (**d**) Normalized (norm.) peak power at $-1,070\,cm^{-1}$ versus normalized total input P_{tot} (black squares, $P_{tot} = 40$–100 μW with $P_1:P_2 = 1{:}2$), P_P change only (orange circles, $P_P = 13$–33 μW with $P_S = 67\,\mu W$) and P_S change only (cyan triangles, $P_S = 27$–67 μW with $P_P = 33\,\mu W$) on a log-log scale. P_S represents the overall power of continuum wavelengths. Dashed lines are linear fitting functions. Error bars represent standard deviations of p-MA CARS signals on five individual quadrumers. (**e**) SECARS spectrum of adenine shows four expected peaks at $-740\,cm^{-1}$ (1), $-950\,cm^{-1}$ (2), $-1,270\,cm^{-1}$ (3) and $-1,450\,m^{-1}$ (4). (**f**) SECARS spectrum of benzocaine shows five expected peaks at $-850\,cm^{-1}$ (1), $-1,160\,cm^{-1}$ (2), $-1,290\,cm^{-1}$ (3), $-1,610\,cm^{-1}$ (4) and $-1,700\,cm^{-1}$ (5).

due to their very large separation (Supplementary Fig. 9 and Supplementary Note 5).

Typical SECARS spectra from three of the measured quadrumers are shown in Fig. 4a (see Supplementary Fig. 10 and Supplementary Note 6 for raw spectra). Two of the spectra shown here correspond extremely closely to the p-MA and adenine SECARS spectra obtained using higher concentrations of each of those respective adsorbate molecules. The 'mixed' spectrum clearly has spectral features that correspond to the spectra of both molecules. At these concentrations, we observed good Raman-band reproducibility among quadrumers with fluctuations in intensities, frequencies and spectral features

(Supplementary Fig. 11), most likely due to different molecular orientations with respect to the analyser and laser polarization, a characteristic that we believe to be a signature of single or a few molecules. In contrast to more concentrated solutions of the two analyte molecules which should always yield a mixed spectrum, we observed, for a series of 49 quadrumers, that the SECARS spectra were dominated by either one analyte or the other, or no molecules detected at all (Fig. 4b). At low concentration, the number of molecules in each probe region should follow a Poisson distribution[7,10] with an expected value μ. We obtained $\mu = 0.89 \pm 0.24$ by fitting the measured data (Supplementary Note 7), close to our estimated value of 1.5. Based on the measured

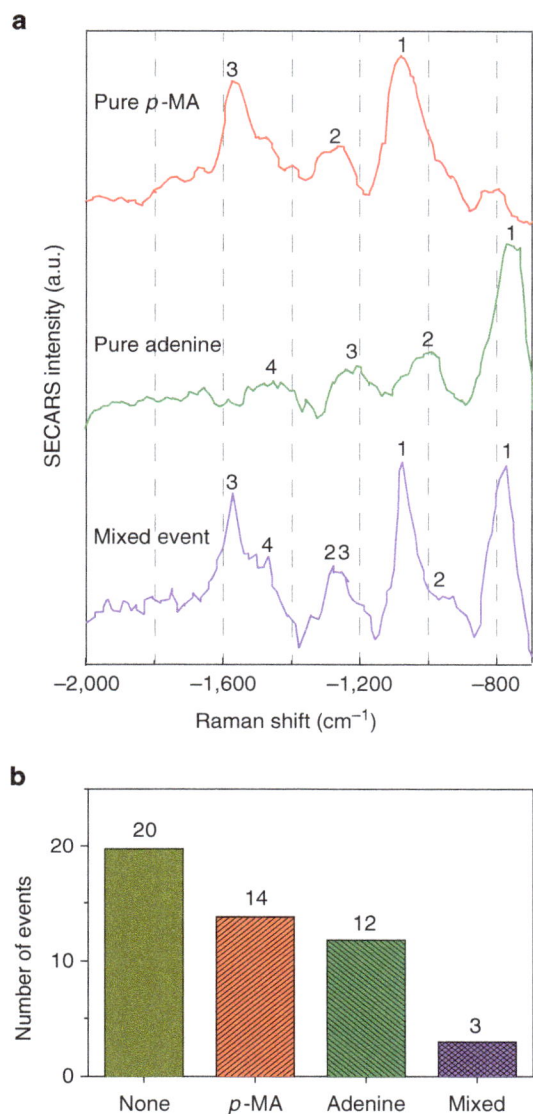

Figure 4 | Single-molecule SECARS detection of bianalytes. (a) Three representative SECARS spectra showing a pure p-MA event (top), a pure adenine event (middle) and a mixed event (bottom). **(b)** Histogram of occurrences of none, pure p-MA, pure adenine and mixed molecules from one sampled array of 49 quadrumers.

signal intensities, our SECARS is $\sim 10^{11}$ times stronger than the spontaneous Raman scattering signal from these molecules (Supplementary Note 8). From our experimental measurements, the SECARS probe region is estimated to be roughly $\sim 167\,nm^2$ in area ($4{,}175\,nm^3$ in volume), only 0.19% of the entire quadrumer surface area (volume), close to the simulation result of $250\,nm^2$ area ($6{,}250\,nm^3$ volume). These observations all provide clear evidence that SECARS obtained in this manner provides single-molecule sensitivity.

Discussion

In conclusion, we have demonstrated SECARS utilizing a Fano-resonant plasmonic quadrumer substrate, which provides an $\sim 10^{11}$ enhancement over spontaneous Raman scattering, and is capable of single-molecule sensitivity with Raman cross-sections σ_{Raman} as low as $\sim 10^{-30}\,cm^2\,sr^{-1}$. These values are among the smallest values for molecules (except perhaps for N_2 and alkane

molecules, which have $\sigma_{Raman} \sim 10^{-31}$ and $10^{-32}\,cm^2\,sr^{-1}$, respectively, at NIR excitations[38,41]). Obtaining single-molecule spectra using a substrate that can be fabricated with standard lithographic methods can, in principle, ultimately open the door to the identification of unknown molecules at the ultimate limits of chemical sensitivity. Such ultimate detection thresholds could have tremendous applicability in fields as diverse as medical diagnostics, forensics, and homeland security.

Methods

SECARS substrates. The quadrumers were fabricated by standard positive-resist electron beam lithography by patterning a 50- or 25-nm thick Au on a 1-nm Ti adhesive layer, which had first been evaporated onto an ultra-smooth fused silica substrate ($1 \times 2\,cm^2$, surface finish 10/5, SVM). The diameters and gaps of the quadrumers were determined by top-view s.e.m. images, and heights were determined by 60° tilted-view s.e.m. images (divided by cos 60°). The standard deviations (error bars) were calculated based on 20 quadrumers. The fabricated quadrumers were arranged in a 7×7 array with 6 μm intercluster distances to avoid coupling effects. The substrate was treated with an oxygen plasma (660 W, Fischione) for 250 s for the removal of residual molecules before experimental use.

SECARS experimental setup. A sample-scanning nonlinear optical microscope was constructed for the single-particle SECARS measurements (Supplementary Fig. 3). An ultrafast Ti:sapphire laser (Mira 900, Coherent Inc.) was used to generate the 800 nm pulses (repetition rate 76 MHz). After passing through a Faraday isolator (EOT Inc.), which allowed only one-way transmission of light, the laser beam was split into two paths. One path formed the pump beam at 800 nm with 5 nm FWHM by two band-pass filters (FB800-10, Thorlabs); the other path generated the Stokes beam at 830–1,050 nm by a nonlinear photonic crystal fibre (Femto-White800, Crystal Fiber A/S) for the Vis-NIR supercontinuum and then passed through a long-pass filter (BLP01-830R-25, Semrock). For each path, the polarization was controlled by a half-wave plate (10RP52-2, Newport) and a linear polarizer (LPVIS100, Thorlabs), and the intensity was attenuated by a neutral density filter (NDL-25C-4, Thorlabs). These two beams were then superimposed spatially using a dichroic mirror (DMSP805, Thorlabs). A motorized linear stage (ILS200CHA, Newport) was inserted into the Stokes path to overlap the pulses in time (pulse durations: pump ~ 200 fs; Stokes ~ 500 fs). The collinear, coherent beams were then focused to ~ 1 μm diameter on an individual quadrumer using a $\times 50/0.8$NA objective (HC PL FLUOTAR, Leica). Substrates were mounted on a three-axis piezoelectric stage (P-545.3R2 NanoXYZ, Physik Instrumente (PI) GmbH) to position the desired quadrumer precisely at the beam focus. The SECARS signal from the molecules was collected by a $60 \times /0.7$NA microscope objective with a variable coverslip correction collar (LUCPlanFL N, Olympus), then transmitted through a pair of short-pass filters (FF01-775/SP-25 and SP01-785RU-25, Semrock) to remove the excitation light. The light was focused onto the entrance slit of a spectrograph (SpectraPro 400i, PI/Acton) by an $f = 40$ mm achromatic doublet lens (AC254-040-A, Thorlabs), and detected by a Si CCD array (PIXIS 100, PI/Acton). The SECARS polarization was analysed by an additional polarizer (LPUV100, Thorlabs). A pair of CCD cameras (Guppy F-146, AVT and UI-2230-C-HQ,IDS) was used to obtain images of the back-scattered and transmitted light to assist in locating the specific quadrumer of interest on the substrate.

Typically, SECARS experiments were performed with a total power of 40–100 μW for ensemble molecules or 600 μW for single molecules, corresponding to a power density of 4–10 kW cm^{-2} or 60 kW cm^{-2} (peak power density 0.13–0.325 GW cm^{-2} or 1.95 GW cm^{-2}). Powers in both cases are smaller than those used in typical SERS experiments (> 100 kW cm^{-2}), but are capable of taking a full SECARS spectrum (-700 to $-2{,}000$ cm^{-1}) within a short time (<1 s).

Ensemble-molecule adsorption. The substrates were soaked in a 10 mM p-MA (4-aminothiophenol, $H_2NC_6H_4SH$, 97%, Aldrich) ethanolic solution, a 1 mM adenine ($C_5H_5N_5$, ≥99%, Sigma) or benzocaine ($C_9H_{11}NO_2$, Sigma) aqueous solutions overnight. A monolayer of p-MA molecules are believed to self-assemble onto the Au quadrumer surface and to bind to the substrate via the strong Au-S bond. Adenine and benzocaine molecules attach to the Au surface due to the Au-NH$_2$ bond. These substrates were then washed thoroughly to remove unattached molecules from the Au substrate before measurement.

Linear optical measurements. Dark-field scattering spectra were obtained by a custom-built microscope with 35° incidence from the substrate normal (Microscope: Axiovert 200 MAT, Zeiss; Objective: 50 × /0.55NA Epiplan-Neofluar, Zeiss; Spectrograph: SP2150, PI/Acton; CCD: PIXIS 400BR, PI/Acton).

Simulation method. Finite-difference-time-domain simulation commercial software (Lumerical FDTD Solutions 7.5.3) was used to obtain the near- and far-field properties of the quadrumer for normal incidence excitation. The top-view geometries were chosen to correspond as closely as possible to the experimentally obtained s.e.m. images of the quadrumer structures, and the heights used in the simulations corresponded to the Au thickness deposited during sample fabrication. A 1.3-nm thick dielectric layer on the side and top surfaces of each disk was used to model the p-MA monolayer. The Johnson and Christy dielectric function[42] was used for Au, $\varepsilon_1 = 2.10$ for the infinite fused silica substrate, and $\varepsilon_2 = 9$ for the p-MA molecules[23]. Charge density plots were obtained by calculating the difference of the normal component of the electric field above and below the Au surface (Gauss's law).

Single-molecule deposition. A volume 25 ml of 200 nM p-MA ethanolic solution and 25 ml of 200 nM adenine aqueous solution were combined and mixed thoroughly. Then 10 µl of such mixture solution was drop-casted onto the SECARS substrate (1×2 cm^2), soaking the entire substrate surface and evaporating in ~ 2 s, such that on average an estimated number of ~ 0.75 p-MA and 0.75 adenine molecules fell into each probe region (~ 250 nm^2 area) of the substrate. This substrate was dried in air overnight to immobilize the probe molecules on the substrate surface, then dried under nitrogen flow before the SECARS measurements.

References

1. Fleischmann, M., Hendra, P. J. & McQuillan, A. J. Raman spectra of pyridine adsorbed at a silver electrode. *Chem. Phys. Lett.* **26**, 163–166 (1974).
2. Jeanmaire, D. L. & Vanduyne, R. P. Surface Raman spectroelectrochemistry.1. Heterocyclic, aromatic, and aliphatic-amines adsorbed on anodized silver electrode. *J. Electroanal. Chem.* **84**, 1–20 (1977).
3. Zhang, R. *et al.* Chemical mapping of a single molecule by plasmon-enhanced Raman scattering. *Nature* **498**, 82–86 (2013).
4. Michaels, A. M., Nirmal, M. & Brus, L. E. Surface enhanced Raman spectroscopy of individual rhodamine 6G molecules on large Ag nanocrystals. *J. Am. Chem. Soc.* **121**, 9932–9939 (1999).
5. Michaels, A. M., Jiang, J. & Brus, L. Ag nanocrystal junctions as the site for surface-enhanced Raman scattering of single Rhodamine 6G molecules. *J. Phys. Chem. B* **104**, 11965–11971 (2000).
6. Xu, H. X., Bjerneld, E. J., Kall, M. & Borjesson, L. Spectroscopy of single hemoglobin molecules by surface enhanced Raman scattering. *Phys. Rev. Lett.* **83**, 4357–4360 (1999).
7. Le Ru, E. C., Meyer, M. & Etchegoin, P. G. Proof of single-molecule sensitivity in surface enhanced Raman scattering (SERS) by means of a two-analyte technique. *J. Phys. Chem. B* **110**, 1944–1948 (2006).
8. Etchegoin, P. G., Meyer, M., Blackie, E. & Le Ru, E. C. Statistics of single-molecule surface enhanced Raman scattering signals: fluctuation analysis with multiple analyte techniques. *Anal. Chem.* **79**, 8411–8415 (2007).
9. Le Ru, E. C., Blackie, E., Meyer, M. & Etchegoin, P. G. Surface enhanced Raman scattering enhancement factors: a comprehensive study. *J. Phys. Chem. C* **111**, 13794–13803 (2007).
10. Dieringer, J. A., Lettan, R. B., Scheidt, K. A. & Van Duyne, R. P. A frequency domain existence proof of single-molecule surface-enhanced Raman spectroscopy. *J. Am. Chem. Soc.* **129**, 16249–16256 (2007).
11. Lim, D. K., Jeon, K. S., Kim, H. M., Nam, J. M. & Suh, Y. D. Nanogap-engineerable Raman-active nanodumbbells for single-molecule detection. *Nat. Mater.* **9**, 60–67 (2010).
12. Kleinman, S. L. *et al.* Single-molecule surface-enhanced Raman spectroscopy of crystal violet isotopologues: theory and experiment. *J. Am. Chem. Soc.* **133**, 4115–4122 (2011).
13. Begley, R. F., Harvey, A. B. & Byer, R. L. Coherent anti-Stokes Raman spectroscopy. *Appl. Phys. Lett.* **25**, 387–390 (1974).
14. Duncan, M. D., Reintjes, J. & Manuccia, T. J. Scanning coherent anti-Stokes Raman microscope. *Opt. Lett.* **7**, 350–352 (1982).
15. Zumbusch, A., Holtom, G. R. & Xie, X. S. Three-dimensional vibrational imaging by coherent anti-Stokes Raman scattering. *Phys. Rev. Lett.* **82**, 4142–4145 (1999).
16. Cheng, J. X., Volkmer, A. & Xie, X. S. Theoretical and experimental characterization of coherent anti-Stokes Raman scattering microscopy. *J. Opt. Soc. Am. B* **19**, 1363–1375 (2002).
17. Shi, K. B., Li, H. F., Xu, Q., Psaltis, D. & Liu, Z. W. Coherent anti-Stokes Raman holography for chemically selective single-shot nonscanning 3D imaging. *Phys. Rev. Lett.* **104**, 093902 (2010).
18. Steuwe, C., Kaminski, C. F., Baumberg, J. J. & Mahajan, S. Surface enhanced coherent anti-Stokes Raman scattering on nanostructured gold surfaces. *Nano Lett.* **11**, 5339–5343 (2011).
19. Liang, E. J., Weippert, A., Funk, J. M., Materny, A. & Kiefer, W. Experimental observation of surface-enhanced coherent anti-Stokes-Raman scattering. *Chem. Phys. Lett.* **227**, 115–120 (1994).
20. Voronine, D. V. *et al.* Time-resolved surface-enhanced coherent sensing of nanoscale molecular complexes. *Sci. Rep.* **2**, 891 (2012).
21. Koo, T. W., Chan, S. & Berlin, A. A. Single-molecule detection of biomolecules by surface-enhanced coherent anti-Stokes Raman scattering. *Opt. Lett.* **30**, 1024–1026 (2005).
22. Zhang, Y., Wen, F., Zhen, Y. R., Nordlander, P. & Halas, N. J. Coherent Fano resonances in a plasmonic nanocluster enhance optical four-wave mixing. *Proc. Natl Acad. Sci. USA* **110**, 9215–9219 (2013).
23. Ye, J. *et al.* Plasmonic nanoclusters: near field properties of the Fano resonance interrogated with SERS. *Nano Lett.* **12**, 1660–1667 (2012).
24. Stockman, M. I. Nanoscience: dark-hot resonances. *Nature* **467**, 541–542 (2010).
25. Gallinet, B., Siegfried, T., Sigg, H., Nordlander, P. & Martin, O. J. F. Plasmonic radiance: probing structure at the angstrom scale with visible light. *Nano Lett.* **13**, 497–503 (2013).
26. Thyagarajan, K., Butet, J. & Martin, O. J. F. Augmenting second harmonic generation using Fano resonances in plasmonic systems. *Nano Lett.* **13**, 1847–1851 (2013).
27. Fan, J. A. *et al.* Fano-like interference in self-assembled plasmonic quadrumer clusters. *Nano Lett.* **10**, 4680–4685 (2010).
28. Ichimura, T., Hayazawa, N., Hashimoto, M., Inouye, Y. & Kawata, S. Tip-enhanced coherent anti-Stokes Raman scattering for vibrational nanoimaging. *Phys. Rev. Lett.* **92**, 220801 (2004).
29. Wang, H. F. *et al.* In vivo and in vitro two-photon luminescence imaging of single gold nanorods. *Proc. Natl Acad. Sci. USA* **102**, 15752–15756 (2005).
30. Durr, N. J. *et al.* Two-photon luminescence imaging of cancer cells using molecularly targeted gold nanorods. *Nano Lett.* **7**, 941–945 (2007).
31. Nagasawa, F., Takase, M., Nabika, H. & Murakoshi, K. Polarization characteristics of surface-enhanced Raman scattering from a small number of molecules at the gap of a metal nano-dimer. *Chem. Commun.* **47**, 4514–4516 (2011).
32. Bosnick, K. A., Jiang, J. & Brus, L. E. Fluctuations and local symmetry in single-molecule rhodamine 6G Raman scattering on silver nanocrystal aggregates. *J. Phys. Chem. B* **106**, 8096–8099 (2002).
33. Lee, K. & Irudayaraj, J. Correct spectral conversion between surface-enhanced raman and plasmon resonance scattering from nanoparticle dimers for single-molecule detection. *Small* **9**, 1106–1115 (2013).
34. Gibson, J. W. & Johnson, B. R. Density-matrix calculation of surface-enhanced Raman scattering for p-mercaptoaniline on silver nanoshells. *J. Chem. Phys.* **124**, 064701 (2006).
35. Patnaik, A. K., Roy, S. & Gord, J. R. Saturation of vibrational coherent anti-Stokes Raman scattering mediated by saturation of the rotational Raman transition. *Phys. Rev. A* **87**, 043801 (2013).
36. Otto, C., Vandentweel, T. J. J., Demul, F. F. M. & Greve, J. Surface-enhanced Raman spectroscopy of DNA Bases. *J. Raman Spectrosc.* **17**, 289–298 (1986).
37. Palafox, M. A. Raman spectra and vibrational analysis for benzocaine. *J. Raman Spectrosc.* **20**, 765–771 (1989).
38. Blackie, E. J., Le Ru, E. C. & Etchegoin, P. G. Single-molecule surface-enhanced Raman spectroscopy of nonresonant molecules. *J. Am. Chem. Soc.* **131**, 14466–14472 (2009).
39. Cheng, J. X., Book, L. D. & Xie, X. S. Polarization coherent anti-Stokes Raman scattering microscopy. *Opt. Lett.* **26**, 1341–1343 (2001).
40. Rigler, R. & Vogel, H. *Single Molecules and Nanotechnology* (Springer, 2008).
41. Murphy, W. F., Fernandezsanchez, J. M. & Raghavachari, K. Harmonic forcefield and Raman scattering intensity parameters of n-butane. *J. Phys. Chem.* **95**, 1124–1139 (1991).
42. Johnson, P. B. & Christy, R. W. Optical constants of noble metals. *Phys. Rev. B* **6**, 4370–4379 (1972).

Acknowledgements

We thank Nche T. Fofang, Amanda Goodman, Fangfang Wen, Linan Zhou, Sandra Whaley Bishnoi and Nathaniel Hogan for helpful discussions. This work was supported by a DoD National Security Science and Engineering Faculty Fellowship (N00244-09-1-0067), the Defense Threat Reduction Agency (HDTRA1-11-1-0040) and the Robert A. Welch Foundation (C-1220 and C-1222).

Author contributions

N.J.H., P.N. and Y.Z. designed the experiments. Y.Z. prepared samples and performed experiments. Y.-R.Z. and Y.Z. analysed the data. Y.Z., N.J.H. and J.K.D. wrote the paper. All authors participated in discussions of this work.

Additional information

A platform for designing hyperpolarized magnetic resonance chemical probes

Hiroshi Nonaka[1], Ryunosuke Hata[1], Tomohiro Doura[1], Tatsuya Nishihara[1], Keiko Kumagai[2], Mai Akakabe[2], Masashi Tsuda[3], Kazuhiro Ichikawa[4] & Shinsuke Sando[1,4]

Hyperpolarization is a highly promising technique for improving the sensitivity of magnetic resonance chemical probes. Here we report [^{15}N, D$_9$]trimethylphenylammonium as a platform for designing a variety of hyperpolarized magnetic resonance chemical probes. The platform structure shows a remarkably long ^{15}N spin–lattice relaxation value (816 s, 14.1 T) for retaining its hyperpolarized spin state. The extended lifetime enables the detection of the hyperpolarized ^{15}N signal of the platform for several tens of minutes and thus overcomes the intrinsic short analysis time of hyperpolarized probes. Versatility of the platform is demonstrated by applying it to three types of hyperpolarized chemical probes: one each for sensing calcium ions, reactive oxygen species (hydrogen peroxide) and enzyme activity (carboxyl esterase). All of the designed probes achieve high sensitivity with rapid reactions and chemical shift changes, which are sufficient to allow sensitive and real-time monitoring of target molecules by ^{15}N magnetic resonance.

[1] INAMORI Frontier Research Center, Kyushu University, 744 Motooka, Nishi-ku, Fukuoka 819 0395, Japan. [2] Science Research Center, Kochi University, Kochi 783 8506, Japan. [3] Center for Advanced Marine Core Research, Kochi University, Kochi 783 8502, Japan. [4] Innovation Center for Medical Redox Navigation, Kyushu University, Fukuoka 812 8582, Japan. Correspondence and requests for materials should be addressed to S.S. (email: ssando@ifrc.kyushu-u.ac.jp).

Considerable effort has long been dedicated to the molecular analysis of living systems. In particular, molecular analysis has recently been attempted for complex systems in cell assembly, tissue, organ and body. Magnetic resonance (MR)-based techniques—MR imaging (MRI) or MR spectroscopy—are the powerful approaches for such *in situ* molecular analysis, and various MR chemical probes (MR probes) have been designed[1]. However, these have an intrinsic limitation for practical applications, namely their low sensitivity.

Hyperpolarization is a highly promising technique for overcoming this limitation[2,3]. The hyperpolarization technique achieves polarization of nuclear spin populations, producing a large enhancement of sensitivity for MR-detectable nuclei. The technique has been applied successfully for *in vitro* or *in vivo* metabolic analyses using stable isotope-enriched natural compounds (metabolites), including *N*-acetylated amino acids, pyruvate, fructose, choline and glucose[4–8].

It was recently demonstrated that the hyperpolarization technique can also be applied to chemical sensors for surveying the chemical status of living systems. In practice, hyperpolarized [^{13}C]bicarbonate[9], [^{13}C]benzoylformic acid[10], [^{13}C, D$_6$] *p*-anisidine[11] and [^{13}C]dehydroascorbate[12,13] have been designed as sensitive MR probes for sensing pH, H_2O_2, HOCl and redox status, respectively. A universal strategy—in other words, the presence of a platform structure for designing hyperpolarized MR probes—can make it easier to develop a variety of hyperpolarized MR sensors.

The importance of a platform structure is obvious, as demonstrated in the design of optical probes. For example, in the case of fluorescent probes, some chromophores work as a platform[14]. A good representative is fluorescein (Fig. 1a). A variety of fluorescent probes have been developed from this fluorophore platform using a well-established strategy (*vide infra*). However, corresponding structures for hyperpolarized MR probes have not yet been realized.

Here we propose [^{15}N, D$_9$]trimethylphenylammonium ([^{15}N, D$_9$]TMPA) as a promising platform structure for designing hyperpolarized MR probes. It achieves improved sensitivity with a remarkably long hyperpolarization lifetime (^{15}N, $T_1 = 816$ s, 14.1 T). The versatile applicability of the platform structure is established by designing three types of hyperpolarized MR probes, one each targeting metal ion (Ca^{2+}), reactive oxygen species (H_2O_2), and enzyme (carboxyl esterase).

Results

Design of platform structure for hyperpolarized MR probes.

Typically, a platform structure in an optical imaging probe is composed of signalling, aromatic and sensing moieties (Fig. 1), where the aromatic unit works as a connector to transmit chemical events on a sensing moiety (R in Fig. 1) to a signalling moiety[15,16]. In the case of fluorescein, benzoic acid and xanthene chromophore act as the aromatic and signalling moieties, respectively (Fig. 1a). Derivatization of benzoic acid (the aromatic moiety) with sensing moieties enables the generation of various signal-on-type fluorescent probes, for example, Fluo-2 for Ca^{2+}[17], DAF-2 for NO[18] and DNAF1 for glutathione *S*-transferase (GST) sensing[19].

In the present case, the signalling moieties are MR-detectable hyperpolarized nuclei (Fig. 1b). When attempting to design a platform for hyperpolarized MR probes, one critical issue is the short lifetime of the hyperpolarized spin state of the nuclei (signalling moieties). For example, the hyperpolarization lifetime of a ^{13}C MR probe is only a few tens of seconds at best, which restricts its application to the analysis of extremely fast kinetic events. Therefore, the challenge is to find a hyperpolarized

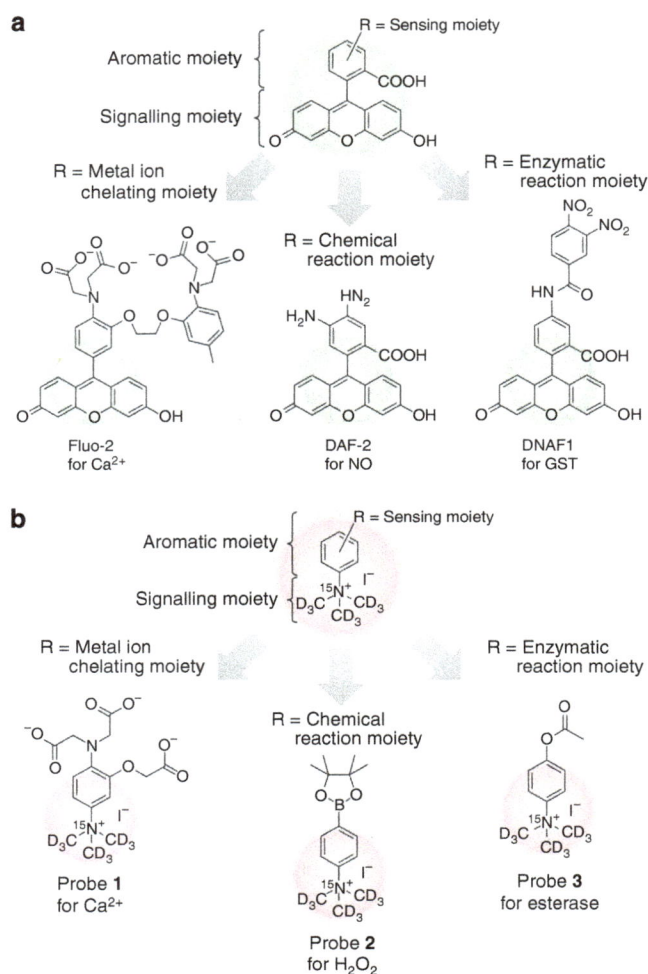

Figure 1 | Platform for designing chemical probes. (**a**) Fluorescein as a platform for designing fluorescent probes. Fluorescein comprises both aromatic and signalling moieties. The fluorescence quantum yield of the signalling moiety can be tuned by the highest occupied molecular orbital (HOMO)/lowest unoccupied molecular orbital (LUMO) level of the aromatic moiety. In three examples—Fluo-2, DAF-2 and DNAF1—the HOMO/LUMO level of the aromatic moiety is changed on the binding or reaction of Ca^{2+}, NO and glutathione *S*-transferase enzyme with the sensing moiety, respectively, leading to light emission from the fluorescent sensors. (**b**) Proposed platform for designing hyperpolarized MR probes. Various hyperpolarized MR probes can be designed by the same strategy used for converting fluorophore platforms to fluorescent sensors, as shown in Fig. 1a. The chemical structures of probes **1-3** used in this study are shown.

nucleus or structure that affords a much longer hyperpolarization lifetime.

The hyperpolarization lifetime is related directly to the spin–lattice relaxation time (T_1)[20]. The T_1 of ^{15}N nuclei in organic compounds is usually longer than those of ^{1}H and ^{13}C nuclei[21]. In addition, this spin–lattice relaxation is caused mainly by dipole–dipole interaction, spin–rotation interaction and chemical shift anisotropy[22–24]. Therefore, typically, ^{15}N nuclei, which have less neighbouring protons in small and rigid structures tend to give a longer T_1 value, achieving a longer hyperpolarization lifetime. Actually, the ^{15}N nucleus of choline (^{15}N(CH$_3$)$_3$CH$_2$CH$_2$OH) has been shown to produce a long hyperpolarization lifetime[7]. With this in mind, we designed [^{15}N]trimethylphenylammonium ([^{15}N]TMPA or [^{15}N]trimethylaniline) as a candidate for the platform structure (Fig. 1b). We anticipated that [15N]TMPA, which has a $-^{15}$N(CH$_3$)$_3$ signalling moiety on an aromatic moiety, might serve as a suitable platform for designing hyperpolarized MR probes.

Long hyperpolarization lifetime of the platform structure. The [15N]TMPA was synthesized from [15N]aniline by nucleophilic displacement with CH$_3$I. The hyperpolarization lifetime of [15N]TMPA was evaluated by measuring the T_1 value (Fig. 2a). The ^{15}N T_1 value of [15N]TMPA was determined as 275 ± 11 s (14.1 T, D$_2$O, 30 °C), which was much longer than that of the practically used [1-^{13}C]pyruvic acid (41 s, 14.1 T, D$_2$O, 30 °C). Interestingly, this value is longer than that of [15N]choline (232 s, 14.1 T, D$_2$O, 30 °C). Reduced interaction with proton (less dipole–dipole interaction) or structural rigidity (less spin–rotation interaction) might explain this longer T_1 value[22–24].

The T_1 value was further extended by deuteration of [15N] TMPA. Non-proton-coupled nuclei tend to show a longer T_1 value because of the lack of dipole–dipole interactions with neighbouring protons. In this sense, deuteration is one of the most straightforward ways to increase the hyperpolarization lifetime[25–28]. We prepared [15N, D$_9$]TMPA, wherein all the methyl protons were replaced with deuterium atoms using CD$_3$I instead of CH$_3$I. As a result, the [15N, D$_9$]TMPA afforded a remarkably long ^{15}N T_1 value of 816 ± 15 s (14.1 T, D$_2$O, 30 °C; 754 ± 23 s in 90% H$_2$O; Fig. 2a), which was 19.9-, 3.5- and 1.3-fold longer than those of [1-^{13}C]pyruvic acid, [15N]choline and [15N, D$_9$]choline,

respectively. To the best of our knowledge, this T_1 value is the longest among the ^{15}N compounds reported to date.

The [15N, D$_9$]TMPA was efficiently hyperpolarized by dynamic nuclear polarization (DNP) using trityl radicals[29]. The sensitivity of the hyperpolarized sample increased and allowed detection of the targeted ^{15}N by a single scan (%$P_{15N} = 2.0\%$, $T = 298$ K, $B_0 = 9.4$ T, 1.5 h polarization). The high sensitivity was obvious when compared with the thermally equilibrated spectrum (Fig. 2b). As little as 10 μM of hyperpolarized [15N, D$_9$]TMPA could be detected (S/N ratio = 3) using a single-scan ^{15}N analysis under our experimental conditions (flip angle = 90°). In addition, because of its remarkably long T_1 value, the hyperpolarized state continued after dissolution of the hyperpolarized sample (stacked spectra; Fig. 2c). These results indicate that deuterated [15N, D$_9$]TMPA has a considerable potential for use as a remarkably long-lived and sensitive hyperpolarization unit.

Hyperpolarized MR probe targeting calcium ions. With the [15N, D$_9$]TMPA platform in hand, we then demonstrated its practical utility by designing new hyperpolarized MR probes. These needed to satisfy the following prerequisites: (1) the probe should have a MR-detectable nucleus with a long T_1 for long hyperpolarization; (2) it should bind/react with the target species rapidly within the hyperpolarization lifetime; and (3) it should induce a sufficiently large chemical shift change upon reaction.

As a first choice, we aimed to develop the hyperpolarized MR probe targeting the calcium ion (Ca^{2+}), a biologically important metal ion[30]. In addition to their biological importance, abnormal Ca^{2+} concentrations in the blood (hyper- or hypocalcemia) are known to be associated with some diseases[31,32]; therefore, the *in situ* analysis and imaging of Ca^{2+} concentrations in the body is potentially useful for an investigation of the mechanism or an early diagnosis of these diseases. We designed MR probe **1** (Fig. 3a), wherein the [15N, D$_9$]TMPA (aromatic and signalling moieties) has been substituted with triacetic acid as a Ca^{2+}-chelating group (sensing moiety)[33]. MR probe **1** was synthesized from the methyl ester of *o*-aminophenol-*N, N, O*-triacetic acid (APTRA), a known Ca^{2+} chelator, in four steps (Supplementary Methods). The absorption analyses confirmed that probe **1** bound to Ca^{2+} rapidly with an affinity of $K_d = 490$ μM (Supplementary Fig. S1a,d), with one-to-one binding stoichiometry (Supplementary Fig. S1b,c) and high selectivity over Mg^{2+} or K$^+$ (Supplementary Fig. S1e).

The sensitivity of MR probe **1** was enhanced dramatically by DNP. As expected, the ^{15}N of MR probe **1** had a long T_1 value (129 ± 22 s, 9.4 T) and the hyperpolarized state of ^{15}N signal was observed by ^{15}N single-scan nuclear magnetic resonance (NMR; 600 s under our experimental conditions, 10 mM of **1**, Supplementary Fig. S2).

The hyperpolarized MR probe **1** worked as a chemical shift-switching Ca^{2+} sensor. Figure 3b shows the single-scan ^{15}N NMR spectra of hyperpolarized MR probe **1** (0.5 mM) in the presence of various concentrations of Ca^{2+} (0–10 mM). The presence of Ca^{2+} induced a ^{15}N chemical shift change (from 49.5 to 51.0 p.p.m.; $\Delta\delta = \sim 1.5$ p.p.m.) in a Ca^{2+} concentration-dependent manner, which was sufficient to be detected by ^{15}N DNP–NMR analysis (Fig. 3b,c). In marked contrast, only a small chemical shift change ($\delta = 0.3$ p.p.m.) was observed in the presence of excess Mg^{2+} (10 mM) (Fig. 3c).

Importantly, the hyperpolarized Ca^{2+} probe worked in biological samples. In blood serum, T_1 value of MR probe **1** was not shortened (142 ± 2 s, 9.4 T, in blood serum containing 50% v/v D$_2$O). Thus, ^{15}N signals of hyperpolarized MR probe **1** (0.5 mM) were detectable in human blood (Fig. 3d). The observed signal could be discriminated clearly from those in blood samples

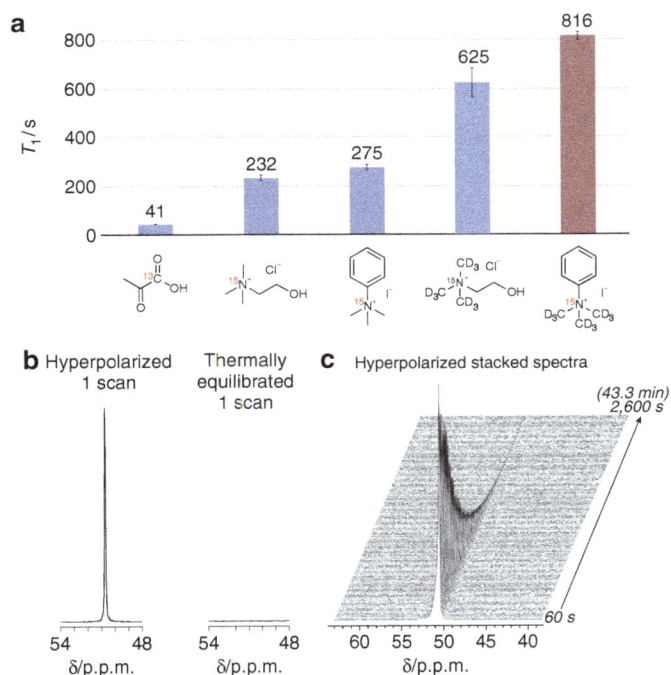

a

Figure 2 | Properties of proposed platform [15N, D$_9$]TMPA. (a) Spin-lattice relaxation time T_1 (14.1 T, D$_2$O, 30 °C) of ^{13}C (450 mM) or ^{15}N (200-300 mM) nuclei of the chemical compound shown at the bottom. Error bars indicate a s.d. of five saturation recovery measurements. **(b)** Single-scan ^{15}N NMR spectra of hyperpolarized (40 s after dissolution) or thermally equilibrated [15N, D$_9$]TMPA (10 mM). **(c)** Single-scan ^{15}N NMR spectra of hyperpolarized [15N, D$_9$]TMPA stacked from ca. 60-2,600 s (every 20 s, 128 times) after dissolution of the hyperpolarized [15N, D$_9$]TMPA (5 mM). The pulse angles for ^{15}N measurements in **b** and **c** were 90° and 13°, respectively.

with a Ca^{2+} excess (10 mM Ca^{2+}, added externally, top spectrum) and a Ca^{2+} deficiency (10 mM EDTA, added externally, bottom spectrum). Estimated from a calibration curve in human serum (Supplementary Fig. S3), the observed ^{15}N signal in blood corresponded to 1.04 mM of Ca^{2+} (typical total Ca concentration in blood (50% v/v) = 1 ~ 1.25 mM). This value was close to that (1.15 mM) determined using a classical optical sensing method for Ca^{2+}. The small difference between the results from MR and optical analyses might be caused by a difference of protocols. In the case of the optical sensing of Ca^{2+} concentration in blood, a purification step is indispensable because the inherent light absorption by blood interferes with the optical measurements, as shown in Fig. 3e (left). In fact, we prepared blood plasma by centrifugation and used it for optical Ca^{2+} sensing. On the other hand, hyperpolarized MR analysis can be carried out in blood directly. This *in situ* (in blood) applicability is an advantage of the present calcium-sensing MR probe.

To show the applicability of the hyperpolarized MR probe **1**, we applied probe **1** for Ca^{2+} imaging in blood (Fig. 3e). The ^{15}N signal of the probe **1** + Ca^{2+} complex was imaged (Fig. 3e).

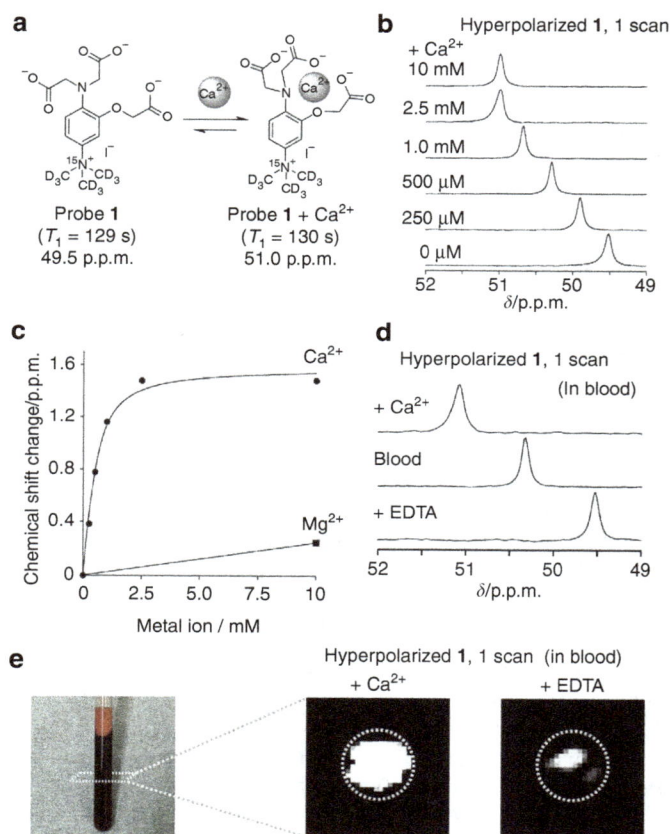

An image with good contrast was obtained in blood samples with a Ca^{2+} excess (8 mM Ca^{2+}, added externally, left image), whereas weak contrast was observed in blood with a Ca^{2+} deficiency (8 mM EDTA, added externally, right image).

These results indicate that the hyperpolarized MR probe **1** works as an *in situ* Ca^{2+} sensor with high sensitivity even in human blood.

Versatility of the platform. The versatility of the platform was confirmed by designing two other hyperpolarized MR probes targeting different molecules but by the same strategy. Probe **2** was designed as a hyperpolarized MR probe targeting H_2O_2 (Fig. 4a), which is one of major disease-related reactive oxygen species[34,35]. H_2O_2 production is associated with endothelial inflammatory responses[35] and the increased production level of H_2O_2 in tumours is correlated with cancer cell growth and malignancy[36]. The probe has an H_2O_2-reactive boronic acid ester

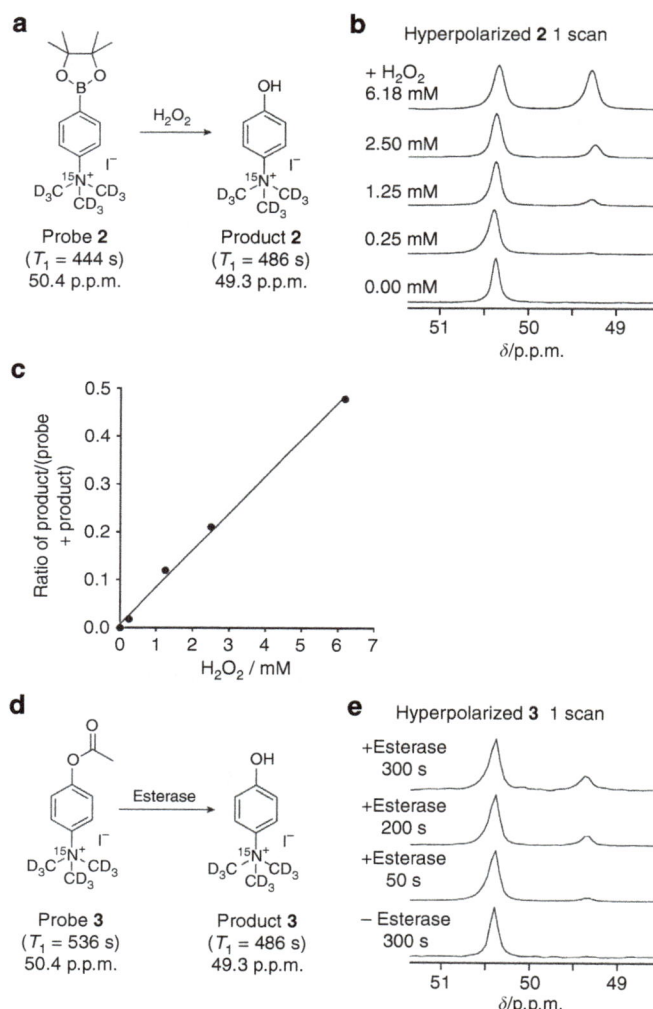

Figure 3 | Hyperpolarized MR probe targeting calcium ions. (a) Ca^{2+} sensing by MR probe **1**. **(b)** Single-scan ^{15}N NMR spectra of hyperpolarized probe **1** (0.5 mM) with various concentrations of Ca^{2+} in HEPES pH 7.4 (40 s after mixing, 30° pulse angle). **(c)** Plot of ^{15}N chemical shift change of hyperpolarized probe **1** (0.5 mM) versus concentrations of Ca^{2+} (circles) and Mg^{2+} (square). **(d)** Single-scan ^{15}N NMR spectra (30° pulse angle) of hyperpolarized probe **1** (0.5 mM) in human blood containing 50% v/v HEPES buffer (middle) without or (top) with 10 mM Ca^{2+} or (bottom) with 10 mM EDTA. **(e)** Single-scan ^{15}N MRI image of hyperpolarized probe **1** (8 mM) in HEPES buffer containing 20% v/v human blood with (left) 8 mM of Ca^{2+} or (right) 8 mM of EDTA. The photograph of Ca^{2+}-added sample is shown in left.

Figure 4 | Hyperpolarized MR probes targeting hydrogen peroxide and carboxyl esterase. (a) H_2O_2 sensing by MR probe **2**. **(b)** Single-scan ^{15}N NMR spectra of hyperpolarized probe **2** (2.5 mM) mixed with various concentrations (0, 0.25, 1.25, 2.50 and 6.18 mM) of H_2O_2 in phosphate buffer pH 7.4 (50 s after mixing, 30° pulse angle). **(c)** Plot of product **2**/(probe **2** + product **2**) peak integral ratios versus concentrations of H_2O_2, $R^2 = 0.996$ for linear fitting. **(d)** Esterase activity sensing by MR probe **3**. **(e)** Single-scan ^{15}N NMR spectra of hyperpolarized probe **3** (10 mM, 15° pulse angle) after mixing with esterase (124 units ml^{-1}, derived from the porcine liver) in PBS (pH 7.4).

(the sensing moiety) on the [^{15}N, D$_9$]TMPA unit (Supplementary Methods)[37]. After reaction with H$_2$O$_2$, the boronic acid was expected to convert to a hydroxyl group and such functional group transformation would induce a chemical shift change of the hyperpolarized ^{15}N to function as a chemical shift-switching MR probe. This proved to be the case. The T_1 values of probe 2 and product 2 were determined as 444 ± 11 and 486 ± 66 s (9.4 T), respectively, which were sufficiently long to be monitored by ^{15}N DNP–NMR spectroscopy. The apparent reaction kinetics were very rapid at $(4.8 \pm 0.4) \times 10^{-3}$ s^{-1} (Supplementary Fig. S4). As shown in the single-scan ^{15}N NMR spectra of Fig. 4b, a new signal of product 2 (49.3 p.p.m.) was observed from single-scan after starting the ^{15}N NMR measurement (corresponding to 50 s after mixing the hyperpolarized probe 2 with 0–6.18 mM of H$_2$O$_2$). Because of almost the same T_1 values of probe and product—that is, almost the same decay rate of hyperpolarized spin state—the signal ratio of the hyperpolarized product to the amount of probe and product was proportional to the concentration of H$_2$O$_2$ (Fig. 4b), displaying a good linear correlation with increasing concentrations (Fig. 4c).

In addition, the platform was applied successfully in designing a hyperpolarized MR probe for analysing carboxyl esterase activity. The carboxyl esterase is a biomarker of cancer[38] and one of the major enzymes related to drug metabolism and pro-drug activation[39]. For example, human carboxyl esterase 2 is commonly expressed in tumour tissues and is correlated with the activation of anticancer drugs[40]. Therefore, detection of carboxyl esterase is biologically and medically significant. Probe 3, with a methyl ester moiety, was designed (Supplementary Methods) as a hyperpolarized MR probe for esterase (Fig. 4d) and incubated with a model carboxyl esterase derived from porcine liver. As with probes 1 and 2, probe 3 also showed a long T_1 value (536 ± 33 s for the probe and 486 ± 66 s for its product at 9.4 T), sufficient enhancement of signal intensity and ^{15}N chemical shift change (1.1 p.p.m.) after reaction with esterase. A new ^{15}N signal of product 3 (49.3 p.p.m.) appeared in the presence of carboxyl esterase (Fig. 4e), in addition to the parent peak of probe 3 (50.4 p.p.m.). This allowed us to detect the presence of carboxyl esterase from the hyperpolarized ^{15}N chemical shift analysis.

Discussion

We propose [^{15}N, D$_9$]TMPA as a suitable platform for designing various hyperpolarized MR probes. The significance of this study can be summarized as follows. First is the proposed platform's high performance. The [^{15}N, D$_9$]TMPA platform achieved good hyperpolarization and a remarkably long hyperpolarization lifetime with the longest T_1 value (816 s, 14.1 T, D$_2$O) among the ^{15}N compounds reported to date. This extended lifetime enabled the detection of the hyperpolarized ^{15}N signal of the platform for several tens of minutes under our experimental conditions, approaching the lifetimes of molecular probes used for positron emission tomography[41]. This overcomes the intrinsic short analysis time of hyperpolarized probes. Given that existing hyperpolarized chemical probes (typically ^{13}C-based) have much shorter T_1 values (≤ 60 s), this long-lived hyperpolarized chemical probe is useful because it allows easy handling, sufficient distribution through the body and long duration measurements of targeted biological events. In addition, the longer hyperpolarization can lower the probe concentration required. This is a distinct advantage of the present platform. The second important aspect of this platform is its ease of incorporation into sensors. It comprises signalling (hyperpolarized ^{15}N) and aromatic (benzene ring) moieties. The platform can be converted to a hyperpolarized ^{15}N MR probe by the same strategy used for designing fluorescent sensors (Fig. 1a), that is, by the simple derivatization of an

aromatic moiety with an appropriate sensing moiety. As various fluorescent probes have already been designed using this strategy, the [^{15}N, D$_9$]TMPA platform has high potential to be diversified to create hyperpolarized MR sensors targeting various biochemical events. The third advantage of [^{15}N, D$_9$]TMPA is its versatility. Three different types of hyperpolarized MR probes were designed successfully from the same platform (Fig. 1b). All of the designed compounds worked as sensitive, selective and fast responsive hyperpolarized MR probes. Further, it was demonstrated that the designed hyperpolarized MR sensor could be utilized for ^{15}N MRI of target biomolecules in blood. These findings demonstrate the considerable potential of [^{15}N, D$_9$]TMPA as a basis for designing a variety of hyperpolarized MR probes.

Although the present research showed the high potential of the platform for generating hyperpolarized MR probes, there are still aspects to be improved. Practical in vivo applications of these probes must await further studies on biostability, toxicity and distribution. However, as demonstrated for fluorescent probes, these factors could be overcome by making improvements to the probes or the platform itself. In fact, preliminary experiments showed that the cytotoxicity and inhibitory activity against acetylcholine esterase could be suppressed markedly by appropriate substitutions to the TMPA platform (probes 1 and 3 showed almost no cytotoxicity at the low mM range, Supplementary Fig. S5). In addition, efforts should be made towards development of a clinical ^{15}N scanner, optimized for the hyperpolarized ^{15}N sensor.

Methods

General information on synthesis. Reagents and solvents were purchased from standard suppliers and used without further purification. Gel permeation chromatography (GPC) was performed on JAIGEL GS310 using a JAI Recycling Preparative HPLC LC-9201. NMR spectra were measured using a Bruker Avance III spectrometer (400 MHz for ^{1}H). Methanol-d$_4$ (3.31 p.p.m.) or D$_2$O (4.79 p.p.m.) was used as the internal standard for ^{1}H NMR. Methanol-d$_4$ (49.0 p.p.m.) and methanol in D$_2$O (49.5 p.p.m.) were used as the internal standard for ^{13}C NMR. Choline chloride-^{15}N (43.4 p.p.m.) was used as the external standard for ^{15}N NMR. Mass spectra were measured using a JEOL JMS-HX110A fast atom bombardment (FAB).

Synthesis of [^{15}N, D$_9$]choline chloride. Potassium carbonate (4.46 g, 32.3 mmol) and [D$_3$]iodomethane (3.12 g, 21.5 mmol) were added to [^{15}N]ethanolamine (334 mg, 5.38 mmol) in dry methanol (15 ml), and the mixture was stirred under nitrogen atmosphere at room temperature for 12 h. After insoluble inorganic salt was removed by filtration, the filtrate was evaporated under reduced pressure. The residue was mixed with small amount of dry methanol, filtered and the filtrate was evaporated. The residue was washed with ethyl acetate:methanol = 10:1 and the remaining solid was collected to give [^{15}N, D$_9$]choline iodide as a pale yellow solid (741 mg, 59%): ^{1}H NMR (CD$_3$OD, 400 MHz) $\delta = 3.60$–3.63 (m, 2H), 4.04–4.07 (m, 2H); ^{13}C NMR (CD$_3$OD, 100 MHz) $\delta = 55.8$, 67.4; ^{15}N NMR (CD$_3$OD, 40 MHz) $\delta = 43.8$. Silver oxide (1.53 g, 6.59 mmol) was added to [^{15}N, D$_9$]choline iodide (741 mg, 3.19 mmol) in dry methanol (10 ml) and the mixture was stirred for 30 min. Solids were removed by filtration. HCl aqueous solution (0.5 M) was added dropwise to the filtrate until pH became 4 and then the solvent was evaporated under reduced pressure. Dry ethanol (10 ml) was added to the residue and insoluble solids were removed by filtration. The solvent was evaporated under reduced pressure from the filtrate to give [^{15}N, D$_9$]choline chloride as a pale yellow solid (147 mg, 31%).^{1}H NMR (D$_2$O, 400 MHz) $\delta = 3.42$–3.44 (m, 2H), 3.96–4.00 (m, 2H); ^{13}C NMR (D$_2$O, 100 MHz) $\delta = 53.1$–54.0 (m), 56.2, 67.7; ^{15}N NMR (D$_2$O, 40 MHz) $\delta = 43.1$; HRMS (FAB): m/z calc. for C$_5$H$_5$D$_9$O^{15}N$^+$ [M − Cl]$^+$ = 114.1611, found = 114.1611.

Synthesis of [^{15}N]TMPA. Iodomethane (330 μl, 5.30 mmol) was added to a solution of [^{15}N]aniline (100 mg, 1.06 mmol) and N,N-diisopropylethylamine (740 μl, 4.25 mmol) in dry dimethylformamide (3 ml). The mixture was stirred at room temperature overnight and evaporated under vacuum. Ethyl acetate was added to the residue resulting in a white precipitate. The resulting precipitate was filtered and purified using GPC (eluent: methanol) to give [^{15}N]TMPA as a white powder (105 mg, 37%): ^{1}H NMR (CD$_3$OD, 400 MHz) $\delta = 3.73$ (d, $J = 0.8$ Hz, 9 H), 7.61–7.70 (m, 3H), 7.96–7.99 (m, 2H); ^{13}C NMR (CD$_3$OD, 100 MHz) $\delta = 56.6$ (d, $J = 5$ Hz), 119.8 (d, $J = 1$ Hz), 130.3, 130.3 (d, $J = 1$ Hz), 147.2 (d, $J = 8$ Hz); ^{15}N NMR (CD$_3$OD, 40 MHz) $\delta = 53.3$; HRMS (FAB): m/z calc. for C$_9$H$_{14}$^{15}N$^+$ [M − I]$^+$ = 137.1097, found = 137.1098.

Synthesis of [^{15}N, D$_9$]TMPA. [D$_3$]Iodomethane (622 µl, 10.0 mmol) was added to a solution of [^{15}N]aniline (188 mg, 2.00 mmol) and *N,N*-diisopropylethylamine (1.39 ml, 8.00 mmol) in dry dimethylformamide (3 ml). The mixture was stirred at room temperature overnight and then at 50 °C overnight. After evaporation under vacuum, ethyl acetate was added to the residue resulting in a white precipitate. The resulting precipitate was filtered and purified using GPC (eluent: methanol) to give [^{15}N, D$_9$]TMPA as a white powder (375 mg, 69%): ^{1}H NMR (CD$_3$OD, 400 MHz) $\delta = 7.62–7.71$ (m, 3H), 8.01–8.04 (m, 2H); ^{13}C NMR (CD$_3$OD, 100 MHz) $\delta = 119.9$ (d, $J = 1$ Hz), 130.3, 130.3 (d, $J = 1$ Hz), 147.0 (d, $J = 8$ Hz); ^{15}N NMR (CD$_3$OD, 40 MHz) $\delta = 52.3$; HRMS (FAB): *m/z* calc. for C$_9$H$_5$D$_9$^{15}N$^+$ [M − I]$^+$ $= 146.1662$, found $= 146.1665$.

T_1 measurements. All T_1 measurements were performed at thermally equilibrated conditions. The T_1 measurements in Fig. 2a were performed using a JEOL ECA 600 (14.1 T, 30 °C) by the saturation recovery method. The T_1 measurements of MR probes **1–3** (Figs 3 and 4) were performed using a Bruker Avance III spectrometer (9.4 T, 25 °C) by the inversion recovery method.

General information on DNP–NMR/MRI measurements. Tris{8-carboxyl-2,2,6,6-tetra[2-(1-hydroxyethyl)]-benzo(1,2-d:4,5-d′)bis(1,3)dithiole-4-yl}methyl sodium salt (Ox63 radical, GE Healthcare) and the ^{15}N-labelled sample were dissolved in a 1:1 solution of D$_2$O (99.9%, D):dimethyl sulfoxide-d6 (99.8%, D; final concentration of Ox63 15 mM). The sample was submerged in liquid helium in a DNP polarizer magnet (3.35 T; HyperSense, Oxford Instruments). The transfer of polarization from the electron spin on the radical to the ^{15}N nuclear spin on the probe was achieved using microwave irradiation at 94 GHz and 100 mW for 1.5 or 3.0 h under 2.8 mbar at 1.4 K. After polarization, samples were dissolved in water containing 0.025% EDTA disodium salt or an appropriate buffer heated to 10 bar. The DNP–NMR measurements of Figs 2b, 3, and 4 were performed using JEOL ECA 300 (7.05 T). The DNP–NMR measurement of Fig. 2c was performed using Japan Redox JXI-400Z spectrometer (9.4 T). Choline chloride-^{15}N (43.4 p.p.m.) was used as the external standard for ^{15}N NMR. The DNP–NMR spectra were obtained using flip angles of 13° (Fig. 2c), 15° (Fig. 4e), 23° (Supplementary Fig. S2), 30° (Figs 3b–d and 4b) or 90° (Fig. 2b). The DNP–MRI measurement of Fig. 3e was performed using Varian 400 MR WB spectrometer (9.4 T).

^{15}N DNP–NMR (time course analysis of [^{15}N, D$_9$]TMPA). The hyperpolarized [^{15}N, D$_9$]TMPA (final concentration 5 mM) dissolved in water containing 0.025% EDTA disodium salt (6 ml). The solution was passed through an anion exchange cartridge (Grace) to remove the remaining Ox63 radical, which affects the hyperpolarization lifetime, and then transferred to a 10-mm NMR tube.

^{15}N DNP–NMR (Ca^{2+} sensing by probe 1). The hyperpolarized probe **1** (final concentration 0.5 mM) was dissolved in various concentrations of Ca^{2+} (final concentrations: 0, 0.25, 0.5, 1.0, 2.5 or 10 mM) or Mg^{2+} (final concentration: 10 mM) in 20 mM HEPES buffer (pH 7.4, 4 ml), and then an aliquot was transferred to a 5-mm NMR tube.

^{15}N DNP–NMR (Ca^{2+} sensing by probe 1 in human blood). The hyperpolarized probe **1** (final concentration 0.5 mM) dissolved in 20 mM HEPES buffer (pH 7.4, 1 ml) was added to human blood (1 ml) with or without externally added Ca^{2+} or EDTA (final concentrations: 10 mM) and an aliquot was transferred to a 5-mm NMR tube.

^{15}N DNP–MRI (Ca^{2+} sensing by probe 1 in human blood). The hyperpolarized probe **1** (final concentration: 8 mM) was dissolved in 20 mM HEPES buffer (pH 7.4, 4 ml) with Ca^{2+} (final concentration: 8 mM) or EDTA-2Na (final concentration: 8 mM). The solution was added to human blood (1 ml) in a 10-mm NMR tube and mixed. DNP–MRI images were acquired with a gradient echo two-dimensional multi-slice acquisition technique (GEMS) with a total acquisition time of 0.64 ms. The excitation pulse was centred at probe **1** + Ca^{2+} complex. Other MR parameters were field of view 40 × 40 mm^2 × 4 mm, matrix size of 32 × 32, 60° radio frequency pulse. In the reconstruction phase, the matrix was zero-filled to 64 × 64.

^{15}N DNP–NMR (H$_2$O$_2$ sensing by probe 2). The hyperpolarized probe **2** (final concentration: 2.5 mM) dissolved in phosphate buffer (pH 7.4, 3.9 ml) was mixed with various concentrations of H$_2$O$_2$ (final concentration: 0–6.18 mM, 100 µl) and an aliquot was transferred to a 5-mm NMR tube. The concentration of H$_2$O$_2$ was determined based on the molar extinction coefficient at 240 nm (43.6 M^{-1} cm^{-1}).

^{15}N DNP–NMR (esterase sensing by probe 3). The hyperpolarized probe **3** (500 µl, final concentration: 10 mM) dissolved in PBS (pH 7.4) was mixed with esterase (Sigma-Aldrich E2884, 62 units, 8 µl) and transferred to a 5-mm NMR tube. The esterase was derived from the porcine liver, which was used as a model esterase.

References

1. Terreno, E., Castelli, D. D., Viale, A. & Aime, S. Challenges for molecular magnetic resonance imaging. *Chem. Rev.* **110**, 3019–3042 (2010).
2. Viale, A. *et al.* Hyperpolarized agents for advanced MRI investigations. *Q. J. Nucl. Med. Mol. Imaging* **53**, 604–617 (2009).
3. Viale, A. & Aime, S. Current concepts on hyperpolarized molecules in MRI. *Curr. Opin. Chem. Biol.* **14**, 90–96 (2010).
4. Wilson, D. M. *et al.* Generation of hyperpolarized substrates by secondary labeling with [1,1-^{13}C] acetic anhydride. *Proc. Natl Acad. Sci. USA* **106**, 5503–5507 (2009).
5. Golman, K., in 't Zandt, R. & Thaning, M. Real-time metabolic imaging. *Proc. Natl Acad. Sci. USA* **103**, 11270–11275 (2006).
6. Keshari, K. R. *et al.* Hyperpolarized [2-^{13}C]-Fructose: a hemiketal DNP substrate for *in vivo* metabolic imaging. *J. Am. Chem. Soc.* **131**, 17591–17596 (2009).
7. Gabellieri, C. *et al.* Therapeutic target metabolism observed using hyperpolarized ^{15}N choline. *J. Am. Chem. Soc.* **130**, 4598–4599 (2008).
8. Meier, S., Jensen, P. R. & Duus, J. Ø. Real-time detection of central carbon metabolism in living Escherichia coli and its response to perturbations. *FEBS Lett.* **585**, 3133–3138 (2011).
9. Gallagher, F. A. *et al.* Magnetic resonance imaging of pH *in vivo* using hyperpolarized ^{13}C-labeled bicarbonate. *Nature* **453**, 940–943 (2008).
10. Lippert, A. R., Keshari, K. R., Kurhanewicz, J. & Chang, C. J. A hydrogen peroxide-responsive hyperpolarized ^{13}C MRI contrast agent. *J. Am. Chem. Soc.* **133**, 3776–3779 (2011).
11. Doura, T., Hata, R., Nonaka, H., Ichikawa, K. & Sando, S. Design of a ^{13}C magnetic resonance probe using a deuterated methoxy group as a long-lived hyperpolarization unit. *Angew. Chem. Int. Ed.* **51**, 10114–10117 (2012).
12. Bohndiek, S. E. *et al.* Hyperpolarized [1-^{13}C]-ascorbic and dehydroascorbic acid: vitamin C as a probe for imaging redox status *in vivo. J. Am. Chem. Soc.* **133**, 11795–11801 (2011).
13. Keshari, K. R. *et al.* Hyperpolarized ^{13}C dehydroascorbate as an endogenous redox sensor for *in vivo* metabolic imaging. *Proc. Natl Acad. Sci. USA* **108**, 18606–18611 (2011).
14. Chang, P. V. & Bertozzi, C. R. Imaging beyond the proteome. *Chem. Commun.* **48**, 8864–8879 (2012).
15. Miura, T. *et al.* Rational design principle for modulating fluorescence properties of fluorescein-based probes by photoinduced electron transfer. *J. Am. Chem. Soc.* **125**, 8666–8671 (2003).
16. Ueno, T. *et al.* Rational principles for modulating fluorescence properties of fluorescein. *J. Am. Chem. Soc.* **126**, 14079–14085 (2004).
17. Minta, A., Kao, J. P. Y. & Tsien, R. Y. Fluorescent indicators for cytosolic calcium based on rhodamine and fluorescein chromophore. *J. Biol. Chem.* **264**, 8171–8178 (1989).
18. Kojima, H. *et al.* Fluorescent indicators for imaging nitric oxide production. *Angew. Chem. Int. Ed.* **38**, 3209–3212 (1999).
19. Fujikawa, Y. *et al.* Design and synthesis of highly sensitive fluorogenic substrates for glutathione S-transferase and application for activity imaging in living cells. *J. Am. Chem. Soc.* **130**, 14533–14543 (2008).
20. Månsson, S. *et al.* ^{13}C imaging—a new diagnostic platform. *Eur. Radiol.* **16**, 57–67 (2006).
21. Gopinath, T. & Veglia, G. Dual acquisition magic-angle spinning solid-state nmr-spectroscopy: simultaneous acquisition of multidimensional spectra of biomacromolecules. *Angew. Chem. Int. Ed.* **51**, 2731–2735 (2012).
22. Lippmaa, E., Saluvere, T. & Laisaar, S. Spin-lattice relaxation of ^{15}N nuclei in organic compounds. *Chem. Phys. Lett.* **11**, 120–123 (1971).
23. Schweitzer, D. & Spiess, H. W. Nitrogen-15 NMR of pyridine in high magnetic fields. *J. Magn. Reson.* **15**, 529–539 (1974).
24. Levy, G. C., Holloway, C. E., Rosanske, R. C., Hewitt, J. M. & Bradley, C. H. Natural abundance nitrogen-15 n.m.r. spectroscopy. Spin-lattice relaxation in organic compounds. *Org. Magn. Reson.* **8**, 643–647 (1976).
25. Allouche-Arnon, H. *et al.* A hyperpolarized choline molecular probe for monitoring acetylcholine synthesis. *Contrast Media Mol. Imaging* **6**, 139–147 (2011).
26. Allouche-Arnon, H., Lerche, M. H., Karlsson, M., Lenkinski, R. E. & Katz-Brull, R. Deuteration of a molecular probe for DNP hyperpolarization - a new approach and validation for choline chloride. *Contrast Media Mol. Imaging* **6**, 499–506 (2011).
27. Sarkar, R. *et al.* Proton NMR of ^{15}N-choline metabolites enhanced by dynamic nuclear polarization. *J. Am. Chem. Soc.* **131**, 16014–16015 (2009).
28. Kumagai, K. *et al.* Synthesis and hyperpolarized ^{15}N NMR studies of ^{15}N-choline-d$_{13}$. *Tetrahedron* **69**, 3896–3900 (2013).
29. Ardenkjær-Larsen, J. H. *et al.* Increase in signal-to-noise ratio of > 10,000 times in liquid-state NMR. *Proc. Natl Acad. Sci. USA* **100**, 10158–10163 (2003).
30. Hofer, A. M. & Brown, E. M. Extracellular calcium sensing and signalling. *Nat. Rev. Mol. Cell Biol.* **4**, 530–538 (2003).

31. Nordenström, E., Katzman, P. & Bergenfelz, A. Biochemical diagnosis of primary hyperparathyroidism: analysis of the sensitivity of total and ionized calcium in combination with PTH. *Clin. Biochem.* **44**, 849–852 (2011).

32. Lumachi, F., Brunello, A., Roma, A. & Basso, U. Cancer-induced Hypercalcemia. *Anticancer Res.* **29**, 1551–1555 (2009).

33. Basarić, N. *et al.* Synthesis and spectroscopic characterisation of BODIPY based fluorescent off-on indicators with low affinity for calcium. *Org. Biomol. Chem.* **3**, 2755–2761 (2005).

34. Spector, A., Ma, W. & Wang, R. The aqueous humor is capable of generating and degrading H_2O_2. *Invest. Ophthalmol. Vis. Sci.* **39**, 1188–1197 (1998).

35. Cai, H. Hydrogen peroxide regulation of endothelial function: origins, mechanisms, and consequences. *Cardiovasc. Res.* **68**, 26–36 (2005).

36. Van de Bittner, G. C., Dubikovskaya, E. A., Bertozzi, C. R. & Chang, C. J. *In vivo* imaging of hydrogen peroxide production in a murine tumor model with a chemoselective bioluminescent reporter. *Proc. Natl Acad. Sci. USA* **107**, 21316–21321 (2010).

37. Miller, E. W., Albers, A. E., Pralle, A., Isacoff, E. Y. & Chang, C. J. Boronate-based fluorescent probes for imaging cellular hydrogen peroxide. *J. Am. Chem. Soc.* **127**, 16652–16659 (2005).

38. Jiang, Y. L. *et al.* A specific molecular beacon probe for the detection of human prostate cancer cells. *Bioorg. Med. Chem. Lett.* **22**, 3632–3638 (2012).

39. Redinbo, M. R. & Potter, P. M. Mammalian carboxylesterases: from drug targets to protein therapeutics. *Drug. Discov. Today* **10**, 313–325 (2005).

40. Xu, G., Zhang, W., Ma, M. K. & McLeod, H. L. Human carboxylesterase 2 is commonly expressed in tumor tissue and is correlated with activation of irinotecan. *Clin. Cancer Res.* **8**, 2605–2611 (2002).

41. Farde, L., Halldin, C., Stone-Elander, S. & Sedvall, G. PET analysis of human dopamine receptor subtypes using ^{11}C-SCH 23390 and ^{11}C-raclopride. *Psychopharmacology* **92**, 278–284 (1987).

Acknowledgements

This work was supported by a NEXT Program from JSPS. We thank Mr T. Abe of Oxford Instruments for helpful discussions and technical assistance for the DNP experiments. We also thank the Network Joint Research Center for Materials and Devices for T_1 and FAB–MS measurements. H.N. thanks the Kato Memorial Bioscience Foundation for financial support. R.H. and T.N. thank JSPS for the fellowship. K.I. was supported by the funding programme 'Creation of Innovation Centers for Advanced Interdisciplinary Research Areas' from JST, commissioned by MEXT.

Author contributions

S.S. conceived the project. H.N. and S.S. designed the experiments. H.N. and R.H. performed all the experiments with the help from T.D., T.N., K.K., M.A., M.T. and K.I. on DNP and NMR measurements. The manuscript was written by H.N. and S.S. and edited by all the co-authors.

Additional information

Radiolysis as a solution for accelerated ageing studies of electrolytes in Lithium-ion batteries

Daniel Ortiz[1,†], Vincent Steinmetz[2], Delphine Durand[3], Solène Legand[3], Vincent Dauvois[3], Philippe Maître[2] & Sophie Le Caër[1,†]

Diethyl carbonate and dimethyl carbonate are prototype examples of eco-friendly solvents used in lithium-ion batteries. Nevertheless, their degradation products affect both the battery performance and its safety. Therefore, it is of paramount importance to understand the reaction mechanisms involved in the ageing processes. Among those, redox processes are likely to play a critical role. Here we show that radiolysis is an ideal tool to generate the electrolytes degradation products. The major gases detected after irradiation (H_2, CH_4, C_2H_6, CO and CO_2) are identified and quantified. Moreover, the chemical compounds formed in the liquid phase are characterized by different mass spectrometry techniques. Reaction mechanisms are then proposed. The detected products are consistent with those of the cycling of Li-based cells. This demonstrates that radiolysis is a versatile and very helpful tool to better understand the phenomena occurring in lithium-ion batteries.

[1] Institut Rayonnement Matière de Saclay, LIDyL et Service Interdisciplinaire sur les Systèmes Moléculaires et les Matériaux UMR 3299 CNRS/CEA SIS2M Laboratoire de Radiolyse, Bâtiment 546, F-91191 Gif-sur-Yvette, France. [2] Laboratoire de Chimie-Physique, UMR 8000 CNRS Université Paris Sud, Faculté des Sciences, Bâtiment 349, F-91405 Orsay, France. [3] CEA/Saclay, DEN/DANS/DPC/SECR/LRMO, F-91191 Gif-sur-Yvette, France. † Present address: Institut Rayonnement Matière de Saclay, NIMBE, UMR 3685 CNRS/CEA, LIONS, Bâtiment 546, F-91191 Gif-sur-Yvette, France. Correspondence and requests for materials should be addressed to S.Le.C. (email: sophie.le-caer@cea.fr).

Lithium-ion batteries (LIBs) are ubiquitous in everyday life as they have very high gravimetric and volumetric energy densities[1-3]. To promote a higher energy storage by future generations[4], major issues in battery developments must be faced and solved, not the least being the characterization of ageing processes[3]. Indeed, failure to do so adequately has led in the past to potentially disastrous safety issues[5] or costly recalls. It is therefore paramount to have an in depth understanding of the underlying ageing phenomena[6-8]. Owing to the complexity of the whole LIB, many factors can induce safety problems but one key factor is the stability of the organic-solvent electrolyte[9-11], which often cannot be studied by conventional thermally activated ageing methods. As a result, ageing studies can be lengthy, costly and usually remain purely qualitative[3,12]. Here, we demonstrate that radiolysis not only allows obtaining in a matter of hours what previously took months or even years to produce[13-17], but also that it allows the quantitative and mechanistic study of these processes. Radiolysis has the potential of significant cost and time savings in the development of new battery electrolytes[18].

This approach entails particularly two important benefits: the time needed to degrade the solvent is shortened (minutes–hours) as compared with the charge/discharge experiments (weeks–months). Second, processes at both short (ps–μs) and long-time scales (seconds–minutes) might also be studied, offering thus an understanding on temporal scales varying over some 12 orders of magnitude.

Commercial electrolytes are normally based on lithium salts in solutions of both linear alkyl carbonates and cyclic alkyl carbonates (Fig. 1). Indeed, the required properties of the electrolyte, such as conductivity or viscosity, can be optimized by combining solvents of different natures[19]. This paper focuses on the study of the stable degradation products formed at long times (seconds–minutes) in diethylcarbonate (DEC) and dimethylcarbonate (DMC) solutions (Fig. 1). Both give similar results; here we will present in detail those on DEC. DMC results are given in Supplementary Figs 1,2 and Supplementary Tables 1,2 and 3.

Gas phase products are identified using a gas chromatography-electron impact-mass spectrometry instrument (GC-EI/MS). A refined quantitative analysis is then performed using both an EI-magnetic sector mass spectrometer (EI/MS) and a Micro-Gas Chromatography (μ-GC) system. Liquid phase is also analysed by GC-EI/MS. Moreover, to further characterize the more complex liquid mixture, electrospray ionization-mass spectrometry (ESI/MS) technique is also employed. As radiolysis implies isomerization and the formation of numerous products, a combination of ESI, Differential Ion Mobility Spectrometry (DIMS) complements usefully the aforementioned mass spectrometry experiments[20-22]. Gas phase Infrared Multiphoton Dissociation (IRMPD) of a selected set of DIMS- and mass-selected ions is used to derive

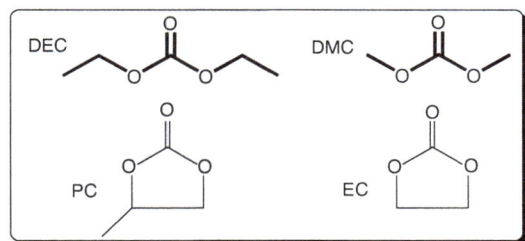

structural information[23-26]. To make the reading easier, a list of abbreviations is available in Supplementary Note 1.

All these complementary techniques enable us to characterize in detail the effect of radiolysis on DEC (and DMC) and to decipher the underlying reaction mechanisms. They are discussed and compared with reaction mechanisms occurring in electrolysis processes. In both cases, the same types of species are produced. Moreover, the detected products are consistent with the ones reported in the literature for the cycling of Li-based cells. This proves that radiolysis is a very useful tool to understand ageing phenomena in LIB.

Results

Gas phase results. Gas decomposition products of irradiated DEC obtained from GC-EI/MS are presented in Fig. 2. Different types of molecules are produced under irradiation: alkanes, alkenes and alkynes (for example, C_2H_6, C_2H_4 and C_2H_2); oxygenated molecules (aldehyde; ether; carboxylic acid). The different retention times of the products formed upon irradiation are given in Supplementary Table 4. It is important to notice that H_2, CH_4 and CO are not detected in the split mode by GC-EI/MS (Fig. 2) but that they are formed and measured by EI/MS and μ-GC.

Having identified the molecules produced upon irradiation, the corresponding radiolytic yields G, which are defined as the amount of formed species per energy unit deposited in the sample, expressed in $\mu mol\,J^{-1}$, can be measured. Main results are displayed both in Fig. 3 and Table 1. We checked that the yields measured with both irradiation setups were the same, within the uncertainty bars.

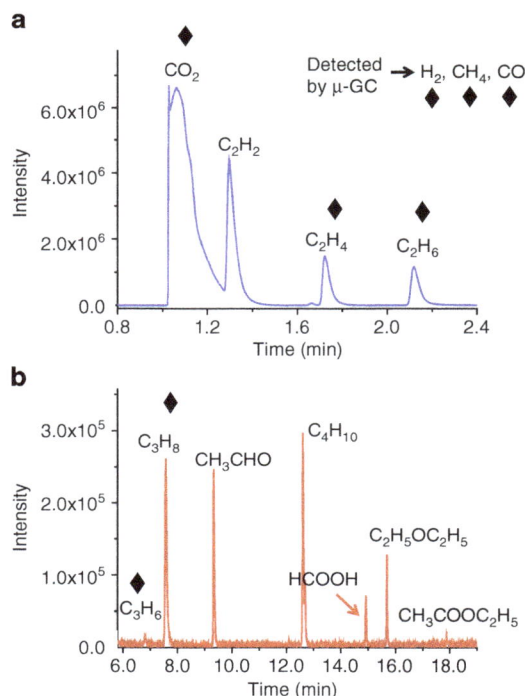

Figure 2 | Gas decomposition products of DEC measured by GC-EI/MS after a 20-kGy irradiation. Gy stands for 'Gray' and corresponds to $1\,J\,kg^{-1}$. Except DEC itself, the part in blue (**a**) corresponds to the most intense peaks between 0.8 and 2.4 min, and the one in red (**b**) to smaller signals between 6 and 18 min. The top right molecules (H_2, CH_4, CO) are identified by μ-GC experiments. The compounds labelled with a diamond are identified in refs 15,17.

Figure 1 | Common alkyl carbonates used in commercial electrolytes. Linear alkyl carbonates (diethylcarbonate, DEC, and dimethylcarbonate, DMC) and cyclic alkyl carbonates (propylene carbonate, PC, and ethylene carbonate, EC). DEC and DMC structures, written in bold, are studied here.

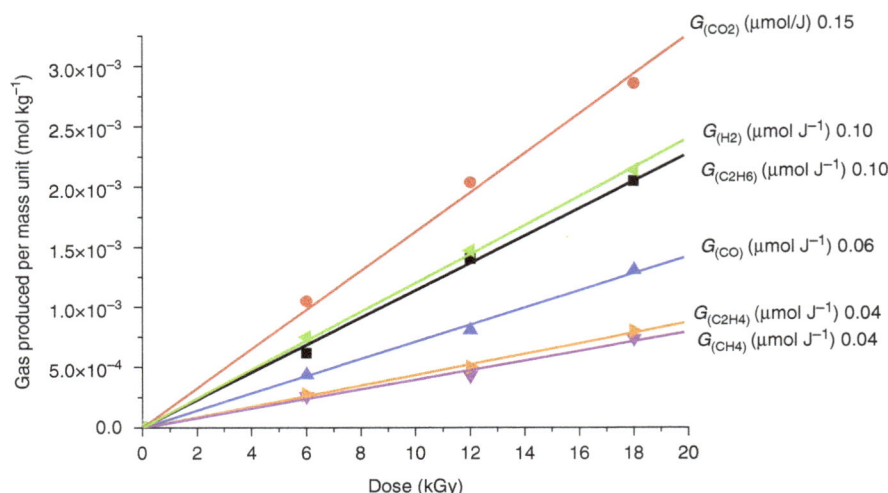

Figure 3 | Evolution of the major DEC degradation products formed in the gas phase. They are obtained with the EI/magnetic sector mass spectrometer (EI/MS). For each gas, the radiolytic yield is given.

Table 1 | Radiolytic yields of the degradation products formed in the gas phase.

	G (μmol J^{-1}) μ-GC	G (μmol J^{-1}) EI/MS
CO_2	0.21	0.15
Ethane (C_2H_6)		0.10
H_2	0.13	0.10
CO	0.05	0.06
Ethylene (C_2H_4)		0.04
Methane (CH_4)	0.08	0.04
Formic acid (HCOOH)		0.03
Diethyl ether ($C_2H_5OC_2H_5$)		0.02
Acetaldehyde (CH_3CHO)		0.02
Propane (C_3H_8)		0.01
Butane (C_4H_{10})		0.01

GC, gas chromatography; EI/MS, electron impact-magnetic sector mass spectrometry. They are expressed in μmol J^{-1} of gas produced after DEC irradiation and measured by both μ-GC (left) and EI/MS (right) techniques. The uncertainty bars are estimated to be 10% for the gas chromatography technique and 20% in the other case. The radiolytic yields can also be expressed in molecule per 100 eV by dividing the value obtained in μmol J^{-1} by the 1.036 10^{-1} factor.

Main decomposition gases (Fig. 3) measured by EI/MS are CO_2, H_2, ethane and CO. In addition, other alkanes, such as CH_4, propane and butane, and oxygenated products, such as acetaldehyde, diethyl ether or formic acid, are detected at lower quantities. Although ethyl acetate, acetylene and propene were detected, their yields were too low to quantify (estimated to be under 0.01 μmol J^{-1}). Knowing the uncertainty bars estimated for EI/MS and gas chromatography (Table 1), the radiolytic yields obtained for CO_2, H_2 and CO are consistent with each other. A discrepancy exists in the case of the methane radiolytic yield. Indeed, in EI/MS experiments, the most abundant fragment of CH_4 is the resulting ion CH_3^+ at m/z 15. This ion is common to almost all identified molecules, leading to a greater uncertainty in this case.

Liquid phase results. Having identified and quantified the compounds formed in the gas phase, the evolution of the liquid phase is then studied in detail. The GC-EI/MS chromatogram obtained for irradiated DEC exhibits numerous peaks between 3 and 30 min (Fig. 4). Among the identified products, three types of

degradation products are evidenced (Fig. 4). A detailed list of the identified products is given in Supplementary Table 5. The first one (symbolized by red circles) corresponds to a linear lengthening of the alkyl carbon chain of DEC. Different alkyl chain lengths ranging from $n = 1$ ($t_r = 4.6$ min) to $n = 3$ ($t_r = 15.7$ min) are detected. Blue squares symbolize compounds in which a C_2H_5-O-CO-O-C_nH_{2n}-CO-C_2H_5 bond is formed (with n equal to zero ($t_r = 20.4$ min) or two ($t_r = 26.5$ min)). Green triangles represent products for which a C-O bond cleavage and a branching in the alkyl chain occur. Finally, the C_2H_5-O-CH_2CH_2-OC_2H_5 ($t_r = 5.6$ min) and C_2H_5-O-CO-O-$\boldsymbol{CH_2CH_2}$-O-C_2H_5 ($t_r = 11.7$ min) molecules, represented as black crosses, are also detected. Unfortunately, the remaining peaks which are not marked in Fig. 4 could not be identified with the NIST library. This led us to use an alternative ionization technique in order to better characterize the degradation products formed in the liquid phase.

A high-resolution ESI/MS spectrum obtained using a Fourier Transform Ion Cyclotron Resonance mass spectrometer (FT-ICR, 7T) for irradiated DEC was recorded (Supplementary Discussion and Supplementary Figs 3–5). Most of the detected peaks are consistent with the compounds obtained previously (Fig. 4 and Supplementary Table 5) but new signals also appear. Among them, the most abundant is the $m/z = 145.091$ ion corresponding to the $[\boldsymbol{C_7H_{12}O_3} + \boldsymbol{H}]^+$ formula. In what follows, a special attention will be paid to the $m/z = 145.091$ ion, hereafter denoted as 145.1, as it was not detected previously. The ion is generated in the gas phase and its structure is unraveled by IRMPD experiments.

IRMPD experiments. In these IRMPD experiments, two dissociation channels are observed at $m/z = 117.1$ (loss of ethylene) and 91.1 (loss of C_4H_6; consistent with Collision Induced Dissociation experiments, see Supplementary Fig. 6). The IRMPD spectrum of the $m/z = 145.1$ ion is dominated by two intense broad bands centred around 1,620 and 1,565 cm^{-1}, respectively (Fig. 5). Two other weaker bands are observed at 1,490 and 1,340 cm^{-1}, respectively. Moreover, DIMS experiments evidence that, at least, three isomers coexist at this m/z ratio (Supplementary Discussion and Supplementary Figs 7 and 8). Two main facts are consistent with this ion co-existence at $m/z = 145.1$ ion: (i) the broad complex shape of the bands observed between 1,500 and 1,660 cm^{-1}; (ii) the fact that each

IRMPD fragment exhibits a specific wavelength dependence (chart *b* of Fig. 5 with the mass-resolved IRMPD spectra of $m/z = 117.1$ and 91.1). Interestingly, the two bands located at 1,565 and 1,525 cm^{-1} are specific of the fragment ion $m/z = 91.1$. To get a structural assignment of the different underlying species, Density functional theory (DFT)-computed infrared absorption

spectra of different isomeric structures were compared with the experimental IRMPD spectra. All the seven structures (S_1–S_7) having the molecular formula $[C_7H_{12}O_3 + H]^+$ and consistent with the expected chemical functions present in the ion and the radiolysis experiments were taken into account (Fig. 6). Depending on their infrared signature, three types of compounds were considered: (i) a carbocation (C_2H_5-O)$_2$-$\mathbf{C^+}$-(OC$_2$H$_3$) (S_1), (ii) protonated carbonates R-O-$\mathbf{COH^+}$-O-R' (S_2–S_5) and (iii) protonated esters R-O-$\mathbf{COH^+}$-R' (S_6–S_7). The calculated infrared spectra of the S_1–S_7 structures are reported in Supplementary Figs 9–12. Comparison of the specific bands centred at 1,565 and 1,525 cm^{-1}, associated with the fragment ion $m/z = 91.1$, matches nicely with the proposed carbocation structure (C_2H_5-O)$_2$-$\mathbf{C^+}$-(OC$_2$H$_3$) (S_1; Supplementary Fig. 10), which is the single structure dominated by two intense bands in this wavenumber range. The bands centred at 1,568 and

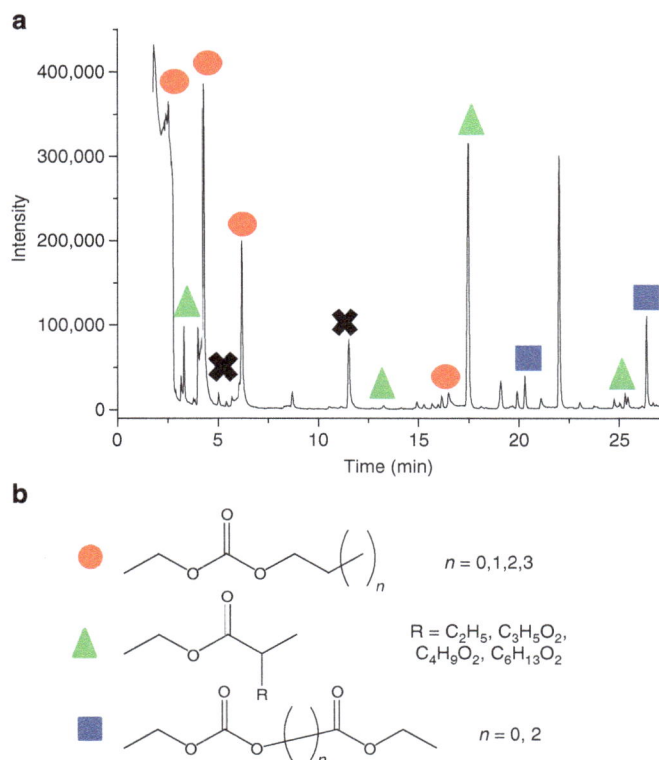

Figure 4 | Liquid decomposition products of DEC measured by GC-EI/MS after a 100-kGy irradiation. (a) The GC-EI/MS chromatogram of the irradiated DEC. **(b)** Different types of degradation products represented with various colours and symbols. Black crosses represent other identified molecules, which do not belong to the above-mentioned types of degradation products.

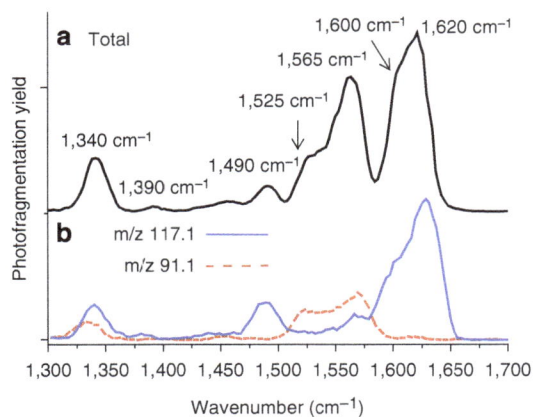

Figure 5 | Total and mass-resolved IRMPD spectra from $m/z = 145.1$ ion. a shows the experimental IRMPD spectrum (black straight line) of the $m/z = 145.1$ ion in the 1,300–1,700 cm^{-1} range. Mass-resolved IRMPD spectra are shown in **b**. Two fragmentation channels are detected at $m/z = 117.1$ (straight blue line) and at $m/z = 91.1$ (dotted red line).

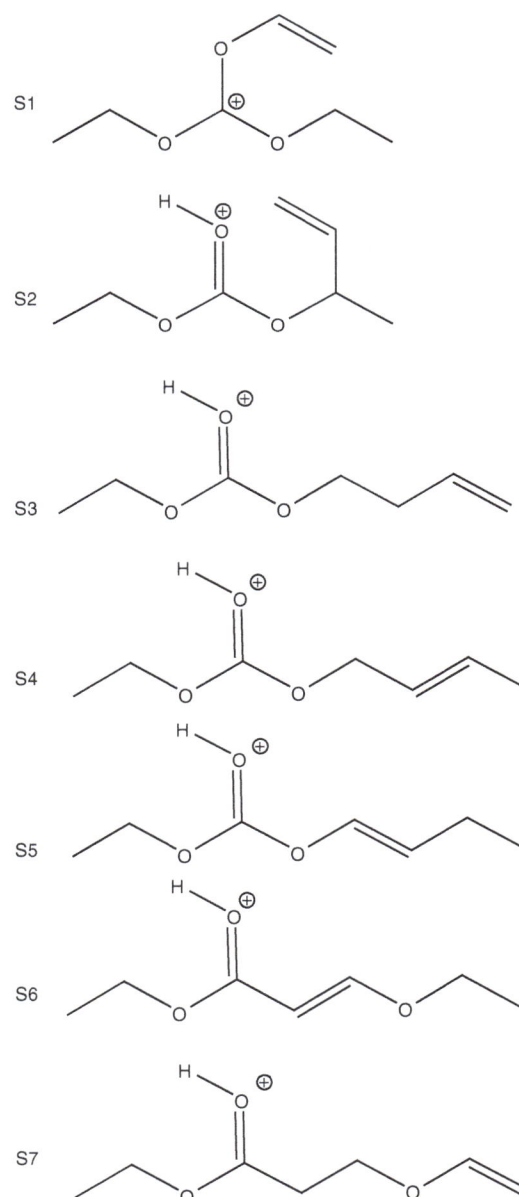

Figure 6 | Considered compounds corresponding to the $[C_7H_{12}O_3 + H]^+$ molecular formula. They were taken into account to elucidate the structure of the $m/z = 145.1$ ion.

1,528 cm^{-1} are assigned to the O–C–O antisymmetric stretching and the C-O-CH = CH$_2$ symmetric stretching bands, respectively. In the DFT-computed spectra of the protonated carbonate structures R-O-**COH**$^+$-O-R' (S$_2$–S$_5$), shown in Supplementary Fig. 11, an efficient isomer differentiation remains complex. All theoretical infrared spectra are dominated by two intense signals calculated between 1,630–1,615 and 1,530–1,515 cm^{-1} and are attributed to the O–C–O antisymmetric stretch and carbonyl C = O symmetric stretching vibrational modes, respectively. As shown in Supplementary Fig. 11, the two computed vibrational modes are close in energy for all structures (S$_2$–S$_5$), which makes an efficient differentiation difficult. Furthermore, Supplementary Fig. 12 shows the theoretical spectra of protonated ester-type structures R-O-**COH**$^+$-R' (S$_6$–S$_7$). Both computed spectra are dominated by the carbonyl C = O symmetrical stretching band, centred between 1,615–1,630 cm^{-1}. Moreover, no structure S$_1$–S$_5$ could account for the experimental 1,490 cm^{-1} band. This band, assigned to the CH$_3$ deformation mode, can be explained by the presence of the S$_6$ structure (C$_2$H$_5$-O-**COH**$^+$-C$_2$H$_2$-O-C$_2$H$_5$) whose calculated infrared spectrum predicts this vibrational mode at 1,493 cm^{-1}. The last band, observed experimentally at 1,340 cm^{-1}, corresponds to a CH bending vibrational mode. However, it does not give any further information as it matches to all calculated structures (S$_1$–S$_7$). These experiments enabled us to confirm structures associated to the m/z = 145.1 ion. The combination of **S$_1$** (C$_2$H$_5$-O)$_2$-**C**$^+$-(OC$_2$H$_3$), of at least one carbonate structure (S$_2$–S$_5$) and of at least the ester structure **S$_6$** (C$_2$H$_5$-O-**COH**$^+$-C$_2$H$_2$-O-C$_2$H$_5$) allows simulating a global infrared spectrum, which is consistent with the complex feature of the experimental spectrum recorded (Supplementary Fig. 13).

Discussion

It is well known that the first effect of ionizing radiation in liquids is to excite and ionize molecules:

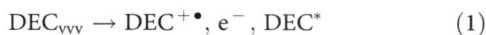

$$DEC_{vvv} \rightarrow DEC^{+\bullet}, e^-, DEC^* \quad (1)$$

The different role of these species is summarized in Fig. 7. First of all, it is important to remark that the static dielectric constant at room temperature of DEC is very low (2.8), meaning that the recombination of the electron with its parent radical cation DEC$^{+\bullet}$ is very fast, as evidenced by pulse radiolysis experiments performed at the picosecond timescale[27]. Moreover, this recombination of DEC$^{+\bullet}$ with the electron leads to the formation of the excited DEC molecule (Fig. 7). It was reported

that the remaining radical cations DEC$^{+\bullet}$, which have not reacted with the electron, may induce a proton transfer reaction to another DEC molecule, leading to the formation of DECH$^+$ and to the neutral species DEC(-H)$^\bullet$ (equation 2):[27]

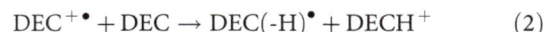

$$DEC^{+\bullet} + DEC \rightarrow DEC(-H)^\bullet + DECH^+ \quad (2)$$

Other reaction channels between the radical cation and the DEC molecule are possible and can be proposed, similarly to the observations performed in the case of irradiated DMC[28]. In an electron spin resonance study (ESR)[28], it was indeed shown that in DEC$^{+\bullet}$, an hydrogen-atom transfer takes place between the –CH$_3$ group (and also between the adjacent –CH$_2$- group) to the carbonyl oxygen, leading to the formation of CH$_2^\bullet$CH$_2$O (C = O$^+$H)OC$_2$H$_5$ and of CH$_3$CH$^\bullet$O(C = O$^+$H)OC$_2$H$_5$, respectively. These species will then react with another DEC molecule (which act as a proton acceptor), and different bond cleavages can take place. Equations 3 and 4 can be proposed:

$$DEC^{+\bullet} + DEC \rightarrow DECH^+ + CH_3CHO + CO_2 + C_2H_5^\bullet \quad (3)$$

$$DEC^{+\bullet} + DEC \rightarrow DECH^+ + C_2H_5O^\bullet + CO_2 + C_2H_4 \quad (4)$$

This leads to the formation of various radicals and molecules such as CO$_2$. In equation 4, the C$_2$H$_5$O$^\bullet$ radical can rearrange to form the CH$_3$C$^\bullet$HOH radical. Moreover, the formation of the ethyl radical (equation 3) can then lead, after H$^\bullet$ atom abstraction from the DEC molecule, to two radicals: CH$_3$C$^\bullet$HOCOOC$_2$H$_5$ and $^\bullet$CH$_2$CH$_2$OCOOC$_2$H$_5$ and to ethane formation. Equations 3 and 4 account for the formation of CH$_3$CHO and of ethylene (Fig. 2) and are presented in the Fig. 7.

The remaining electrons, which have not reacted with the radical cation, will mainly solvate, leading to the formation of the e_{DEC}^- species. Once they are solvated, and similar to reactions written in the case of DMC and ethylmethylcarbonate[29,30], dissociative electron attachment can take place, leading to different bond cleavages and to the formation of various radicals such as C$_2$H$_5^\bullet$, which can then abstract an H$^\bullet$ atom from the DEC molecule (see above). This reductive path is shown in the Fig. 7. Nevertheless, pulse radiolysis experiments[27] indicate that both reaction pathways, reductive and oxidative (Fig. 7), are minor under our experimental conditions. The various products reported in the present study are then mainly attributed to the reactivity of the DEC* molecule. Moreover, and similar to the observations performed in irradiated dodecane[31], radiolysis can also form the excited radical cation, which can also lead to the

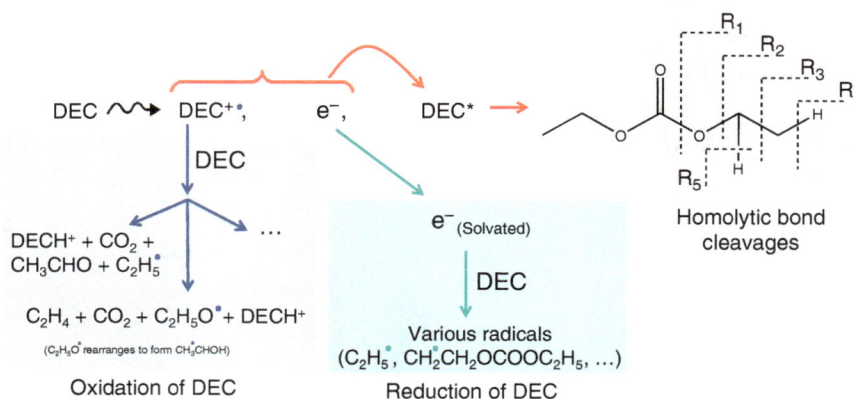

Figure 7 | Different reactive channels of the radiolytic processes. The main channel, represented with the red colour, leads to the formation of the excited DEC molecule[27], in which various homolytic bond cleavages (R1,…, R5) take place. The nature and amount of gases formed imply that the R2, R4 and R5 bond cleavages are the preferential ones. Among the possible reaction pathways, the reduction of DEC is indicated in the turquoise box[29] and its oxidation is given in the blue box[28]. In battery applications, oxidation is observed at the positive electrode when the cell is overcharged at high voltages, and reduction takes place at the negative electrode.

excited DEC molecule. It is important to point out here that the same type of behaviour is expected in the case of DMC. These excited molecules can form directly small molecules such as CO, CO_2 or H_2..., and also lead to different homolytic bond cleavages (Fig. 7). For instance, bond cleavage R_1 leads to the following radicals:

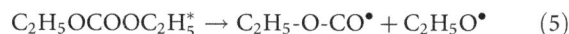

$$C_2H_5OCOOC_2H_5^* \rightarrow C_2H_5\text{-}O\text{-}CO^\bullet + C_2H_5O^\bullet \qquad (5)$$

Similar equations can be written for R_2–R_5 (Fig. 7). Let us point out that radicals arising from C–H bond cleavage may also be secondary radicals formed after $H^\bullet$ atom abstraction from ethyl radicals, for example. The formation of the decomposition products arises then from different possible reaction mechanisms from the excited state: (i) direct production of small molecules such as CO, CO_2 and H_2.... from the excited state of DEC. The high CO_2 yield measured, for example, cannot be explained by consecutive R_1 and R_2 bond cleavages, which are not very probable, but by a direct production from the excited state. The same explanation accounts for the high CO and CO_2 yields measured in irradiated DMC (Supplementary Table 2). The difference in the relative proportions of gases formed from DMC and DEC implies that the reaction pathways from their respective excited state are different; (ii) radical recombination: the formation of diethylether can for instance arise from the recombination of radicals issued from R_1 and R_2. This recombination explains also the lengthening and branching of the alkyl chain of DEC (red circles and green triangles in Fig. 4); (iii) proton abstraction of radicals from the solvent (DEC) molecules, which are the most abundant ones. Therefore, this process is more probable than the previous one. For instance, the ethyl radical can abstract an hydrogen atom from the DEC molecule, leading to the formation of ethane and to the DEC(-H)$^\bullet$ radical. A similar reaction mechanism will explain the formation of H_2 from $H^\bullet$ radicals.

The higher ethane yield measured as compared with methane (Table 1) implies then that R_2 is preferred over R_3 (Fig. 7). Ethylene can be formed by two different pathways: (i) an initial cleavage of C–H group via R5 (Fig. 7) could lead to an O–C bond cleavage, producing the $CH^\bullet CH_3$ radical, which could rearrange to form ethylene and the $C_2H_5^\bullet$ radical by pathway R2 followed by an H-atom abstraction. The nature and amount of gases measured from DEC imply that the preferential bond cleavages achieved are the C(=O)O–C and the C–H ones. The same trends are observed in DMC, and the C(=O)O–C bond cleavage leads to $CH_3^\bullet$ radicals and then to methane formation. This is consistent with previous work, which has observed the $CH_3^\bullet$ and $CH_3OCOOCH_2^\bullet$ radicals using ESR spectroscopy in irradiated frozen DMC[29]. Last, the high H_2 yields measured here are in line with the dihydrogen yields determined for similar carbonyl compounds such as ketones[32] and esters ($G_{H2} = 0.099\,\mu mol\,J^{-1}$ and $G_{H2} = 0.079\,\mu mol\,J^{-1}$, respectively)[33,34].

Results obtained on the liquid phase evidences that the use of different analytical techniques is required for a detailed identification. Both 'soft' and 'hard' ionization techniques give consistent and complementary information. Although EI sorts out direct structural determination based on fragmentation patterns, ESI avoids fragmentation and generates intact degradation products. Moreover, ESI was important to unravel non-identified species in the chromatogram (Fig. 4), especially unsaturated ones. The ESI/IRMPD experiments have enabled to detect three main types of compounds: (i) a carbocation (S_1), (ii) protonated carbonate R-O-COH^+-O-R' and (iii) protonated ester R-O-COH^+-R' (Fig. 6). For instance, the degradation product identified at $t_r = 6.4$ min (Fig. 4 and Supplementary Table 5) can convert into the S_3–S_5 structures (Fig. 6) after

subsequent irradiation. However, the formation of the carbocation (S_1) can only be explained by a 1,3-sigmatropic rearrangement process between S_1 (C_2H_5-O)$_2$-C^+-(OC_2H_3) and S_2 (C_2H_5-O-COH^+-O-CH(CH=CH_2)-CH_3) structures, taking place within the mass spectrometer during the electrospray ionization. The S_1 structure is then not due to radiation chemistry.

As above mentioned, our experimental data together with previous pulse radiolysis experiments[27] indicate that the reactivity in the radiolysis experiments is mainly due to the excited state of diethylcarbonate (DEC*), leading then to various homolytic cleavages and to the formation of small molecules. Nevertheless, other reactive channels (implying the electron and the radical cation) exist and are summarized in Fig. 7. Let us point out that radiolysis leads to the formation of both the products arising from solvent reduction and from solvent oxidation.

In the field of batteries, the oxidative decomposition of the solvent is observed at the positive electrode when the battery is overcharged at high voltages, whereas reduction reactions of the solvent occur at the negative electrode. Even if numerous species take part in the complex surface chemistry on electrodes, the formation of the DEC$^{+\bullet}$ radical cation[35] in overcharged cells and electron transfer at the negative electrode will lead to the formation of various radicals as described in both oxidative and reductive pathways of Fig. 7. The same intermediates (radicals) and the same molecules (CO_2...) can then be found in the electrolysis experiments and in the radiolysis of DEC (Fig. 7). So, even if the molecules will not be formed in the same amounts in both techniques, radiolysis gives a precious insight into the processes taking place.

As mentioned before, one of the main goals of this study was to compare the species formed under irradiation with those formed by electrolysis. This comparison is depicted in Fig. 2, in which the gases detected in our study and in previous works on overcharged cells are labelled with a diamond[15,17]. Even if the solvents commonly studied in the field of LIBs are usually composed by mixtures of both linear and cyclic carbonates[36], the major gases detected (H_2, CO, CO_2, CH_4, C_nH_{2n+2} or C_nH_{2n} (refs 15,17)) are the same, and the main oxidative and reductive gases (CO_2, H_2...) are measured and quantified in the present work. Our study evidences also the presence of minor oxygenated compounds (ester, ether..., Fig. 2). This can be linked to the possibility of radiation chemistry to deliver a significant amount of energy in a reasonable time, then enabling to evidence the formation of minor molecules, which is not easily accessible in batteries studies. In the liquid phase, Tarascon et al.[12–14,16,37,38] have identified different sets of families in linear alkyl carbonates/ EC-LiPF$_6$ systems, also detected in our radiolysis experiments (Fig. 4 and Supplementary Fig. 14). These degradation products are based on: (i) $(CH_2$-CH_2-O)$_n$ structure such as C_2H_5-O-C_2H_4-O-C_2H_5, (ii) (C_2H_5-O-CO-[O-$C_2H_4]_n$-O-C_2H_5) as C_2H_5-O-CO-O-C_2H_4-O-C_2H_5 or (iii) (C_2H_5-O-CO-[O-$C_2H_4]_n$-CO-O-C_2H_5), which are detected by GC-EI/MS (Supplementary Table 5). Last, it is possible, with the radiolysis tool, to study separately the reactivity of each solvent used in batteries, without or with LiPF$_6$, and then to focus on different mixtures. Therefore, the role of each solvent and of the salt can be understood in details.

Radiolysis enabled us to generate stable degradation products of neat linear alkyl carbonates used in commercial LIBs. This entails three important advantages: (i) the time needed to degrade the electrolyte is shortened as compared with standard charge/ discharge experiments, (ii) both short and long time-scale phenomena can be studied and (iii) the possibility to study the role of each solvent without/with salt. Our attention was first focused on both neat DEC and DMC. In the gas phase, H_2, CO, CO_2, alkanes ($C_nH_{2n+2})_{n=1-4}$, ethylene, acetylene and different kinds of oxygenated molecules were identified. In the liquid phase, more complex to analyse, different chain lengths and branching in the alkyl groups, as well as $-(CH_2\text{-}CH_2\text{-}O)_n$ or $(C_2H_5\text{-}O\text{-}CO\text{-}[O\text{-}C_2H_4]\text{-}O\text{-}C_2H_5)$ based structures have been found. The use of a different ionization technique (ESI as compared with EI) has also allowed identifying other decomposition products, especially unsaturated carbonates. The detected products are consistent with the ones reported in the literature for the cycling of Li-based cells.

Moreover, our experiments evidence the critical importance of the excited states of linear alkyl carbonates in radiolysis. These excited states can lead directly to the formation of small molecules (CO, CO_2, H_2...) in different proportions in DEC and in DMC, and to different bond cleavages forming various radicals. Once radicals are formed, radical recombination and proton abstraction from the solvent account for the detected compounds.

In conclusion, we show that the 'radiolysis approach' used herein can provide a fast overview of the electrolyte decomposition phenomenon. Indeed, our results lead to similar sets of molecules as electrolysis. Radiolysis can then be very useful for a rapid screening of anti-ageing properties of new electrolytes.

Methods

Chemicals and sample preparation. Anhydrous grade DEC, DMC and potassium chunks (in mineral oil) were obtained from Sigma-Aldrich. To remove all water traces in the solvent, the solvent was pre-treated with potassium under argon atmosphere. The solution was then distilled under argon atmosphere in a flask containing a molecular sieve, which was before dried 24 h at 300 °C. The water amount was measured by a coulometric Karl–Fischer titrator and was never higher than 100 p.p.m. Before irradiation, the samples were degassed during 30 min by argon bubbling and placed in a Pyrex glass ampoule. They were then outgassed at approximately 3 mbar and subsequently filled with 1.5 bar of argon 6.0. This operation was repeated three times.

Irradiation experiments. A Gammacell 3000 with a ^{137}Cs source was used. The dose rate (5.1 ± 0.2 Gy min^{-1}, with 1 Gy = 1 J kg^{-1}) was determined using the Fricke dosimeter[39]. The total dose received by the sample was about 20 kGy, which is achieved here in 2–3 days.

Moreover, to get a significant amount of degradation products in the liquid phase, irradiations were also performed using the electron pulses of a Titan Beta, Inc. accelerator (10 MeV electrons with a pulse duration of 10 ns (ref. 40)), which delivers a higher dose to the sample than the Gammacell in a short time. We checked for these organic liquids that the degradation products obtained by these two irradiation setups are the same. A dose rate of 25 Gy per pulse was determined using the Fricke dosimeter[39]. To avoid a macroscopic heating of the sample during irradiation, the repetition rate was set to 2 Hz, for which the temperature of the sample remains below 40 °C as required in batteries. The 100-kGy dose is then delivered to the sample in roughly 30 min.

Gas phase analytical methods. H_2, CH_4, CO and CO_2 gases were quantified by gas chromatography (μ-GC-R3000, SRA instrument) using ultra-high purity helium as a carrier gas[41]. Moreover, to fully identify the degradation products formed in the gas phase, Gas Chromatography Mass Spectrometry (GC-MS) experiments were performed with an Agilent 6890 GC system interfaced with an Agilent 5973 MS equipped with an EI source, and a quadrupole mass analyser. The mass range is 4–160. Helium is used as the vector gas with a flow rate of 2 ml min^{-1}. More details concerning the GC–MS apparatus are given in ref. 42. Finally, a gas mass spectrometer with a direct inlet equipped with an EI ionization source and a magnetic sector for mass analyser was used for a quantitative analysis (EI/MS). The mass range goes from 1 to 200 amu and the detection limit is about 1 p.p.m.

Liquid phase analytical methods. GC-EI/MS experiments were carried out using a Waters GCT Premier-Time-of-Flight (TOF) mass spectrometer. Degradation products were separated with a (25 m × 0.25 mm) CP Sil 5 CB capillary column. The initial and final temperatures were 60 °C and 280 °C, respectively, with a temperature rate of 3 °C min^{-1}. Helium was used as the vector gas with an inlet initial flow regulated at 1 ml min^{-1}. The ion source was operated at 180 °C with an electron energy of 70 eV. EI spectra were obtained in the 10–800 m/z range.

For the electrospray-mass spectrometry (ESI-MS) experiments, the solutions were prepared by mixing 100 μl of alkyl carbonate, 1 ml of H_2O/MeOH (40:60) and 2 μl of formic acid (98%). Solutions were infused with a syringe pump at a flow of 5 μl min^{-1} and a 5.5-kV voltage was applied to the capillary entrance. High-resolution mass spectra were obtained using a Fourier Transform Ion Cyclotron

Resonance Spectrometer (Bruker FT-ICR, 7T). MS/MS Collision-Induced Dissociation spectra were also recorded for elucidating the structure of selected ions.

More direct structural information on mass-selected ions could also be derived from their IRMPD spectra recorded in the 1,200–1,800 cm^{-1} wavenumber range. These experiments were performed with a quadrupole ion trap mass spectrometer (Bruker Esquire 3000+) equipped with an electrospray ion source, which is coupled with IR lasers[23]. Tunable mid-infrared radiation produced by the free electron laser of CLIO (Centre Laser Infrarouge d'Orsay) was used[43]. The average laser power was on the order of 1,000 mW by setting the electron energy to 45 MeV. Upon resonant vibrational excitation, dissociation of the mass-selected ion is induced through the IRMPD mechanism. The infrared spectra are obtained by plotting the IRMPD efficiency as a function of wavenumber. Structural information can be obtained by comparing IRMPD spectra to the IR absorption spectra calculated for different isomeric structures of the ion.

Computational details. DFT calculations were carried out with the Gaussian 03 package[44]. Geometry optimizations and the harmonic vibrational frequencies were computed by combining the B3LYP[45,46] functional with the 6–311 ++ G** basis set. The theoretical vibrational frequencies were then scaled by a factor of 0.98 (refs 47,48). Each calculated band was convoluted assuming a Gaussian function having a full-width at half-maximum of 20 cm^{-1}.

References

1. Wang, Q. *et al.* Molecular wiring of insulators: charging and discharging electrode materials for high-energy Lithium-ion batteries by molecular charge transport layers. *J. Am. Chem. Soc.* **129**, 3163–3167 (2007).
2. Armstrong, A. R. *et al.* Synthesis of tetrahedral LiFeO$_2$ and its behavior as a cathode in rechargeable lithium batteries. *J. Am. Chem. Soc.* **130**, 3554–3559 (2008).
3. Tarascon, J. M. & Armand, M. Issues and challenges facing rechargeable Lithium batteries. *Nature* **414**, 359–367 (2001).
4. Hu, Y.-Y. *et al.* Origin of additional capacities in metal oxide Lithium-ion battery electrodes. *Nat. Mater.* **12**, 1130–1136 (2013).
5. Bhattacharyya, R. *et al.* In situ NMR observation of the formation of metallic lithium microstructures in lithium batteries. *Nat. Mater.* **9**, 504–510 (2010).
6. Broussely, M. *et al.* Main aging mechanisms in Li ion batteries. *J. Power Sources* **146**, 90–96 (2005).
7. Vetter, J. *et al.* Ageing mechanisms in Lithium-ion batteries. *J. Power Sources* **147**, 269–281 (2005).
8. Fong, R., Vonsacken, U. & Dahn, J. R. Studies of lithium intercalation into carbons using nonaqueous electrochemical-cells. *J. Electrochem. Soc.* **137**, 2009–2013 (1990).
9. Jia-Yan, L., Wang-Jun, C., Ping, H. & Yong-Yao, X. Raising the cycling stability of aqueous lithium-ion batteries by eliminating oxygen in the electrolyte. *Nat. Chem.* **2**, 760–765 (2010).
10. Zonghai, C. *et al.* New class of nonaqueous electrolytes for long-life and safe lithium-ion batteries. *Nat. Commun.* **4**, 1513 (2013).
11. Bridel, J.-S. *et al.* Decomposition of ethylene carbonate on electrodeposited metal thin film anode. *J. Power Sources* **195**, 2036–2043 (2010).
12. Gachot, G. *et al.* Deciphering the multi-step degradation mechnanisms of carbonate-based electrolyte in Li batteries. *J. Power Sources* **178**, 409–421 (2008).
13. Gireaud, L. *et al.* Identification of Li battery electrolyte degradation products through direct synthesis and characterization of alkyl carbonate salts. *J. Electrochem. Soc.* **152**, A850–A857 (2005).
14. Gireaud, L. *et al.* Mass spectrometry investigations on electrolyte degradation products for the development of nanocomposite electrodes in lithium ion batteries. *Anal. Chem.* **78**, 3688–3698 (2006).
15. Kumai, K. *et al.* Gas generation mechanism due to electrolyte decomposition in commercial Lithium-ion cell. *J. Power Sources* **81**, 715–719 (1999).
16. Laruelle, S. *et al.* Identification of Li-based electrolyte degradation products through DEI and ESI High-Resolution Mass Spectrometry. *J. Electrochem. Soc.* **151**, A1202–A1209 (2004).
17. Yoshida, H. *et al.* Degradation mechanism of alkyl carbonate solvents used in lithium-ion cells during initial charging. *J. Power Sources* **68**, 311–315 (1997).
18. Spotheim-Maurizot, M., Mostafavi, M., Douki, T. & Belloni, J. *Radiation Chemistry from Basics to Applications in Material and Life Sciences* Ch.7 (EDP Sciences, 2012).
19. Xu, K. Nonaqueous liquid electrolytes for Lithium-based rechargeable batteries. *Chem. Rev.* **104**, 4303–4417 (2004).
20. Isenberg, S. L., Armistead, P. M. & Glish, G. L. Optimization of peptide separations by differential ion mobility spectrometry. *J. Am. Soc. Mass Spectrom* **25**, 1592–1599 (2014).
21. Kanu, A. B. *et al.* Ion mobility-mass spectrometry. *J. Mass Spectrom.* **43**, 1–22 (2008).
22. Guevremont, R. High-field asymmetric waveform ion mobility spectrometry: a new tool for mass spectrometry. *J. Chromat. A* **1058**, 3–19 (2004).

23. Mac Aleese, L. *et al.* Mid-IR spectroscopy of protonated leucine methyl ester performed with an FTICR or a Paul type ion-trap. *Int. J. Mass Spectrom.* **249**, 14–20 (2006).

24. MacAleese, L. & Maitre, P. Infrared spectroscopy of organometallic ions in the gas phase: from model to real world complexes. *Mass Spectrom. Rev.* **26**, 583–605 (2007).

25. Simon, A. *et al.* Fingerprint vibrational spectra of protonated methyl esters of amino acids in the gas phase. *J. Am. Chem. Soc.* **129**, 2829–2840 (2007).

26. Lemaire, J. *et al.* Gas phase infrared spectroscopy of selectively prepared ions. *Phys. Rev. Lett.* **89** (2002).

27. Torche, F. *et al.* Picosecond pulse radiolysis of the liquid diethyl carbonate. *J. Phys. Chem. A* **117**, 10801–10810 (2013).

28. Ganghi, N. S., Rao, D. N. R. & Symons, M. C. R. Radical cations of organic carbonates, trimethyl borate and methyl nitrate - A Radiation Electron-Spin-Resonance Study. *J. Chem. Soc. Faraday Trans. I* **82**, 2367–2376 (1986).

29. Shkrob, I. A., Zhu, Y., Marin, T. W. & Abraham, D. Reduction of carbonate electrolytes and the formation of solid-electrolyte interface (SEI) in Lithium-ion batteries. 1. spectroscopic observations of radical intermediates generated in one-electron reduction of carbonates. *J. Phys. Chem. C* **117**, 19255–19269 (2013).

30. Shkrob, I. A., Zhu, Y., Marin, T. W. & Abraham, D. Reduction of carbonate electrolytes and the formation of solid-electrolyte interface (SEI) in Lithium-Ion Batteries. 2. Radiolytically induced polymerization of ethylene carbonate. *J. Phys. Chem. C* **117**, 19270–19279 (2013).

31. Kondoh, T. *et al.* Femtosecond pulse radiolysis study on geminate ion recombination in n-dodecane. *Rad. Phys. Chem.* **80**, 286–290 (2011).

32. Matsui, M. & Imamura, M. Radiation chemical studies with cyclotron beams. 3. Heavy-ion radiolysis of liquid aliphatic-ketones. *Bull. Chem. Soc. Jpn* **47**, 1113–1116 (1974).

33. Ausloos, P. & Trumbore, C. N. Radiolysis of CH3COOCH3 and CH3COOCD3 by cobalt-60 gamma-rays. *J. Am. Chem. Soc.* **81**, 3866–3871 (1959).

34. Hall, K. L., Bolt, R. O. & Carroll, J. K. *Radiation Effects on Organic Materials* Vol. 104 (Academic, 1963).

35. Matsuta, S. *et al.* Electron-spin-resonance study of the reaction of electrolytic solutions on the positive electrode for lithium-ion secondary batteries. *J. Electrochem. Soc.* **148**, A7–A10 (2001).

36. Sasaki, T. *et al.* Formation mechanism of alkyl dicarbonates in Li-ion cells. *J. Power Sources* **150**, 208–215 (2005).

37. Dedryvere, R. *et al.* Characterization of lithium alkyl carbonates by X-Ray photoelectron spectroscopy: experimental and theoretical Study. *J. Phys. Chem. B* **109**, 15868–15875 (2005).

38. Dedryvere, R. *et al.* XPS identification of the organic and inorganic components of the electrode/electrolyte interface formed on a metallic cathode. *J. Electrochem. Soc.* **152**, A689–A696 (2005).

39. Fricke, H. & Hart, E. J. in *Radiation Dosimetry* (eds Attix, F. H. & Roesch, W. C.) Vol. 2, 167–232 (Academic, 1966).

40. Mialocq, J. C., Hickel, B., Baldacchino, G. & Juillard, M. The radiolysis project of CEA. *J. Chim. Phys* **96**, 35–43 (1999).

41. Fourdrin, C. *et al.* Water radiolysis in exchanged-montmorillonites: the H_2 production mechanisms. *Environ. Sci. Technol.* **47**, 9530–9537 (2001).

42. Le Caer, S. *et al.* Modifications under irradiation of a self-assembled monolayer grafted on a nanoporous silica glass: a solid-state NMR characterization. *J. Phys. Chem. C* **116**, 4748–4759 (2012).

43. Prazeres, R. *et al.* Two-colour operation of a free-electron laser and applications in the mid-infrared. *Eur. Phys. J. D* **3**, 87–93 (1998).

44. Gaussian 03 (Gaussian, Inc., Wallingford, CT, 2004).

45. Becke, A. D. Density-functional thermochemistry 3. the role of exact exchange. *J. Chem. Phys.* **98**, 5648–5652 (1993).

46. Lee, C. T., Yang, W. T. & Parr, R. G. Development of the Colle-Salvetti correlation energy formula into a functional of the electron density. *Phys. Rev. B* **37**, 785–789 (1988).

47. Halls, M. D. & Schlegel, H. B. Comparison of the performance of local, gradient-corrected, and hybrid density functional models in predicting infrared intensities. *J. Chem. Phys.* **109**, 10587–10593 (1998).

48. Halls, M. D., Velkovski, J. & Schlegel, H. B. Harmonic frequency scaling factors for Hartree-Fock, S-VWN, B-LYP, B3-LYP, B3-PW91 and MP2 with the Sadlej pVTZ electric property basis set. *Theo. Chem. Accounts* **105**, 413–421 (2001).

Acknowledgements

We acknowledge CEA's DSM Energie program for funding, and Dr Jean-Christophe P. Gabriel for scientific and editorial comments. This work was also supported by a public grant from the 'Laboratoire d'Excellence Physics Atom Light Mater' (LabEx PALM) overseen by the French National Research Agency (ANR) as part of the 'Investissements d'Avenir' program (reference: ANR-10-LABX-0039). Financial support from the national FT-ICR network (FR 3624 CNRS) for conducting the research is also gratefully acknowledged. We thank Dr Jean-Claude Berthet for his help in drying the alkyl carbonates and also Dr Fabienne Testard for the Karl–Fischer measurements. Dr Jean-Frédéric Martin and Dr Pascal Mailley are gratefully acknowledged for precious help and fruitful discussions. We also thank the CLIO team (Dr Jean Michel Ortega, Catherine Six, Gilles Perilhous and Jorge Viera) as well as Oscar Hernandez for their help during the IRMPD experiments. Finally, we thank Professor Mehran Mostafavi and Professor Jacqueline Belloni for fruitful discussions.

Author contributions

D.O. and S.le.C. conceived, designed and performed the experiments, analysed the results and wrote the paper. P.M. and V.S. helped in the IRMPD experiments and contributed to their interpretation. D.D., S.L. and V.D. helped during the GC–MS experiments.

Additional information

18

Understanding silicate hydration from quantitative analyses of hydrating tricalcium silicates

Elizaveta Pustovgar[1], Rahul P. Sangodkar[2], Andrey S. Andreev[3], Marta Palacios[1], Bradley F. Chmelka[2], Robert J. Flatt[1] & Jean-Baptiste d'Espinose de Lacaillerie[1,3]

Silicate hydration is prevalent in natural and technological processes, such as, mineral weathering, glass alteration, zeolite syntheses and cement hydration. Tricalcium silicate (Ca_3SiO_5), the main constituent of Portland cement, is amongst the most reactive silicates in water. Despite its widespread industrial use, the reaction of Ca_3SiO_5 with water to form calcium-silicate-hydrates (C-S-H) still hosts many open questions. Here, we show that solid-state nuclear magnetic resonance measurements of ^{29}Si-enriched triclinic Ca_3SiO_5 enable the quantitative monitoring of the hydration process in terms of transient local molecular composition, extent of silicate hydration and polymerization. This provides insights on the relative influence of surface hydroxylation and hydrate precipitation on the hydration rate. When the rate drops, the amount of hydroxylated Ca_3SiO_5 decreases, thus demonstrating the partial passivation of the surface during the deceleration stage. Moreover, the relative quantities of monomers, dimers, pentamers and octamers in the C-S-H structure are measured.

[1] Institute for Building Materials, Department of Civil, Environmental and Geomatic Engineering, ETH Zürich 8093, Switzerland. [2] Department of Chemical Engineering, University of California, Santa Barbara, California 93106, USA. [3] Soft Matter Science and Engineering Laboratory, UMR CNRS 7615, ESPCI Paris, PSL Research University, 10 rue Vauquelin, Paris 75005, France. Correspondence and requests for materials should be addressed to J.-B.d.E.d.L. (email: jean-baptiste.despinose@espci.fr).

Since Le Chatelier[1], it is well understood that Portland cement hydration is initiated by the dissolution of calcium silicate monomers in water, followed by the precipitation of less soluble layered calcium-silicate-hydrates (C-S-H), in which silicate ions condense to form short chains. However, despite two centuries of widespread applications and a century of detailed study, the molecular mechanisms behind the kinetic stages of hydration (that is, induction, acceleration and deceleration) are still debated. Similar kinetic stages are observed in various heterogeneous hydration processes occurring during mineral weathering[2,3], glass alteration[4,5] and hydrothermal syntheses. For example, although hydrothermal zeolite syntheses under alkaline aqueous conditions proceeds over different timescales[6], the effective reaction rates in cementitious and zeolite systems exhibit similar distinct stages (induction, acceleration and deceleration), and are governed by several coupled parameters varying in space and time near the liquid–solid interface. This situation is thus extremely complex to describe accurately. An added difficulty is that for porous materials such as cement or zeolites, interfacial energy contributes to the stabilization of nanoscale intermediates, which are typically challenging to characterize. For Portland cement in particular, the lack of quantitative experimental data obtained with sufficient time resolution has precluded the validation of existing models aimed at explaining the complex kinetics of cement hydration.

Similar to the homogeneous versus heterogeneous pathways dichotomy in zeolite crystallization mechanisms[7], two landmark competing theories have been proposed to explain the early-age time dependence of the rate of tricalcium silicate (Ca_3SiO_5) hydration, the principal component in commercial Portland cements responsible for the development of mechanical strength[8–10]. The first theory proposes that early-age hydration products form a diffusion barrier on the surfaces of Ca_3SiO_5 particles, thus affecting subsequent reactions of the underlying non-hydrated core[11]. The second theory[12–14] suggests that the early-age time-dependence of the rate of hydration is determined by the rate of Ca_3SiO_5 dissolution and by a change in the associated rate limiting step from etch pit formation to step retreat, which is a mechanism also often invoked in the geochemical literature on natural weathering[15,16]. The relevance of these theories to silicate hydration can be examined by understanding the molecular compositions and structures of species at the solid–liquid interfaces during the early stages of hydration. Similar questions are raised in heterogeneous catalysis and geochemistry; however, Portland cement hydration faces the additional complexity that the main product, C-S-H, is not only poorly crystalline but also nanostructured with variable stoichiometry and silicate coordinations[17,18]. These challenges have been previously addressed partially through numerical modelling of hydration reactions at Ca_3SiO_5 surfaces[19,20] and of the local structure and disorder of the resulting hydration products[21]. Nevertheless, these models suffer from a lack of experimental support at the molecular level.

Here, solid-state NMR measurements of triclinic [29]Si-enriched Ca_3SiO_5 hydration are used to determine the transient molecular-level compositions at silicate surfaces and the interactions between silicate species, hydroxyl groups and water molecules, which influence the rates of hydration reactions. The isotopic [29]Si enrichment provides significantly enhanced NMR signal sensitivity that can be used to monitor the structures of the hydrates *in situ* during the hydration process, as a function of hydration time. In addition, [29]Si enrichment enabled two-dimensional (2D) through-bond (*J*-mediated) NMR measurements that are sensitive to [29]Si-O-[29]Si covalent bonding. They are used to crucially provide detailed information on the local atomic-level compositions, structures and site connectivities in

hydrated silicate species, here C-S-H. These analyses shed new insights on the origin of rate limiting steps and the kinetics of silicate polymerization at the solid–liquid interface during Ca_3SiO_5 hydration.

Results

Experimental approach. To the seminal approach of [29]Si enrichment by Brough *et al.*[22], we added for the first time the sophistication of carefully controlled structure and granulometry of the Ca_3SiO_5 particles (see Supplementary Methods) and hydration reaction conditions (see Supplementary Notes 1 and 2). Indeed the surface structure and area of the Ca_3SiO_5 particles strongly affect their reactivity, which must be carefully controlled to ensure meaningful results[23]. For example, the high surface area of the synthesized [29]Si-enriched Ca_3SiO_5 ($4.4\,m^2\,g^{-1}$, see Supplementary Methods) allowed $\sim 90\%$ of the silicate hydration process to be monitored in 24 h of NMR spectrometer time, without external acceleration. In this way, subtle and unique quantitative information pertinent to hydration mechanisms can be obtained non-invasively and with a time resolution of 30 min

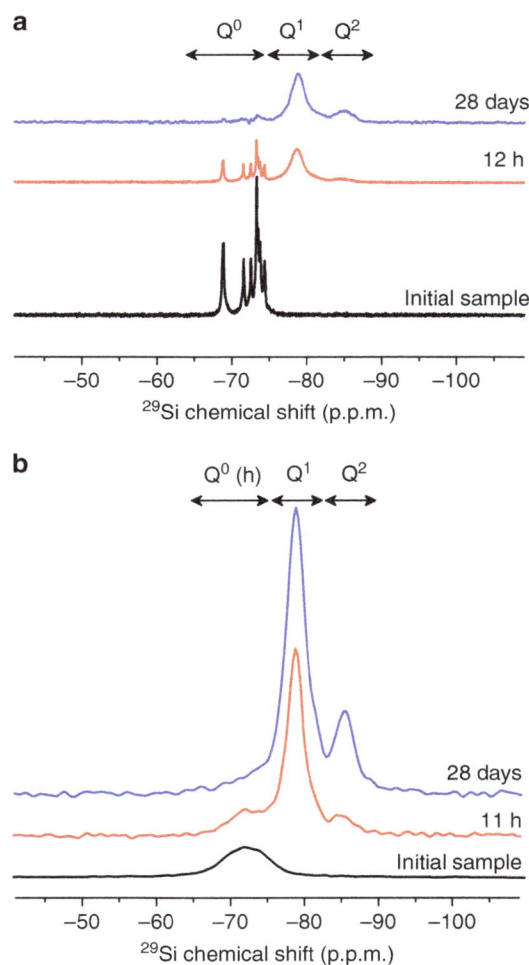

Figure 1 | Dynamics of silicate hydrates formation studied *in situ* by ^{29}Si NMR. (**a**) ^{29}Si MAS NMR and (**b**) {^{1}H}^{29}Si CPMAS NMR spectra of [29]Si-enriched triclinic Ca_3SiO_5 sample in its initial non-hydrated state (in black) and after hydration for 11 or 12 h (in red) and 28 days (in blue). [29]Si resonances from isolated silicate (Q^0) species in non-hydrated Ca_3SiO_5, hydroxylated surface Q^0 (Q^0(h)) species and polymerized calcium-silicate-hydrates (Q^1 and Q^2) are clearly resolved and can be quantified as a function of time.

(measurement time for the NMR spectra). Consequently, the progress of the hydration reaction could be accurately and quantitatively correlated to the corresponding ^{29}Si speciation. In addition, ^{29}Si enrichment allows NMR measurements to be performed on samples without the need for conventional water removal schemes for quenching the hydration process[24], which otherwise often disrupt the fragile microstructure of the C-S-H or may detrimentally alter chemical composition. Representative one-pulse ^{29}Si and ^{1}H{^{29}Si} cross-polarization (CP) magic-angle-spinning (MAS) NMR spectra are presented in Fig. 1a,b, respectively, for non-hydrated and hydrated Ca_3SiO_5. In anhydrous triclinic Ca_3SiO_5 which exhibits long-range crystalline order and well-defined local atomic ^{29}Si environments, eight distinct and narrow (<0.5 p.p.m. full-width at half maximum (FWHM)) ^{29}Si signals are resolved between -68 and -75 p.p.m. corresponding to anhydrous Q^0 species (Supplementary Fig. 2). In contrast, in hydration products, the ^{29}Si resonances are broad (3–4 p.p.m. FWHM) with signals centred at -72, -79 and -85 p.p.m. from silanol Q^0(h), hydrated Q^1 and hydrated Q^2 silicate species, respectively (Fig. 1). The last two species are associated with the C-S-H structure (Q^n refers to silicon atoms that are covalently bonded via bridging oxygen atoms to $0 \leq n \leq 4$ other silicon atoms[25]). These molecular-level insights of the local silicate structures in Ca_3SiO_5 hydration products (C-S-H) are consistent with previous ^{29}Si NMR (refs 26,27), ^{17}O NMR (ref. 28), X-ray and neutron scattering results[18] for C-S-H.

The degree of silicate hydration is determined by quantitative in situ ^{29}Si NMR analyses and forms the crux of our results, which are summarized in Fig. 2. These results are in close agreement with the degree of silicate hydration as established by independent isothermal calorimetric measurements, which reveal the successive stages of initial dissolution, induction, acceleration and deceleration (Fig. 2b) during the silicate hydration process. This comparison crucially establishes the accuracy of the quantitative ^{29}Si NMR results acquired during Ca_3SiO_5 hydration, and indicates that the hydration process is negligibly altered by factors such as the MAS conditions of the NMR experiment (see Supplementary Notes 1 and 2). This detailed

time-resolved, in situ, quantitative NMR analysis answers three central questions about Ca_3SiO_5 hydration: the molecular origin of the reduced apparent solubility of Ca_3SiO_5 during the induction period, the possible 'switch' from one type of hydration products to another between the acceleration and deceleration period, and the relative proportions of silica oligomers in the final C-S-H structure.

Induction period. The apparent solubility of Ca_3SiO_5 during the induction period of hydration has been reported to be lower compared with pristine anhydrous Ca_3SiO_5 (refs 11–13). This reduced apparent solubility has been proposed to arise from the deposition of a layer of hydration products (the metastable barrier hypothesis)[11] or from surface hydroxylation[12,13]. The molecular compositions at the Ca_3SiO_5 surface during this induction period

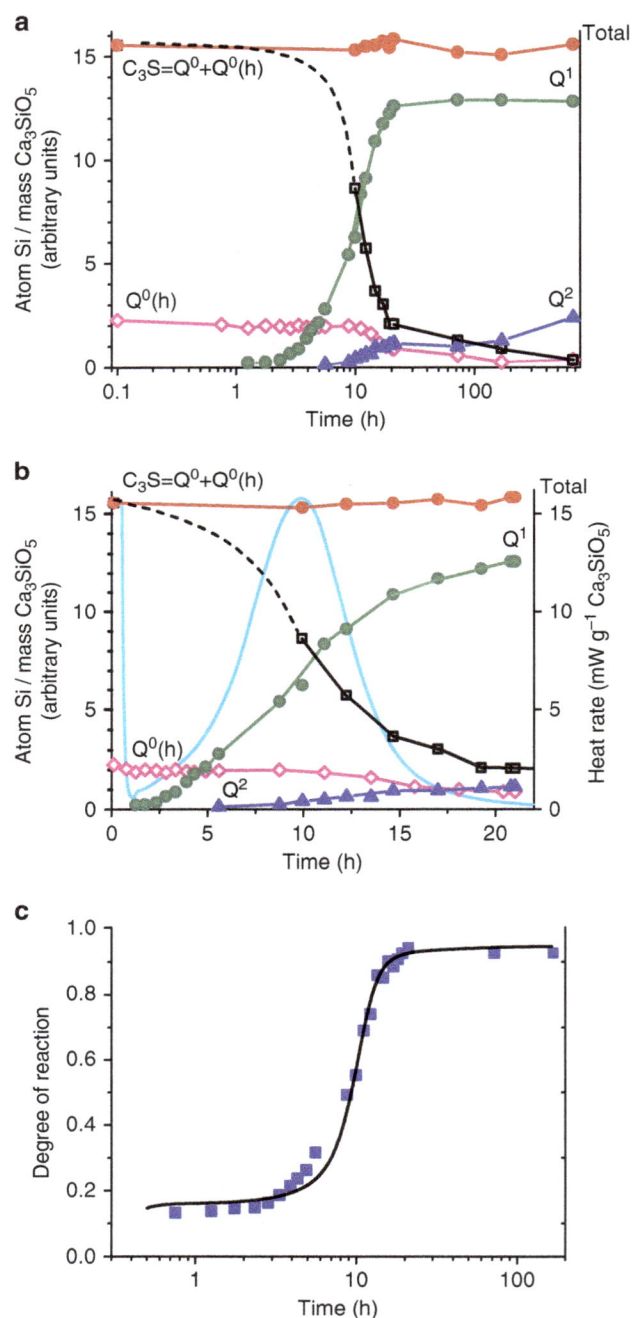

Figure 2 | Quantitative monitoring of silicate speciation during the hydration of ^{29}Si-enriched triclinic Ca_3SiO_5. (a) The quantities of different ^{29}Si silicate species as established by ^{29}Si MAS and {^{1}H}^{29}Si CPMAS NMR measurements for hydration times up to 28 days (see Supplementary Note 1). The quantities, normalized to the initial amount of Ca_3SiO_5, of anhydrous Q^0 (in black), hydroxylated Q^0(h) (in pink), hydrated Q^1 (in green), hydrated Q^2 (in blue) and total silicate species (in red) resulting from this analysis are as shown. (b) Comparison of the quantities of different ^{29}Si silicate species and the reaction heat flow rate determined by isothermal calorimetry (cyan line) for Ca_3SiO_5 up to 24 h of hydration. Based on the heat released in the calorimetry measurements, four stages in the hydration process can be identified: first a brief exothermic peak during the first few minutes (<15 min) corresponding to initial dissolution of Ca_3SiO_5, then a short (15 min–2 h) induction period during which no significant heat is released, followed by a peak corresponding to the acceleration period (2–10 h), and finally the deceleration period (>10 h) associated with decreasing rate of heat release (Supplementary Fig. 8). (c) Comparison of the degree of silicate hydration determined independently by ^{29}Si MAS and {^{1}H}^{29}Si CPMAS NMR quantitative analyses (squares) and isothermal calorimetry results (black line), which are in close agreement. The fact that the total amount of Si atoms remains constant, within the uncertainties of the measurements, over the entire hydration period (28 days) establishes the accuracy of the associated quantitative NMR methods and analyses. Details of these analyses are included in the Supplementary Note 1.

(as determined by the NMR analyses presented here) points towards the latter scenario. The $^{29}Si\{^1H\}$ CPMAS NMR measurements of the initial sample (that is, non-hydrated) (Fig. 1b) establish the presence of Q^0 silicate species in proximity to protons (henceforth labelled Q^0(h)) on Ca_3SiO_5 particle surfaces, even before contact with bulk water. Although previous studies have reported the presence of similar Q^0(h) silicate species at the surfaces of 'anhydrous' Ca_3SiO_5 particles[22,26], it has not been largely publicized nor quantitatively analysed. The 2D $^{29}Si\{^1H\}$ heteronuclear correlation (HETCOR) NMR spectrum of the same sample of non-hydrated Ca_3SiO_5 (Fig. 3) exhibits correlated intensities between the ^{29}Si signal at -72 p.p.m. from Q^0(h) species and unresolved ^1H signals around 1.3 and 0.9 p.p.m. from –SiOH and -CaOH moieties, thereby establishing the close molecular-level proximities of surface Q^0(h) species to at least one type of such ^1H moieties. In addition, the absence of resonances characteristic of polymerized hydration products (that is, Q^1 and Q^2 species), establishes that the reaction of surface silicate species in non-hydrated Ca_3SiO_5 with atmospheric moisture results solely in the formation of hydroxylated Q^0(h) species at particle surfaces, within the sensitivity limits of the measurement. In other words, no separate hydrate phase forms at this stage, it is solely the Ca_3SiO_5 particle near-surface which is hydroxylated.

From a crystal chemistry perspective, the Ca_3SiO_5 particle surface is unlikely to be inert when exposed to atmospheric water vapour. Specifically, Ca_3SiO_5 is an ionic crystal of Ca^{2+} cations with oxide and monomeric silicate anions ($3Ca^{2+} \cdot O^{2-} \cdot SiO_4^{4-}$) (refs 19,29). There is a strong ionization of the atoms ($+1.5$ on Ca^{2+} and -1.5 on O^{2-}) (ref. 19) and consequently Ca_3SiO_5 acts as a basic oxide that readily yields hydroxide ions when reacting with water,

$$O^{2-} + H_2O \rightarrow 2 OH^- \qquad (1)$$

Therefore, one expects OH^- to replace oxide ions on the particle surfaces. However, replacement of one O^{2-} by two OH^- would yield a heterogeneous distribution of local atomic environments at the Ca_3SiO_5 surface, due to the different sizes and formal charges of these anions. Indeed the Q^0(h) ^{29}Si NMR resonance of the initial sample is very broad (Fig. 1b), reflecting a wide distribution of local ^{29}Si environments. In summary, the ^{29}Si

NMR analyses reveal that near-surface ^{29}Si species on Ca_3SiO_5 particles are predominantly hydroxylated and that negligible quantities of polymerized silicate hydration products form (within the sensitive detection limits of the measurements), a result consistent with previous force-field atomistic simulations[19]. Overall, hydroxylated Q^0 (h) species are predominant at particle surfaces during the induction period and expected to result in the reduced apparent solubility of Ca_3SiO_5, compared with pristine anhydrous Ca_3SiO_5 whose level of hydroxylation is lower.

Acceleration stage. With the progress of Ca_3SiO_5 hydration, the monomeric Q^0 silicate species polymerize to form oligomeric units of C-S-H. As shown in Fig. 2, while the population of hydroxylated Q^0(h) species remains constant, the populations of Q^1 species increase significantly during the acceleration stage (~ 2–10 h). Compared with the induction stage (<2 h), the ^{29}Si polymerization during the acceleration stage results predominantly in the formation of Q^1 species (dimers) at early times, and a combination of Q^1 and Q^2 species (for example, pentamers and octamers) at later time (10–20 h). In particular, the population of Q^1 species increases approximately linearly with the progress of hydration (Fig. 2b) across the entire acceleration stage, consistent with the formation of predominantly dimeric C-S-H units. No significant change nor in the silicon second coordination sphere of the hydration products nor in their rate of formation could be detected at this stage.

Deceleration stage. The data in Fig. 2 indicate that at the end of the acceleration stage (after ~ 10 h in the present case) greater quantities of long (>2 silicate tetrahedra) C-S-H chains containing Q^2 species are formed compared with dimeric C-S-H units (without Q^2). Although the amounts of Q^2 species increase progressively after the hydration peak (~ 20 h), the population of Q^1 species remains approximately constant, which indicates the formation of longer silicate chains besides the dimers. By comparison, the amount of Q^0(h) species remains constant for several hours (~ 10 h) during the induction and acceleration stages, it subsequently decreases just when, according to isothermal calorimetry, the Ca_3SiO_5 hydration slows down, that is during the so-called deceleration stage. This observation provides

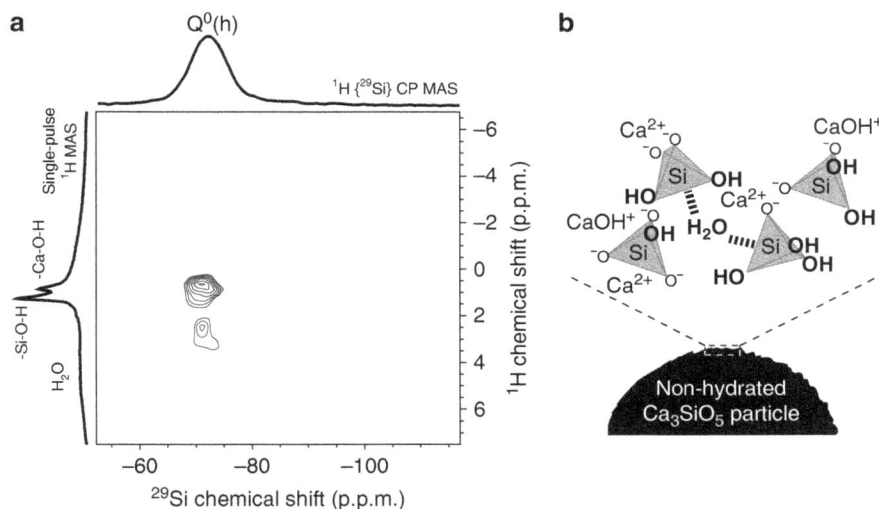

Figure 3 | Proton to silicon signal intensity correlations on the initial non-hydrated ^{29}Si-enriched triclinic Ca_3SiO_5. (**a**) The 2D $\{^1H\}$ ^{29}Si HETCOR NMR spectrum shows intensity correlations between ^{29}Si and ^1H signals that result from molecular proximity between ^{29}Si and ^1H nuclei. ^{29}Si CPMAS and ^1H MAS 1D spectra are shown along the horizontal and vertical axis of the 2D spectrum. The chemical shift of ^{29}Si is detected (horizontal dimension), while chemical shift of ^1H is recorded in the indirect (vertical) dimension. (**b**) The right inset schematizes the protonated moieties detected on the Ca_3SiO_5 surfaces.

important insights regarding the debate on the origin of the deceleration period. While some previous studies suggest that the deceleration period results from coverage of Ca_3SiO_5 particles by hydration products[30], others claim that hydration initially results in products forming a low-density structure, the subsequent densification of which corresponds to the beginning of the deceleration stage[31,32]. Our analyses suggest that compared with the acceleration period that is associated with the formation of predominantly dimeric C-S-H units, the deceleration period corresponds to the formation of greater relative fractions of C-S-H units with longer chain lengths. Such increasing extents of silicate polymerization might possibly be accompanied by an increased density of the C-S-H that consequently would present a diffusion barrier for mass transport and, thus, slow the rate of hydration reaction, consistent with the deceleration stage. This alone is not conclusive as it could either support the view according to which the deceleration would be based indeed on the filling of an ultra-low-density gel[33] or the one based on an inhibition of hydration by hydrates themselves[34], impinging on each other's growth[35,36]. Nevertheless, the decrease of the amount of near-surface $Q^0(h)$ species population at the onset of the deceleration period reflects a proportional decrease of the average surface area available to drive hydration by silicate dissolution. The decrease of the particles surface area as revealed here by NMR supports strongly the conclusions of recent modelling studies[37], namely that the deceleration stage results from the reduction of the average particle surface area available for reaction due to increasing surface coverage of the Ca_3SiO_5 particles by hydration products. This conclusion is also supported by the fact that at 7 days 5% of the Ca_3SiO_5 has not yet hydrated, bringing support to a coverage and passivation of its surface by deposited hydrates. Moreover, the long period during which $Q^0(h)$ remains constant suggests that during dissolution, the surface decrease due to the reduction in particle size is compensated by roughening (opening of etch pits and step retreat)[38]. In other words dissolution does not simply proceed by shrinking of the core of the particles, but also by etching.

Final C-S-H structure. The atomic site interconnectivities of different silicate species can be used to elucidate the molecular structures and lengths of silicate chains in the C-S-H. Such detailed insights can be obtained by using solid-state 2D J-mediated $^{29}Si\{^{29}Si\}$ correlation NMR techniques[39] that probe J-coupled $^{29}Si-O-^{29}Si$ spin pairs and have been previously applied to establish silicate framework connectivities in a variety of heterogeneous materials[40-43]. Previously, Brunet et al.[44] have conducted 2D dipolar-mediated $^{29}Si\{^{29}Si\}$ NMR measurements that rely on through-space $^{29}Si-^{29}Si$ dipolar couplings and which yield information on the molecular-level proximities of different ^{29}Si moieties in synthetic C-S-H. However, such measurements cannot be used to directly establish the covalent connectivity among different ^{29}Si moieties in the C-S-H structure. In contrast, by relying on through-bond J-interactions associated with $^{29}Si-O-^{29}Si$ moieties (J-interactions between ^{29}Si spin pairs separated by more than two covalent bonds are negligibly small and consequently expected to be below the detection limits of the 2D J-mediated $^{29}Si\{^{29}Si\}$ NMR measurement.), 2D J-mediated $^{29}Si\{^{29}Si\}$ double-quantum (DQ) correlation NMR measurements provide detailed insights regarding the tetrahedral site connectivity in the C-S-H chains. Notably, the 2D J-mediated $^{29}Si\{^{29}Si\}$ NMR spectrum of hydrated ^{29}Si-labelled Ca_3SiO_5 shown in Fig. 4b provides significantly enhanced ^{29}Si resolution, compared with the single-pulse ^{29}Si MAS spectrum (Fig. 4a), and unambiguously establishes distinct $^{29}Si-O-^{29}Si$ covalent connectivities in the silicate chains.

The 2D J-mediated $^{29}Si\{^{29}Si\}$ NMR spectrum (Fig. 4b) exhibits three well separated regions of correlated intensities in the Q^1 (approximately -79 p.p.m.) and Q^2 (approximately -85 p.p.m.) chemical shift ranges along the single-quantum (SQ)–DQ $y = 2x$ line, and two pairs of cross-correlated peaks between the Q^1 and Q^2 chemical shifts ranges. The broad continuous distribution of correlated chemical shifts in the 2D $^{29}Si\{^{29}Si\}$ spectrum between signals at -82 and -87 p.p.m. in the ^{29}Si SQ dimension are attributed to different $^{29}Si-O-^{29}Si$ Q^2 moieties, consistent with the structural disorder of C-S-H. Interestingly, the spectrum reveals narrow (0.6 p.p.m. FWHM) ridges of intensity correlations that are parallel to the SQ–DQ line. Such features typically arise from structural disorder on length scales (>1 nm) that are larger than the distances between the $^{29}Si-^{29}Si$ spin pairs (or also due to anisotropy in the magnetic susceptibility)[45]. The presence of such poor long-range structural order is consistent with the broad distributions of local ^{29}Si environments that are associated with the heterogeneous nature of the C-S-H. Nevertheless, careful analysis of the 2D spectrum distinguishes discrete correlated signal intensities that are resolved to greater than a tenth of a p.p.m. Specifically, a strong correlated intensity (labelled i) between the ^{29}Si signals centred at -84.8 and -85.4 p.p.m. in the SQ dimension and at -170.2 p.p.m. in the DQ dimension (Supplementary Fig. 11) unambiguously establishes the presence of two chemically distinct Q^2 ^{29}Si species that are covalently bonded through a shared bridging oxygen atom. The different isotropic ^{29}Si chemical shifts of these distinct Q^2 species likely reflect differences in the number and types of species in the C-S-H interlayer (calcium ions or proton moieties such OH groups or water molecules) that are in close (<1 nm) molecular-level proximity to the non-bridging oxygen atoms of the four-coordinate silicate units. Indeed, the different electronegativities of Ca^{2+} and H^+ result in different ^{29}Si nuclear shielding, as shown by recent density functional theory calculations[46]. These molecular-level differences in the Q^2 species are shown in the schematic diagram (Fig. 4, inset) of a postulated structure of C-S-H that is consistent with the observed 2D NMR correlations (as well as previous experimental[28,18] and modelling analyses[17,47]). Although the Q^{2L} resonances (the four-coordinate Q^2 silicate units that are positioned away from the interlayer space between two C-S-H chains, as shown in the inset in Fig. 4) are not resolved in the spectrum, the external ridges of the Q^2 correlation spot correspond to correlated intensity between the ^{29}Si SQ signals of the two Q^2 silicate species at -85.4 and -84.8 p.p.m. with the ^{29}Si SQ signals from the Q^{2L} species to which they are, respectively, bound. Within this hypothesis and with the constraint that the DQ frequency must be the sum of the SQ frequencies, two additional correlations can be identified for the Q^2 species at SQ signals -85.4 and -84.8 p.p.m. at DQ signals approximately -168.9 p.p.m. (ii) and -168.1 p.p.m. (iii), respectively, thus establishing the presence of two distinct Q^{2L} species with SQ signals at -83.5 and -83.1 p.p.m. Furthermore, the same ^{29}Si SQ signals at -85.4 and -84.8 p.p.m. from the two Q^2 silicate species are also separately correlated with ^{29}Si signals centred around -79 p.p.m. (iv, v) (DQ $\simeq -164$ p.p.m.) from Q^1 species, further corroborating that these Q^2 species are indeed chemically distinct. Therefore, analyses of the 2D J-mediated $^{29}Si\{^{29}Si\}$ spectrum establish the occurrence of oligomeric silicate units with two distinct Q^2 and two distinct Q^{2L} species in the C-S-H structure.

The partially resolved pair correlated intensities (ix–xii) in the range of -77 to -80 p.p.m. reveal the presence of different types of Q^1 silicate species associated with at least four distinct dimeric C-S-H units. These results are further corroborated by differences in the spin–spin (T_2) relaxation-time behaviours of the associated ^{29}Si Q^1 species, which were exploited to provide improved ^{29}Si resolution by using one-dimensional (1D) T_2-filtered ^{29}Si MAS measurements (Supplementary Fig. 12). In combination, the different pair correlated intensities establish the presence of dimeric units (*ix–xii*) and C-S-H chains that consist of two distinct Q^1-Q^2 (*iv*, *v*) and Q^2-Q^{2L} (*ii*, *iii*) connectivities and at least one Q^2-Q^2 (*i*) connectivity. To accommodate this diversity of atomic connectivity revealed by the 2D ^{29}Si{^{29}Si} NMR measurements, the C-S-H structure must contain a linear chain of at least eight four-coordinated silicate units (that is, an octamer). A similar analysis of pair correlated intensities *vi–viii* indicate the presence of pentameric C-S-H units, as discussed in the Supplementary Note 4. This result is supported by recent studies using density functional theory that have evaluated the relative stabilities of linear C-S-H units of different chain lengths and proposed the presence of stable octameric units[48], for which no direct experimental evidence has previously been available.

The relative populations of ^{29}Si silicate species associated with C-S-H units of different chain lengths (for example, dimers and octamers) are determined based on the enhanced ^{29}Si resolution afforded by the 2D ^{29}Si{^{29}Si} NMR spectrum. Specifically, the single-pulse ^{29}Si MAS spectrum shown in Fig. 5a can be simulated by using the peak positions of ^{29}Si signals as established by the 2D ^{29}Si{^{29}Si} NMR spectrum and the relative fractions of Q^1, Q^2 and Q^{2L} species associated with C-S-H units of different chain lengths (for example, $Q^2/Q^1 = 2$, $Q^2/Q^{2L} = 2$ for octamer as shown in Fig. 5b). Such an analysis yields estimates of 44, 7 and 42% ($\pm$ 4%) for the relative populations of ^{29}Si silicate engaged in octameric, pentameric and dimeric units, respectively. These values correspond to 20 mole% octamers, 5 mole% pentamers and 75 mole% dimers in the C-S-H. The salient result is, thus, that despite the fact that the average chain length is 5, pentamers are actually a minority feature. Such distributions of chain lengths are consistent with previous studies that have reported mean chain lengths for C-S-H, which suggest the presence of pentamers and octamers, in addition to dimers[22,49–51]. It must be understood that the high amount of octamers was obtained here in a relatively short hydration times (1.5 month) compared with what would be required in a usual cement paste. Specifically, the use of pure tricalcium silicate, the high surface area (4.4 m^2 g^{-1}) of the non-hydrated sample and the water-to-solids ratio (0.8) used in this study are expected to result in relatively fast hydration kinetics and a faster precipitation of C-S-H. The end result is a higher extent of hydration and silicate cross-linking. Interestingly, the analysis also indicates that small quantities of monomeric ^{29}Si

silicate species, such as hydroxylated Q^0(h) ($5 \pm 1\%$) and anhydrous Q^0 ($2 \pm 1\%$), are present even after hydration of Ca$_3$SiO$_5$ for 1.5 months at 25 °C. These monomers likely arise from remnants of surface hydroxylation of Ca$_3$SiO$_5$ particles or are components of the C-S-H structure, which is consistent with recent numerical modelling results[47].

Discussion

The carefully synthesized ^{29}Si-enriched sample enables, for the first time, 2D *J*-mediated (through ^{29}Si-O-^{29}Si bonds) ^{29}Si{^{29}Si} NMR measurements that provide detailed insights regarding the different silicate species, their respective site connectivities, and relative populations, especially for previously unidentified discrete silicate moieties in the C-S-H. Consequently, the lengths of C-S-H

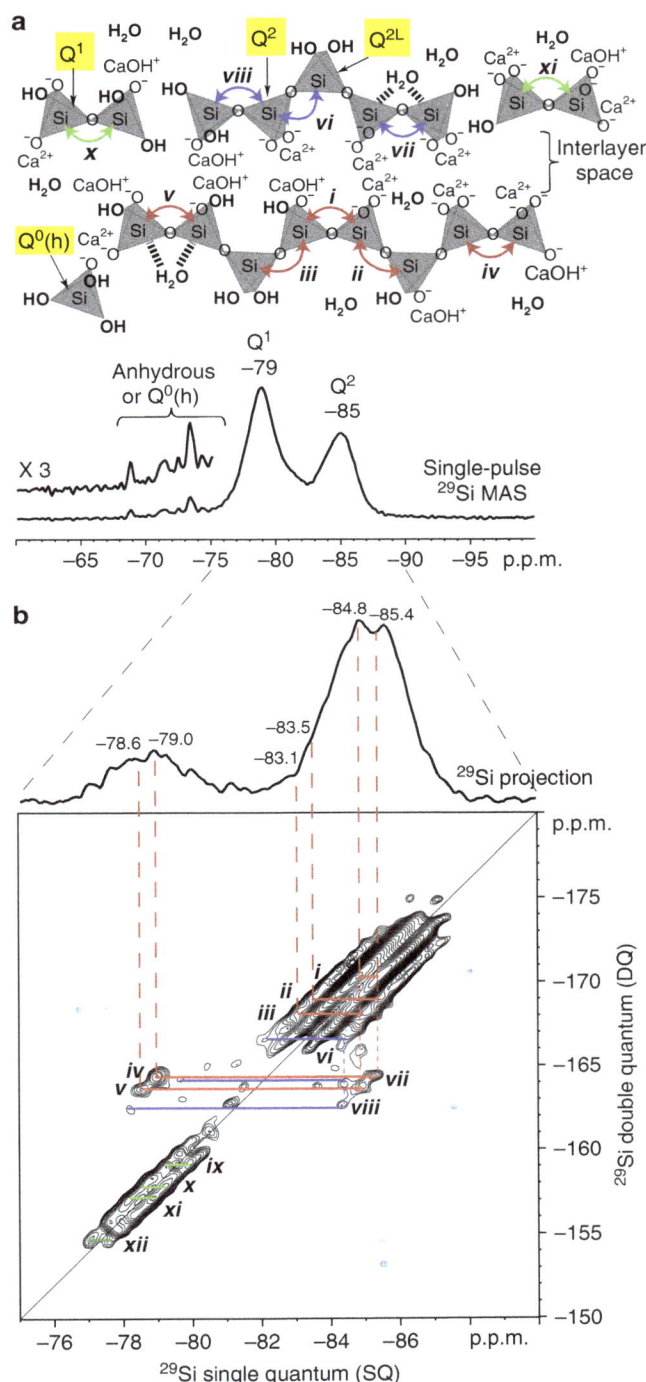

Figure 4 | Molecular structures and silicate site connectivities in partially polymerized calcium-silicate-hydrates. (**a,b**) Solid-state (**a**) 1D single-pulse ^{29}Si MAS and (**b**) 2D *J*-mediated ^{29}Si{^{29}Si} correlation NMR spectra of hydrated (1.5 month, 25 °C) ^{29}Si-enriched triclinic Ca$_3$SiO$_5$. The lowest contour lines in the 2D spectrum are 9% of the maximum signal intensity. The 'double-quantum' filter used to acquire the spectrum in **b** enables selective detection of pairs of signals (*i, j*) from distinct ^{29}Si nuclei that are covalently bonded. Consequently, the 2D spectrum exhibits intensity correlations between ^{29}Si signals at distinct frequencies (ω_i, ω_j) from ^{29}Si-O-^{29}Si spin pairs (*i, j*) in the horizontal SQ dimension (isotropic ^{29}Si chemical shifts) and at the sum of these frequencies ($\omega_i + \omega_j$) in the vertical DQ dimension. Therefore, correlated intensities at these specific positions in the 2D spectrum unambiguously establish the presence of covalently bonded ^{29}Si silicate species corresponding to the distinct isotropic ^{29}Si chemical shifts. The inset in **a** shows a schematic diagram of the different silicate moieties present in the calcium-silicate-hydrates with double-headed arrows indicating the *J*-interactions in ^{29}Si-O-^{29}Si species that are established by the intensity correlations in the 2D spectrum, specifically from dimeric (green), pentameric (blue) or octameric (red) units. For sake of clarity, the calcium layers are not represented.

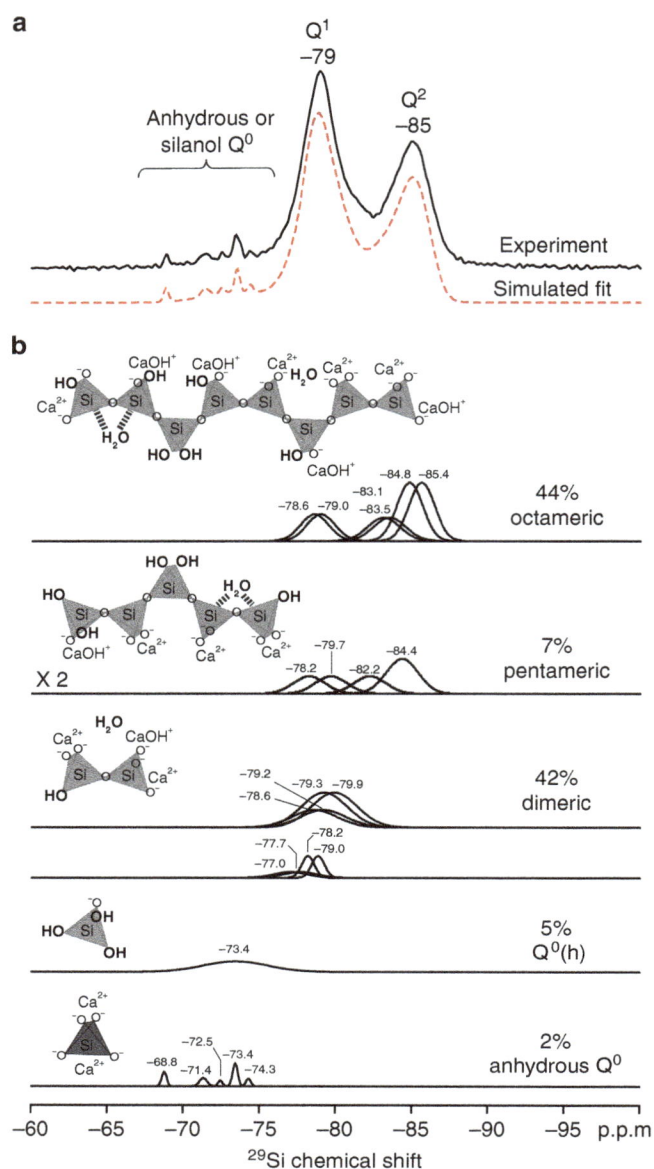

Figure 5 | Relative populations of ^{29}Si silicate species in hydrated triclinic Ca$_3$SiO$_5$. (a) Solid-state 1D single-pulse ^{29}Si MAS spectrum (black) of hydrated (1.5 month, 25 °C) ^{29}Si-enriched Ca$_3$SiO$_5$ and corresponding simulated fit (red) to the spectrum based on the signal decompositions shown in **b**. (b) Signal decompositions and relative populations of the different ^{29}Si moieties that comprise anhydrous Q^0, Q^0(h) and octameric, pentameric and dimeric C-S-H units, which contribute to the simulated fit (red) in **a**. Insets in **b** show schematic diagrams of the possible types of C-S-H units and the associated silicate moieties.

chains and the relative populations of associated silicate species are determined, which can be used to evaluate the validity of molecular models for Portland cement hydration that have been previously proposed in the literature[17,21,47]. This opens new perspective for understanding the complex molecular-level mechanical properties of C-S-H.

Solid-state ^{29}Si NMR measurements of ^{29}Si-enriched triclinic Ca$_3$SiO$_5$ also enable the transient silicate speciation and polymerization in the developing C-S-H structure to be monitored and quantified as a function of hydration time, especially during the crucial induction, acceleration and deceleration stages. Importantly, hydroxylated monomeric (Q^0(h)) silicate species can be detected

and quantified by using ^{29}Si{^{1}H} CPMAS NMR measurements to monitor changes in surface composition with the progress of hydration. The NMR results presented here establish that non-hydrated Ca$_3$SiO$_5$ particle surfaces predominantly consist of hydroxylated Q^0 silicate species with negligible quantities of Q^1 and Q^2 hydration products, including for the pre-induction and induction stages of the hydration process. Such detailed insights of silicate-water mixtures have heretofore been challenging and often infeasible to determine by other characterization techniques due to the low absolute quantities, complicated structures and poor long-range order of the hydroxylated surface species. Compared with the induction period, the onset of silicate polymerization (that is, Q^1 and/or Q^2 species) during hydration corresponds to the formation of dimeric units in C-S-H during the acceleration stage, consistent with previous cement literature. Interestingly, during the deceleration stage the hydration rate reduces (at a hydration level of 50%) before any significant reduction of the Q^0(h) populations are observed at the Ca$_3$SiO$_5$ surface. This corresponds to a relatively fast decrease in the reaction rate compared with the rate of reduction of the hydroxylated species available for reaction at the surface, which indicates that part of the surface is likely covered by C-S-H products. These results are consistent with previous studies that suggest that the rate of hydration is controlled by the surface coverage of C-S-H species during the deceleration stage[37]. Calculations based on a shrinking core model (hydration reaction slows down due to consumption of the particles) indicate that for monodispersed spherical particles, a decrease in volume by a factor of 0.5 would be accompanied by a decrease in surface area by a factor 0.63 ($2^{-2/3}$). Ca$_3$SiO$_5$ particles are neither spherical nor monodisperse but the present NMR results are definitely not compatible with a shrinking core model. Consequently, the surface area available for reaction is clearly modified by the surface roughness produced by dissolution driven etching of the surface[38].

The relations directly observed here for the first time between surface passivation and etching phenomena on the one hand and the succession of the induction, acceleration and deceleration stages of hydration of Ca$_3$SiO$_5$ on the other hand, provide new understanding for the occurrence of this complex kinetic behaviour actually observed in a variety of silicate systems. Ca$_3$SiO$_5$, because of its high reactivity, constitutes an interesting model for understanding long term silicate hydration processes occurring during geochemical weathering or hydrothermal synthesis[23].

Methods

NMR spectroscopy. The ^{1}H and ^{29}Si NMR isotropic chemical shifts were referenced to tetramethylsilane using tetrakis(trimethylsilyl)silane [((CH$_3$)$_3$Si$_4$)Si] as a secondary standard[52]. All measurements were performed using zirconia MAS rotors and at room temperature. Solid-state 1D ^{29}Si NMR experiments were carried out using a Bruker Avance-III 500 spectrometer (magnetic field 11.7 T). Magic-angle-spinning (MAS) spectra were measured using a Bruker MAS NMR probe with 4 mm rotors, at spinning frequencies of 7 kHz, and without decoupling. The single-pulse ^{29}Si MAS spectra were acquired with a π/2 pulse length of 6 μs, a recycle delay of 1,000 or 100 s, and 64 or 16 scans for the ^{29}Si-enriched non-hydrated and hydrated Ca$_3$SiO$_5$ samples, respectively. {^{1}H}^{29}Si CPMAS NMR spectra were recorded using a ^{1}H rf power of 93 kHz, a contact time of 5 ms, and recycle delay of 10 s. The number of scans was 184 for hydrated Ca$_3$SiO$_5$ samples and 2,000 for non-hydrated sample. Hartmann–Hahn matching was ensured by a ramp on the ^{29}Si rf field intensity. 2D {^{1}H}-^{29}Si heteronuclear dipolar correlation (HETCOR) experiments were conducted on a Bruker Avance-700 (16.4 T) spectrometer at ambient temperature, under 4 kHz MAS conditions, with a 7 ms CP contact time, recycle delay of 10 s and 66 t_1 increments of 50 μs each. Solid-state 2D J-mediated ^{29}Si{^{29}Si} DQ correlation NMR experiments were conducted using the refocused-INADEQUATE technique[39] and a 18.8 T Bruker AVANCE-III NMR spectrometer. The experiments were conducted under conditions of 12.5 kHz MAS using a Bruker 3.2 mm H-X double resonance probehead. The 2D ^{29}Si{^{29}Si} spectrum was acquired using a 2.5 μs ^{1}H π/2 pulse, 3.5 ms contact time for ^{29}Si{^{1}H} CP, 6.0 μs ^{29}Si π/2 pulses, SPINAL-64 ^{1}H decoupling[53], 152 t_1 increments, an incremental step size of 80 μs, a recycle delay of 2 s and 3,072 scans for each t_1 increment, which corresponds to an experimental time of 260 h (~11 days).

Hydration experiments. Paste for *in situ* NMR measurements was prepared by mixing 0.3 g of non-hydrated ^{29}Si-enriched Ca$_3$SiO$_5$ and 0.24 g of ultrapure water in a cylindrical 2 ml plastic vial for 3 min using a vortex mixer (Analog, VWR) at 2,500 r.p.m. With the help of a syringe and needle, part this paste was introduced as such in the zirconia MAS rotor thus enabling the acquisition of the NMR spectra during the reaction and avoiding any possible microstructural changes caused by the commonly used drying techniques[24]. After 6 h of hydration, the paste was removed from the ZrO$_2$ rotor to prevent its hardening inside the rotor, and the NMR measurements were continued on the part of the sample previously set aside and stored in the closed vial at room temperature. The kinetics of ^{29}Si-enriched Ca$_3$SiO$_5$ hydration were measured by isothermal calorimetry using a TAM Air microcalorimeter at 23 °C. One gram of ^{29}Si-enriched Ca$_3$SiO$_5$ was mixed with 0.8 g of ultrapure water under identical conditions as for samples prepared for NMR measurements. The paste was immediately sealed in a glass ampoule and placed in the isothermal calorimeter. The degree of reaction of ^{29}Si-enriched Ca$_3$SiO$_5$ was calculated by dividing the cumulative heat released at a certain time by the enthalpy of the hydration reaction of Ca$_3$SiO$_5$ ($-520\,\mathrm{J\,g}^{-1}$ Ca$_3$SiO$_5$) (refs 54,55). Additional details of synthesis, Ca$_3$SiO$_5$ characterization and NMR quantitative analysis are reported in the Supplementary Methods and Supplementary Note 1.

References

1. Le Chatelier, H. *Recherches Expérimentales sur la Constitution des Mortiers Hydrauliques* (Doctoral thesis, Faculté des Sciences de Paris, 1887).
2. Nugent, M. A., Brantley, S. L., Pantano, C. G. & Maurice, P. A. The influence of natural mineral coatings on feldspar weathering. *Nature* **395**, 588–591 (1998).
3. Casey, W. H., Westrich, H. R., Banfield, J. F., Ferruzzi, G. & Arnord, W. G. Leaching and reconstruction at the surface of dissolving chain-silicate minerals. *Nature* **366**, 253–256 (1993).
4. Conradt, R. Chemical durability of oxide glasses in aqueous solutions. *J. Am. Ceram. Soc.* **91**, 728–735 (2008).
5. Cailleteau, C. *et al.* Insight into silicate-glass corrosion mechanisms. *Nat. Mater.* **7**, 978–983 (2008).
6. Cundy, C. S. & Cox, P. A. The hydrothermal synthesis of zeolites: history and development from the earliest days to the present time. *Chem. Rev.* **103**, 663–702 (2003).
7. Serrano, D. & Van Grieken, R. Heterogenous events in the crystallization of zeolites. *J. Mater. Chem.* **11**, 2391–2407 (2001).
8. Bullard, J. W. *et al.* Mechanisms of cement hydration. *Cement Concrete Res.* **41**, 1208–1223 (2011).
9. Taylor, H. F. W. *Cement Chemistry* (Thomas Telford, 1997).
10. Bullard, J. W. & Flatt, R. J. New insights into the effect of calcium hydroxide precipitation on the kinetics of tricalcium silicate hydration. *J. Am. Ceram. Soc.* **93**, 1894–1903 (2010).
11. Gartner, E. M. & Jennings, H. M. Thermodynamics of calcium silicate hydrates and their solutions. *J. Am. Ceram. Soc.* **70**, 743–749 (1987).
12. Juilland, P., Gallucci, E., Flatt, R. J. & Scrivener, K. S. Dissolution theory applied to the induction period in alite hydration. *Cement Concrete Res.* **40**, 831–844 (2010).
13. Nicoleau, L., Nonat, A. & Perrey, D. The di- and tricalcium silicate dissolutions. *Cement Concrete Res.* **47**, 14–30 (2013).
14. Nicoleau, L., Schreiner, E. & Nonat, A. Ion-specific effects influencing the dissolution of tricalcium silicate. *Cement Concrete Res.* **59**, 118–138 (2014).
15. Lasaga, A. C. & Luttge, A. Variation of crystal dissolution rate based on a dissolution stepwave model. *Science* **291**, 2400–2404 (2001).
16. Arvidson, R. S., Ertan, I. E., Amonette, J. E. & Luttge, A. Variation in calcite dissolution rates: A fundamental problem? *Geochim. Cosmochim. Acta* **67**, 1623–1634 (2003).
17. Richardson, I. G. Tobermorite/jennite- and tobermorite/calcium hydroxide-based models for the structure of C–S–H: applicability to hardened pastes of tricalcium silicate, β-dicalcium silicate, Portland cement, and blends of Portland cement with blast-furnace slag, metakaolin, or silica fume. *Cement Concrete Res.* **34**, 1733–1777 (2004).
18. Allen, A. J., Thomas, J. J. & Jennings, H. M. Composition and density of nanoscale calcium-silicate-hydrate in cement. *Nat. Mater.* **6**, 311–316 (2007).
19. Mishra, R. K., Flatt, R. J. & Heinz, H. Force field for tricalcium silicate and insight into nanoscale properties: cleavage, initial hydration, and adsorption of organic molecules. *J. Phys. Chem. C* **117**, 10417–10432 (2013).
20. Thomas, J. J. *et al.* Modeling and simulation of cement hydration kinetics and microstructure development. *Cement Concrete Res.* **41**, 1257–1278 (2011).
21. Qomi, M. J. A. *et al.* Combinatory molecular optimization of cement hydrates. *Nat. Commun.* **5**, 4960 (2014).
22. Brough, A. R., Dobson, C. M., Richardson, I. G. & Groves, G. W. *In situ* solid-state NMR studies of Ca$_3$SiO$_5$: hydration at room temperature and at elevated temperatures using ^{29}Si enrichment. *J. Mater. Sci.* **29**, 3926–3940 (1994).
23. Fischer, C., Arvidson, R. S. & Lüttge, A. How predictable are dissolution rates of crystalline material? *Geochim. Cosmochim. Acta* **98**, 177–185 (2012).
24. Zhang, J. & Scherer, G. W. Comparison of methods for arresting hydration of cement. *Cement Concrete Res.* **41**, 1024–1036 (2011).
25. Engelhardt, G. & Michel, D. *High-Resolution Solid-State NMR of Silicates and Zeolites* (John Wiley & Sons, 1987).
26. Bellmann, F., Damidot, D., Möser, B. & Skibsted, J. Improved evidence for the existence of an intermediate phase during hydration of tricalcium silicate. *Cement Concrete Res.* **40**, 875–884 (2010).
27. Rawal, A. *et al.* Molecular silicate and aluminate species in anhydrous and hydrated cements. *J. Am. Chem. Soc.* **132**, 7321–7337 (2010).
28. Cong, X. & Kirkpatrick, R. J. 17O MAS NMR investigation of the structure of calcium silicate hydrate gel. *J. Am. Ceram. Soc.* **79**, 1585–1592 (1996).
29. Durgun, E., Manzano, H., Kumar, P. V. & Grossman, J. C. The characterization, stability, and reactivity of synthetic calcium silicate surfaces from first principles. *J. Phys. Chem. C* **118**, 15214–15219 (2014).
30. Garrault, S., Behr, T. & Nonat, A. Formation of the C–S–H layer during early hydration of tricalcium silicate grains with different sizes. *J. Phys. Chem. B* **110**, 270–275 (2006).
31. Kumar, A., Bishnoi, S. & Scrivener, K. L. Modelling early age hydration kinetics of alite. *Cement Concrete Res.* **42**, 903–918 (2012).
32. Gonzalez-Teresa, R., Dolado, J. S., Ayuela, A. & Gimel, J. C. Nanoscale texture development of C-S-H gel: a computational model for nucleation and growth. *Appl. Phys. Lett.* **103**, 234105 (2013).
33. Ioannidou, K., Pellenq, R. J. M. & Del Gado, E. Controlling local packing and growth in calcium-silicate-hydrate gels. *Soft Matter* **10**, 1121–1133 (2014).
34. Bishnoi, S. & Scrivener, K. L. Studying nucleation and growth kinetics of alite hydration using µic. *Cement Concrete Res.* **39**, 849–860 (2009).
35. Tzschichholz, F. & Zanni, H. Global hydration kinetics of tricalcium silicate cement. *Phys. Rev. E* **64**, 016115 (2001).
36. Garrault, S., Finot, E., Lesniewska, E. & Nonat, A. Study of C-S-H growth on C3S surface during its early hydration. *Mater. Struct.* **38**, 435–442 (2005).
37. Bullard, J. W., Scherer, G. W. & Thomas, J. J. Time dependent driving forces and the kinetics of tricalcium silicate hydration. *Cement Concrete Res.* **74**, 26–34 (2015).
38. Nicoleau, L. & Bertolim, M. A. Analytical model for the alite (C3S) dissolution topography. *J. Am. Ceram. Soc.* http://dx.doi.org/10.1111/jace.13647 (2015).
39. Lesage, A., Bardet, M. & Emsley, L. Through-bond carbon-carbon connectivities in disordered solids by NMR. *J. Am. Chem. Soc.* **121**, 10987–10993 (1999).
40. Fyfe, C. A. & Brouwer, D. H. Optimization, standardization, and testing of a new NMR method for the determination of zeolite host-organic guest crystal structures. *J. Am. Chem. Soc.* **128**, 11860–11871 (2006).
41. Cadars, S., Brouwer, D. H. & Chmelka, B. F. Probing local structures of siliceous zeolite frameworks by solid-state NMR and first-principles calculations of ^{29}Si-O-^{29}Si scalar couplings. *Phys. Chem. Chem. Phys.* **11**, 1825–1837 (2009).
42. Köster, T. K.-J. *et al.* Resolving the different silicon clusters in Li$_{12}$Si$_7$ by ^{29}Si and ^{6,7}Li solid-state NMR spectroscopy. *Angew. Chem. Int. Ed. Engl.* **50**, 12591–12594 (2011).
43. Shayib, R. M. *et al.* Structure-directing roles and interactions of fluoride and organocations with siliceous zeolite frameworks. *J. Am. Chem. Soc.* **133**, 18728–18741 (2011).
44. Brunet, F., Bertani, P., Charpentier, T., Nonat, A. & Virlet, J. Application of 29Si homonuclear and 1H-29Si heteronuclear NMR correlation to structural studies of calcium silicate hydrates. *J. Phys. Chem. B* **108**, 15494–15502 (2004).
45. Cadars, S., Lesage, A. & Emsley, L. Chemical shift correlations in disordered solids. *J. Am. Chem. Soc.* **127**, 4466–4476 (2005).
46. Rejmak, P., Dolado, J. S., Stott, M. J. & Ayuela, A. ^{29}Si NMR in cement: a theoretical study on calcium silicate hydrates. *J. Phys. Chem. C* **116**, 9755–9761 (2012).
47. Pellenq, R. *et al.* A realistic molecular model of cement hydrates. *Proc. Natl Acad. Sci. USA* **106**, 16102–16107 (2009).
48. Ayuela, A. *et al.* Silicate chain formation in the nanostructure of cement-based materials. *J. Chem. Phys.* **127**, 164710 (2007).
49. Chen, J. J. *et al.* Solubility and structure of calcium silicate hydrate. *Cement Concrete Res.* **34**, 1499–1519 (2004).
50. Kulik, D. A. Improving the structural consistency of C-S-H solid solution thermodynamic models. *Cement Concrete Res.* **41**, 477–495 (2011).
51. Richardson, I. G. Model structures for C-(A)-S-H (I). *Acta Crystallogr. B* **70**, 903–923 (2014).
52. Hayashi, S. & Hayamizu, K. Chemical shift standards in high-resolution solid-state NMR (1) 13C, 29Si, and 1H nuclei. *Bull. Chem. Soc. Jpn* **64**, 685–687 (1991).
53. Fung, B. M., Khitrin, A. K. & Ermolaev, K. An improved broadband decoupling sequence for liquid crystals and solids. *J. Magn. Reson.* **142**, 97–101 (2000).

54. Thomas, J. J., Jennings, H. M. & Chen, J. J. Influence of nucleation seeding on the hydration mechanisms of tricalcium silicate and cement. *J. Phys. Chem. C* **113**, 4327–4334 (2009).

55. Damidot, D. & Nonat, A. C3S hydration in diluted and stirred suspensions: (I) study of two kinetic steps. *Adv. Cem. Res.* **6**, 27–35 (1994).

Acknowledgements

This research was supported by the Commission for Technology and Innovation (CTI project number 15846.1), the US Federal Highway Administration (FHWA) under agreement No. DTFH61-12-H-00003 and by Halliburton, Inc. (Any opinions, findings and conclusions or recommendations expressed in this publication are ours and do not necessarily reflect the view of the US Federal Highway Administration). Funding was also provided by the program 'Germaine de Staël' and a mobility fellowship from the French Embassy in Bern. We thank Dr R. Verel (ETH Zürich), Dr M Plötze (ETH Zürich) and S. Mantellato (ETH Zürich) for their assistance in the 1D ^{29}Si NMR, XRD and BET surface area measurements, respectively. D. Marchon (ETH Zürich) is also thanked for her support in the development of the synthesis protocol of Ca_3SiO_5. The solid-state 2D J-mediated ^{29}Si{^{29}Si} correlation NMR measurements were conducted using the UCSB Materials Research Laboratory (MRL) Shared Experimental Facilities that are supported by the MRSEC Program of the US National Science Foundation under Award No. DMR 1121053; a member of the NSF-funded Materials Research Facilities Network (www.mrfn.org).

Authors contributions

E.P. was the main investigator. She developed the synthesis of the Ca_3SiO_5 samples, designed and carried out the characterization and calorimetry studies. J.-B.d.E.d.L., R.J.F. and M.P. designed the project. B.F.C. proposed the 2D J-mediated and relaxation NMR experiments. E.P., A.S.A., J.-B.d.E.d.L. and R.P.S. performed the NMR experiments. All authors contributed to the analyses of the results and the writing of the manuscript.

Additional information

Competing financial interests: The authors declare no competing financial interests.

Noble metal-comparable SERS enhancement from semiconducting metal oxides by making oxygen vacancies

Shan Cong[1,*], Yinyin Yuan[1,2,*], Zhigang Chen[1], Junyu Hou[3], Mei Yang[1], Yanli Su[3], Yongyi Zhang[1], Liang Li[2], Qingwen Li[1], Fengxia Geng[3] & Zhigang Zhao[1]

Surface-enhanced Raman spectroscopy (SERS) represents a very powerful tool for the identification of molecular species, but unfortunately it has been essentially restricted to noble metal supports (Au, Ag and Cu). While the application of semiconductor materials as SERS substrate would enormously widen the range of uses for this technique, the detection sensitivity has been much inferior and the achievable SERS enhancement was rather limited, thereby greatly limiting the practical applications. Here we report the employment of non-stoichiometric tungsten oxide nanostructure, sea urchin-like $W_{18}O_{49}$ nanowire, as the substrate material, to magnify the substrate-analyte molecule interaction, leading to significant magnifications in Raman spectroscopic signature. The enrichment of surface oxygen vacancy could bring additional enhancements. The detection limit concentration was as low as 10^{-7} M and the maximum enhancement factor was 3.4×10^5, in the rank of the highest sensitivity, to our best knowledge, among semiconducting materials, even comparable to noble metals without 'hot spots'.

[1] Key Lab of Nanodevices and Applications, Suzhou Institute of Nano-Tech and Nano-Bionics, Chinese Academy of Sciences (CAS), Suzhou 215123, China. [2] Key Laboratory for Ultrafine Materials of Ministry of Education, School of Materials Science and Engineering, East China University of Science and Technology, Shanghai 200237, China. [3] College of Chemistry, Chemical Engineering and Materials Science, Soochow University, Suzhou 215123, China. * These authors contributed equally to this work. Correspondence and requests for materials should be addressed to Z.Z. (email: zgzhao2011@sinano.ac.cn).

The surface-enhanced Raman scattering (SERS) effect is a surface-sensitive technique characterized by an increase in the Raman intensity by orders of magnitude for molecules adsorbed on particular surfaces such as rough metals with respect to that expected from the same number of non-adsorbed molecules in solution or the gas phase[1,2]. The enhanced sensitivity makes the technique serve as one of the most powerful analytical tools for the unequivocal identification of chemical and biological analytes and additionally in numerous other fields, such as electrochemistry, surface science, catalysis, chemical and biomolecular sensing and so on[1,3,4]. While the exact mechanism for the enhancement effect is still a matter of debate, two widely accepted theories have been adopted in most cases to explain the phenomenon[5]. The primary one is electromagnetic mechanism predicting that the electric field is magnified when excitation takes place within localized surface plasmon resonances of substrate materials, leading to enhancement factors (EFs) up to the order of 10^6. The induced enhancement of local field by plasmonic coupling is called electromagnetic 'hot spots'. The other is a chemical mechanism proposing the formation of charge-transfer complexes between the chemisorbed species and the substrate materials and obtaining enhancement when the excitation frequency is in resonance with the charge-transfer transition with EFs around 10–100. SERS experiments have been essentially dominated by adsorbates on rough metallic surfaces, especially noble and alkali metals such as Au, Ag and Cu, because their plasmon resonance frequencies locate within excitation wavelength ranges commonly used in Raman spectroscopy. While precise tuneable position of adjacent metallic nanoparticles is a key factor to bring about strong plasmonic coupling, fabricating perfect plasmonic nanostructures with high density 'hot spots' to achieve high enhancements usually requires delicate procedures and high cost. In addition, noble metals typically show poor stability and biocompatibility, and therefore the search of other alternative materials for working as SERS substrate becomes an urgent task. A few semiconducting materials have been proven to show Raman enhancement, for example, InAs/GaAs quantum dots[6], CuTe nanocrystals[7], Cu_2O nanospheres[8] and TiO_2 nanostructures[9,10], in which charge transfer at the semiconductor–analyte interface plays a major role in Raman scattering enhancement. The TiO_2 SERS substrate offers not only a biocompatibility, but also a greater chemical and mechanical stability against variation in pH or temperature of the environment[9]. The effective utilization of semiconductor SERS-active substrates would greatly expand the applications of SERS in many fields, such as direct monitoring of interfacial chemical reactions on individual nanoparticles[11]. However, one significant problem of metal-free SERS-active substrates is that the EFs were typically quite low, usually in the range of 10–10^2, far from sufficient for applications in biological and biomedical analysis and diagnosis. It is consequently one formidable challenge but greatly desirable to find efficient semiconductor SERS substrate materials and obtain EFs comparable to noble metals.

The surface states of substrate material and its interaction with analyte molecule is a key to the SERS effect. Among the numerous methods in tuning the surface states of semiconducting oxide materials, adjusting oxygen deficiency represents one main strategy and may work as an efficient and simple way to achieve the goal of high sensitivity comparable to that of noble metals. Tungsten oxide materials, a traditional semiconductor, have attracted considerable attention because they possess distinctive physical and chemical properties, which endow them to be effective candidates in a wide range of applications, covering photocatalysts, gas sensors and electrochromic devices[12–14]. Importantly, rich phases of substoichiometric composition (WO_{3-x}) can be obtained by the reduction and formation of various types of defect structure, such as $WO_{2.72}$ ($W_{18}O_{49}$), $WO_{2.8}$ (W_5O_{14}), $WO_{2.92}$ ($W_{25}O_{73}$) and so on[15–17], which makes the system work as an ideal platform for investigating the effect of vacancies or defects on SERS phenomenon. $WO_{2.72}$ ($W_{18}O_{49}$) with the largest vacancy has been reported as the only oxide that can be isolated in a pure form, which contains tungsten ions of mixed valency[16,18].

Herein, employing vacancy-containing $W_{18}O_{49}$ as the substrate material, we achieve greatly enhanced SERS effect for the first time on function-rich tungsten oxide material, and the enhancement is further improved by creating surface deficiencies, which gives SERS EF as high as 3.4×10^5 and 100–10,000 times higher than the previously reported values for most other semiconductor SERS-active substrates, and even comparable to noble metals without 'hot spots'.

Results

Sample characterizations. $W_{18}O_{49}$ was synthesized using a previously reported approach, WCl_6 dissolved in ethanol followed by hydrothermal reaction at 180 °C for 24 h (ref. 13). Phase and purity of the sample was verified by X-ray diffraction characterizations as shown in Fig. 1a. The hydrothermally synthesized sample was crystallized in a monoclinic phase of $W_{18}O_{49}$ ($P2/m$, JCPDS no. 84-1516) with lattice constants refined to be $a = 18.31(2)$, $b = 3.839(8)$, $c = 14.00(1)$ Å and $\beta = 115.19(9)°$. The relatively high intensities and narrowing exhibited by (010) and (020) peaks suggest that the crystals preferably grew along the b axis as a result of retarded growth along the close-packed (010) planes. For comparison study, tungsten trioxide, WO_3, was acquired by annealing the $W_{18}O_{49}$ sample in air. The transformed WO_3 was also of high purity with all reflections perfectly fitted in stoichiometric monoclinic (no. 14, $P2_1/n$) tungsten trioxide with cell dimensions $a = 7.305(5)$, $b = 7.523(4)$, $c = 7.689(4)$ Å and $\beta = 90.90(9)°$. To obtain more information about the intrinsic nature of the samples, ultraviolet–visible absorption spectra of the two samples were collected (Fig. 1b). Distinct from the absorption to 500 nm normally observed in tungsten oxide materials, $W_{18}O_{49}$ sample exhibited an obvious blue shift of the absorption edge. In addition, an absorption tail beyond the absorption edge in the visible and near-infrared region was observed, up to the measurement limit, 800 nm, which can be considered to be closely related with the free electrons and/or oxygen deficiency-induced small polarons[15], providing clear evidence that the as-synthesized $W_{18}O_{49}$ sample contains a large number of oxygen vacancies. Morphology of the samples were characterized by scanning electron microscope observations, which suggested that the as-grown $W_{18}O_{49}$ sample comprised of thin nanowires with average length estimated to be about several micrometres while width in the range of 10–20 nm (Fig. 1c). The nanowires were entangled featuring a prickly, spherical and sea urchin-like morphology. High-resolution transmission electron microscopy on one single nanowire discerned clear lattice fringes belonging to (010) planes of monoclinic $W_{18}O_{49}$ with an interplanar spacing of ~0.38 nm (inset in Fig. 1c), which undoubtedly indicates that the nanowires preferably grow along the [010] direction, in good agreement with X-ray diffraction observations. After annealing in air and oxidizing into WO_3, the morphology was featureless, showing irregular particles of several micrometres (Fig. 1d).

The defect structure of $W_{18}O_{49}$ is illustrated in Fig. 2, in comparison with the perfect structure of WO_3, projected along the [010] direction. The crystal structure of WO_3 consists of slabs of corner-sharing WO_6 octahedra (ReO$_3$-type), which have an infinite extension in two dimensions (ac plane) and a finite, characteristic width in the [010] direction. The slabs are mutually

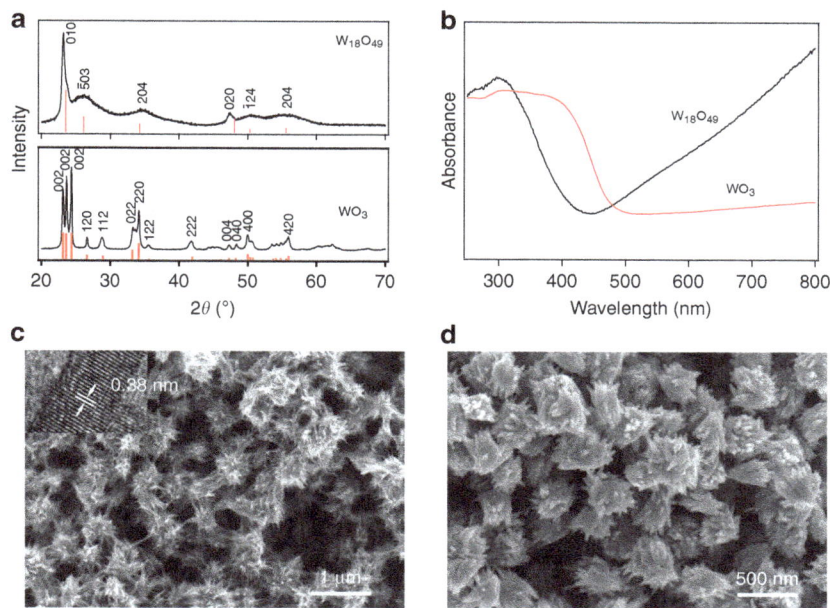

Figure 1 | Characterization of W$_{18}$O$_{49}$ and annealed WO$_3$ samples. (**a**) X-ray diffraction patterns of as-obtained W$_{18}$O$_{49}$ sample and annealed WO$_3$ with all reflections perfectly indexed; (**b**) ultraviolet–visible profile comparison between W$_{18}$O$_{49}$ sample and annealed WO$_3$, showing a blue shift and an obvious absorption tail beyond the edge that may arise from the presence of oxygen vacancies; (**c,d**) scanning electron microscopy images for W$_{18}$O$_{49}$ and WO$_3$ samples, respectively. Inset in (**c**): high-resolution transmission electron microscopy image on one single nanowire illustrating clear lattice fringe of 0.38 nm, which suggested that the nanowire growth was along the [010] direction.

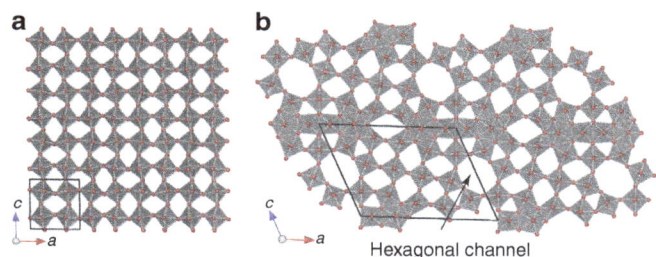

Figure 2 | Structure illustration for WO$_3$ and W$_{18}$O$_{49}$ in [010] projection. (**a**) The crystalline WO$_3$ structure consists of stacking of infinite corner-sharing WO$_6$ octahedra layers, while (**b**) the defects in W$_{18}$O$_{49}$ structure lead to the formation of hexagonal channels.

linked, forming channels between the octahedra. In contrast, the reduction of WO$_3$ to W$_{18}$O$_{49}$ leads to structural changes of the ReO$_3$-type structure and the formation of tungsten pentagonal columns causes the arising of additional hexagonal channels[19]. The presence of oxygen defects may enrich the surface states of semiconductor and enhance the interaction affinity between the adsorbent and adsorbate, providing promises for enhancements in Raman signals.

Raman enhancement for W$_{18}$O$_{49}$ sample. The Raman enhancement behaviour of the materials was examined using laser dye Rhodamine 6G (R6G) as a Raman probe, the molecular structure of which is provided in inset of Fig. 3a. R6G is a strongly fluorescent xanthene derivative that shows a molecular resonance effect when excited into its visible absorption band. Figure 3a shows the Raman spectra of R6G (10^{-6} M) with the excitation wavelength being 532.8 nm on substrates deposited with W$_{18}$O$_{49}$ nanowires, annealed WO$_3$, along with bare SiO$_2$/Si. While no clear signals related with R6G molecule were discerned on bare

SiO$_2$/Si substrate and WO$_3$ except the fluorescence background, a substantial Raman enhancement was observed on W$_{18}$O$_{49}$, and four characteristic bands of R6G centred at 612, 773, 1,360 and 1,650 cm^{-1}, named as P1, P2, P3 and P4, were clearly detected, which suggests that W$_{18}$O$_{49}$ could work as SERS-active substrate. Raman characterizations on bare W$_{18}$O$_{49}$ in the absence of probe molecules were also performed, in which only the signals related to O–W–O bending and W–O stretching modes were found (Supplementary Fig. 1), further confirming that the P1, P2, P3 and P4 bands originated from R6G molecules. P1 and P2 can be assigned to in-plane and out-of-plane bending motions of carbon and hydrogen atoms of the xanthenes skeleton, respectively; P3 and P4 correspond to aromatic C–C stretching vibration modes[20]. Raman spectra at four different R6G concentrations (the range was selected according to adsorption isotherms in Supplementary Fig. 2), decreasing from 10^{-4}, 10^{-5} and 10^{-6} to 10^{-7} M, were collected, showing obvious intensity enhancement until an extreme dilute solution was used, 10^{-7} M, in which just detectable signal was obtained (Fig. 3b). Therefore, the detection limit for W$_{18}$O$_{49}$ material can be determined to be 10^{-7} M. Such a low detection limit enables a high sensitivity towards trace amounts of analyte molecules, and makes it possible for a systematic investigation of different vibration modes in the SERS profiles.

Two signature bands, P1 and P3, were selected for calculating EFs based on the magnification of Raman intensity compared with that on bare substrate (details in Supplementary Methods)[21–23], the variation of which was depicted in Fig. 3c as a function of R6G analyte concentrations (C_{R6G}). It is clear that the greater enhancement occurs for P1 at each concentration, which was found to be surprisingly high as 1.9×10^5 with the R6G concentration of 10^{-6} M, about 100–10,000 times higher than that previously reported for most other semiconductor SERS-active substrates (Supplementary Table 1). Furthermore, the intensity variation was not uniform and instead strongly dependent on the band identity. The selective enhancement of the

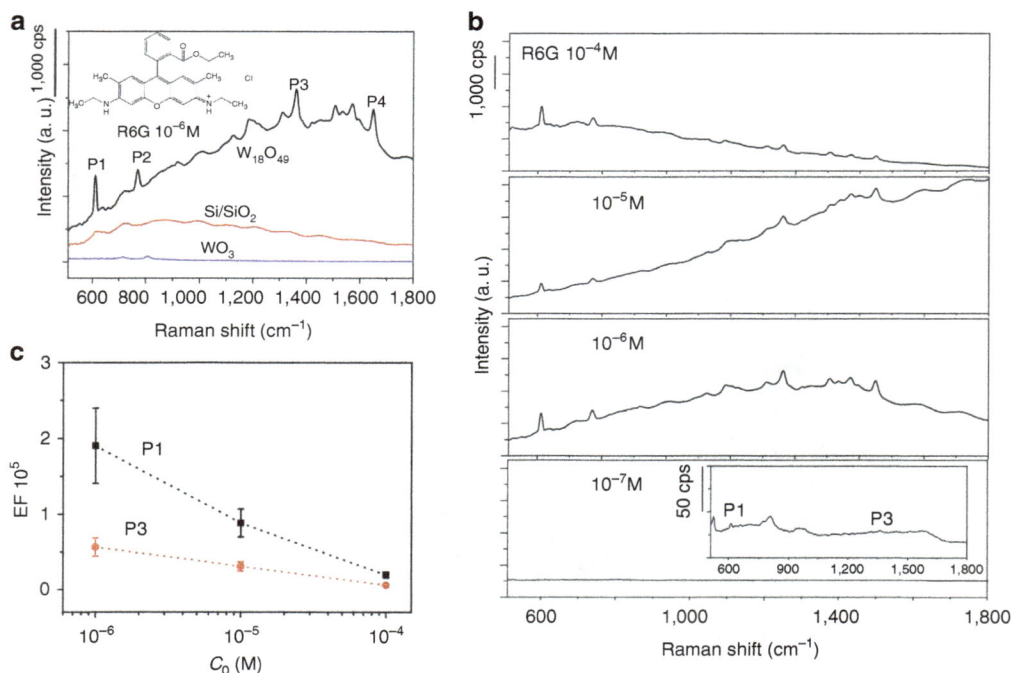

Figure 3 | SERS properties. (a) Raman profile of R6G (10^{-6} M) on substrates deposited with $W_{18}O_{49}$ sample compared with that for WO_3 and bare SiO_2/Si substrate. Inset: molecule structure of R6G. **(b)** Raman spectra collected for $W_{18}O_{49}$ at four different concentrations, 10^{-4}, 10^{-5}, 10^{-6} and 10^{-7} M, suggesting the detection limit was as low as 10^{-7} M (Inset: with narrowed y scale for 10^{-7} M). **(c)** The statistical evolution of EF as a function of R6G concentration plotted in logarithmic scale, with the analysis carried out over 30 different regions per sample. The Raman enhancement typically increased with decreasing concentrations and selective enhancement occurred with different bands.

different vibration modes indicates that the SERS on $W_{18}O_{49}$ sample largely depends on the distinct binding and geometry of R6G molecules on the $W_{18}O_{49}$ surface, which is a character of the chemical enhancement mechanism. Specifically, the change of P1 was more obvious than that for P3. The relative lower intensity exhibited by P3, especially at low concentrations, indicates that the long axis plane of R6G molecule was not parallel to the sample surface or the aromatic rings were separated by chemical groups including C–H bond[24]. With dilution of the R6G solutions, the signal intensity of P3 gradually increases, suggesting the gradual decrease of tilt angles of R6G molecules on the substrate. The emergence of greatly enhanced SERS sensitivity for $W_{18}O_{49}$ compared with WO_3 can probably be attributed to the presence of oxygen vacancies in $W_{18}O_{49}$ structure (*vide ante*, Fig. 2).

Surface vacancy related Raman enhancement. To confirm the role of oxygen vacancy played in SERS, additional surface vacancy was deliberately created by annealing the $W_{18}O_{49}$ sample in Ar/H_2 atmosphere[25]. After the treatment at 300 °C for 1 h, the crystalline phase of $W_{18}O_{49}$ and the sea urchin-like morphology were still well kept (Fig. 4a). X-ray photoelectron spectroscopy was used to check the surface states, and Fig. 4b displays the spectra of W4f core levels for pristine $W_{18}O_{49}$ and the Ar/H_2-treated samples. The W4f core-level spectra could be deconvoluted into three doublets (W^{6+}, W^{5+} and W^{4+}) for all samples. On the fitting analysis, it was found that the atomic percentage of W^{6+}, W^{5+} and W^{4+} for pristine $W_{18}O_{49}$ was 54.1, 30.4 and 15.5%, roughly agreeing the documented values[26]. By contrast, the percentage for W^{5+} increased to 35.4 and 47.5%, W^{4+} to 21.4 and 13.9% for Ar and H_2 thermal-treated samples, respectively. The increased percentage for reduced tungsten, W^{4+} and W^{5+}, is likely to be related with the creation of

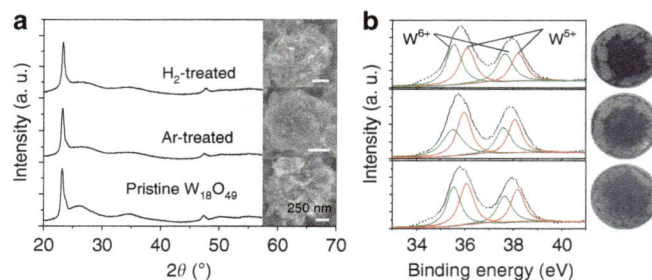

Figure 4 | Characterization of $W_{18}O_{49}$ samples with modulated surface oxygen vacancies. (a) X-ray diffraction patterns of Ar- and H_2-treated samples along with that for as-prepared $W_{18}O_{49}$ sample for comparison purpose, indicating that the crystalline phase remained during the annealing. Inset: the corresponding SEM images showing the morphology was also unchanged. **(b)** XPS spectra of W4f core levels for $W_{18}O_{49}$ samples after treatment in Ar and H_2 atmosphere at 300 °C and pristine $W_{18}O_{49}$ sea urchin-like aggregates. Right: corresponding optical images of the three samples, showing colour change from pale blue for pristine $W_{18}O_{49}$ to cyan and deep blue for Ar- and H_2-treated samples.

positively charged oxygen vacancies with accompanying charge-compensating electrons[27]. With increasing oxygen vacancy, the intensity of the absorption tail in the visible and near-infrared region increased (Supplementary Fig. 3), consistent with the recorded results[28]. The sample colour changed from pale blue for pristine $W_{18}O_{49}$ to cyan and deep blue for Ar- and H_2-treated samples, respectively, confirming the increment of oxygen deficiencies present in the examined samples.

The Raman spectra of R6G with concentration of 1×10^{-6} M for the two treated samples are given in Fig. 5a along with the spectrum of pristine $W_{18}O_{49}$ sample for comparison. Both treated samples clearly manifested the four characteristic peaks of R6G molecule. Neither a new peak nor an obvious shift of the characteristic peak was detected, indicating that no formation or vanish of chemical bonds between R6G and the substrates were induced on oxygen vacancy alternation. Notably, the Ar- and H_2-treated samples were found to display further Raman intensity increase on some degree compared with pristine $W_{18}O_{49}$ sample, corroborating the fact that oxygen vacancy may help increase Raman detect sensitivity. The Raman intensity of the four feature bands for all the three samples were summarized in Fig. 5b, and it is obvious that the Raman enhancement was in the order of H_2-treated $W_{18}O_{49}$ > Ar-treated $W_{18}O_{49}$ > pristine $W_{18}O_{49}$. The P1 band of H_2-treated $W_{18}O_{49}$ displayed the greatest enhancement, and the corresponding EF was evaluated to be 3.4×10^5. Such high EF has been rarely observed in semiconducting materials, even comparable to that for noble metals without hot spots[20], which may come from the combined function of intrinsic and deliberately created oxygen deficiencies.

It seems the presence of oxygen vacancies in semiconducting oxide materials, either intrinsic ones or deliberately created on the surface, could bring significant Raman enhancement for R6G molecule. To examine the interaction between tungsten oxide materials with adsorbate molecule, ultraviolet–visible spectrum of R6G molecules deposited on $W_{18}O_{49}$ was collected in comparison with those for neat R6G and $W_{18}O_{49}$, which is presented in Fig. 6a. The spectrum for the hybrid features occurrence of a new band with an optical absorption onset at $\sim 580\,nm$. The band locates between the band–band transition absorption edge for $W_{18}O_{49}$, 420 nm, and the photoabsorption threshold for R6G dye, 670 nm, providing clear evidence for the efficient charge transfer between R6G and $W_{18}O_{49}$ material. This observation of high Raman enhancement on $W_{18}O_{49}$ may be based on the charge-transfer mechanism in an adsorbate–semiconductor system.

Moreover, oxygen vacancy may play an irreplaceable role in enriching the surface states of semiconductor to provide magnified affinity for the adsorbent–adsorbate interaction, which results in a further increase to the Raman signals for the probe molecules adsorbed on the semiconductor substrate surface through charge transfer involved in a CM mechanism. Providing with efficient charge transfer processes between the matching energy levels of adsorbed probe molecules and the semiconductor, both the polarizability tensor and the electron density distribution of the molecule would be modified, leading to the observation of non-totally symmetric SERS modes. As shown in Fig. 6b, R6G as a typical SERS probe molecule has the highest-occupied molecular orbital and lowest-unoccupied molecular orbital levels at -5.70 and $-3.40\,eV$, respectively[29]. The valence band and conduction band of semiconductor $W_{18}O_{49}$ locate at -7.71 and $-5.11\,eV$ (ref. 30), respectively, with oxygen vacancy-associated electronic state (V_O) well separated from the bottom of the CB, located at about 0.5–1.0 eV below the conduction band minimum[31]. It can be expected that contributions from several types of thermodynamically feasible charge transfer resonance may be related to the overall Raman enhancement in our $W_{18}O_{49}$–R6G system at an excitation of 532 nm, including molecule resonance (μ_{mol}) of R6G, exciton resonance (μ_{ex}) of $W_{18}O_{49}$ defect states, and the photon induced charge transfer resonance (μ_{PICT}) together with the ground-state charge transfer resonance (μ_{GSCT}) from matched energy level between $W_{18}O_{49}$ and R6G molecules. These resonances will lead to a magnification of Raman scattering cross-section.

In fact, defect state $|V\rangle$ stemming from oxygen vacancies in the lattice of semiconductor has a crucial contribution to the molecular polarizability tensor via the vibronic coupling with molecular ground state $|I\rangle$ and molecular excited states $|K\rangle$, besides the conduction band state $|S\rangle$ and valence band state $|S'\rangle$ of semiconductor (details are illustrated in Supplementary Methods). Under the incident laser frequency higher than the molecular resonance ($\omega_0 > \omega_{IK}$), the polarizability tensor $\alpha_{\sigma\rho}$ can be expressed by a simple formula $\alpha_{\sigma\rho} = A + B + C$[32], where A represents the contribution of the molecular resonance. B and C represent the contributions from the photoinduced charge

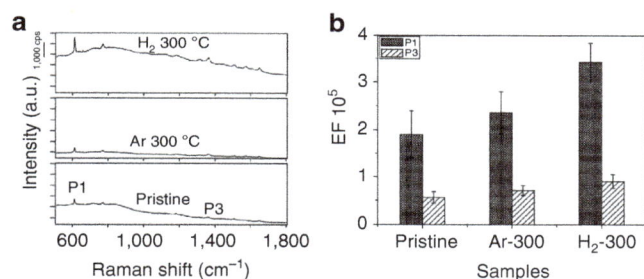

Figure 5 | Improved SERS properties of $W_{18}O_{49}$ samples after Ar/H_2 annealing treatment. (a) Raman signals of R6G molecule on pristine $W_{18}O_{49}$ and the samples after annealing treatment (in Ar/H_2 at 300 °C for 1 h). The tested concentration of R6G was 1×10^{-6} M. (b) A comparison of Raman EF for the two respective vibration modes P1 and P3. Data reported in this histogram resulted from Raman spectra acquired over 30 different regions per sample and provide an indication of the EF for each Raman mode. The H_2-treated sample shows the greatest enhancement at band P1, the EF for which was evaluated to be 3.4×10^5.

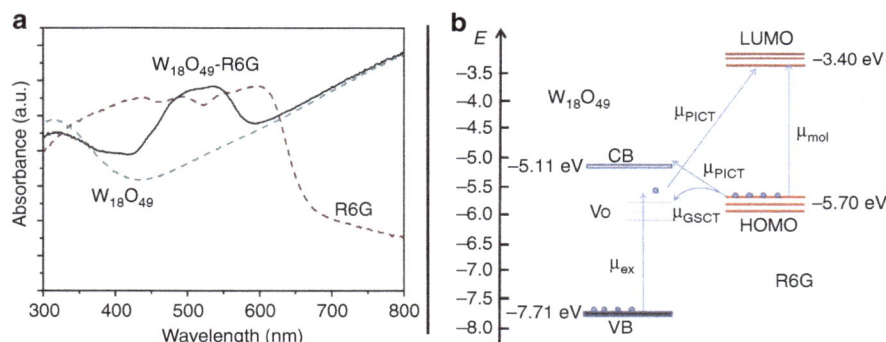

Figure 6 | Charge transfer between R6G and $W_{18}O_{49}$. (a) Absorption spectra for R6G on $W_{18}O_{49}$ compared with neat $W_{18}O_{49}$ and R6G dye. (b) Energy-level diagram of R6G on oxygen-deficit $W_{18}O_{49}$ measured in a vacuum.

transfer of the molecule-to-semiconductor and semiconductor-to-molecule, respectively, which are both related to the defect states in the semiconductor. It can be evidenced that additional possible resonant contribute to the total enhancement in defect-rich semiconductor–molecule system, compared with its non-defect counterpart. When these predicted multiplicatively coincide, large EFs are expected (Supplementary Fig. 4). Thus, the observed high SERS enhancement indicates that modulation of the surface vacancy of semiconductor substrate is a simple but effective means in design of a molecule-semiconductor system with high SERS enhancement.

Discussion

In summary, metal-free SERS substrate with highly sensitive detection has been successfully fabricated employing non-stoichiometric tungsten oxide, $W_{18}O_{49}$ as an example, which achieved a limit detection as low as 10^{-7} M and EF up to 3.4×10^5, an outstanding observation for semiconducting materials and even comparable to noble metal without 'hot spots'. The oxygen vacancy played a critical role in amplifying the spectroscopic signatures of probe molecules, since tungsten trioxide, WO_3, only gave extremely weak signals. Moreover, the artificial creation of oxygen deficiencies by annealing the material substrate in inert or reducing atmosphere (Ar/H_2) brought about further Raman enhancements, providing unambiguous evidence that the presence of oxygen vacancy, either intrinsic or post-created ones, could help magnify the Raman signals. The extremely high SERS sensitivity on semiconducting materials can probably be attributed to the presence of oxygen deficiencies, either intrinsic or on the surface, and the resultant strengthened interaction with probe molecules via vibronic coupling. These observations provide important clues in the future strategy design of efficient semiconducting SERS substrate, and may bring important fundamental advance in material science and chemistry.

Methods

Synthesis of sea urchin-like $W_{18}O_{49}$ nanowires and WO_3 nanoparticles on Si/SiO₂.

$W_{18}O_{49}$ nanowires in sea urchin-like morphology on Si/SiO₂ substrate were prepared according to a recently published hydrothermal approach[13]. In brief, WCl_6 (0.099 g) was dissolved in 30 ml absolute ethanol, which was transferred to a Teflon-lined stainless steel autoclave holding several horizontally oriented Si/SiO₂ substrates. The autoclave was then sealed and heated at 180 °C for 12 h. After finishing the reaction, the substrates were rinsed thoroughly with absolute ethanol and dried naturally at room temperature before use. The $W_{18}O_{49}$ nanowires on substrate were transformed to the corresponding trioxide form, WO_3, by annealing in air. The temperature was set at 500 °C and the sample was kept at the temperature for 1 h.

Modulating surface vacancy states.

Modulation of the surface oxygen vacancy states was achieved through annealing the as-prepared $W_{18}O_{49}$ nanowires by heating in Ar/H_2 atmosphere at 300 °C for 1 h. The velocity of Ar/H_2 was fixed at 400 and 200 ml min⁻¹, respectively.

Raman measurement.

To study the Raman enhancement effect by the tungsten oxide materials, R6G dissolved in deionized water was employed to be the probe molecule because R6G has a large Raman scattering cross-section at the laser excitation wavelength applied for the SERS experiments, 532.8 nm (ref. 33). The tested concentration ranged from 10^{-4} to 10^{-7} M. Stock solution of 10^{-3} M was initially made, and solutions of other concentrations were obtained by successive dilution by factors of 10^1 or 10^2. After dropping an aliquot of the respective solutions on the substrate and drying for at least 5 h, Raman spectra were subsequently collected on a high-resolution confocal Raman spectrometer (LabRAM HR-800) using the same instrumental settings for ready comparisons. The excitation wavelength was 532.8 nm and a ×50 L objective was used to focus the laser beam. The spectra were acquired for 30 s with three accumulations and the laser power was maintained at 0.3 mW with an average spot size of 1 µm in diameter in all acquisitions. For each sample, Raman spectra from different areas were collected, and the signal intensity was averaged for final analysis, from which relative s.d. values for EFs were estimated (Supplementary Methods and Supplementary Figs 5–8).

References

1. Chang, R. K. & Furtak, T. E. *Surface Enhanced Raman Scattering* (Plenum Press, 1982).
2. Campion, A. & Kambhampati, P. Surface-enhanced Raman scattering. *Chem. Soc. Rev.* **27**, 241–250 (1998).
3. Nie, S. & Emory, S. R. Probing single molecules and single nanoparticles by surface-enhanced Raman scattering. *Science* **275**, 1102–1106 (1997).
4. Kneipp, J., Kneipp, H. & Kneipp, K. SERS-a single-molecule and nanoscale tool for bioanalytics. *Chem. Soc. Rev.* **37**, 1052–1060 (2008).
5. Lombardi, J. R. & Birke, R. L. A unified view of surface-enhanced Raman scattering. *Acc. Chem. Res.* **42**, 734–742 (2009).
6. Quagliano, L. G. Observation of molecules adsorbed on III-V semiconductor quantum dots by surface-enhanced Raman scattering. *J. Am. Chem. Soc.* **126**, 7393–7398 (2004).
7. Li, W. *et al.* CuTe nanocrystals: shape and size control, plasmonic properties, and use as SERS probes and photothermal agents. *J. Am. Chem. Soc.* **135**, 7098–7101 (2013).
8. Jiang, L. *et al.* Surface-enhanced Raman scattering spectra ofadsorbates on Cu₂O nanospheres: charge-transfer andelectromagnetic enhancement. *Nanoscale* **5**, 2784–2789 (2013).
9. Musumeci, A. *et al.* SERS of semiconducting nanoparticles (TiO₂ hybrid composites). *J. Am. Chem. Soc.* **131**, 6040–6041 (2009).
10. Qi, D., Lu, L., Wang, L. & Zhang, J. Improved SERS sensitivity on plasmon-free TiO₂ photonic microarray by enhancing light-matter coupling. *J. Am. Chem. Soc.* **136**, 9886–9889 (2014).
11. Li, L. *et al.* Metal oxide nanoparticle mediated enhanced Raman scattering and its use in direct monitoring of interfacial chemical reactions. *Nano Lett.* **12**, 4242–4246 (2012).
12. Abe, R., Takami, H., Murakami, N. & Ohtani, B. Pristine simple oxides as visible light driven photocatalysts: highly efficient decomposition of organic compounds over platinum-loaded tungsten oxide. *J. Am. Chem. Soc.* **130**, 7780–7781 (2008).
13. Tian, Y. *et al.* Synergy of $W_{18}O_{49}$ and polyaniline for smart supercapacitor electrode integrated with energy level indicating functionality. *Nano Lett.* **14**, 2150–2156 (2014).
14. Cong, S., Tian, Y., Li, Q., Zhao, Z. & Geng, F. Single-crystalline tungsten oxide quantum dots for fast pseudocapacitor and electrochromic applications. *Adv. Mater.* **26**, 4260–4267 (2014).
15. Manthiram, K. & Alivisatos, A. P. Tunable localized surface plasmon resonances in tungsten oxide nanocrystals. *J. Am. Chem. Soc.* **134**, 3995–3998 (2012).
16. Remškar, M. *et al.* W₅O₁₄ nanowires. *Adv. Funct. Mater.* **17**, 1974–1978 (2007).
17. Sundberg, M. The crystal and defect structures of W₂₅O₇₃, a member of the homologous series WₙO₃ₙ₋₂. *Acta Crystallogr. Sec. B* **32**, 2144–2149 (1976).
18. Guo, C. *et al.* Morphology-controlled synthesis of $W_{18}O_{49}$ nanostructures and their near-infrared absorption properties. *Inorg. Chem.* **51**, 4763–4771 (2012).
19. Viswanathan, K., Brandt, K. & Salje, E. Crystal structure and charge carrier concentration of $W_{18}O_{49}$. *J. Solid State Chem.* **36**, 45–51 (1981).
20. Hildebrandt, P. & Stockburger, M. Surface-enhanced resonance Raman spectroscopy of Rhodamine 6G adsorbed on colloidal silver. *J. Phys. Chem.* **88**, 5935–5944 (1984).
21. Le Ru, E. C., Blackie, E., Meyer, M. & Etchegoin, P. G. Surface enhanced Raman scattering enhancement factors: a comprehensive study. *J. Phys. Chem. C* **111**, 13794–13803 (2007).
22. Hsiao, W.-H. *et al.* Surface-enhanced Raman scattering imaging of a single molecule on urchin-like silver nanowires. *ACS Appl. Mater. Interfaces* **3**, 3280–3284 (2011).
23. Lal, S., Grady, N. K., Goodrich, G. P. & Halas, N. J. Profiling the near field of a plasmonic nanoparticle with Raman-based molecular rulers. *Nano Lett.* **6**, 2338–2343 (2006).
24. Yang, H. *et al.* Comparison of surface-enhanced Raman scattering on graphene oxide, reduced graphene oxide and graphene surfaces. *Carbon* **62**, 422–429 (2013).
25. Naldoni, A. *et al.* Effect of nature and location of defects on bandgap narrowing in black TiO₂nanoparticles. *J. Am. Chem. Soc.* **134**, 7600–7603 (2012).
26. Jeon, S. & Yong, K. Synthesis and characterization of tungsten oxide nanorods from chemical vapor deposition-grown tungsten film by low-temperature thermal annealing. *J. Mater. Res.* **23**, 1320–1326 (2008).
27. He, X. W. *et al.* Memristive properties of hexagonal WO₃ nanowires induced by oxygen vacancy migration. *Nanoscale Res. Lett.* **8**, 50 (2013).
28. Xi, G. *et al.* Ultrathin $W_{18}O_{49}$ nanowires with diameters below 1 nm: synthesis, near-infrared absorption, photoluminescence, and photochemical reduction of carbon dioxide. *Angew. Chem. Int. Ed.* **51**, 2395–2399 (2012).
29. Ling, X. *et al.* Can graphene be used as a substrate for Raman enhancement? *Nano Lett.* **10**, 553–561 (2010).
30. Rawal, S. B. *et al.* Design of visible-light photocatalysts by coupling of narrow bandgap semiconductors and TiO₂: effect of their relative energy band positions on the photocatalytic efficiency. *Catal. Sci. Technol.* **3**, 1822–1830 (2013).

31. Wang *et al.* Semiconductor-to-metal transition in WO_{3-x}: nature of the oxygen vacancy. *Phys. Rev. B* **84,** 073103 (2011).
32. Lombardi *et al.* Theory of surface-enhanced Raman scattering in semiconductors. *J. Phys. Chem. C* **118,** 11120–11130 (2014).
33. Shim, S., Stuart, C. M. & Mathies, R. A. Resonance raman cross-sections and vibronic analysis of Rhodamine 6G from broadband stimulated Raman spectroscopy. *ChemPhysChem* **9,** 697–699 (2008).

Acknowledgements

This work was supported by the National Natural Science Foundation of China (51372266), the Natural Science Foundation of Jiangsu Province (BK20130348) and Suzhou Industrial Science and Technology Programm (ZXG201426). F.G. acknowledges support from the National Natural Science Foundation of China (51402204), Thousand Young Talents Program and Jiangsu Specially-Appointed Professor Program. Y.Z. and Q.L. acknowledge the support of the National Natural Science Foundation of China (51202282) and National Basic Research Program by Ministry of Science and Technology (2011CB932600).

Author contributions

Z.Z. and F.G. conceived the project and designed the experiments. S.C., Y.Y., Z.C., J.H., M.Y. and Y.S. performed material synthesis, structural characterization and Raman measurements. S.C., Y.Z., L.L., Q.L., F.G. and Z.Z. analysed the data. S.C., F.G. and Z.Z. co-wrote the paper. All authors discussed the results and commented on the manuscript.

Additional information

Competing financial interests: The authors declare no competing financial interests.

Identification of phases, symmetries and defects through local crystallography

Alex Belianinov[1,2], Qian He[3], Mikhail Kravchenko[1,2], Stephen Jesse[1,2], Albina Borisevich[1,3] & Sergei V. Kalinin[1,2]

Advances in electron and probe microscopies allow 10 pm or higher precision in measurements of atomic positions. This level of fidelity is sufficient to correlate the length (and hence energy) of bonds, as well as bond angles to functional properties of materials. Traditionally, this relied on mapping locally measured parameters to macroscopic variables, for example, average unit cell. This description effectively ignores the information contained in the microscopic degrees of freedom available in a high-resolution image. Here we introduce an approach for local analysis of material structure based on statistical analysis of individual atomic neighbourhoods. Clustering and multivariate algorithms such as principal component analysis explore the connectivity of lattice and bond structure, as well as identify minute structural distortions, thus allowing for chemical description and identification of phases. This analysis lays the framework for building image genomes and structure–property libraries, based on conjoining structural and spectral realms through local atomic behaviour.

[1] Institute for Functional Imaging of Materials, Oak Ridge National Laboratory, Oak Ridge, Tennessee 37831, USA. [2] The Center for Nanophase Materials Sciences, Oak Ridge National Laboratory, Oak Ridge, Tennessee 37831, USA. [3] Materials Sciences and Technology Division, Oak Ridge National Laboratory, Oak Ridge, Tennessee 37831, USA. Correspondence and requests for materials should be addressed to A.B. (email: belianinova@ornl.gov) or to A.B. (email: albinab@ornl.gov) or to S.V.K. (email: sergei2@ornl.gov).

The introduction of scattering techniques in the beginning of twentieth century by the Braggs has paved the way for probing the structure of matter on the atomic scales[1]. Early milestones include structure identification of simple crystalline substances as well as DNA, with recent advances encompassing small-angle scattering, radial distribution function analysis, inelastic scattering methods and surface diffraction and ptychography[2]. Despite the broad variety of scattering techniques, the basic principle—analysis of the structure factor—or equivalently a pair correlation function averaged over the probing volume remained invariant since the early days of the Braggs team. Furthermore, operating in the reciprocal space, natural to the scattering-based techniques, forged the way many generations of condensed matter scientists think. In fact, working in k-space to explore elementary excitations and normal modes can be considered a classical approach to physics. Typically, these surface or lattice descriptions are based on the periodicity of the system in real space and are intrinsically linked to the underlying symmetry. Unsurprisingly, cases where such description fails, including quasicrystals, nanoscale phase separation in strongly correlated oxides[3,4], morphotropic materials and relaxors[5–7]; remain a topic of much scientific excitement. For all these cases, the knowledge of the structure factor alone is insufficient to reconstruct the lattice of the material.

The progress in high-resolution, real space imaging techniques such as (scanning) transmission electron microscopy (STEM)[8–10] and scanning tunnelling microscopy (STM)[11,12] have allowed direct imaging of atomic columns (STEM) and surface atomic structures (STM). From the beginning of twenty-first century, the resolution (more specifically, information limit) of these methods has steadily risen to a level where minute displacement of atoms, from idealized high symmetry positions, can be visualized and quantified with high veracity. The examples in the field of aberration corrected (S)TEM include direct imaging of ferroelectric polarization[13–16], octahedral tilts[17,18] and chemical expansion strains[19]. Another example is high-resolution STM, allowing direct visualization of octahedral tilts[20], surface strains[21], complex structural reconstructions[22] and Jahn–Teller distortion fields[23]. In this manner, not only atomic structure but also subatomic order parameter fields can be visualized.

Typically, such image-based analyses are based on implicit, *a priori*, assumptions of the macroscopic symmetry of the system. These approaches fail when multiple crystallographic phases and/ or extended defects are present. Finding atoms without a reference to a global lattice is a general particle search problem, a well explored area with multiple available algorithms. That said, adaptation of these algorithms to an atom search is non-trivial, especially when multiple atom types are present. Other approaches circumvent atom finding by analysis of image segments that contain features of interest to extract relevant information[24,25]. However, contrast-based image analysis methods are prone to error propagation, especially in the case of lower-quality images, significant computational time cost for large images and multiple image arrays, and rather extensive user involvement and expertise. Notably, steps have been taken to ameliorate these complications through the use of correcting algorithms including para-meters such as orientation of the detector and environmental distortions[26]. These image processing techniques are quite powerful and ensure maximal data veracity before the analysis is initiated. Conjoined with such a powerful and extensive suite of image processing software, contrast-based methods have achieved impressive results that extend into the realm of three-dimensional reconstruction and internal structure mapping[27].

Nevertheless, once all atoms are found, a bigger challenge arises: without the global lattice as an intuitive vehicle for interpretation of the local structural data, we need to find completely new ways to categorize, analyse and interpret. Here we aim to explore whether a universal physical description of the system, including local and global symmetries, phases, and topological defects, can be built up only from the local information, obviating the overall lattice structure[28]. We propose an approach based on the multivariate statistical analysis of the coordination spheres of individual atoms, made up by an array of values that represent a variety of metrics between an atom and its nearest or next nearest neighbours, to reveal preferential structures and symmetries. We test this approach on a mixed-phase system with a variety of nearest neighbour environments and show how a framework for interpretation of this new type of structural data can be developed.

Results

Algorithm workflow. We define the chemical neighbourhood of the atom via the number and identity of the nearest neighbours. The types of atoms with statistically different chemical neigh-bourhoods and their spatial distribution define chemical com-position, that is, phases. If the chemical neighbourhoods are related by point symmetry operations, such as rotation or mirror symmetry, this defines physical ferroic variants. Note that clas-sical definitions of phase and ferroic variants further rely on the presence of translation symmetries, and below we demonstrate the development of a local picture and discuss possible pathways for global description.

The schematic of our near-atomic neighbourhood-based approach is illustrated in Fig. 1. Before the image enters the analysis workflow, it is lightly preprocessed to remove jitter noise and normalize contrast. As a first step of analysis, we identify all atoms in the image. Classically, this is achieved by overlaying a coarse ideal atomic lattice based on periodicity or Fast Fourier Transform (FFT) filtering and then relaxing the ideal lattice until an appropriate level of fit is found. In the case shown in Fig. 2a, this global approach is not applicable due to the lack of long-range periodicity across the image, and hence a local approach is used. The local approach works by first approximating a typical atomic shape (Fig. 1a). This is performed either manually or automatically by identifying a set, or a representative member in the image. Following the target identification, we perform a correlation of the target across the entire image over each pixel, by sliding the target window over the image. This produces a correlation map where locations of the target are strongly highlighted and areas that do not match the target are suppressed. The correlated surface is then thresholded to remove any small artefacts, resulting in a binary image where all areas are zero, except those that correlate strongly with the target (Fig. 1b). The binary image is processed further using a size histogram, to remove any point artefacts; we also found this procedure well suited for images with varying intensity backgrounds, as size of the object will vary with the background. The centroids of the final, disconnected objects in the image are identified through standard image processing functions available in Matlab Image Processing Toolbox (Fig. 1c). With the centroids identified, position refinement and the multivariate analysis steps can be launched (Fig. 1d–f).

As a model system, we have chosen mixed oxide Mo–V–M–O (M = Nb, Ta, Te and/or Sb), which is currently a promising catalyst for many industrially important reactions, such as propane (amm)oxidation[29]. In this system, the following phases are noted for their catalytic performance: the orthorhombic M1 phase with a space group Pba2 and a pseudo-hexagonal M2 phase with a space group Pmm2. The M1 phase, containing Mo_6O_{21}-

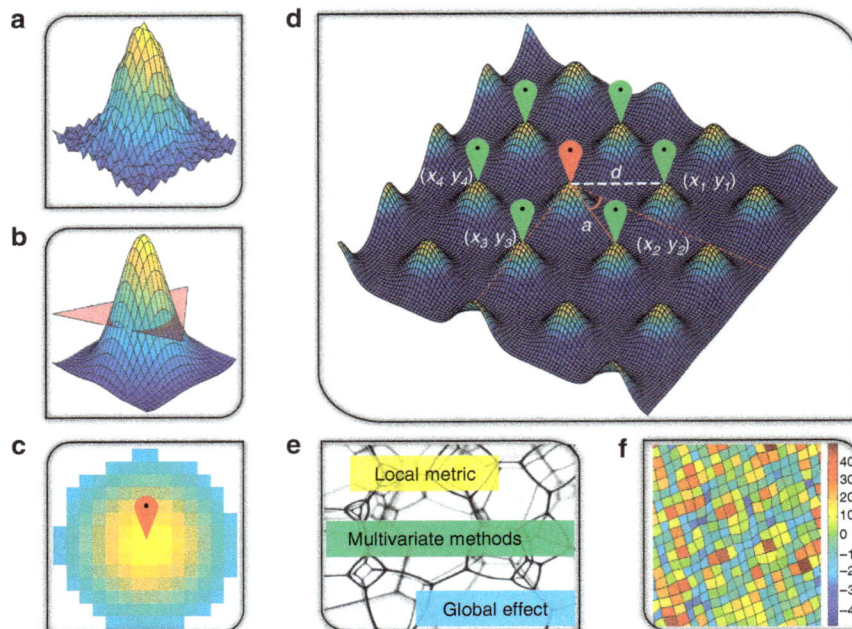

Figure 1 | Operational workflow of atom finding and subsequent analysis. (**a**) A representative atom (target) extracted from an image or supplied by the user. (**b**) Threshold the atom after correlation analysis. (**c**) Finding the centre of the atom in a binary image after a threshold in **b**. (**d**) A central atom (labelled no. 1) and six neighbours with distance and angle metric assignments shown. (**e**) Artistic representation of converting image data into vector data. At this step atomic centres from and their neighbourhoods are compiled into a single array on which multivariate analysis is performed. (**f**) Visualization of the multivariate analysis results.

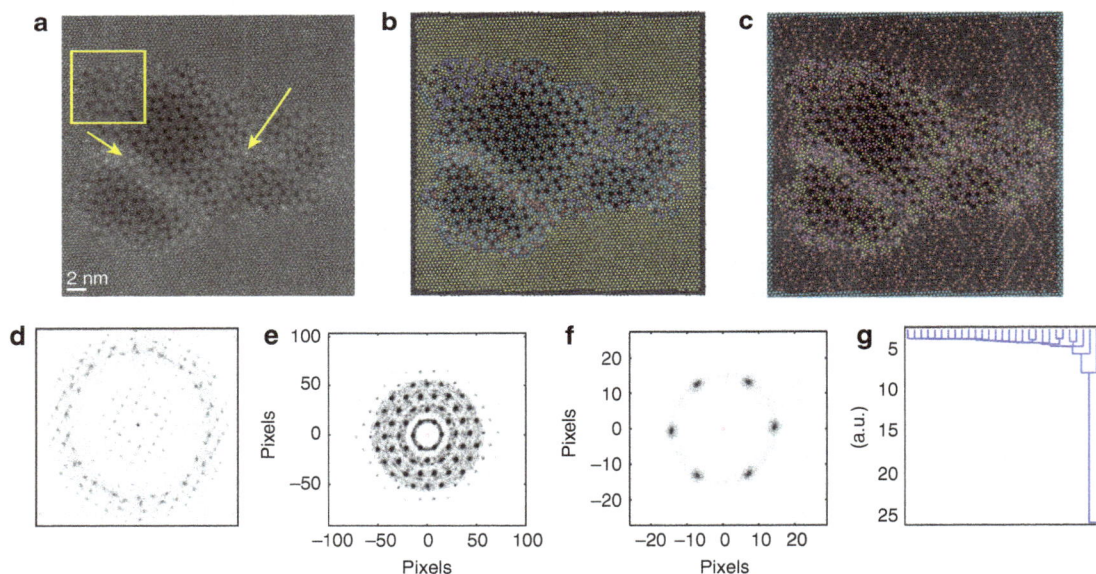

Figure 2 | Two-phase Mo–V–Te–Ta oxide. (**a**) M1 and M2 mixed-phase STEM image. (**b**) k-means clustering results for six neighbours, sorted by distance metric. (**c**) k-means clustering results for six neighbours, sorted by angle metric. (**d**) FFT of image in **a**. (**e**) Fifty member neighbourhood of the image in **a**. (**f**) Six member neighbourhood of the image in **a**. (**g**) Dendrogram for the six neighbour, sorted by distance metric, with the y axis signifying the cluster separation in the hierarchical tree.

type pentagonal units and doublet heptagonal channels, is the main active phase for paraffin activation, while the M2 phase, containing only hexagonal channels, has a possible synergistic effect when used together with the M1 by improving the reaction selectivity[30]. We recently discovered that these two phases can actually form coherent interfaces and intergrowths, suggesting new directions for catalyst improvement[31]. STEM images were taken with relatively fast scanning speed (that is, 1 μs per pixel).

The displayed image were the summation of multiple (that is, 30) fast sequentially scanned images aligned via cross-correlation. In this way, the image artefacts[26] due to scan noise, possible beam damage and drift can be minimized while the signal-to-noise ratio is maintained. A representative image showing the coexistence of M1 and M2 phases is shown in Fig. 2a. Note that the image contains multiple clearly visible regions with different crystalline ordering separated by a boundary (emphasized by yellow arrows) that contains lattice elements from both regions (highlighted by a yellow rectangle).

Using defined atomic shape as a template at the first step of the processing flow offers a distinct advantage in being able to differentiate sublattices as well as tilts or other contrast-based features. The correlation step is sensitive to small details of the supplied shape as well as its size, maximizing selectivity of finding the member of interest. Finally, flooding and histogram binning allow any remaining small features, such as edge artefacts and intensity imbalances, to be removed. Once atom positions are identified, the centres are refined with sub-pixel precision. Since the approximate centres are already known from the centroid identification, they are used as a seed fit for a two-dimensional (2D) Gaussian function that determines the best fit in the radial area of an atom. The maximum of the smooth function is extracted, resulting in an accurate measurement of the centre point of an atom with a higher precision than native resolution.

Implemented statistical framework. This analysis yields absolute positions of each atom in an image, as well as local descriptors such as column intensity and peak width determined at the refinement stage. For the task of nearest neighbour environment identification, we are only using atomic positions. For each atom, i, in the image, we construct a near-coordination sphere as an array $N_i = ([x_1, y_1]..., [x_j, y_j])$, where (x_j, y_j) is the position coordinate of the jth nearest neighbour. The number of nearest neighbours, or the search radius, can be defined separately and are chosen depending on the analysis. In the simplest case, neighbours are chosen based on dominant symmetry, for example, 6 for hexagonal lattice or 4/8 for cubic lattice. When the number of defined neighbours exceeds the available nearest neighbours, the next nearest neighbours are included. In the case when the search radius is used, the returned number of neighbours varies for each atom, due to vacancies (in the case of STM images), image edges or different coordination numbers.

At the first step, we explore statistical properties of neighbour distributions. Shown in Fig. 2e,f are statistical distributions for defined neighbourhoods of 50 and 6 neighbours, respectively. For a large number of neighbours, the derived distribution effectively represents the 2D pair correlation function illustrating the global periodicity in the image. For a smaller number of components, the average structure of the nearer chemical neighbourhood is revealed. In both cases, the maxima in the distribution correspond to preferential inter atomic distances in a hexagonal lattice. However, in the Fig. 2f, additional intermediate points are observed that do not fall into the hexagonal maxima. This is due to the local environments in the image that do not follow the same symmetry. Note that while six nearest neighbours were used for the analysis, this approach can be extended for more remote neighbours and incorporate multiple sublattices. In addition, the central atom to neighbour relationship can be further explored by classification of the members in the coordination sphere by arranging them by length to the centre or angle of the bonds and so on. That is, using the N_i vector as a descriptor for a particular behaviour of interest, as illustrated in Fig. 2b,c.

Once the set of N_i vectors is assembled, the data object can be analysed as a multispectral data set via multivariate statistical methods. To identify the chemical structure of material, we perform clustering analysis of local neighbourhoods, effectively establishing the types of chemical environments. Here we utilize a k-means clustering algorithm to divide i points (or their corresponding N_i vectors) into K clusters so that the within cluster sum of squares is minimized (Equation 1).

$$\arg \min_S \sum_{i=1}^{K} \sum_{x_j \in S_i} \left\| x_j - \mu_i \right\|^2, \tag{1}$$

here μ_i is the mean of all points in S_i. We use the square Euclidean distance with each centroid being the component wise median of the points in a given cluster. The clustering is performed as a function of number of clusters, K, and the quality of separation, which can be represented as a dendrogram (explained in detail below), allowing a range of optimal number of clusters to be determined. Thus, determined clusters define the groups of atoms with specific chemical neighbourhoods that can be further positioned in real space and corresponding configurations can be explored through direct visualization, classical correlation function and Fourier transform methods.

Shown in Fig. 2g is a dendrogram for the image shown in Fig. 2a based on a classification of the distance to the central atom in a six neighbour case. A dendrogram plot illustrates hierarchical cluster arrangement in a top-down approach, where all observations are grouped into a single cluster initially and are recursively separated down the hierarchy. This is achieved by establishing a distance metric between observations and linkage criteria used to find the dissimilarity of clusters as a function of pairwise distances. It then follows that vertical axis in the dendrogram plot represents the distance between the two data points being connected (by whatever metric of choice), and the largest drops indicate major changes in data organization. On the basis of dendrogram for Fig. 2g, the strongest cluster separation occurs for two, three and four clusters with more than four clusters being the limit of significant optimization gain. We have found that a large percentage of total clusters identify strong outliers, that is, points found at the edges of the image, which while rigorously correct does not add to the understanding of the material being imaged. Therefore, we chose to omit the atoms that lie on the image boundary as centre atoms; however, their positions are still utilized as neighbours for the atoms further inward in the image. The results of four cluster separation based on the distance metric are shown in Fig. 2b, and for four clusters with an angle metric in Fig. 2c. Rotation is accounted for by always placing the first neighbour atom in the same location relative to the centre and filling the rest in a clockwise manner. Figure 3a–d shows where the atoms from each of the clusters are located on the initial image, with an accompanying FFT (Fig. 3(I–IV)) of the cluster points illustrating the symmetry of their relative distribution, as well as a 2D histogram of the nearest neighbour environment for atoms within the cluster.

Note that the analysis clearly distinguishes different areas of the image based on the similarity of chemical neighbourhoods of their constituent atoms. The coordination environments in Fig. 3a,c,d exist only within the central grain. The component in Fig. 3d forms a clearly visible region at the boundary between the grain and the outer matrix, characterized by least long-range order as is evident from the FFT in Fig. 3 IV; the Fig. 3b defines the matrix. Broadly, components in Fig. 3a–c have clear long-range periodicity in space, corresponding to specific sites within the unit cells of the respective phases. Note that in this case, the spatial periodicity of the individual clusters does not follow from, or contribute to classification, since the latter relies purely on the properties of nearest neighbourhood and does not contain any information regarding the long-range order in the system; rather,

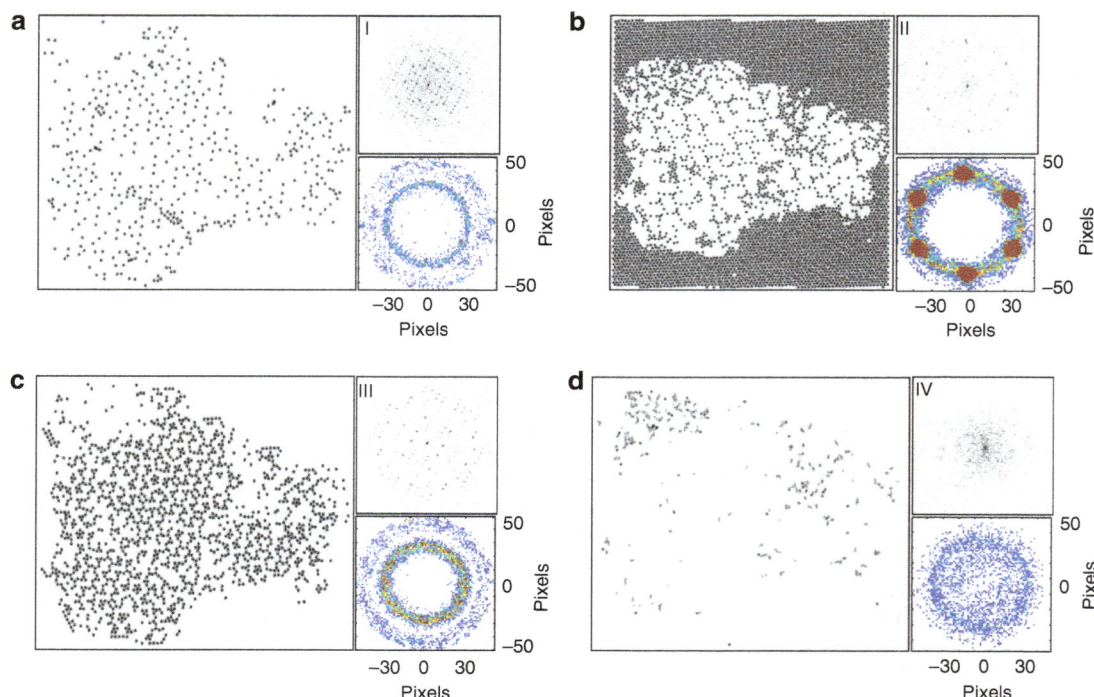

Figure 3 | Individual *k*-means clusters for image in Fig. 2, distance metric. (a) Cluster 1 spatial distribution with (I) FFT of the distribution and a 2D histogram of neighbours of atoms in the cluster. **(b)** Cluster 2 spatial distribution with (II) FFT of the distribution and a 2D histogram of neighbours of atoms in the cluster. **(c)** Cluster 3 spatial distribution with (III) FFT of the distribution and a 2D histogram of neighbours of atoms in the cluster. **(d)** Cluster 4 spatial distribution with (IV) FFT of the distribution and a 2D histogram of neighbours of atoms in the cluster.

we are using FFT as a post processing approach allowing us to differentiate between periodic and non-periodic classes in the initial image.

The relative distribution of the clusters can also be viewed in the form of the colour map as shown in Fig. 2b,c. Examination of the original image (Fig. 2a) and the image with overlaid cluster information (Fig. 2b,c) side by side makes apparent several characteristic patterns. First, the regions of single-phase M2 matrix, several subgrains within the central M1 grain and clearly visible amorphous boundaries separating these regions are distinguished. Second, a region emerges that is comprised of closely located atoms with similar local environments not found in either M1 or M2 phase, which can be tentatively associated with the emergence of a distinct third-phase region. While the first conclusion is also apparent from the visual examination of the initial image, the second one is not, demonstrating the advantages of the statistical analysis of the local neighbourhoods for analysing internal phase composition and structure of partially ordered phases in real space.

We further extend the multivariate approach to explore minute deviations of the internal structure in a single-phase region. Shown in Fig. 4 is a STEM image of a crystalline region of the M2 phase. The corresponding Fourier transform and nearest neighbour distributions for 50 neighbours and 6 neighbours are shown in Fig. 4d–f, respectively. Note the high degree of crystallinity in the material as reflected in the FFT. Interestingly, the neighbourhood histogram shows the internal structure with peaks having x, linear, and dot-like shapes as seen in more detail in Fig. 4e,f. This histogram is a clear indication that M2 phase can be viewed as a simpler hexagonal structure with several small distortions that vary in a periodic fashion from one primitive cell to another, forming a superstructure. Therefore, the dot-like spots delineate the unit cell for the superstructure (a multiple of the primitive cell), and the distorted spots carry information about

the symmetry of the distortions on specific sites within this larger unit cell. The corresponding length and angle *k*-means clustered images are shown in Fig. 4b,c with the individual clusters shown in the Supplementary Fig. 1. Notice the clear delineation of the sites of different symmetry within the phase, as well as a clearly visible antiphase boundary, which is not at all apparent in the original image and hardly evident from the raw data in Fig. 4a.

The presence of the fine structure of the neighbour distribution shown in Fig. 4f suggests the presence of minuscule distortions from ideal symmetry of the primitive cell giving rise to the M2 crystalline structure. To visualize these physical behaviours, we analyse these distortions using the principal component analysis (PCA) of the neighbourhood vector to separate statistically significant deformation of the nearest neighbourhood. PCA[32–36] is used to convert **N** observations into a superposition of orthogonal, linearly uncorrelated eigenvectors w_j shown in equation (2).

$$\mathbf{N}_i = a_{ij}w_j, \qquad (2)$$

where a_{ij} are expansion coefficients (PCA loadings). The eigenvectors w_j and the corresponding eigenvalues λ_k are found from the singular value decomposition of the covariance matrix, $\mathbf{C} = \mathbf{N}\mathbf{N}^T$, where **N** is the matrix of all experimental data points N_{ij}. That is, the rows of **N** correspond to individual atoms, and columns correspond to x and y components of radius vectors to nearest neighbours. The eigenvectors w_j are orthogonal and are arranged such that corresponding eigenvalues are placed in descending order, $\lambda_1 > \lambda_2 > ....$ by variance.

Shown in Fig. 5a–f are eigenvectors (represented as deviations from the average 6 neighbour shape, located in the upper left corner) and their corresponding loading maps. The scree plot is shown in Supplementary Fig. 2. The first eigenvector is the average, and the loading map shows the residual intensity of all six components. To visualize the higher eigenmodes, we plot

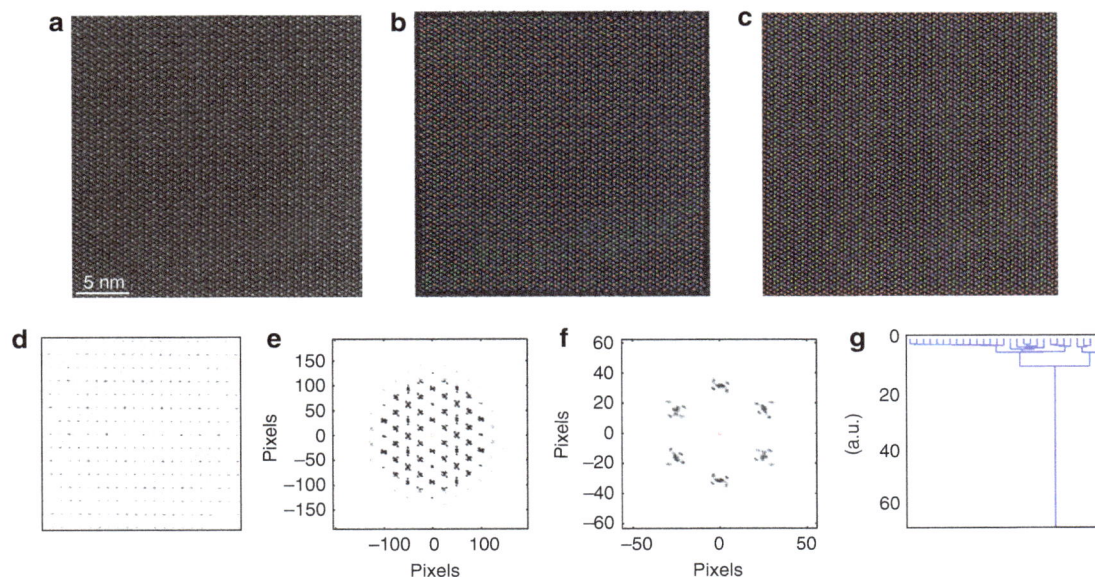

Figure 4 | Single M2 phase Mo–V–Te oxide. (**a**) M2 phase STEM image. (**b**) *k*-means clustering results for six neighbours, sorted by distance metric. (**c**) *k*-means clustering results for 6 neighbours, sorted by angle metric. (**d**) FFT of image in **a**. (**e**) Fifty member neighbourhood of the image in **a**. (**f**) Six member neighbourhood of the image in **a**. (**g**) Dendrogram for the six neighbour, sorted by angles metric, with the *y* axis signifying the cluster separation in the hierarchical tree.

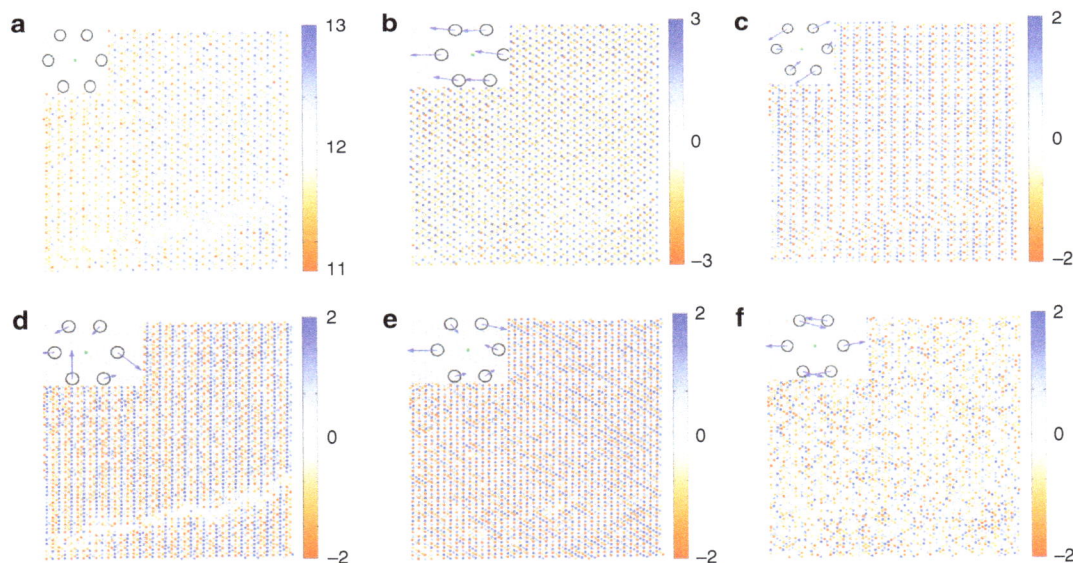

Figure 5 | PCA loadings and vectors for the image in Fig. 4a. (**a**) First eigenvector, in the upper left, and a corresponding loading map. (**b**) Second eigenvector, in the upper left, and a corresponding loading map. (**c**) Third eigenvector, in the upper left, and a corresponding loading map. (**d**) Fourth eigenvector, in the upper left, and a corresponding loading map. (**e**) Fifth eigenvector, in the upper left, and a corresponding loading map. (**f**) Sixth eigenvector, in the upper left, and a corresponding loading map.

them as deformation of the average, represented as vectors of deformation from the ideal lattice positions of the neighbours. We further note that these statistical normal modes do not have well defined physical meaning. Practically, they reveal spatial frequencies present in the image from which symmetry can be inferred, allowing the interpreter to ascribe a likely physical interpretation to some earlier PCA components. For example, the second eigenmode corresponds to the uniform shift of the coordinate sphere along one of the principal directions of the primitive hexagonal cell. In this case, the displacement of the

entire nearest neighbour sphere is clearly equivalent to the displacement of the central atom in the opposite direction, a polar distortion. In a more general case, the vector sum of the shifts of the nearest neighbour atoms will determine whether the collective distortion being considered is polar or non-polar in nature (non-zero vector sum versus zero vector sum, respectively). The loading map reflects the previously reported antipolar structure characteristic of the M2 phase; interestingly, the antiphase boundary is not apparent on the loading map, suggesting that the antipolar structure is not altered by the presence of the boundary. The third eigenvector corresponds to a symmetric deformation of the neighbourhood similar to a shear mode. The corresponding map is almost uniform, but close examination shows that the contrast of the map exhibits the shift associated with the antiphase boundary, even though no contrast alteration is associated with the boundary itself. The fourth, fifth and sixth components are more difficult to interpret, as the distortions appear to be very complex. The fifth component is somewhat reminiscent of the rotational transformation of the third. Interestingly, the loading maps associated with the third and forth components show very clear contrast at the antiphase boundary, suggesting that these distortions might be characteristic of the frustrated environment of that defect. In contrast, the loading map for the fifth component, similarly to the second, shows no contrast shift at the antiphase boundary. Information of this type could be tremendously useful when determining polar character of different extended defects.

Discussion

We have implemented a locality-based analysis of complex materials from high veracity atomically resolved images to explore chemistry and physics at the nanoscale, grounded in the analysis of atomic neighbourhoods. Unlike the classical, symmetry-based descriptions, our approach utilizes local bond characteristics including the structure of the coordination sphere and bonding type. For materials with a significantly varying chemical neighbourhood, this analysis allows identification of the uniform phase regions, as well as clear delineation of unknown phases and structural defects. The Fourier analysis of individual cluster components allows associated symmetries to be revealed.

In the single-phase region, clustering analysis allows decomposition of the system into the elementary sublattices. In this case, additional opportunities are opened by the PCA of local neighbourhoods, defining the statistical normal modes of the system. Again, this statistical description illustrates the predominant statistically significant distortions and ranks them in terms of relevant prevalence.

In general, we believe that this approach paves the way for full information recovery in high-resolution imaging such as electron and scanning probe microscopies, as well as allows for classification and automatic identification of materials. Subsequent effort will be aimed at development of the identification of the repeated statistically defined units based on graph partitioning of underlying lattice, creating a basis for development of image genomes and further development of structure-property correlative libraries based on STEM–EELS and STM–STS data.

Methods

Sample. The Mo–V–M oxides were prepared by hydrothermal synthesis or slurry evaporation as previously reported[37,38]. Ammonium paramolybdate, telluric acid, antimony trioxide, vanadium (IV) sulfate, niobium (V) oxalate hexahydrate and tantalum (V) ethoxide were used as precursors. All operations, preparation and stirring of the solution, were performed at 353 K except Sb system at 373 K. The slurry was introduced into the Teflon inner tube of a stainless steel autoclave. In the case of slurry evaporation, this slurry was dried overnight in the oven at 383 K. The autoclave was sealed and heated at 448 K for 48 h. After hydrothermal synthesis, the dark blue powder obtained was washed, filtered with distilled water (200 ml) and dried at 353 K for overnight. Then, the dried powder catalyst was calcined under ultrahigh purified nitrogen flow (50 ml min^{-1}) at 873 K for 2 h before use. In the case of catalyst containing Sb, the obtained solid was preheated in furnace with air at 573 K for 4 h before calcined under ultrahigh purified nitrogen flow at 873 K for 2 h. The nominal compositions (molar ratio) of the catalysts involved in the work are as follows: Mo–V–Te–Ta oxide shown in Fig. 2, Mo:V:Te:Ta = 1:0.3:0.17:0.12; Mo–V–Te oxide shown in Fig. 4, Mo:V:Te = 1:0.75:0.75.

Electron microscopy imaging. To make specimen for electron microscopy, the catalyst sample was embedded in a resin, and sectioned by microtome as ~50-nm slices[39]. These specimens were introduced into a holey-carbon-coated Cu grid. The

HAADF-STEM imaging was performed on UltraSTEM 200 (operated at 200 kV) in Oak Ridge National Laboratory. The inner angle of the High-Angle Annular Dark-Field (HAADF) detector is around 63 mrad. To minimize the beam damage and specimen drift, the images used for analysis were the sum images of 20–30 fast scanned frames (1 µs per pixel and ~20 pA probe current) stacked with cross-correlation algorithm. Gatan Digitalmicrograph was used for image acquisition, and all the images are 32 bit in depth. No further image processing was performed before PCA, Independent Component Analysis (ICA) and k-mean clustering analysis. The pixel size for the original image in Fig. 2 is 0.27 Å and in Fig. 4 is 0.13 Å.

References

1. Coontz, R., Fahrenkamp-Uppenbrink, J., Lavine, M. & Vinson, V. Going from Strength to Strength. *Science* **343**, 1091 (2014).
2. Hruszkewycz, S. O. *et al.* Quantitative nanoscale imaging of lattice distortions in epitaxial semiconductor heterostructures using nanofocused X-ray Bragg projection ptychography. *Nano Lett.* **12**, 5148–5154 (2012).
3. Dagotto, E. Complexity in strongly correlated electronic systems. *Science* **309**, 257–262 (2005).
4. Dagotto, E., Hotta, T. & Moreo, A. Colossal magnetoresistant materials: the key role of phase separation. *Phys. Rep.* **344**, 1–153 (2001).
5. Woodward, D. I., Knudsen, J. & Reaney, I. M. Review of crystal and domain structures in the PbZrxTi1-xO3 solid solution. *Phys. Rev. B* **72**, 104110 (2005).
6. Rao, W. F., Wuttig, M. & Khachaturyan, A. G. Giant Nonhysteretic Responses of Two-Phase Nanostructured Alloys. *Phys. Rev. Lett.* **106**, 105703 (2011).
7. Vugmeister, B. E. Polarization dynamics and formation of polar nanoregions in relaxor ferroelectrics. *Phys. Rev. B* **73**, 174117 (2006).
8. Crewe, A. V. Scanning electron microscopies—is high resolution possible. *Science* **154**, 729 (1966).
9. Pennycook, S. J. & Nellist, P. D. *Scanning Transmission Electron Microscopy: Imaging and Analysis* (Springer, 2011).
10. Ardenne, M. v. Das Elektronen-Rastermikroskop. Praktische Ausführung. *Z. Tech. Phys.* **19**, 407–416 (1938).
11. Binnig, G., Rohrer, H., Gerber, C. & Weibel, E. 7X7 Reconstruction on Si(111) Resolved in Real Space. *Phys. Rev. Lett.* **50**, 120–123 (1983).
12. Binnig, G. & Rohrer, H. Scanning tunneling microscopy. *Helv. Phys. Acta* **55**, 726–735 (1982).
13. Jia, C. L. *et al.* Atomic-scale study of electric dipoles near charged and uncharged domain walls in ferroelectric films. *Nat. Mater.* **7**, 57–61 (2008).
14. Chang, H. J. *et al.* Atomically Resolved Mapping of Polarization and Electric Fields Across Ferroelectric/Oxide Interfaces by Z-contrast Imaging. *Adv. Mater.* **23**, 2474 (2011).
15. Nelson, C. T. *et al.* Spontaneous vortex nanodomain arrays at ferroelectric heterointerfaces. *Nano Lett.* **11**, 828–834 (2011).
16. Chisholm, M. F., Luo, W. D., Oxley, M. P., Pantelides, S. T. & Lee, H. N. Atomic-scale compensation phenomena at polar interfaces. *Phys. Rev. Lett.* **105**, 197602 (2010).
17. Borisevich, A. *et al.* Mapping octahedral tilts and polarization across a domain wall in BiFeO(3) from Z-contrast scanning transmission electron microscopy image atomic column shape analysis. *Acs Nano* **4**, 6071–6079 (2010).
18. Jia, C. L. *et al.* Oxygen octahedron reconstruction in the SrTiO(3)/LaAlO(3) heterointerfaces investigated using aberration-corrected ultrahigh-resolution transmission electron microscopy. *Phys. Rev. B* **79**, 081405(R) (2009).
19. Kim, Y.-M. *et al.* Probing oxygen vacancy concentration and homogeneity in solid-oxide fuel-cell cathode materials on the subunit-cell level. *Nat. Mater.* **11**, 888–894 (2012).
20. Li, Q. *et al.* Atomically resolved spectroscopic study of Sr2IrO4: Experiment and theory. *Sci. Rep.* **3** (2013).
21. Maksymovych, P., Sorescu, D. C. & Yates, J. T. Gold-adatom-mediated bonding in self-assembled short-chain alkanethiolate species on the Au(111) surface. *Phys. Rev. Lett.* **97**, 146103 (2006).
22. Hamers, R. J., Tromp, R. M. & Demuth, J. E. Surface electronic structure of Si (111)-(7 × 7) resolved in real space. *Phys. Rev. Lett.* **56**, 1972–1975 (1986).
23. Gai, Z. *et al.* Chemically induced Jahn–Teller ordering on manganite surfaces. *Nat. Commun.* **5** (2014).
24. Sarahan, M. C., Chi, M., Masiel, D. J. & Browning, N. D. Point defect characterization in HAADF-STEM images using multivariate statistical analysis. *Ultramicroscopy* **111**, 251–257 (2011).
25. Lu, P. & Gauntt, B. D. Structural mapping of disordered materials by nanobeam diffraction imaging and multivariate statistical analysis. *Microsc. Microanal.* **19**, 300–309 (2013).
26. Jones, L. & Nellist, P. D. Identifying and correcting scan noise and drift in the scanning transmission electron microscope. *Microsc. Microanal.* **19**, 1050–1060 (2013).
27. Jones, L., MacArthur, K. E., Fauske, V. T., van Helvoort, A. T. J. & Nellist, P. D. Rapid estimation of catalyst nanoparticle morphology and atomic-coordination

by high-resolution Z-contrast electron microscopy. *Nano Lett.* **14**, 6336–6341 (2014).

28. Keen, D. A. & Goodwin, A. L. The crystallography of correlated disorder. *Nature* **521**, 303–309 (2015).

29. Shiju, N. R. & Guliants, V. V. Recent developments in catalysis using nanostructured materials. *Appl. Catal. A* **356**, 1–17 (2009).

30. Holmberg, J., Grasselli, R. K. & Andersson, A. Catalytic behaviour of M1, M2, and M1/M2 physical mixtures of the Mo–V–Nb–Te–oxide system in propane and propene ammoxidation. *Appl. Catal.* **270**, 121–134 (2004).

31. He, Q., Woo, J., Belianinov, A., Guliants, V. V. & Borisevich, A. Y. Better catalysts through microscopy: mesoscale M1/M2 intergrowth in molybdenum-vanadium based complex oxide catalysts for propane ammoxidation. *ACS Nano* **9**, 3470–3478 (2015).

32. Bosman, M., Watanabe, M., Alexander, D. T. L. & Keast, V. J. Mapping chemical and bonding information using multivariate analysis of electron energy-loss spectrum images. *Ultramicroscopy* **106**, 1024–1032 (2006).

33. Bonnet, N. in *Advances in Imaging and Electron Physics* Vol. 114 (eds Hawkes, P. W.) (Elsevier Academic Press Inc., 2000).

34. Bonnet, N. Multivariate statistical methods for the analysis of microscope image series: applications in materials science. *J. Microsc.* **190**, 2–18 (1998).

35. Jesse, S. & Kalinin, S. V. Principal component and spatial correlation analysis of spectroscopic-imaging data in scanning probe microscopy. *Nanotechnology* **20**, 085714 (2009).

36. Belianinov, A. *et al.* Big data and deep data in scanning and electron microscopies: deriving functionality from multidimensional data sets. *Adv. Struct. Chem. Imaging* **1**, 6 (2015).

37. Watanabe, N. & Ueda, W. Comparative study on the catalytic performance of single-phase Mo – V – O-based metal oxide catalysts in propane ammoxidation to acrylonitrile. *Ind. Eng. Chem. Res.* **45**, 607–614 (2006).

38. Nguyen, T. T., Deniau, B., Baca, M. & Millet, J. M. M. Synthesis and monitoring of MoVSbNbO oxidation catalysts using V K and Sb L1-edge xanes spectroscopy. *Top. Catal.* **54**, 650–658 (2011).

39. Yu, J., Woo, J., Borisevich, A., Xu, Y. & Guliants, V. V. A combined HAADF STEM and density functional theory study of tantalum and niobium locations in the Mo–V–Te–Ta(Nb)–O M1 phases. *Catal. Commun.* **29**, 68–72 (2012).

Acknowledgements

Research for A. Bel., Q. H., M. K., A. Bor., S. V. K., was supported by the US Department of Energy, Basic Energy Sciences, Materials Sciences and Engineering Division. Research for SJ was sponsored by Laboratory Directed Research and Development Program of Oak Ridge National Laboratory, managed by UT-Battelle, LLC, for the U.S. Department of Energy. This research was conducted at the Center for Nanophase Materials Sciences, which is sponsored at Oak Ridge National Laboratory by the Scientific User Facilities Division, Office of Basic Energy Sciences, U. S. Department of Energy.

Author contributions

A.Be. analysed the data and prepared the manuscript. Q.H. provided images and sample data. M.K. refined analysis software. S.J. provided analysis technical knowledge and prepared figures. A.Bo. provided electron microscopy expertise and oversaw imaging. S.V.K. conceived the concept and contributed to manuscript writing.

Additional information

Gln40 deamidation blocks structural reconfiguration and activation of SCF ubiquitin ligase complex by Nedd8

Clinton Yu[1,*], Haibin Mao[2,*], Eric J. Novitsky[3], Xiaobo Tang[2], Scott D. Rychnovsky[3], Ning Zheng[2] & Lan Huang[1]

The full enzymatic activity of the cullin-RING ubiquitin ligases (CRLs) requires a ubiquitin-like protein (that is, Nedd8) modification. By deamidating Gln40 of Nedd8 to glutamate (Q40E), the bacterial cycle-inhibiting factor (Cif) family is able to inhibit CRL E3 activities, thereby interfering with cellular functions. Despite extensive structural studies on CRLs, the molecular mechanism by which Nedd8 Gln40 deamidation affects CRL functions remains unclear. We apply a new quantitative cross-linking mass spectrometry approach to characterize three different types of full-length human Cul1–Rbx1 complexes and uncover major Nedd8-induced structural rearrangements of the CRL1 catalytic core. More importantly, we find that those changes are not induced by Nedd8(Q40E) conjugation, indicating that the subtle change of a single Nedd8 amino acid is sufficient to revert the structure of the CRL catalytic core back to its unmodified form. Our results provide new insights into how neddylation regulates the conformation and activity of CRLs.

[1] Department of Physiology and Biophysics, University of California, Irvine, California 92697, USA. [2] Department of Pharmacology and Howard Hughes Medical Institute, University of Washington, Seattle, Washington 98195, USA. [3] Department of Chemistry, University of California, Irvine, California 92697, USA. * These authors contributed equally to this work. Correspondence and requests for materials should be addressed to N.Z. (email: nzheng@uw.edu) or to L.H. (email: lanhuang@uci.edu).

Cullin-RING ubiquitin ligases (CRLs) represent a super-family of multi-subunit E3 ubiquitin ligases comprised of a cullin-RING catalytic core and adaptor proteins that mediate the recruitment of protein substrates[1–6]. Eight cullin family proteins (Cul1, Cul2, Cul3, Cul4A/B, Cul5, Cul7, Cul9 and APC2) are found in humans, each functioning as a scaffold on which a variety of CRLs are assembled. The SCF/CRL1 (Skp1–Cul1–F-box protein) complex represents the prototypical CRL E3, which uses Cul1–Rbx1 as the catalytic core[2,7,8]. The Cul1 scaffold binds the Skp1 adaptor and the Rbx1 RING subunit at its N-terminal and C-terminal domains, respectively. Skp1 in turn docks F-box proteins, which are substrate receptors that confer substrate specificity to the SCF, while the RING-finger domain of Rbx1 engages ubiquitin-charged E2, mediating the transfer of ubiquitin to the F-box protein-bound substrate. A reconstructed structure model of the SCF based on crystal structures of several overlapping sub-complexes reveals an elongated E3 platform, in which the F-box protein is separated from the Rbx1-bound E2 by a ~ 50-Å distance[9].

Covalent conjugation of ubiquitin-like protein Nedd8 (that is, neddylation) to a specific Lysine (Lys720) of Cul1 has been shown to promote both E2 recruitment and subsequent ubiquitin transfer, thereby stimulating the E3 activity of SCF ligases[2,10–13]. Although the intact neddylated Cul1–Rbx1 complex remains recalcitrant to crystallization, crystal structures of a truncated C-terminal domain of Cul5 in complex with Rbx1 have shed light on the effects of neddylation on the conformation of the cullin-RING catalytic core[14]. In the unneddylated form, the $Cul5^{CTD}$–Rbx1 complex adopts a 'closed' conformation in which the RING-finger domain of Rbx1 is nestled within a hydrophobic pocket of $Cul5^{CTD}$. Upon neddylation, the RING-finger domain of Rbx1 is released from the pocket, deemed the 'open' state, but remains tethered by its N-terminus to Cul5, presumably allowing the extended RING-finger to sample the three-dimensional (3D) space around Cul5. This conferred flexibility has been proposed to enable Rbx1 to close the distance between substrate and E2, facilitating the transfer of ubiquitin from E2 to substrate protein.

Notably, the cycle-inhibiting factors (Cifs) found in many pathogenic Gram-negative bacteria can irreversibly deamidate a specific glutamine residue (Gln40) of Nedd8 and convert it to glutamate[15]. This Q40E modification has no effect on cullin neddylation, but can effectively abolish the E3 activity of CRLs and affect proper cullin deneddylation by the COP9 signalosome[15–18]. These observations raise an intriguing question as to how the subtle change of a single Nedd8 amino acid is able to negate the effect of neddylation in remodelling the ~ 100-kDa CRL catalytic core. In the structure of the neddylated $Cul5^{CTD}$–Rbx1 complex, Gln40 of Nedd8 is close to the isopeptide bond between Nedd8 and Cul5 and partially sandwiched between the two proteins. The amide group in the Gln40 side chain, however, is exposed to the solvent and does not participate in any hydrogen bond interactions[14]. The molecular mechanism by which Nedd8 Gln40 deamidation alters CRL functions remains elusive.

Recently, cross-linking mass spectrometry (XL-MS) has risen as a powerful method to study protein–protein interactions and characterize the structure of large protein complexes[19–28]. In comparison with X-ray crystallography or NMR, XL-MS approaches have much less restriction on sample preparation due to its sensitivity, flexibility and versatility, and are capable of capturing the dynamic states of large, heterogeneous protein structures. By stabilizing transient interactions, chemical cross-linking preserves various structural states of dynamic complexes, yielding a representation that describes the average state of a protein complex and providing a complementary set of structural data different from that obtained from rigid state data analyses such as X-ray crystallography. Recently, we have developed a new

class of cross-linkers, that is, sulfoxide-containing MS-cleavable cross-linking reagents, to enable simplified and unambiguous identification of cross-linked peptides using multistage tandem mass spectrometry $(MS^n)^{29–31}$. These new types of cross-linkers are robust and reliable, and have been successfully applied to define protein–protein interactions both in vitro[22,29,30] and in vivo[30]. To establish a robust quantitative XL-MS (QXL-MS) platform to study dynamic protein complexes, we have then developed a pair of stable isotope-labelled amine reactive cross-linkers (that is, d_0- and d_{10}-labelled dimethyl-disuccinimidyl sulfoxide (DMDSSO)), which allow simultaneous identification and quantitation of cross-linked peptides[31]. In combination with quantitative analysis, XL-MS can determine dynamic conversion between the average states of protein complexes under different conditions.

Here we employ this DMDSSO-based QXL-MS strategy to define the structural changes of full-length Cul1–Rbx1 modified by either wild-type Nedd8 or its Q40E mutant, which is the product of Gln40 deamidation. Quantitative similarities and differences in cross-linked peptide abundances can be attributed to the changes in protein complex structures under different conditions, as the occurrences of spatially proximal amino-acid residues suited for cross-linking is directly dependent on the 3D structural conformation of these complexes. Our results have provided new insights on how Nedd8 modification impacts the topology of Cul1–Rbx1 and the effect of Nedd8 Gln40 deamidation on the structure of the activated CRL core.

Results

Reconstitution of SCF E3 activity with intact proteins.
To enhance the solubility and stability of Cul1, we removed two short segments (see details in Methods) of Cul1 which were not visible in the crystal structure of Cul1–Rbx1 complex (PDB: 1LDJ)[32], drastically improving protein behaviour. This truncated Cul1 and $Rbx1^{16–108}$ were co-expressed and purified from Escherichia coli. The purified Cul1–Rbx1 was conjugated to wild-type Nedd8 to yield neddylated Cul1–Rbx1 complex (Nedd8$\sim$Cul1–Rbx1). The Q40E mutant Nedd8, in which Gln40 was replaced with Glu40 to mimic deamidated Nedd8, can also be efficiently conjugated to Cul1–Rbx1 to form Nedd8(Q40E)$\sim$Cul1–Rbx1. Both neddylated Cul1–Rbx1 samples were purified with affinity-tagged Nedd8 after neddylation reaction to remove the unmodified species.

To understand ubiquitin ligase activities of different forms of Cul1–Rbx1, we have employed two in vitro ubiquitination assays: free ubiquitin chain assembly and ubiquitination of cryptochrome 2 (CRY2). CRY2, a key regulator of circadian rhythm, is a well-characterized substrate of the SCF^{FBXL3} ubiquitin ligase[33–36]. In both assays, we used Cdc34, the canonical E2 of Cul1. Consistent with previous reports, Cul1–Rbx1 can promote substrate-independent free ubiquitin chain assembly, while Nedd8$\sim$Cul1–Rbx1 significantly enhanced the reaction kinetics (Fig. 1a)[37]. In contrast, Nedd8(Q40E)$\sim$Cul1–Rbx1 only exhibited comparable activity to unneddylated Cul1–Rbx1, which was much weaker than that of Nedd8$\sim$Cul1–Rbx1. This is consistent with the discovery that deamidation of Q40 in Nedd8 abolishes the ligase activity of neddylated Cul1–Rbx1 (refs 15,16).

We further confirmed this observation with an in vitro ubiquitination assay of CRY2. As shown in Fig. 1b, polyubiquitin chains were formed on CRY2 in the presence of Nedd8$\sim$Cul1–Rbx1. In contrast, neither unneddylated nor Nedd8(Q40E)-modified Cul1–Rbx1 complexes were able to catalyse ubiquitination of CRY2. This further confirms that Nedd8$\sim$Cul1–Rbx1 is the only active form and that deamidation of Q40 in Nedd8 can

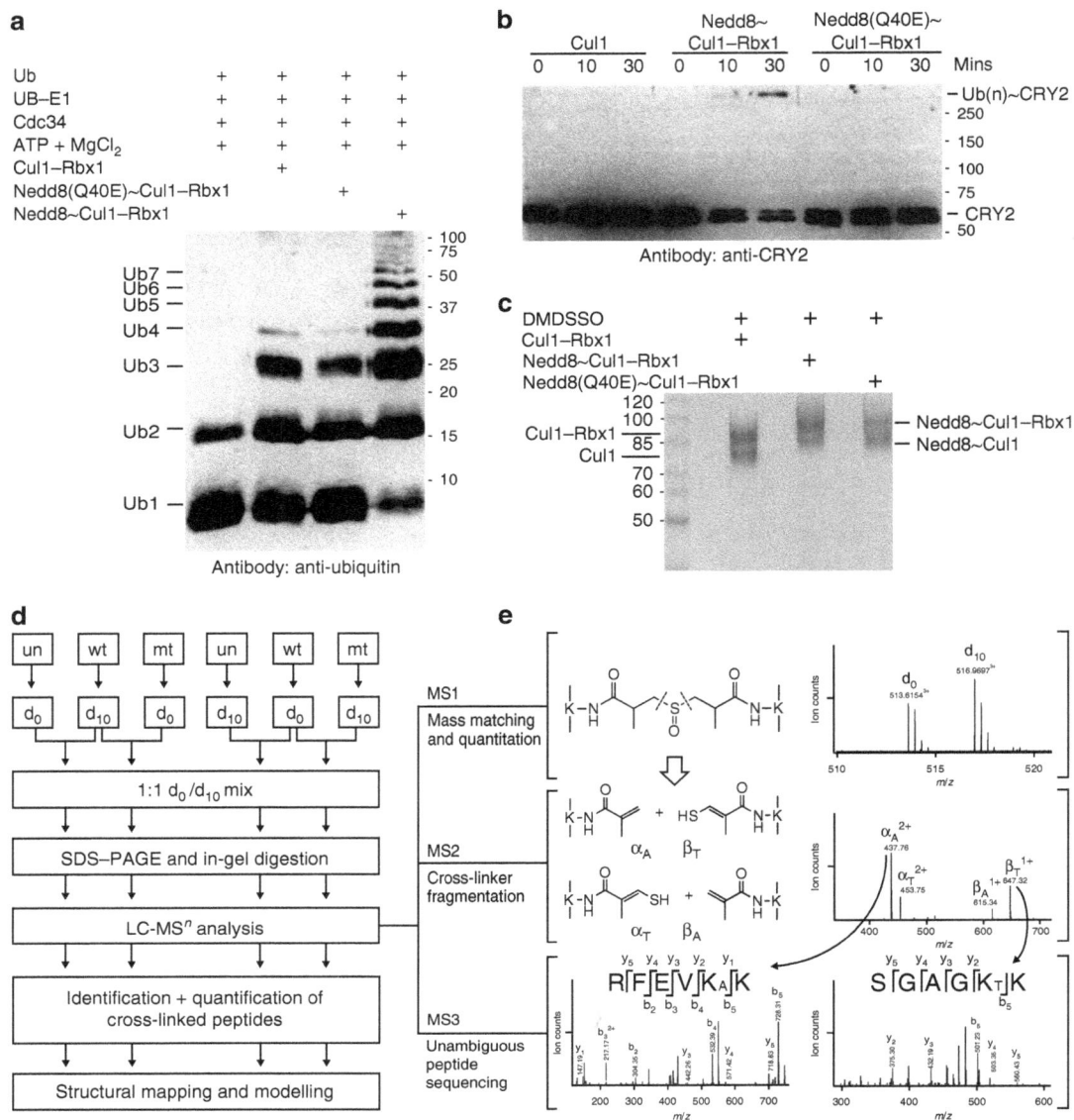

Figure 1 | Biochemical assays for ubiquitin ligase activity and general quantitative XL-MS experimental workflow. (a) Comparisons of ubiquitin ligase activities of different Cul1-Rbx1 variants on free ubiquitin chain assembly. Synthesized unanchored polyubiquitin chains were detected by anti-ubiquitin western blot. Highly efficient ubiquitin synthesis was only detected in the presence of Nedd8-Cul1-Rbx1 (lane 4). For visual clarity, ubiquitin polymers are simply abbreviated as Ub(n) (for example, Ub2 as ubiquitin dimer, Ub3 as ubiquitin trimer and so on). **(b)** Comparisons of ubiquitin ligase activities of different Cul1-Rbx1 variants on CRY2 ubiquitination. Ubiquitination reactions were quenched at indicated time points. Ubiquitinated CRY2 was detected using an anti-CRY2 antibody. Successful ubiquitination of CRY2 occurs only in the presence of Nedd8-Cul1-Rbx1 (lane 6). **(c)** SDS-PAGE analysis of DMDSSO cross-linked Cul1-Rbx1, Nedd8 ~ Cul1-Rbx1 and Nedd8(Q40E) ~ Cul1-Rbx1 complexes. **(d)** d_0/d_{10}-DMDSSO-based quantitative XL-MS workflow for identifying and quantifying cross-linked peptides of Cul1-Rbx1 complexes. The three types of Cul1-Rbx1 complexes, that is, un, unneddylated; wt, wild-type neddylated; mt, mutant Q40E-neddylated, were first cross-linked by DMDSSO separately, two of which were then selected for mixing before SDS-PAGE. Four types of mixing were made to obtain sufficient pairwise comparison among the three samples. Gel bands representing cross-linked protein complexes were subsequently excised and in-gel digested before LC-MS^n analysis for identification and quantification. **(e)** Representative MS^n analysis of DMDSSO interlinked peptides. MS^1 spectrum shows the detection of a pair of d_0-DMDSSO and d_{10}-DMDSSO cross-linked peptides (m/z 513.6154^{3+} and m/z 516.9697^{3+}), whose spectral relative abundance ratio is used for quantitation. MS^2 analysis of the d_0-DMDSSO interlinked peptides α-β (m/z 513.6154^{3+}) yielded two peptide fragment pairs: $α_A/β_T$ (m/z 437.76^{2+}/647.32^{1+}) and $α_T/β_A$ (m/z 453.75^{2+}/615.34^{1+}), confirming its cross-link type as an interlink. Subsequent MS^3 analyses of the $α_A$ (m/z 437.76^{2+}) and $β_T$ (647.32^{1+}) ions produced series of y and b ions that enabled unambiguous identification of $α_A$ as RFEVK$_A$K of Rbx1 and $β_T$ as SGAGK$_T$K of Rbx1. Integration of the MS^n (that is, MS^1, MS^2 and MS^3) data has confirmed the d_0-DMDSSO cross-linked peptide as an intrasubunit interlink between K19 and K25 of Rbx1. K_A, alkene modified lysine; K_T, unsaturated thiol modified lysine.

in fact abrogate its activity. Compared with the free ubiquitin chain assembly assay, the polyubiquitin chain synthesis on CRY2 was highly processive, as the ubiquitinated CRY2 band was observed at the top of the gel as shown by western blot analysis.

Taken together, our results have demonstrated neddylation is essential for the activation of the Cul1-Rbx1 complex in protein ubiquitination. Importantly, distinct functional disparity between Nedd8 ~ Cul1-Rbx1 and Nedd8(Q40E) ~ Cul1-Rbx1 have been further validated, and only the former is the active E3 ligase for protein ubiquitination.

QXL-MS strategy. To understand molecular details underlying the functional differences between different forms of Cul1–Rbx1 complexes, we have employed a QXL-MS strategy based on a newly developed pair of stable isotope-coded MS-cleavable cross-linkers, d_0-DMDSSO and d_{10}-DMDSSO[31] (Supplementary Fig. 1a,b), to examine the structural similarities and dissimilarities between these complexes. Concurrent usage of these two cross-linking reagents enables quantitative comparisons between the 3D structures of protein complexes under various conditions. To establish the QXL-MS workflow for comprehensive structural comparisons among unneddylated (un), wild-type neddylated (wt) and Q40E mutant neddylated (mt) Cul1–Rbx1 complexes, cross-linking conditions were first optimized through *in vitro* cross-linking of the three protein complexes with various concentrations of either d_0-DMDSSO or d_{10}-DMDSSO for different amounts of reaction times. Cross-linking efficiency was then evaluated by separating the resulting cross-linked products using one-dimensional SDS–polyacrylamide gel electrophoresis (SDS–PAGE). As shown in Fig. 1c, the resulting cross-linked products correspond well to respective molecular weights of these three complexes with $\sim 50\%$ cross-linking efficiency. In addition, we have determined that d_0- and d_{10}-DMDSSO reacted with Cul1–Rbx1 complexes with similar efficiency as illustrated in Supplementary Fig. 1c, also reflected in previous testing on standard proteins[31]. These results demonstrate that d_0- and d_{10}-labelled DMDSSO are well suited for QXL-MS analysis of these protein complexes.

To enable sufficient comparisons among the three different types of protein complexes with the minimal number of samples for analysis, we have strategically selected Nedd8 ~ Cul1–Rbx1 as the cross-sample reference in the pairwise comparison experiments. As illustrated in Fig. 1d, d_{10}-DMDSSO cross-linked Nedd8 ~ Cul1–Rbx1 was mixed with d_0-DMDSSO cross-linked Cul1–Rbx1 or d_0-DMDSSO cross-linked Nedd8(Q40E) ~ Cul1–Rbx1, followed by SDS–PAGE separation. The regions corresponding to expected cross-linked complexes were in-gel digested and the resulting peptides were subjected to liquid chromatography (LC)-MS^n analysis. DMDSSO cross-linked peptides were identified unambiguously based on MS^n data, that is, MS^1, MS^2 and MS^3, as previously described[29–31]. Representative MS^n analyses of d_0-DMDSSO and d_{10}-DMDSSO interlinked peptides α–β (*m/z* 513.6154^{3+} and *m/z* 516.9697^{3+}, respectively) are shown (Fig. 1e; Supplementary Fig. 2). As shown in Fig. 1e, MS^2 analysis of the d_0-DMDSSO interlinked peptide α–β yielded two expected fragment pairs α_A/β_T (*m/z* 437.76^{2+}/647.32^{1+}) and α_T/β_A (*m/z* 453.75^{2+}/615.34^{1+}), which are characteristic of DMDSSO interlinked peptides[31], confirming the type of cross-link observed here. Subsequent MS^3 analysis of the α_A fragment (*m/z* 437.76^{2+}) produced a series of y and b ions that enabled its unambiguous identification as ^{21}RFEVK$_A$K^{26} of Rbx1 with K25 modified with DMDSSO alkene remnant. Similarly, MS^3 analysis of the β_T fragment (*m/z* 647.32^{1+}) identified its sequence unambiguously as ^{15}SGAGK$_T$K^{20} of Rbx1 with K19 modified with the unsaturated thiol remnant. Together with MS^1 mass matching, we confidently determined this d_0-DMDSSO cross-linked peptide as an intraprotein interlink between K19 and K25 of Rbx1. Similar MS^n analysis of the same peptide cross-linked with d_{10}-DMDSSO (*m/z* 516.9697^{3+}) further confirms and identifies the intrasubunit K–K linkage within Rbx1 (Supplementary Fig. 2). On the basis of the identical fragmentation patterns of d_0- and d_{10}-DMDSSO cross-linked peptides, our results demonstrate that d_0- and d_{10}-DMDSSO

contain the same functionality and characteristics required for the unambiguous identification of their respective cross-linked peptides by MS^n analysis. Therefore, identification of either of the d_0- or d_{10}- cross-linked peptide pairs in each pairwise experiment would allow us to quantify differences in their relative abundances.

To quantify the identified d_0- and d_{10}-DMDSSO cross-linked peptides, we then determined the relative abundance ratios of corresponding peptide pairs based on their MS^1 spectral intensities. The same pairwise comparison experiments were repeated using reversed cross-linker treatments (that is, d_0-DMDSSO cross-linked Nedd8 ~ Cul1–Rbx1 was mixed with d_{10}-DMDSSO cross-linked Cul1–Rbx1 or d_{10}-DMDSSO cross-linked Nedd8(Q40E) ~ Cul1–Rbx1) to rule out cross-linking bias due to reagent deuteration (Fig. 1d).

Mapping XL-MS data to Cul1–Rbx1 complexes. The current structural model of unneddylated Cul1–Rbx1 is described in Fig. 2a, based on a previously reported crystal structure of full-length human Cul1–Rbx1 (PDB 1LDJ)[32]. In this structure, the N-terminal domain of Cul1 consists of three helical repeats (repeat 1, 2 and 3), each comprising five α-helices. These three repeats pack consecutively to form a long stalk-like shape. The Cul1 C-terminal domain (CTD) is composed of a four-helix bundle (4HB), an α/β domain and two copies of the winged-helix motif (WHA and WHB). The 4HB connects the N-terminal domain to CTD and organizes other subdomains in the CTD. It packs with the α/β domain and the long H29 helix, which connects WHA and WHB. The α/β domain and the N-terminal β-strand of Rbx1 form an intermolecular five-stranded β-sheet. One face of the WHB interacts with the long H29 helix and the 4HB, and the other contacts the RING domain of Rbx1. This compact architecture has been proposed to represent the 'closed' conformation of the Cul1–Rbx1 complex (Fig. 2a)[14].

Although there is no high-resolution structure available for the Nedd8 ~ Cul1–Rbx1 complex, the crystal structure of Nedd8 ~ Cul5^{CTD}–Rbx1 was previously resolved[14]. By threading the Cul1^{CTD} sequence into the neddylated Cul5 structure, we have derived a homology model of Nedd8 ~ Cul1–Rbx1 (Fig. 2b). Similar to the structure of Nedd8 ~ Cul5^{CTD}–Rbx1 (ref. 14), this model shows that neddylation has minor effects on the structures of individual subdomains, but induces marked rearrangements in their relative positions. The H29 helix rotates about 45°, which changes the WHB position relative to the 4HB and α/β domain. The repositioning of WHB abolishes the interaction between the WHB and the Rbx1 RING domain and frees the latter from the Cul1 scaffold. Nedd8 contacts the WHB to stabilize this 'open' state of the Cul1–Rbx1 complex. Two orientations for the RING, resulting from crystal packing, are observed in the crystal structure of Nedd8 ~ Cul5^{CTD}–Rbx1, indicating that the relative position of the RING domain and cullin scaffold are very promiscuous in solution. Therefore, we have generated two structural models of Nedd8 ~ Cul1–Rbx1 to describe the two different RING conformations (I and II) in the complex, and their overlays are illustrated in Fig. 2b. As shown, RING I (in yellow) is more proximal to the Cul1 scaffold, whereas RING II (in grey) is more distal. It is noted that the orientations of the Cul1 scaffold and Nedd8 remain the same in both RING conformations (Fig. 2b).

To further elucidate the structures of Cul1–Rbx1 complexes, we focused on the identification and quantification of interlinked peptides as they are most informative in describing residue proximity and interaction contacts in 3D structures. With our XL-MS strategy, we have identified a total of 68 unique interlinked d_0/d_{10} peptide pairs from eight replicate sets

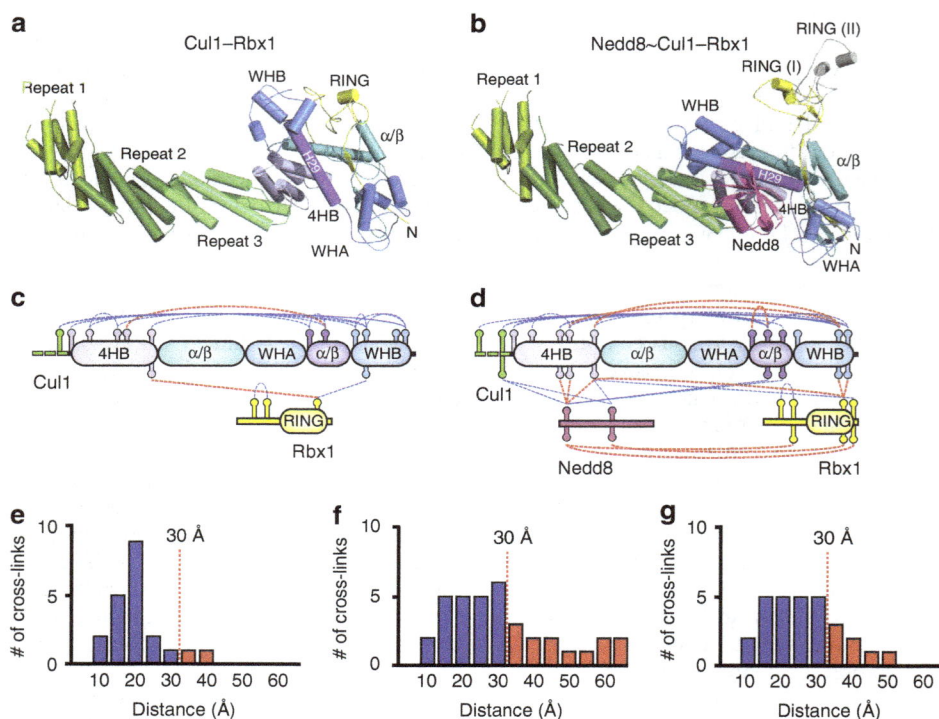

Figure 2 | Mapping cross-link data onto current structural models of Cul1–Rbx1 complexes. (a) The known structure of unneddylated Cul1–Rbx1 complex. **(b)** The overlay of the two homology models of neddylated Cul1–Rbx1 derived from Nedd8~Cul5^{CTD}-Rbx1 structure, depicting two conformations of the Rbx1 RING domain with I in yellow and II in grey. On the basis of the identified interlinked peptides, the cross-link maps were generated for **(c)** Cul1–Rbx1 and **(d)** Nedd8~Cul1–Rbx1 complexes. Note: linkages between residues with spatial distances below 30 Å are shown in blue-dotted lines, while those above 30 Å in red-dotted lines, correlating with colour-coded bar graphs in **e–g**. **(e)** The distribution plot of identified linkages versus their spatial distances between interlinked lysines in Cul1–Rbx1 structure. **(f)** The distribution plot of identified linkages versus their spatial distances between interlinked lysines in Nedd8~Cul1–Rbx1 structure models. **(g)** The distribution plot of identified linkages involving only Cul1 and Nedd8 versus their spatial distances between interlinked lysines in Nedd8~Cul1–Rbx1 structure models.

of comparison experiments (Supplementary Table 1), representing 27 intraprotein and 17 interprotein linkages. To correlate our XL-MS data with the current structural models of Cul1–Rbx1 complexes, we first generated K–K linkage maps of Cul1–Rbx1 and Nedd8~Cul1–Rbx1 based on the interlinks identified from each sample, as shown in Fig. 2c,d. It is important to note that our structural models of Nedd8~Cul–Rbx1 with either RING (I) or RING (II) conformations have the same cross-link maps as shown in Fig. 2d. Interestingly, 23 of 25 intraprotein Cul1 K–K linkages are localized in Cul1^{CTD} regions that interact with Rbx1 and Nedd8. In addition, 5 and 10 linkages represent interprotein interactions of Cul1 with Rbx1 or Nedd8, respectively. Collectively, extensive interactions among the three proteins were detected for us to evaluate the structural differences between the various forms of Cul1–Rbx1 complexes.

Next, we have mapped the identified cross-linked residues onto the structural models of Cul1–Rbx1 and Nedd8~Cul1–Rbx1, respectively, and calculated the distances between α-carbons (C_α–C_α distance) of cross-linked lysines using the molecular visualization software PyMOL. Considering the lengths of the DMDSSO (11 Å) and lysine side chains as well as backbone dynamics, the theoretical upper limit for the C_α–C_α distance between DMDSSO cross-linked lysine residues is ~30 Å, suggesting that lysines within distance <30 Å can be preferably cross-linked by DMDSSO. To examine the distance constraints of identified cross-links, we have plotted the distance distribution of the Cul1–Rbx1 cross-link data set (Fig. 2e). Ninety per cent of cross-links satisfy the distance cutoff of 30 Å, indicating a good

correlation with the current known structure of Cul1–Rbx1. However, when plotting Nedd8~Cul1–Rbx1 cross-link data to either of our homology-derived models, only 64% of cross-links (23/36) are within the desired distance constraint (Fig. 2f). In fact, the cross-links outside the cutoff predominantly represent interactions among Nedd8, Rbx1 and the C-terminal domain of Cul1. As Rbx1 is suspected to be mobile in previous publications[14], we then excluded 6 Rbx1-associated cross-links, thus yielding 30 remaining cross-links describing interactions within and between Nedd8 and Cul1 proteins. As a result, ~73% of linkages (22/30) fall within our expected distance constraints (Fig. 2g), with the 8 outliers all representing cross-links that involve either Nedd8 or the winged-helix domains of Cul1^{CTD}. This discrepancy could be explained by either the inaccuracy of the Nedd8~Cul1–Rbx1 structural model in those regions or a highly dynamic topology associated with the 'open' conformation.

Quantitation of DMDSSO cross-linked peptides. Generally, the likelihood of forming a cross-link between two given lysine residues is dependent on multiple factors. One of the important aspects is the 3D spatial distance between cross-linkable lysines. In addition, the relative orientations of proteins and their subdomains in different conformations under compared conditions can influence the relative reactivity of lysine residues. For instance, lysine residues localized in buried or protected regions would have decreased solvent and cross-linker accessibility compared with flexible, unprotected regions. Moreover, certain conformations could potentially influence the electronic environments of lysine residues by positioning them to form salt-bridge interactions with nearby acidic residues, decreasing their relative reactivity. Therefore, a combination of multiple factors could ultimately be responsible for the differences in observed spectral abundances of cross-linked peptides. Nonetheless, comparative analysis using QXL-MS strategies can unravel conformational changes of protein complexes under different conditions[24,38]. Of the total 68 unique interlinks identified in this work, 41 were identified at least in three biological replicates—our minimum requirement for reproducibility—representing 26 unique and high-confidence K–K linkages that were used for quantitative structural comparisons. Among them, there are two linkages associated with K720 of Cul1, which is the neddylation site and therefore covalently modified in the two neddylated Cul1–Rbx1 complexes, but free in unmodified Cul1–Rbx1 complex. As such, the two identified interlinked peptides associated with K720 of Cul1 were only detected in Cul1–Rbx1 complex, and were excluded from further analyses. The final list of 24 unique and quantifiable K–K linkages used for assessing structural changes of Cul1–Rbx1 complexes is summarized in Table 1. As shown, 13 were intraprotein (12 Cul1–Cul1 and 1 Rbx1–Rbx1) and 11 were interprotein (3 Cul1–Rbx1, 7 Cul1–Nedd8 and 1 Rbx1–Nedd8) interlinks. Among them, 15 linkages exhibited significant changes ($\geq$ 4-fold) in their relative abundances and suggested structural differences in different samples, while the remaining 9 displayed marginal changes (< 2-fold), indicating those interaction regions are relatively stable.

Comparison of Cul1–Rbx1 and Nedd8 ~ Cul1–Rbx1 complexes. Existing structural models have suggested that unneddylated Cul1–Rbx1 adopts a 'closed' conformation, while neddylated Cul1–Rbx1 exists in an 'open' state, as represented in Fig. 3a,b. To determine the structural effects of neddylation in the context of the full-length proteins, we examined intraprotein interlinks identified within Cul1 and Rbx1, respectively. For the 12 intraprotein interlinks identified within Cul1, 6 of them (that is, Cul1$^{K410-K743}$, Cul1$^{K417-K689}$, Cul1$^{K468-K693}$, Cul1$^{K472-K689}$, Cul1$^{K472-K693}$ and Cul1$^{K701-K708}$) exhibited below twofold difference between Cul1–Rbx1 and Nedd8 ~ Cul1–Rbx1 complexes, suggesting that there were no substantial structural reorientations between these cross-linked lysine residues upon neddylation (Table 1). In consistence with the cross-linking data, all of their C$_\alpha$–C$_\alpha$ distances are within 30 Å in the current Cul1–Rbx1 and Nedd8 ~ Cul1–Rbx1 complex models. For example, the relative spectral abundance ratio of the Cul1$^{K472-K693}$ linkage in unneddylated and neddylated Cul1–Rbx1 is ~ 1, and their respective C$_\alpha$–C$_\alpha$ distances are 16.0 and 16.4 Å (Fig. 3c). Although the 4HB and α/β domains containing these two residues become closer (Fig. 3a,b), the overall 3D spatial distance of these two residues has minimal change, thus leading to comparable cross-linking efficiency.

In contrast, the remaining six intraprotein interlinks of Cul1 had at least fourfold difference in their relative abundance ratios, in which five interlinks (that is, Cul1$^{K337-K750}$, Cul1$^{K410-K750}$, Cul1$^{K464-K693}$, Cul1$^{K464-K743}$ and Cul1$^{K693-K743}$) were detected much more dominantly in Nedd8 ~ Cul1–Rbx1 complex and one (that is, Cul1$^{K431-K472}$) significantly abundant in unneddylated Cul1–Rbx1 (Table 1; Fig. 3). These differences indicate that the

---

Table 1 | Comparative linkage profiles for SCF core ligases as determined by LC-MS[n].

Linkages identified						Mapped distances (Å)		Quantitative ratio*		
Linkage	Cul1	Cul1	Rbx1	Rbx1	Nedd8	Cul1-Rbx1	wtNedd8 ~ Cul1-Rbx1	Cul1-Rbx1	wtNedd8 ~ Cul1-Rbx1	mtNedd8 ~ Cul1-Rbx1
Cul1—Cul1	K337	K750				37.1	26.5	0.05	1.00	0.20
	K410	K720				38.1	21.7	1.00	0.01	0.01
	K410	K743				17.4	22.5	0.85	0.67	1.00
	K410	K750				24.9	21.0	0.14	1.00	0.36
	K417	K689				18.8	24.8	1.00	0.60	0.91
	K431	K472				9.7	9.3	1.00	0.24	0.50
	K464	K693				15.8	11.9	0.36	1.00	0.41
	K464	K743				15.7	26.3	0.05	1.00	0.22
	K468	K693				13.7	12.2	1.00	0.49	0.85
	K472	K689				16.3	10.7	1.00	0.87	0.90
	K472	K693				15.7	16.4	0.94	0.87	1.00
	K693	K743				24.8	26.6	0.03	1.00	0.17
	K701	K708				11.1	10.5	0.69	1.00	0.92
Cul1-Rbx1[†]	K493		K89			30.2	40.8 (I)/55.2 (II)	1.00	0.67	0.77
	K720		K89			11.7	72.8 (I)/82.8 (II)	1.00	0.00	0.00
	K743		K89			26.8	59.2 (I)/66.0 (II)	0.06	1.00	0.10
	K750		K89			22.6	54.5 (I)/64.9 (II)	0.11	1.00	0.05
Rbx1-Rbx1			K19	K25		18.1	---	1.00	0.62	0.71
Cul1-Nedd8	K410				K6	—	17.0	0.01	1.00	0.22
	K464				K6	—	30.6	0.01	1.00	0.12
	K468				K6	—	35.3	0.05	1.00	0.23
	K493				K6	—	41.0	0.04	0.07	1.00
	K493				K48	—	29.9	0.01	0.11	1.00
	K693				K6	—	28.7	0.00	1.00	0.11
	K701				K6	—	19.6	0.01	1.00	0.13
Rbx1[†]-Nedd8			K89		K48	—	65.0 (I)/74.3 (II)	0.02	0.96	1.00

LC-MS[n], liquid chromatography-multistage tandem mass spectrometry; SCF, Skp1-Cul1-F-box protein.
— Denotes distance undeterminable due to missing residues in structure/model.
*Spectral abundances normalized to the highest value for each linkage (per row).
†Mapped distances calculated for both Rbx1 RING reconformation I and II (Fig. 2b).

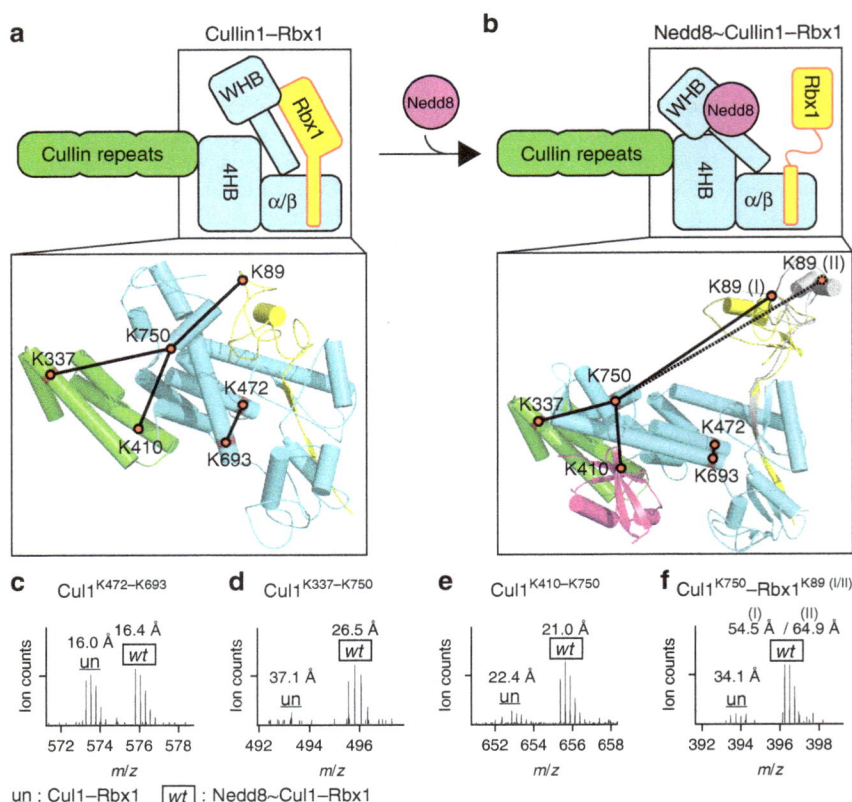

Figure 3 | Quantitative analysis of K–K linkages to determine neddylation-dependent structural changes in the Cul1–Rbx1 complex. Structural representation of (**a**) unneddylated Cul1–Rbx1 in the 'closed' state, (**b**) neddylated Cul1–Rbx1 in the 'open' conformation, in which K89 (I) and K89 (II) represent K89 position in Rbx1 RING (I) (in yellow) or (II) (in grey) conformations, respectively. The insets display the mapping of four selected interlinks onto the structures of Cul1–Rbx1 complexes, whose MS[1] spectra are displayed as follows: (**c**) Cul1$^{K472-K693}$, (**d**) Cul1$^{K337-K750}$; (**e**) Cul1$^{K410-K750}$; (**f**) Cul1^{K750}–Rbx1$^{K89(I/II)}$. These d_O/d_{10}-DMDSSO cross-linked peptide pairs measured in MS[1] were used to determine their relative abundance ratios between unneddylated and neddylated Cul1–Rbx1 complexes for quantitative analysis (Table 1).

two complexes feature substantial structural differences in regions containing the cross-linked lysines. In this study, two lysine residues that are proximal (<30 Å) would have higher chance of being captured by DMDSSO cross-linkers, which in turn increases the spectral abundance compared with one between lysine residues that are spatially distant (>30 Å). Figure 3d displays the MS spectrum of the Cul1$^{K337-K750}$ interlink, and quantitative analysis revealed that this interaction occurs much more favourably in Nedd8 ~ Cul1–Rbx1 than Cul1–Rbx1, on an average of 20:1. From the current models, the C_α–C_α distances between K337 and K750 of Cul1 were calculated to be 36.3 and 26.5 Å based on the Cul1–Rbx1 structure and our Nedd8 ~ Cul1–Rbx1 model, respectively. These calculated C_α–C_α distances fall outside of and within the distance that can be cross-linked by DMDSSO, which are in agreement with the increased spectral abundance of this interlink in Nedd8 ~ Cul1–Rbx1 compared with Cul1–Rbx1.

Similarly, the intraprotein interlink Cul1$^{K410-K750}$ was calculated to be on an average of approximately seven times more abundant in wild-type neddylated than unneddylated forms (Table 1; Fig. 3e). However, distinct from the Cul1$^{K337-K750}$ interlink, the C_α–C_α distances of K410 and K750 in the current models were determined to be 22.4 and 21.0 Å, respectively. Despite this similarity in their calculated proximities, the differential spectral abundance suggests that their cross-link is obstructed in unneddylated Cul1. In addition to distance, cross-linked peptide spectral abundance is influenced by the relative orientation of the lysine pair and their surroundings. In fact,

K410 and K750 of unneddylated Cul1 point away from each other and are separated by other residues, which can presumably impede the cross-linking reaction. Similarly, an additional four Cul1 intraprotein interlinks (that is, Cul1$^{K464-K693}$, Cul1$^{K464-K743}$, Cul1$^{K693-K743}$ and Cul1$^{K431-K472}$) have no apparent correlation between their relative spectral abundance ratios and respective C_α–C_α distance (<30 Å). However, most of them can be rationalized based on the structural environment of the lysine residues in the context of the current structure models (Supplementary Fig. 3). The Cul1$^{K693-K743}$ interlink represents the only noticeable outlier, which is >30-folds more abundant in Nedd8 ~ Cul1–Rbx1 than Cul1–Rbx1, albeit a similar C_α–C_α distance in the two models. K693 and K743 are located on the H29 helix and the WHB domain, which together act as a single rigid body. Their preferred cross-links in the neddylated Cul1–Rbx1 cannot be explained without significant changes of the current model of Nedd8 ~ Cul1–Rbx1. Overall, our results suggest that certain regions in the Cul1 scaffold, including the structural elements where those lysines are located, likely undergo profound structural reorientations in response to neddylation.

To further dissect the impact of neddylation, we examined the three unique Cul1–Rbx1 interprotein K–K linkages identified here (Table 1; Figs 3 and 4). Two cross-links, Cul1^{K743}–Rbx1^{K89} and Cul1^{K750}–Rbx1^{K89}, were quantitatively determined to have spectral abundances on an average of 10-fold higher in neddylated Cul1–Rbx1 compared with their unneddylated counterparts. Mapping of Cul^{K743}–Rbx1^{K89} and Cul^{K750}–Rbx1^{K89} to the Cul1–Rbx1 structure determines their

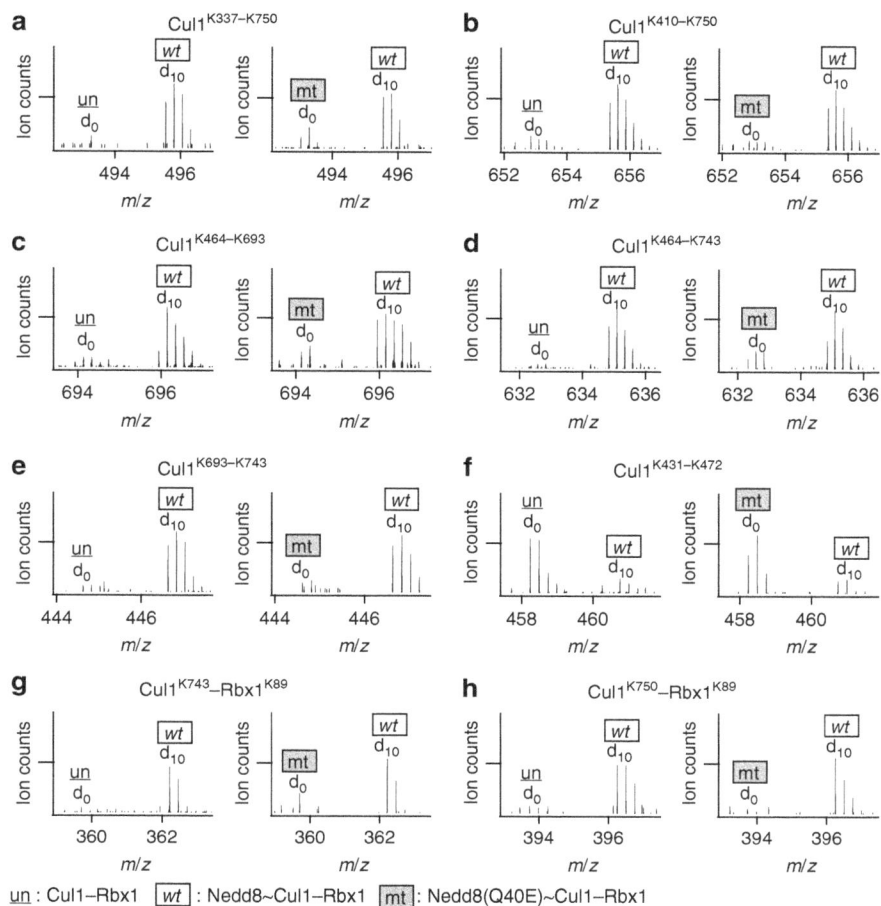

Figure 4 | Deciphering the structural dynamics of Cul1–Rbx1 complexes using QXL-MS. Eight selected K-K linkages are presented to describe conformational changes in the three types of Cul1–Rbx1 complexes. Two sets of pairwise comparison results (that is, un (d_0) versus wt (d_{10}) and mt (d_0) versus wt (d_{10})) are displayed for each selected cross-link, in which both the un and mt forms were cross-linked by d_0-DMDSSO and the wt form was cross-linked by d_{10}-DMDSSO. MS[1] spectra of (**a**) Cul1$^{K337-K750}$, (**b**) Cul1$^{K410-K750}$, (**c**) Cul1$^{K464-K693}$, (**d**) Cul1$^{K464-K743}$, (**e**) Cul1$^{K693-K743}$, (**f**) Cul1$^{K431-K472}$, (**g**) Cul1^{K743}-Rbx1^{K89} and (**h**) Cul1^{K750}-Rbx1^{K89}. The sequences of these cross-links are summarized in Supplementary Table 1. The relative spectral abundance of cross-links (d_0:d_{10}) measured during MS analysis describes the cross-linkability of lysine residues in 3D structure.

C_α–C_α distances to be 31.9 and 34.1 Å, respectively, just outside the range covered by DMDSSO and accounting for their low cross-linking abundances. However, when mapped to our homology-derived models of neddylated Cul1–Rbx1 with either RING (I) or RING (II) conformations, those same interlinks yielded C_α–C_α distances of 59.2 Å (I)/66.0 Å (II) and 57.5 Å (I)/64.9 Å (II), respectively (Table 1; Fig. 3f), even more unlikely to be cross-linked by DMDSSO. Instead, these unusual cross-links must be explained by either the structural flexibility of the 'open-state' conformation exhibited by Nedd8 ∼ Cul1–Rbx1 or a geometry different from the current model. In the crystal structure of Nedd8 ∼ Cul5^{CTD}-Rbx1, neddylation causes the globular RING domain of Rbx1 to eject from the WHB while remaining tethered to the Cul5^{CTD} by its N-terminal sequence. As a result, Rbx1 is free to sample the 3D space above Cul5^{CTD}. Our observations on the three Cul1–Rbx1 interprotein interlinks indicate that such a dynamic topology is plausible and can account for the formation of linkages with lysine residues that are too distant to be cross-linked in the unneddylated complex.

Interestingly, mapping of the identified K-K linkages between Cul1 and Nedd8 to the homology-derived Nedd8 ∼ Cul1–Rbx1 model shows that three out of seven cross-linking events were calculated to bridge C_α–C_α distances >30 Å, with two more above 28.5 Å (Table 1). This suggests that the position of Nedd8

relative to 4HB and α/β subdomains of Cul1, which comprise the majority of the Cul1–Nedd8 interlinks, may not be accurate in the current model. On one hand, the crystal structure of the Nedd8 ∼ Cul5^{CTD}-Rbx1 complex, which our Nedd8 ∼ Cul1–Rbx1 model was based on, might represent snapshots of an otherwise dynamic scaffold in addition to the flexibly linked Rbx1 RING domain. On the other hand, it remains possible that neddylation may cause different conformational changes on different cullins. Therefore, a more accurate structure model of the Nedd8 ∼ Cul1–Rbx1 complex is needed to explain all comparative cross-links between the free and modified Cul1–Rbx1 assembly.

Effects of Nedd8 deamidation on Cul1–Rbx1. To investigate the structural mechanism underlying deamidation of Nedd8, we conducted pairwise structural comparisons between Nedd8 ∼ Cul1–Rbx1 (wt) and Nedd8(Q40E) ∼ Cul1–Rbx1 (mt) complexes using the same DMDSSO-based QXL-MS strategy as described above. Similar to the previous results obtained from the comparison between unneddylated and wild-type neddylated Cul1–Rbx1 complexes, the six Cul1 interlinks (that is, Cul1$^{K410-K743}$, Cul1$^{K417-K689}$, Cul1$^{K468-K693}$, Cul1$^{K472-K689}$, Cul1$^{K472-K693}$ and Cul1$^{K701-K708}$) and one Rbx1 interlink (that is, Rbx1$^{K19-K25}$)

displayed non-significant changes (<2-fold) when comparing their spectral abundances in wt-neddylated and mt-neddylated $\sim$ Cul1–Rbx1 complexes (Table 1). Interestingly, the five interlinks of Cul1 (that is, Cul1$^{K337-K750}$, Cul1$^{K410-K750}$, Cul1$^{K464-K693}$, Cul1$^{K464-K743}$ and Cul1$^{K693-K743}$) that were found primarily in Nedd8 $\sim$ Cul1–Rbx1 compared with unneddylated Cul1–Rbx1 were also difficult to detect in Nedd8(Q40E) $\sim$ Cul1–Rbx1 (Fig. 4a–e). Furthermore, the Cul1$^{K431-K472}$ interlink, which was found to be much more abundant in unneddylated Cul1–Rbx1 than Nedd8 $\sim$ Cul1–Rbx1, was also detected more intensely in Nedd8 (Q40E) $\sim$ Cul1–Rbx1 (Fig. 4f). Overall, the relative abundance ratios of these core CRL interlinks are similar when comparing Nedd8 $\sim$ Cul1–Rbx1 to both Cul1–Rbx1 and Nedd8(Q40E) $\sim$ Cul1–Rbx1, respectively, suggesting that covalent attachment of Nedd8(Q40E) to Cul1 did not result in the same conformational changes in Cul1 as the wild-type Nedd8 modification.

This observation is also supported by the identification of interlinked peptides between Cul1 and Rbx1. The Cul1^{K743}–Rbx1^{K89} and Cul1^{K750}–Rbx1^{K89} interprotein interlinks were primarily detected in Nedd8 $\sim$ Cul1–Rbx1, but not in Cul1–Rbx1 or Nedd8(Q40E) $\sim$ Cul1–Rbx1 (Fig. 4g,h). Collectively, the quantitative MS profiles of the identified Nedd8(Q40E) $\sim$ Cul1–Rbx1 linkages are much more similar to those of unneddylated Cul1–Rbx1.

The structural dissimilarities in the two types of neddylated Cul1–Rbx1 complexes are further confirmed by Cul1–Nedd8 linkage comparisons. We have identified seven unique interprotein K–K linkages between Cul1 and Nedd8 as summarized in Table 1. Interestingly, all of the Cul1–Nedd8 intersubunit interlinks had relative abundance ratios indicating significant differences (≥ 4-fold) between Nedd8 $\sim$ Cul1–Rbx1 and Nedd8(Q40E) $\sim$ Cul1–Rbx1 (Fig. 5). Among them, five cross-links (Nedd8^{K6}–Cul1^{K410}, Nedd8^{K6}–Cul1^{K464}, Nedd8^{K6}–Cul1^{K468}, Nedd8^{K48}–Cul1^{K693} and Nedd8^{K6}–Cul1^{K701}) were only detected in Nedd8 $\sim$ Cul1–Rbx1, while the remaining cross-links (Nedd8^{K6}–Cul1^{K493} and Nedd8^{K48}–Cul1^{K493}) were only measured in Nedd8(Q40E) $\sim$ Cul1–Rbx1. In particular, K6 of Nedd8 and K493 of Cul1 are localized to opposite sides of the Cul1 scaffold in the Nedd8 $\sim$ Cul1–Rbx1 model, resulting in a C_α–C_α distance > 30 Å. Therefore, this Nedd8^{K6}–Cul1^{K493} cross-link preferably detected in Nedd8(Q40E) $\sim$ Cul1–Rbx1 further suggests that the Q40E mutation imparts a large degree of influence on the position of Nedd8 in relation to the Cul1 scaffold. Taken together, our results have demonstrated that Nedd8(Q40E) cannot induce the same structural effect on Cul1–Rbx1 as wild-type Nedd8, and the overall conformation of Nedd8(Q40E) $\sim$ Cul1–Rbx1 is much more similar to that of unneddylated Cul1–Rbx1 (Fig. 5a–f).

Discussion

We have developed an effective QXL-MS workflow based on our previously developed pair of isotope-labelled (that is, d_0 and d_{10}) MS-cleavable DMDSSO[31] to characterize the structural differences and similarities of three Cul1–Rbx1 complexes. This approach allows us to quantitatively assess neddylation-dependent conformational changes within the Cul1–Rbx1 complex and gain insights into the molecular basis underlying its activation mechanism. In this work, we have demonstrated that DMDSSO reagents are well suited to quantitatively compare protein complexes as they cross-link proteins with similar efficiency and the relative spectral intensity ratios of d_0- and d_{10}-DMDSSO cross-linked peptides are indicative of their respective abundances in the two compared samples. Thus, these isotope-coded cross-linkers can be used orthogonally to study differential protein structures by characterizing their intraprotein and interprotein interlinked peptides to describe protein interactions associated with conformational changes.

With this QXL-MS approach, we have reproducibly quantified 24 unique intraprotein and interprotein lysine–lysine linkages within the three different forms of Cul1–Rbx1 complexes (that is, Cul1–Rbx1, Nedd8 $\sim$ Cul1–Rbx1 and Nedd8(Q40E) $\sim$ Cul1–Rbx1). Although a substantial amount of cross-link data correlates well with existing models, several cross-links have spatial distances outside the desired range and cannot be rationalized based on current structure models. While our results generally support the homology model of Nedd8 $\sim$ Cul1–Rbx1 derived from the Nedd8 $\sim$ Cul5^{CTD}–Rbx1 crystal structure[14], a more accurate description of neddylation-induced conformational changes of the Cul1 scaffold calls for the necessity of a better defined model.

Independent of such a model, multiple pairwise comparisons have revealed that the molecular structure of Nedd8(Q40E)-modified Cul1–Rbx1 is very similar to that of its unmodified form, but significantly different from wild-type Nedd8-modified Cul1–Rbx1, indicating that Gln40 in Nedd8 is critical for the structural stability of neddylated Cul1–Rbx1. In the structure of Nedd8–Cul5^{CTD}–Rbx1, Gln40 is proximal to the isopeptide bond between Nedd8 and Cul5 (ref. 14) and may interact with the cullin scaffold to stabilize its active conformation. On the other hand, the unmodified CRL catalytic core adopts a rigid, thermodynamically stable 'closed' structure, lacking ligase activity for polyubiquitination of substrates. During neddylation, CRL interacts with neddylation machinery and shifts to a flexible, 'open' conformation with the extending RING-finger domain[14,39]. The neddylated CRL remains in its active state until Nedd8 is removed (deneddylation). We propose that Gln40 in Nedd8 can interact with amino-acid residues in cullin through weak interactions, such as hydrogen bonds and electrostatic interactions, which are responsible for stabilization of the 'open' state. Deamidation of Gln40 abolishes or weakens these interactions such that the CRL switches back to its thermodynamically more stable 'closed' state. On the basis of our cross-link data, we have proposed schematic models representing neddylation-dependent conformational changes in the Cul1–Rbx1 complex by wild-type or mutant Nedd8 (Fig. 5g). As illustrated, wt-neddylation leads to the 'open' conformation of the CRL core in which Rbx1 is free to rotate as previously shown by crystallography[14]. In contrast, mt-neddylation of Cul1 prevents switching from the inactive to active state by maintaining the 'closed' structure of the CRL with Rbx1 embedded. To verify this hypothesis, an experimental structure of neddylated full-length CRL will be required.

In summary, we have successfully applied our recently developed MS-cleavable, stable isotope-labelled cross-linkers d_0-DMDSSO and d_{10}-DMDSSO to quantitatively study structural differences in Cul1–Rbx1 complexes in response to neddylation. Such structural characterization has previously been hindered using conventional structural tools because of their large sizes (over 100 kDa) and dynamic conformations. Our QXL-MS approach enables us to quantitatively compare multiple lysine interlinks in three types of full-length Cul1–Rbx1 complexes. Comparing these cross-linkage profiles, we found that neddylation can induce large structural rearrangements of the Cul1–Rbx1 complex, which are partially consistent with structural models obtained with truncated and neddylated Cul5–Rbx1 complex. Our results also indicates Nedd8(Q40E)-conjugated Cul1–Rbx1 has a similar structure as that of free Cul1–Rbx1, answering the puzzle of how a subtle change of a single Nedd8 amino acid, Gln^{40}, can abolish the activity of the much larger CRL complex. Given the speed and accuracy of the approach, we expect that our

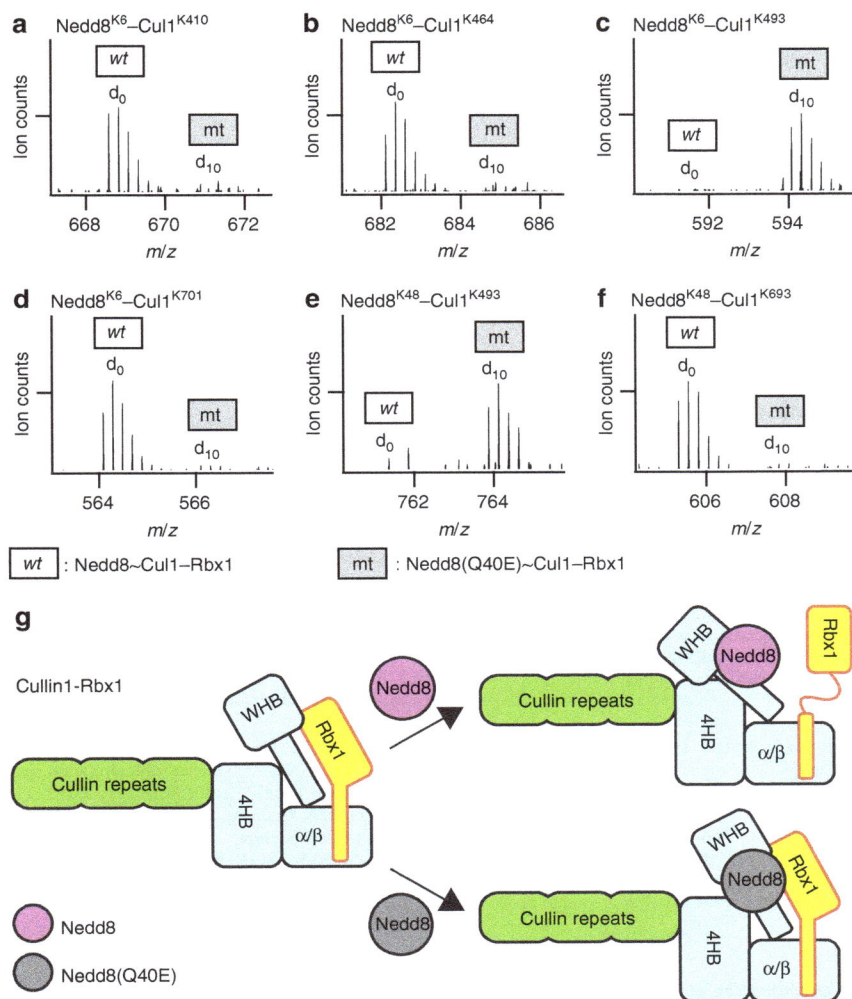

Figure 5 | Elucidation of structural dissimilarities between wt- and mt-neddylated Cul1–Rbx1 complexes by quantitative analysis of Cul1–Nedd8 interlinks. MS1 spectra of (**a**) Nedd8^{K6}–Cul1^{K410}, (**b**) Nedd8^{K6}–Cul1^{K464}, (**c**) Nedd8^{K6}–Cul1^{K493}, (**d**) Nedd8^{K6}–Cul1^{K701}, (**e**) Nedd8^{K48}–Cul1^{K493} and (**f**) Nedd8^{K48}–Cul1^{K693}. The mt form was d_0-DMDSSO cross-linked, and the wt form was d_{10}-DMDSSO cross-linked. (**g**) Proposed models for Nedd8-dependent conformational changes of the Cul1–Rbx1 complex.

QXL-MS strategy will enable us to perform future studies in characterizing E2–E3 interactions and further dissect the action mechanism of CRLs during protein ubiquitination. In addition, our work has paved the way for adapting QXL-MS methods for elucidating dynamic structures of proteins and protein complexes in the future.

Methods

Materials and reagents. General chemicals were purchased from Fisher Scientific or VWR international. Sequencing grade-modified trypsin was purchased from Promega (Fitchburg, WI).

Preparation of Cul1-Rbx1 protein complexes. The heterodimeric NEDD8-activating enzyme APPBP1–Uba3 was prepared similarly as before[40]. Briefly, APPBP1 was subcloned into a modified pGEX4T1 (Amersham Biosciences) vector containing a glutathione S-transferase (GST) tag followed by a Tobacco etch virus (TEV) protease cleavage site, while Uba3 was subcloned into a modified pET15b (Novagen) vector containing a chloramphenicol resistance cassette. GST–APPBP1 and Uba3 were co-expressed in BL21(DE3) (Novagen) and purified by glutathione-affinity chromatography. After TEV cleavage, the APPBP1–Uba3 complex was further purified by anion exchange and gel filtration.

Nedd8 and the Nedd8-conjugating enzyme Ubc12 were subcloned into the same pGEX4T1 vector. Both were expressed in E. coli BL-21(DE3) cells and purified by glutathione-affinity and anion-exchange chromatography. In this study, we used a truncated version of Nedd8 ending at glycine 76, representing its mature form.

Two short unstructured segments in the N-terminus of Cul1 (residues 1–12 and 58–81) were removed from full-length human Cul1 to form Cul1$^{\Delta N}$ (referred to here as Cul1). Both Cul1 and Rbx1$^{16-108}$ were fused with an N-terminal His$_6$ tag followed by a TEV cleavage site and co-expressed in BL-21(DE3). The complex was first purified by a Ni^{2+} sepharose-affinity column (GE Healthcare) and further purified by cation-exchange and gel filtration chromatography after TEV cleavage. To prepare Nedd8 ∼ Cul1–Rbx1, 10 μM purified Cul1–Rbx1 was neddylated with 10 μM GST–Nedd8 in the presence of 0.2 μM APPBP1–Uba3 and 0.5 μM Ubc12 for 1 h at 4 °C. Nedd8 ∼ Cul1–Rbx1 was then separated from free Cul1–Rbx1 by a glutathione-affinity column. After TEV cleavage, Nedd8 ∼ Cul1–Rbx1 was eluted off the column and further purified by cation-exchange and gel filtration chromatography. Nedd8(Q40E)-modified Cul1–Rbx1 was purified similarly to the wild type.

Full-length human ubiquitin-activating enzyme Ube1 was expressed as a GST fusion protein in High Five insect cells using the Bac-to-Bac baculovirus expression system (Invitrogen). Insect cells were collected 48–72 h post infection and lysed, followed by glutathione-affinity chromatography. Recombinant human Cdc34 was overexpressed and purified from E. coli by a similar approach as the Nedd8 purification. Recombinant untagged ubiquitin (Ub) was expressed in BL21(DE3). After sonication and centrifugation, cleared lysate was adjusted to 3.5% perchloric acid. After precipitated protein was removed by centrifugation, ubiquitin in the supernatant was further purified by cation-exchange chromatography and dialysis against 20 mM Tris, pH 8.0 thoroughly.

Ubiquitination assays. For free ubiquitin chain synthesis assay, a mixture containing 100 μM Ub, 0.3 μM UBE1 and 1.0 μM Cdc34 was incubated with 0.4 μM Cul1–Rbx1 variants in a reaction buffer of 50 mM Tris-HCl, 200 mM NaCl, 2 mM ATP and 10 mM MgCl$_2$, pH8.0. After incubation at 37 °C for 4 h, the reaction mixtures were resolved by a 15% SDS–PAGE gel and transferred onto a nitrocellulose membrane, which was incubated overnight with a mouse monoclonal anti-ubiquitin antibody (Sigma-Aldrich, #U0508) at 1:2,500 dilution. The membrane was washed and incubated with horseradish peroxidase-linked ECL-anti mouse IgG (GE Healthcare, #NA931V) for 1 h. Free ubiquitin chains were visualized using SuperSignal West Pico Chemiluminescent Substrate (Pierce Biotechnology, #34080).

For the CRY2 ubiquitination assay, a reaction mixture containing 0.2 µM CRY2–FBXL3–SKP1 complex[36], 70 µM Ub, 0.15 µM UBE1 and 1.5 µM Cdc34 was incubated with 0.4 µM Cul1–Rbx1 variants in a reaction buffer of 40 mM Tris-HCl (pH 7.5), 2 mM dithiothreitol, 5 mM MgCl$_2$ and 2 mM ATP. The reactions were carried out at 37 °C and quenched at different time points by adding SDS–PAGE loading buffer, then analysed by western blot with a rabbit anti-CRY2 antibody(LifeSpan BioSciences, Inc. #LS-C6229) at 1:1,000 dilution and horseradish peroxidase-linked ECL-anti rabbit IgG (GE Healthcare, #NA934V).

DMDSSO cross-linking and digestion of Cul1-Rbx1 complexes. Purified complexes were diluted to 4 µM in 20 mM HEPES (pH 7.5) and reacted with d$_0$- or d$_{10}$-DMDSSO in a molar ratio of 1:25 (protein: cross-linker) for 45 min at room temperature and quenched with excess ammonium bicarbonate. Cross-linked proteins were then separated by SDS–PAGE and visualized by Coomassie blue. Bands corresponding to cross-linked complexes were excised, reduced with tris (2-carboxy-ethyl) phosphine for 30 min at room temperature and alkylated with chlor-oacetamide for 30 min at room temperature in dark, and then digested with trypsin at 37 °C overnight. Peptide digests were extracted, concentrated and reconstituted in 3% ACN/2% formic acid before LC-MS^n analysis. To allow quantitative pairwise complex comparisons, individually cross-linked proteins are strategically mixed, for example, d$_0$-DMDSSO cross-linked Cul1-Rbx1 with d$_{10}$-DMDSSO cross-linked Nedd8 ∼ Cul1-Rbx1, at a 1:1 ratio and subjected to subsequent analysis together as outlined above.

Liquid chromatography-multistage tandem mass spectrometry. DMDSSO cross-linked peptides were analysed by LC-MS^n utilizing an LTQ-Orbitrap XL MS (Thermo Fisher, San Jose, CA) coupled on-line with an Easy-nLC 1,000 (Thermo Fisher, San Jose, CA) as previously described[29,31]. Each MS^n experiment consists of one MS scan in FT mode (350–1,400 m/z, resolution of 60,000 at m/z 400) followed by two data-dependent MS2 scans in FT mode (resolution of 7,500) with normalized collision energy at 20% on the top two MS peaks with charges at 3 + or up, and three MS3 scans in the LTQ with normalized collision energy at 35% on the top three peaks from each MS2.

Data analysis, identification and quantification of cross-linked peptides.
Monoisotopic masses of parent ions and corresponding fragment ions, parent ion charge states, and ion intensities from LC-MS^2 and LC-MS^3 spectra were extracted using an in-house software based on the Raw_Extract script from Xcalibur v2.4 (Thermo Scientific)[29–31]. MS^3 data were subjected to a developmental version of Protein Prospector (v. 5.10.10) for database searching, using Batch-Tag against a limited database containing recombinant Cul1, Rbx1, Nedd8 and Nedd8(Q40E) sequences with mass tolerances for parent ions and fragment ions set as ± 20 p.p.m. and 0.6 Da, respectively. Trypsin was set as the enzyme with five maximum missed cleavages allowed. A maximum of five variable modifications were also allowed, including protein N-terminal acetylation, methionine oxidation, N-terminal conversion of glutamine to pyroglutamic acid, asparagine deamidation and cysteine carbamidomethylation. In addition, three defined modifications on uncleaved lysines and free protein N-termini were also selected: alkene (A: C$_4$H$_4$O, + 68 Da; or A*: C$_4$$_{-1}D_5$O, +73 Da), sulfenic acid (S: C$_4$H$_6$O$_2$S, +118 Da; or S*: C$_4$H$_1$D$_5$O$_2$S, +123 Da) and unsaturated thiol (T: C$_4$H$_4$OS, +100 Da; or T*: C$_4$H$_{-1}$D$_5$OS, +105 Da) modifications, due to remnant moieties of d$_0$- (that is, A, S and T) or d$_{10}$-DMDSSO (that is, A*, S* and T*), respectively. It is noted that the sulfenic acid moiety often undergoes dehydration to become a more stable and dominant unsaturated thiol moiety (that is, T, + 100 Da or T*, + 105 Da) as previously described[29–31]. Initial acceptance criteria for peptide identification required a reported expectation value ≤ 0.1.

Integration of MS^n data was carried out using the in-house program LinkHunter, a revised version of the previously written Link-Finder program, to validate and summarize cross-linked peptides[22,29]. Basically, monoisotopic masses of parent ions measured in MS^1 scans for those putative interlinked peptides are required to match the sum of the two MS^2 cross-linked fragment ions that have been sequenced in MS^3.

Only the identified interlinked DMDSSO cross-linked peptides were subjected for subsequent manual quantitation as only interlinked peptides provide the most useful information on protein structures. Using Skyline (v. 2.5.06157; https://skyline.gs.washington.edu), we have determined the spectral abundances of all individually identified cross-linked peptides in each pairwise comparison, and the calculated relative abundance of d$_0$/d$_{10}$ cross-linked peptides. This allows determining the relative occurrence of the identified K–K linkages across all purified complexes. All linkages were then mapped onto existing Cul1–Rbx1 crystal structure (PDB: 1LDJ), as well as the derived homology model of Nedd8 ∼ Cul1–Rbx1

(ref. 32), to compare experimentally derived ratios of occurrence of K–K linkages to the C$_\alpha$–C$_\alpha$ distances as determined by structural models.

References

1. Fang, S., Lorick, K. L., Jensen, J. P. & Weissman, A. M. RING finger ubiquitin protein ligases: implications for tumorigenesis, metastasis and for molecular targets in cancer. *Semin. Cancer Biol.* **13**, 5–14 (2003).
2. Petroski, M. D. & Deshaies, R. J. Function and regulation of cullin-RING ubiquitin ligases. *Nat. Rev. Mol. Cell Biol.* **6**, 9–20 (2005).
3. Nalepa, G., Rolfe, M. & Harper, J. W. Drug discovery in the ubiquitin-proteasome system. *Nat. Rev. Drug Discov.* **5**, 596–613 (2006).
4. Zimmerman, E. S., Schulman, B. A. & Zheng, N. Structural assembly of cullin-RING ubiquitin ligase complexes. *Curr. Opin. Struct. Biol.* **20**, 714–721 (2010).
5. Sarikas, A., Hartmann, T. & Pan, Z. Q. The cullin protein family. *Genome Biol.* **12**, 220 (2011).
6. Duda, D. M. *et al.* Structural regulation of cullin-RING ubiquitin ligase complexes. *Curr. Opin. Struct. Biol.* **21**, 257–264 (2011).
7. Deshaies, R. J. SCF and Cullin/Ring H2-based ubiquitin ligases. *Annu. Rev. Cell Dev. Biol.* **15**, 435–467 (1999).
8. Cardozo, T. & Pagano, M. The SCF ubiquitin ligase: insights into a molecular machine. *Nat. Rev. Mol. Cell Biol.* **5**, 739–751 (2004).
9. Zheng, N., Wang, P., Jeffrey, P. D. & Pavletich, N. P. Structure of a c-Cbl-UbcH7 complex: RING domain function in ubiquitin-protein ligases. *Cell* **102**, 533–539 (2000).
10. Podust, V. N. *et al.* A Nedd8 conjugation pathway is essential for proteolytic targeting of p27Kip1 by ubiquitination. *Proc. Natl Acad. Sci. USA* **97**, 4579–4584 (2000).
11. Read, M. A. *et al.* Nedd8 modification of cul-1 activates SCF(beta(TrCP))-dependent ubiquitination of IkappaBalpha. *Mol. Cell. Biol.* **20**, 2326–2333 (2000).
12. Amir, R. E., Iwai, K. & Ciechanover, A. The NEDD8 pathway is essential for SCF(beta -TrCP)-mediated ubiquitination and processing of the NF-kappa B precursor p105. *J. Biol. Chem.* **277**, 23253–23259 (2002).
13. Saha, A. & Deshaies, R. J. Multimodal activation of the ubiquitin ligase SCF by Nedd8 conjugation. *Mol. Cell* **32**, 21–31 (2008).
14. Duda, D. M. *et al.* Structural insights into NEDD8 activation of cullin-RING ligases: conformational control of conjugation. *Cell* **134**, 995–1006 (2008).
15. Cui, J. *et al.* Glutamine deamidation and dysfunction of ubiquitin/NEDD8 induced by a bacterial effector family. *Science* **329**, 1215–1218 (2010).
16. Jubelin, G. *et al.* Pathogenic bacteria target NEDD8-conjugated cullins to hijack host-cell signaling pathways. *PLoS Pathog.* **6**, e1001128 (2010).
17. Morikawa, H. *et al.* The bacterial effector Cif interferes with SCF ubiquitin ligase function by inhibiting deneddylation of Cullin1. *Biochem. Biophys. Res. Commun.* **401**, 268–274 (2010).
18. Toro, T. B., Toth, J. I. & Petroski, M. D. The cyclomodulin cycle inhibiting factor (CIF) alters cullin neddylation dynamics. *J. Biol. Chem.* **288**, 14716–14726 (2013).
19. Leitner, A. *et al.* Probing native protein structures by chemical cross-linking, mass spectrometry, and bioinformatics. *Mol. Cell. Proteomics* **9**, 1634–1649 (2010).
20. Chen, Z. A. *et al.* Architecture of the RNA polymerase II-TFIIF complex revealed by cross-linking and mass spectrometry. *EMBO J.* **29**, 717–726 (2010).
21. Herzog, F. *et al.* Structural probing of a protein phosphatase 2A network by chemical cross-linking and mass spectrometry. *Science* **337**, 1348–1352 (2012).
22. Kao, A. *et al.* Mapping the structural topology of the yeast 19S proteasomal regulatory particle using chemical cross-linking and probabilistic modeling. *Mol. Cell. Proteomics* **11**, 1566–1577 (2012).
23. Leitner, A. *et al.* The molecular architecture of the eukaryotic chaperonin TRiC/CCT. *Structure* **20**, 814–825 (2012).
24. Schmidt, C. *et al.* Comparative cross-linking and mass spectrometry of an intact F-type ATPase suggest a role for phosphorylation. *Nat. Commun.* **4**, 1985 (2013).
25. Erzberger, J. P. *et al.* Molecular architecture of the 40SeIF1eIF3 translation initiation complex. *Cell* **158**, 1123–1135 (2014).
26. Shi, Y. *et al.* Structural characterization by cross-linking reveals the detailed architecture of a coatomer-related heptameric module from the nuclear pore complex. *Mol. Cell. Proteomics* **13**, 2927–2943 (2014).
27. Lasker, K. *et al.* Molecular architecture of the 26S proteasome holocomplex determined by an integrative approach. *Proc. Natl Acad. Sci. USA* **109**, 1380–1387 (2012).
28. Zeng-Elmore, X. *et al.* Molecular architecture of photoreceptor phosphodiesterase elucidated by chemical cross-linking and integrative modeling. *J. Mol. Biol.* **426**, 3713–3728 (2014).

29. Kao, A. *et al.* Development of a novel cross-linking strategy for fast and accurate identification of cross-linked peptides of protein complexes. *Mol. Cell. Proteomics* **10**, M110.002212 (2011).

30. Kaake, R. M. *et al.* A new in vivo cross-linking mass spectrometry platform to define protein-protein interactions in living cells. *Mol. Cell. Proteomics* **13**, 3533–3543 (2014).

31. Yu, C., Kandur, W., Kao, A., Rychnovsky, S. & Huang, L. Developing new isotope-coded mass spectrometry-cleavable cross-linkers for elucidating protein structures. *Anal. Chem.* **86**, 2099–2106 (2014).

32. Zheng, N. *et al.* Structure of the Cul1-Rbx1-Skp1-F boxSkp2 SCF ubiquitin ligase complex. *Nature* **416**, 703–709 (2002).

33. Busino, L. *et al.* SCFFbxl3 controls the oscillation of the circadian clock by directing the degradation of cryptochrome proteins. *Science* **316**, 900–904 (2007).

34. Godinho, S. I. *et al.* The after-hours mutant reveals a role for Fbxl3 in determining mammalian circadian period. *Science* **316**, 897–900 (2007).

35. Siepka, S. M. *et al.* Circadian mutant Overtime reveals F-box protein FBXL3 regulation of cryptochrome and period gene expression. *Cell* **129**, 1011–1023 (2007).

36. Xing, W. *et al.* SCF(FBXL3) ubiquitin ligase targets cryptochromes at their cofactor pocket. *Nature* **496**, 64–68 (2013).

37. Wu, K., Chen, A. & Pan, Z. Q. Conjugation of Nedd8 to CUL1 enhances the ability of the ROC1-CUL1 complex to promote ubiquitin polymerization. *J. Biol. Chem.* **275**, 32317–32324 (2000).

38. Schmidt, C. & Robinson, C. V. A comparative cross-linking strategy to probe conformational changes in protein complexes. *Nat. Protoc.* **9**, 2224–2236 (2014).

39. Scott, D. C. *et al.* Structure of a RING E3 trapped in action reveals ligation mechanism for the ubiquitin-like protein NEDD8. *Cell* **157**, 1671–1684 (2014).

40. Huang, D. T. & Schulman, B. A. Expression, purification, and characterization of the E1 for human NEDD8, the heterodimeric APPBP1-UBA3 complex. *Methods Enzymol.* **398**, 9–20 (2005).

Acknowledgements

We wish to thank members of the Huang and Zheng laboratories for their help during this study, especially Alex Huszagh in the Huang lab for his help in data analysis. We would like to thank Professor A.L. Burlingame, Drs Robert Chalkley, Shenheng Guan and Peter Baker at UCSF for using Protein Prospector. This work was supported by National Institutes of Health grants RO1GM074830 to L.H. and R01GM106003 to L.H. and S.R. N.Z. is an Howard Hughes Medical Institute investigator. Eric Novitsky was supported by an institutional Chemical and Structural Biology Training Grant predoctoral fellowship (T32-GM10856).

Author contributions

C.Y. designed and carried out QXL-MS experiments and data analyses, and prepared all figures and tables. H.M. prepared purified protein samples and performed ubiquitination assay. X.T. contributed to ubiquitination assay. E.J.N. and S.D.R. synthesized cross-linking reagents; L.H. and N.Z. conceived the study and directed the research. C.Y., H.M., N.Z. and L.H. wrote the paper.

Additional information

Competing financial interests: The authors declare no competing financial interests.

22

Hot electron-induced reduction of small molecules on photorecycling metal surfaces

Wei Xie[1] & Sebastian Schlücker[1]

Noble metals are important photocatalysts due to their ability to convert light into chemical energy. Hot electrons, generated via the non-radiative decay of localized surface plasmons, can be transferred to reactants on the metal surface. Unfortunately, the number of hot electrons per molecule is limited due to charge–carrier recombination. In addition to the reduction half-reaction with hot electrons, also the corresponding oxidation counter-half-reaction must take place since otherwise the overall redox reaction cannot proceed. Here we report on the conceptual importance of promoting the oxidation counter-half-reaction in plasmon-mediated catalysis by photorecycling in order to overcome this general limitation. A six-electron photocatalytic reaction occurs even in the absence of conventional chemical reducing agents due to the photoinduced recycling of Ag atoms from hot holes in the oxidation half-reaction. This concept of multi-electron, counter-half-reaction-promoted photocatalysis provides exciting new opportunities for driving efficient light-to-energy conversion processes.

[1] Physical Chemistry I, Faculty of Chemistry and Center for Nanointegration Duisburg-Essen (CENIDE), University of Duisburg-Essen, Universitätsstr. 5, Essen 45141, Germany. Correspondence and requests for materials should be addressed to S.S. (email: sebastian.schluecker@uni-due.de).

Noble metal nanoparticles are nanoscale sources of light, heat and electrons[1]. This can be exploited in numerous applications such as plasmon-assisted optical microscopy and spectroscopy (light)[2], photothermal therapy (heat)[3] and plasmon-driven redox reactions in chemistry (electrons)[4–6].

Upon resonant illumination with light in the visible range, the free electron gas of a metal nanoparticle is externally driven to perform oscillations at its plasma eigenfrequency. This resonant electronic excitation is called localized surface plasmon. There are two fundamentally different decay channels for localized surface plasmons: a radiative channel, in which the nanoparticle acts as a nanoantenna (scattering), and a non-radiative channel (absorption). The absorption of photons may lead to the generation of heat by electron–phonon coupling or to the generation of charge carriers by the excitation of electron–hole pairs[7]. In the latter case, the generated high-energy electrons can be transferred from the surface of the metal nanoparticle to an adjacent electron acceptor such as a semiconductor or a molecule[8–11].

The radiative channel is the physical basis of plasmon-assisted light scattering. Chemically relevant information on molecules adsorbed on the surface of metal nanoparticles can be obtained via surface-enhanced Raman scattering (SERS)[12–14], which inherits the high molecular specificity of Raman spectroscopy, but with signal levels that are orders of magnitude higher[15,16]. In heterogeneous catalysis, SERS has been used as a surface-selective tool for label-free monitoring of chemical reactions[17,18]. Proof-of-concept studies were focused on monitoring the metal nanoparticle-catalyzed reduction of aromatic nitro compounds by hydride reagents[19–21]. So far, Ag nanoparticles have not been able to catalyse this reaction under otherwise same conditions.

Interestingly, we discovered that even in the absence of hydride reagents, this six-electron reduction reaction can take place on the surface of Ag nanoparticles in the presence of protons and halide ions. Our accidental observation has only been possible by employing highly plasmonically active Ag superstructures for label-free monitoring of the reaction by in situ SERS spectroscopy since otherwise the product could not have been identified. Central elements in this plasmon-mediated reduction reaction are high-energy hot electrons and halide ions. The hot electrons are transferred to the adsorbed molecules in the reduction half-reaction, while the hot holes are responsible for the oxidation counter-half-reaction. Halide ions are a key component in this redox reaction since they form photosensitive silver halides. The photoinduced dissociation of the silver halides acts as the oxidation half-reaction to compensate the hot holes and facilitate the electron transfer to the adsorbate. Importantly, it this recycling of surface silver atoms which drives the overall redox reaction.

Results

Core–satellite Ag superstructures.
Figure 1a shows a multiple-step method for the synthesis of core–satellite Ag superstructures. Silver nanoparticle cores with a diameter of ~ 100 nm are first encapsulated by an ultrathin silica shell. The silica surface is then functionalized with thiol groups. Silver nanoparticle satellites (~ 25 nm) are assembled onto the large core via Ag–S bonds. Silver superstructures with dozens of satellites can be synthesized in large scale (Fig. 1b,c).

High plasmonic activity is required for the generation of hot electrons for redox chemistry (non-radiative plasmon decay) and for SERS reaction monitoring (radiative plasmon decay). Although silver is one of the most plasmonically active metals, Ag nanoparticle monomers do not have sufficiently high scattering cross-sections for SERS monitoring (Supplementary Figs 1–3 and Supplementary Note 1). In contrast, our rationally designed core/satellite superstructures (Supplementary Fig. 4)

exhibit significantly larger absorption and scattering cross-sections (Fig. 1d). Computer simulations predict a local electric field enhancement of $|E| \sim 90$ upon resonant excitation of the Ag superstructure (Fig. 1e and Supplementary Fig. 5). This corresponds to a SERS enhancement factor[15] (EF) of $|E|^4 \sim 6.62 \times 10^7$. Experimentally, the very high plasmonic activity of the Ag superstructures is demonstrated in single-particle SERS experiments (Supplementary Fig. 6). Figure 1f shows the enhanced Raman spectrum of 4-mercaptobenzoic acid (4-MBA) on a single Ag superstructure on a silicon wafer. The Raman intensity of dominant peaks from 4-MBA (1,076 and 1,587 cm^{-1}) is about 10 times stronger than the first-order phonon peak from the silicon substrate at ~ 520 cm^{-1}. This single-particle SERS activity enables quantitative and label-free reaction monitoring.

Hot electron reduction. The Ag superstructures were coated with a self-assembled monolayer (SAM) of 4-nitrothiophenol (4-NTP). The reduction of the educt 4-NTP to the product 4-aminothiophenol (4-ATP) by sodium borohydride has been used as a model reaction in previous proof-of-concept studies to test the catalytic activity of noble metal nanoparticles such as Au, Pt and Pd[19–23]. In contrast, Ag nanoparticles cannot catalyse this hydride (H$^-$) reduction reaction (Supplementary Figs 7 and 8 and Supplementary Note 2). Surprisingly, we detected the SERS signal of the reaction product 4-ATP when the Ag core/satellite superstructures were suspended in aqueous HCl (Fig. 2a). This unexpected finding is exciting since it demonstrates that even in the absence of a chemical hydride agent, the reduction on much cheaper Ag surfaces is now possible. Formally, the hydride equivalent in this photocatalytic reaction is provided by a proton and two electrons: H$^- =$ H$^+ + 2$e$^-$. In control experiments we tested other acids such as H$_2$SO$_4$. However, in this case only the SERS signal of the educt 4-NTP and no contributions from the product 4-ATP were detected, indicating that the reduction did not occur. This led us to the hypothesis that not only acidic conditions, but also the counter anion is relevant. We tested this by adding NaCl to the Ag superstructures suspended in H$_2$SO$_4$ and indeed the SERS signal of the product 4-ATP was detected. In a negative control experiment, we added only aqueous NaCl solution, that is, without acid, to the Ag superstructures. As expected, no SERS signal of the product 4-ATP was observed. We therefore concluded that both protons and chloride anions are required for this photocatalytic reduction. While protons are obviously needed as the hydrogen source in this reaction, we initially did not have an explanation for the role of the chloride anions.

The generation of hot electrons is in principle possible with all plasmonic materials upon resonant excitation by light. Tuning the laser excitation wavelength away from λ_{max} of the plasmon peak leads to lower reduction activity (Supplementary Fig. 9 and Supplementary Note 3). We therefore assumed that this unexpected photocatalytic route on Ag, involving hot electrons but no chemical reducing agent, may also occur on Au surfaces. In order to test this, we synthesized hybrid Ag core–Au satellite superstructures since the Au satellites might donate their hot electrons to the SAM on their surface. It is important to mention that this is only possible for the satellites (Au), but not for the core (Ag): the large Ag core is isolated by an inert ultrathin glass as a dielectric spacer, which prevents the chemisorption of molecules onto the surface of the core as well as the transfer of hot electrons to them[24]. To our surprise, no reduction reaction was observed for the Ag core–Au satellite superstructures (Supplementary Fig. 10), neither in aqueous HCl nor aqueous H$_2$SO$_4$ (Fig. 2b and Supplementary Note 4). This finding indicates that not only the plasmonic but also the chemical properties of the metal (Supplementary Figs 11 and 12), in

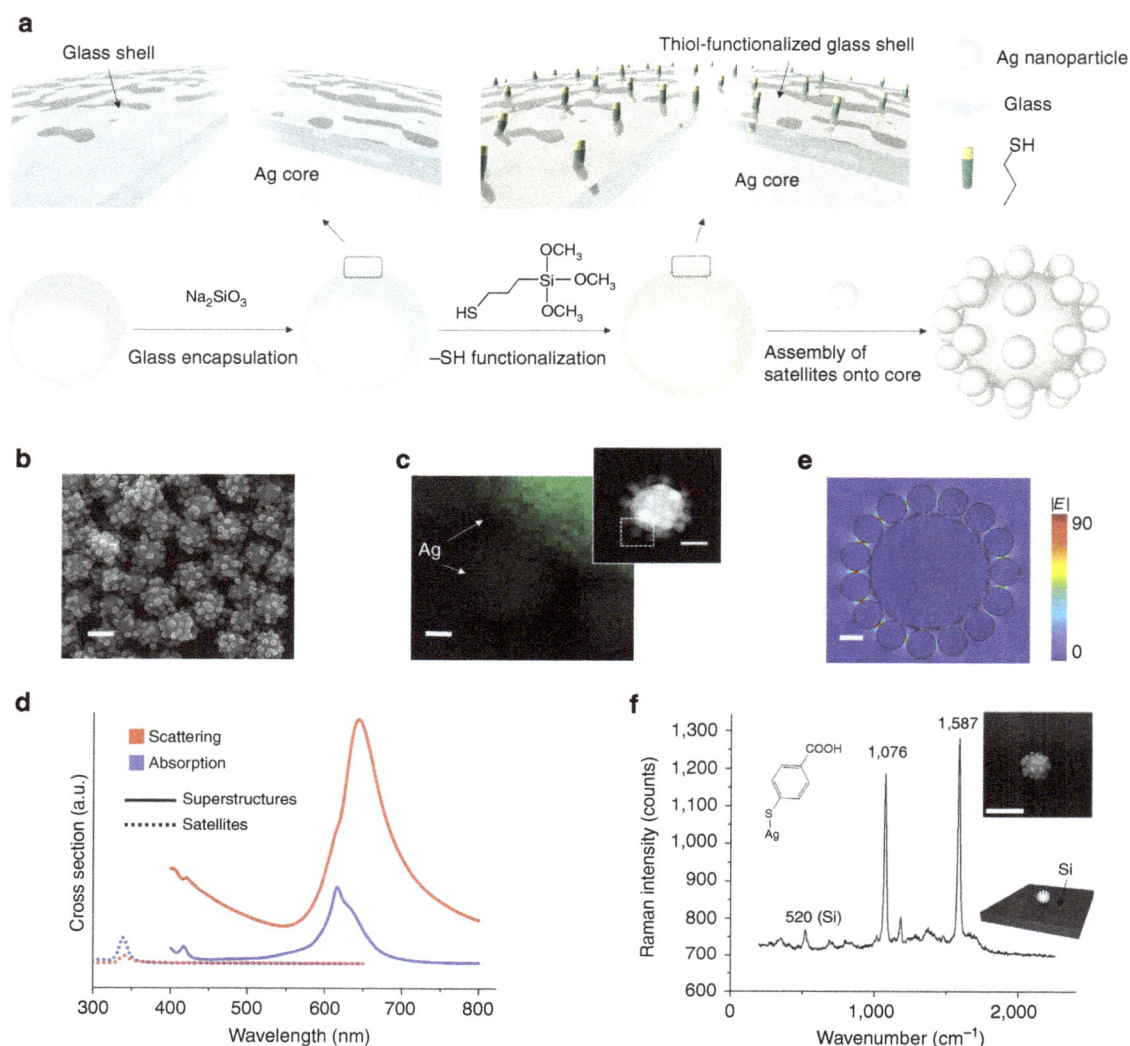

Figure 1 | Synthesis and characterization of highly plasmonically active Ag core–satellite superstructures. (**a**) Scheme of the synthesis: 100 nm Ag cores are encapsulated with an ultrathin glass shell and functionalized with thiol groups. Ag satellite nanoparticles (25 nm) are assembled onto the shell-isolated Ag core. (**b**) SEM image of the Ag superstructures (scale bar, 100 nm). (**c**) EDS element map (Ag) of the superstructure (scale bar, 10 nm) and the corresponding STEM image (scale bar, 50 nm). (**d**) Calculated absorption and scattering cross-sections of a 25 nm Ag satellite compared with the Ag superstructure. The cross-sections in both channels, absorption for efficient generation of hot electron–hole pairs and scattering for chemical monitoring with molecular specificity by SERS, are much larger in the superstructure. (**e**) Finite element method simulation of the incident electric field amplitude $|E|$ distribution on resonant excitation at 632.8 nm (scale bar, 20 nm). (**f**) SERS spectrum of 4-MBA from a single Ag superstructure on a Si wafer shown in the SEM image (top right; scale bar, 200 nm). The strong SERS peaks of 4-MBA at 1,076 and 1,587 cm^{-1} (laser power 0.6 mW and intergration time 36.5 ms; first-order phonon peak of silicon at ~ 520 cm^{-1}) demonstrate the single-particle SERS activity of the Ag superstructures.

particular Ag, are relevant for driving this photocatalytic reaction. In order to find out whether a chloride-specific effect is present, we also tested other halide anions.

Counter-half-reaction. Silver chloride has a very low solubility in water ($K_{SP} = 1.77 \times 10^{-10} \, M^2$ at room temperature) and, more importantly, it is a photosensitive salt that has been used in photography since the early 19th century. The proposed reaction mechanism of this photocatalytic reaction, which is in agreement with all experimental findings, is shown in Fig. 3. First, hot electron–hole pairs (e^-/h^+) are generated on the Ag super-structure upon resonant light excitation:

$$Ag \xrightarrow{h\nu} Ag^+ + e^- \qquad (1)$$

The hot electrons can be transferred to the molecules chemi-sorbed on the Ag surface. In the absence of chloride anions the

number of the hot electrons is not sufficient for driving the reaction due to the high charge–carrier recombination rate. In the presence of Cl$^-$ the hot holes ($h^+ = Ag^+$) combine with Cl$^-$ ions and form photosensitive AgCl on the Ag surface, which is then decomposed via photodissociation:

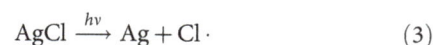

$$Ag^+ + Cl^- \rightarrow AgCl \downarrow \qquad (2)$$

$$AgCl \xrightarrow{h\nu} Ag + Cl\cdot \qquad (3)$$

In this case the oxidation half-reaction is the electron transfer from Cl$^-$ to the Ag catalyst and more hot electrons can be provided for the reduction half-reaction. The Cl$^-$ ions are therefore required for recycling Ag atoms as the electron donors from the corresponding hot holes ($h^+ = Ag^+$). This photo-recycling is a necessary condition since overall six electrons are

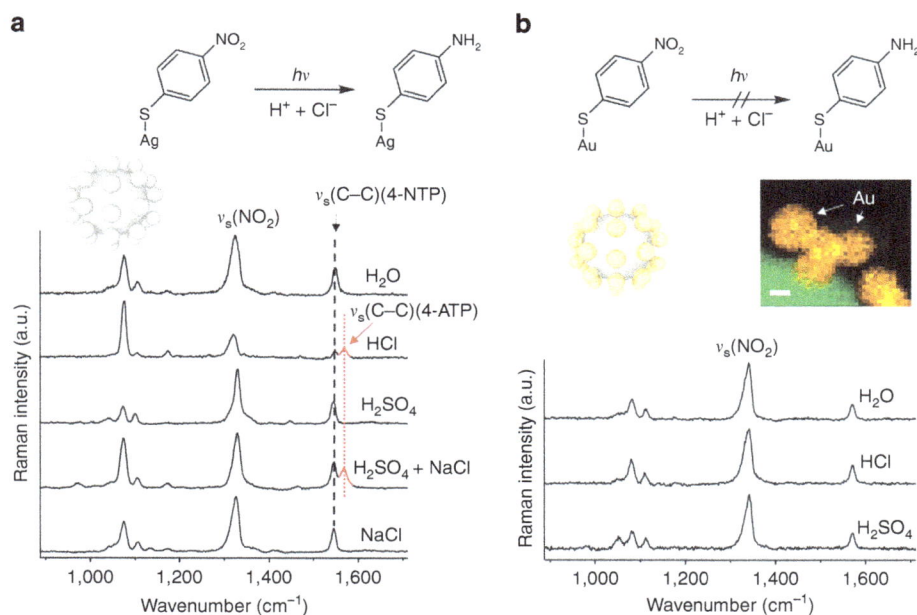

Figure 2 | Hot electron-induced reduction from 4-NTP to 4-ATP on Ag versus on Au surfaces. (**a**) SERS spectra of 4-NTP on the surface of plasmonic Ag superstructures in different environments. 4-ATP, the product of the hot electron-induced reaction, can be detected (the C–C and C–S stretching bands of 4-ATP appear at ~1,590 and ~1,080 cm^{-1}, respectively) only in the presence of both H$^+$ and Cl$^-$. (**b**) SERS control experiments using the ~25 nm Au satellites of the superstructure. The inset (middle right) shows the element mapping of both Au and Ag (scale bar, 10 nm). It is important to mention that the 4-NTP educt molecules can only adsorb on the satellite particle surface because the Ag core is isolated by an ultrathin glass shell. No signal from 4-ATP was detected on the Au surface, indicating that the reduction cannot proceed on Au nanoparticles even in the presence of hot electrons. The hot holes remaining on the Au surface limit the number of hot electrons that can be transferred to the 4-NTP molecules. See Fig. 3 for proposed reaction mechanism.

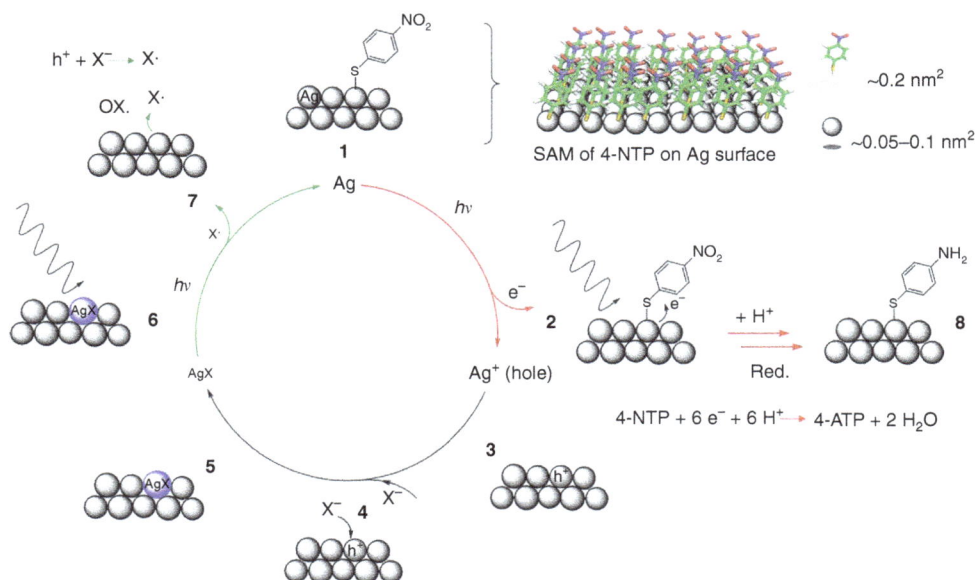

Figure 3 | Counter-half-reaction-promoted hot electron reduction. Schematic illustration of the excitation of hot electrons for the reduction of 4-NTP to 4-ATP. Halide anions compensate the hot holes via the oxidation counter-half-reaction, thereby recycling the plasmonic Ag surface. When a SAM of 4-NTP molecules covers the Ag surface (**1**), each molecule occupies about 2–4 surface Ag atoms (see top right of the figure). However, six electrons per 4-NTP molecule (**2**) are required to complete the reduction to 4-ATP (**8**). Thus, the Ag superstructure (**3**) has to gain electrons from outside via the oxidation half-reaction (labelled with green in the reaction circle). Due to the very low solubility of the corresponding silver halides, the halide anions have a very strong affinity to the hot holes (Ag$^+$) (**4**) and lead to the formation of photosensitive AgX (**5**), which undergoes a photodissociation on the Ag surface (**6**). Thus, the oxidation half-reaction (**7**) is established due to this formation and subsequent decomposition of the photoactive AgX. In this case the hot holes are compensated and thereby recycle the Ag surface for the hot electron-induced reduction half-reaction (labelled in red).

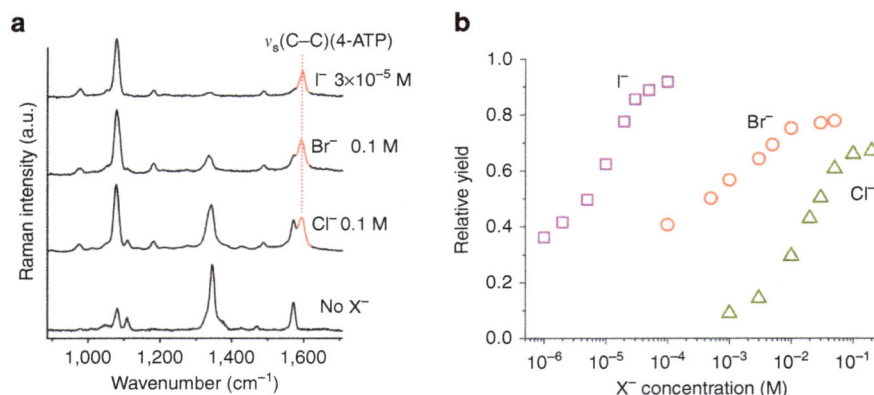

Figure 4 | Hot electron-induced reduction in the presence of halide anions. (**a**) SERS spectra of the reactions on plasmonic Ag superstructures using different halide anions (Cl$^-$, Br$^-$ and I$^-$) to compensate hot holes (h$^+$ = Ag$^+$). Iodide shows the strongest activity for the reaction even at very low concentrations. (**b**) Relative contribution of the product (4-ATP) to the SERS spectra from the reaction suspension as a function of halide anion concentration. The higher activity of I$^-$ is because the corresponding silver halide has a lower solubility K_{sp} (AgI = 8.52×10^{-17} M^2 < AgBr = 5.4×10^{-13} M^2 < AgCl = 1.77×10^{-10} M^2), i.e. a higher tendency to form solid AgX on the Ag surface, and a higher photosensitivity (AgI > AgBr > AgCl) in the counter-half-reaction for the recycling of the Ag surface atoms (Ag$^+$ → Ag).

formally needed to drive this reduction reaction:

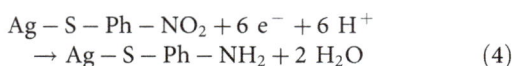

$$Ag - S - Ph - NO_2 + 6\ e^- + 6\ H^+$$
$$\rightarrow Ag - S - Ph - NH_2 + 2\ H_2O \qquad (4)$$

A single 4-NTP molecule in the SAM occupies about 0.2 nm^2 Ag surface area[25], where only 2–4 Ag atoms are present (Fig. 3). Because of the fast charge–carrier recombination it is not possible that the Ag surface can provide enough hot electrons for the entire reduction alone, that is, without recruiting additional electrons from the environment.

We confirmed the role of photosensitive Ag salts for photo-recycling of Ag atoms as electron donors in quantitative control experiments using silver halides (Supplementary Figs 13–15). According to the proposed reaction mechanism, Br$^-$ and I$^-$ should be even more active than Cl$^-$ in this reaction due to their lower water solubility ($K_{SP} = 5.4 \times 10^{-13}$ and 8.52×10^{-17} M^2 respectively) and the higher photosensitivity[26] of their corresponding silver salts (AgBr and AgI). We found that with the same concentration of Br$^-$ and Cl$^-$, more 4-NTP molecules are reduced to 4-ATP in the Br$^-$-containing solution (Fig. 4a). In particular I$^-$ shows an extremely high activity, even at much lower concentrations than those used for Cl$^-$ and Br$^-$ (30 µM versus 100 mM). The relative contribution of the product 4-ATP to the SERS spectra calculated according to the characteristic Raman bands of 4-NTP and 4-ATP at 1,573 and 1,591 cm^{-1}, respectively, at different X$^-$ concentrations, are shown in Fig. 4b. The activity of X$^-$ for the hot electron-induced redox chemistry follows the trend I$^-$ > Br$^-$ > Cl$^-$, which directly correlates with the solubility and photosensitivity of their corresponding Ag salts. In negative control experiments we also used other anions such as SO$_4^{2-}$ and PO$_4^{3-}$ (Supplementary Fig. 16). The corresponding Ag salts have a higher water solubility and/or low photosensitivity compared with the Ag halides and cannot compensate the hot holes by photorecycling of Ag atoms and therefore exhibit no photocatalytic activity in this hot electron-induced reaction.

The proposed oxidation half-reaction in Fig. 3 (**6**→**7**) is the photodissociation of AgX (photography). This recovers Ag atoms which then induce reduction by hot electrons (**1**→**2**). Alternatively, AgX may be directly converted to Ag$^+$ (hole) + X· + e$^-$ (**6**→**3**) and no hot electrons are required for recycling. In this particular reaction (4-NTP to 4-ATP) it is not possible to disentangle both processes since AgX is required for

photorecycling, while metallic Ag is necessary for SERS monitoring. The characterization of the Ag superstructures after reaction can be found in Supplementary Figs 17 and 18. Therefore, we tested if hot electrons from the photorecycling Ag surface can be used in other redox reactions which do not require SERS monitoring. Results in Supplementary Fig. 19 show the catalytic activity of Ag nanoparticles in the photoreduction of yellow [Fe(III)(CN)$_6$]$^{3-}$ to colourless [Fe(II)(CN)$_6$]$^{4-}$. Also here the halide anion efficacy trended with I$^-$ > Br$^-$ > Cl$^-$. In contrast, AgX alone (without metallic Ag) cannot catalyse the same reaction in the control experiments (Supplementary Fig. 20), indicating the key role of the photorecycling metallic Ag/AgX surface in the reduction process. These results further support the proposed general reaction mechanism.

Discussion

We report on the discovery of a counter-half-reaction promoted reduction chemistry on silver surfaces without the need of chemical reducing agents. Key elements for driving this reaction are hot electrons, protons and halide anions. Hot electrons are generated from silver and transferred to molecules adsorbed on the metal surface. Protons serve as the hydrogen source. Halide ions are required for photorecycling the electron-donating silver atoms (Ag) from hot holes (h$^+$ = Ag$^+$) after photodissociation of the insoluble silver halides present on the Ag surface (Supplementary Discussion). A series of control experiments demonstrates that without this photorecycling counter-half-reaction the six-hot electron reduction cannot proceed. The discovery was enabled by the use of *in situ* SERS spectroscopy, which allowed us to directly identify the reaction product on-line and in a label-free approach. This unexpected photocatalytic route paves the way for exciting new opportunities for light-to-energy conversion schemes employing silver as a significantly cheaper catalyst compared with other noble metals such as Au, Pt and Pd.

Methods

Synthesis of monodisperse Ag nanoparticles. A modified method from ref. 27 was used to synthesize Ag nanoparticles. To synthesize ∼25 nm small Ag nanoparticles, 23 ml glycerol was mixed with 27 ml water in a 100 ml flask and heated up to 104 °C under vigorous magnetic stirring. Sodium citrate (10 mg) was dissolved in 1 ml water and added to solution in the flask. One millilitre of 0.5% AgNO$_3$ aqueous solution was subsequently added. After 1 h, the reaction system was cooled down and yellow brown colloid was collected. The obtained ∼25 nm Ag nanoparticles were used as the seeds to produce ∼100 nm Ag nanoparticles. In a 250 ml flask, 580 mg of polyvinylpyrrolidone K30 was added to 138 ml water together with 23 ml glycerol. Seeds colloid (2.5 ml; 25 nm Ag nanoparticles) was added under stirring. After 20 s, 1.15 ml of a mixture of 20 mg AgNO$_3$, 1 ml water and 220 µl ammonium hydroxide (25%) solution, was added to the flask together with 92 ml aqueous solution containing 36.8 mg ascorbic acid. The growth finished after 1 h. A milky white colloid was obtained. Fresh Ag nanoparticles are preferred for the synthesis of the plasmonic superstructures.

Synthesis of Ag superstructures. The 100 nm Ag nanoparticles were used as cores of the superstructures. The surface of these large Ag nanoparticles was first coated with 3-mercaptopropyltrimethoxysilane (MPTMS) by adding 10 µl ethanolic MPTMS solution (10%) to 1 ml Ag colloid. This mixture was incubated in a temperature-controlled shaker (Eppendorf, Thermomixer comfort) at 50 °C for 1 h (750 r.p.m.). Then the nanoparticles were washed two times with water and resuspended in 1 ml water. An aqueous 0.054% sodium silicate solution (6 µl) was added to the colloid and the mixture was incubated at 90 °C for 1 h (750 r.p.m.). After centrifugation the ultrathin glass shell-isolated Ag cores were resuspended in 1 ml isopropanol. An aqueous ammonium hydroxide solution (50 µl, 25%) and ethanolic MPTMS solution (5 µl, 10%) were added together to the obtained glass shell-isolated Ag core suspension and then incubated at room temperature. After 30 min, the products were washed two times with isopropanol and resuspended in 1 ml isopropanol. The obtained suspension was mixed with 1 ml of 25 nm Ag satellite nanoparticles and then sonicated (ultrasonic bath Elmasonic P 60 H, 60% power) at room temperature for 20 min. Unbounded small satellite nanoparticles were removed after overnight incubation. For removal of the small satellite particles the mixture was centrifuged at relative centrifugal force (RCF) 600g for 10 min and the supernatant was discarded. The superstructures were resuspended in 1 ml water. After repeating the washing steps two times the final products were resuspended in 100 µl water.

Hot electron-induced reduction from 4-NTP to 4-ATP. Ten microlitre of 10 mM ethanolic 4-NTP solution was added to 1 ml aqueous colloidal suspension of the Ag core/satellite superstructures with an ultrathin silica shell around the Ag core. The mixture was incubated overnight at room temperature to form a SAM of 4-NTP on the surface of the small satellites. Then the Ag superstructures were washed two times with water and resuspended in 1 ml water. To perform the hot electron-induced reaction with protons and halide anions, aqueous sulfuric acid and sodium or potassium halide, respectively, were added to 1 ml of the colloidal super-structures, with a final volume of 2 ml and proton concentration of 1 M. The 632.8 nm line from a He–Ne laser was employed for initiating the redox reaction and for the excitation of the SERS spectra. A WITec Alpha 300R microscope equipped with a monochromator (600 grooves per mm grating) and an EM-CCD (Andor) was used for the detection of the SERS spectra.

Materials. Silver nitrite, sodium citrate, L-ascorbic acid, sodium silicate, MPTMS, tetraethylorthosilicate (TEOS), 4-ATP, sodium borohydride, ammonium hydroxide (25%), isopropanol, sulfuric acid, hydrochloric acid, phosphoric acid, sodium chloride, sodium bromide and potassium iodide were purchased from Sigma/Aldrich/Fluka. 4-NTP was purchased from Fluorochem. 4-MBA was purchased from TCI. Glycerol was purchased from AppliChem. Polyvinyl-pyrrolidone K30 was purchased from Carl Roth. All chemical reagents were used as received without further purification.

References

1. Hartland, G. V. Optical studies of dynamics in noble metal nanostructures. *Chem. Rev.* **111**, 3858–3887 (2011).
2. Lal, S. *et al.* Tailoring plasmonic substrates for surface enhanced spectroscopies. *Chem. Soc. Rev.* **37**, 898–911 (2008).
3. Jain, P. K., Huang, X. H., El-Sayed, I. H. & El-Sayed, M. A. Noble metals on the nanoscale: optical and photothermal properties and some applications in imaging, sensing, biology, and medicine. *Acc. Chem. Res.* **41**, 1578–1586 (2008).
4. Linic, S., Christopher, P. & Ingram, D. B. Plasmonic-metal nanostructures for efficient conversion of solar to chemical energy. *Nat. Mater.* **10**, 911–921 (2011).
5. Christopher, P., Xin, H. L. & Linic, S. Visible-light-enhanced catalytic oxidation reactions on plasmonic silver nanostructures. *Nat. Chem.* **3**, 467–472 (2011).
6. Lee, J. *et al.* Plasmonic photoanodes for solar water splitting with visible light. *Nano Lett.* **12**, 5014–5019 (2012).
7. Manjavacas, A., Liu, J. G., Kulkarni, V. & Nordlander, P. Plasmon-induced hot carriers in metallic nanoparticles. *ACS Nano* **8**, 7630–7638 (2014).
8. Mukherjee, S. *et al.* Hot electrons do the impossible: plasmon-induced dissociation of H$_2$ on Au. *Nano Lett.* **13**, 240–247 (2013).
9. Ingram, D. B. & Linic, S. Water splitting on composite plasmonic-metal/semiconductor photoelectrodes: evidence for selective plasmon-induced formation of charge carriers near the semiconductor surface. *J. Am. Chem. Soc.* **133**, 5202–5205 (2011).
10. Seh, Z. W. *et al.* Janus Au-TiO$_2$ photocatalysts with strong localization of plasmonic near-fields for efficient visible-light hydrogen generation. *Adv. Mater.* **24**, 2310–2314 (2012).
11. Giugni, A. *et al.* Hot-electron nanoscopy using adiabatic compression of surface plasmons. *Nat. Nanotechnol.* **8**, 845–852 (2013).
12. Haynes, C. L., McFarland, A. D. & Van Duyne, R. P. Surface-enhanced Raman spectroscopy. *Anal. Chem.* **77**, 338A (2005).
13. Lee, H. M., Jin, S. M., Kim, H. M. & Suh, Y. D. Single-molecule surface-enhanced Raman spectroscopy: a perspective on the current status. *Phys. Chem. Chem. Phys.* **15**, 5276–5287 (2013).
14. Nie, S. M. & Emory, S. R. Probing single molecules and single nanoparticles by surface-enhanced Raman scattering. *Science* **275**, 1102–1106 (1997).
15. Le Ru, E. C. & Etchegoin, P. G. in *Principles of Surface-Enhanced Raman Spectroscopy* (Elsevier, 2009).
16. Schlücker, S. Surface-enhanced Raman spectroscopy: concepts and chemical applications. *Angew. Chem. Int. Ed.* **53**, 4756–4795 (2014).
17. van Schrojenstein Lantman, E. M. *et al.* Catalytic processes monitored at the nanoscale with tip-enhanced Raman spectroscopy. *Nat. Nanotechnol.* **7**, 583–586 (2012).
18. Sun, M. T. & Xu, H. X. A novel application of plasmonics: plasmon-driven surface-catalyzed reactions. *Small* **8**, 2777–2786 (2012).
19. Xie, W., Herrmann, C., Kömpe, K., Haase, M. & Schlücker, S. Synthesis of bifunctional Au/Pt/Au core/shell nanoraspberries for in situ SERS monitoring of platinum-catalyzed reactions. *J. Am. Chem. Soc.* **133**, 19302–19305 (2011).
20. Huang, J. F. *et al.* Site-specific growth of Au-Pd alloy horns on Au nanorods: a platform for highly sensitive monitoring of catalytic reactions by surface-enhanced Raman spectroscopy. *J. Am. Chem. Soc.* **135**, 8552–8561 (2013).
21. Xie, W., Walkenfort, B. & Schlücker, S. Label-free SERS monitoring of chemical reactions catalyzed by small gold nanoparticles using 3D plasmonic superstructures. *J. Am. Chem. Soc.* **135**, 1657–1660 (2013).
22. Joseph, V. *et al.* Characterizing the kinetics of nanoparticle-catalyzed reactions by surface-enhanced raman scattering. *Angew. Chem. Int. Ed.* **51**, 7592–7596 (2012).
23. Zhang, Q. F., Blom, D. A. & Wang, H. Nanoporosity-enhanced catalysis on subwavelength Au nanoparticles: a plasmon-enhanced spectroscopic study. *Chem. Mater.* **26**, 5131–5142 (2014).
24. Li, J. F. *et al.* Shell-isolated nanoparticle-enhanced Raman spectroscopy. *Nature* **464**, 392–395 (2010).
25. Mohri, N., Inoue, M., Arai, Y. & Yoshikawa, K. Kinetic study on monolayer formation with 4-aminobenzenethiol on a gold surface. *Langmuir* **11**, 1612–1616 (1995).
26. Greenwood, N. N. & Earnshaw, A. in *Chemistry of the Elements* (Butterworth-Heinemann, 1997).
27. Steinigeweg, D. & Schlücker, S. Monodispersity and size control in the synthesis of 20–100 nm quasi-spherical silver nanoparticles by citrate and ascorbic acid reduction in glycerol–water mixtures. *Chem. Commun.* **48**, 8682–8684 (2012).

Acknowledgements

This work was supported by the German Research Foundation (Grand No. XI 123/1-1). We thank B. Walkenfort for FEM simulations and single-particle scattering measurements, and M. König for preparing framed Si wafers. We also thank J.E. Wissler and J.H. Yoon for discussions and suggestions.

Author contributions

W.X. and S.S. designed the experiments. W.X. performed the experiments. W.X. and S.S. discussed the results and wrote the manuscript. S.S. supervised the project.

Additional information

23

A wearable chemical–electrophysiological hybrid biosensing system for real-time health and fitness monitoring

Somayeh Imani[1,*], Amay J. Bandodkar[2,*], A.M. Vinu Mohan[2], Rajan Kumar[2], Shengfei Yu[1], Joseph Wang[2] & Patrick P. Mercier[1]

Flexible, wearable sensing devices can yield important information about the underlying physiology of a human subject for applications in real-time health and fitness monitoring. Despite significant progress in the fabrication of flexible biosensors that naturally comply with the epidermis, most designs measure only a small number of physical or electrophysiological parameters, and neglect the rich chemical information available from biomarkers. Here, we introduce a skin-worn wearable hybrid sensing system that offers simultaneous real-time monitoring of a biochemical (lactate) and an electrophysiological signal (electrocardiogram), for more comprehensive fitness monitoring than from physical or electrophysiological sensors alone. The two sensing modalities, comprising a three-electrode amperometric lactate biosensor and a bipolar electrocardiogram sensor, are co-fabricated on a flexible substrate and mounted on the skin. Human experiments reveal that physiochemistry and electrophysiology can be measured simultaneously with negligible cross-talk, enabling a new class of hybrid sensing devices.

[1] Department of Electrical and Computer Engineering, University of California, San Diego, La Jolla, California 92093-0407, USA. [2] Department of Nanoengineering, University of California, San Diego, La Jolla, California 92093-0448, USA. * These authors contributed equally to this work. Correspondence and requests for materials should be addressed to J.W. (email: josephwang@ucsd.edu) or to P.P.M. (email: pmercier@ucsd.edu).

Wearable sensors present an exciting opportunity to measure human physiology in a continuous, real-time and non-invasive manner[1,2]. Recent advances in hybrid fabrication techniques have enabled the design of wearable sensing devices in thin, conformal form factors that naturally comply with the smooth curvilinear geometry of human skin, thereby enabling intimate contact necessary for robust physiological measurements[1,3,4]. Development of such epidermal electronic sensors has enabled devices that can monitor respiration rate[5–7], heart rate[8,9], electrocardiograms[4,10–12], blood oxygenation[13], skin temperature[14,15], bodily motion[16–20], brain activity[21–23] and blood pressure[24,25]. To date, most systems have targeted only a single measurement at a time, and most such sensors measure only physical and electrophysiological parameters, significantly limiting monitoring and diagnostic opportunities. For example, the human body undergoes complex physiological changes during physical activities such as exercise[26,27], and monitoring the physiologic effect of physical activity can be important for a wide variety of subjects ranging from athletes to the elderly[28–30]. However, current wearable devices that only measure heart rate, motion and electrocardiogram provide an incomplete picture of the complex physiological changes taking place. As a result, further progress in the area of wearable sensors must include new, relevant sensing modalities, and must integrate these different modalities into a single platform for continuous, simultaneous sensing of multiple parameters relevant to a wide range of conditions, diseases, health and performance states.

Inclusion of chemical measurements can provide extremely useful insights not available from physical or electrophysiological sensors[31]. Chemical information can be conventionally acquired via clinical labs or point-of-care devices[32–34]; unfortunately, such approaches do not support continuous, real-time measurements, therefore limiting their utility to applications where stationary, infrequent tests are sufficient. While recent work, including our own, has demonstrated that chemicals such as electrolytes and metabolites can be measured continuously using epidermal electronics on the skin[35–38], or through non-invasive monitoring of other body fluids[38–40], these devices measure only a single parameter at once, and are not integrated with other sensing modalities. Recently, Gao et al.[41] demonstrated a wearable patch that can simultaneously track levels of metabolites and electrolytes in human sweat. However, electrophysiology sensors were not included, and such multimodal sensor fusion is crucial to obtain a more comprehensive knowledge about a wearer's well-being.

Here, we introduce a wearable device that can simultaneously measure chemical and electrophysiological parameters in the form factor of a single epidermal patch. The hybrid wearable, termed here as a Chem–Phys patch, comprises a screen-printed three-electrode amperometric lactate biosensor and two electrocardiogram electrodes, enabling concurrent real-time measurements of lactate and electrocardiogram. When used in physical-exertion monitoring, electrocardiogram measurements can help monitor heart health and function, while sweat lactate can be used to track an individual's performance and exertion level, and is also an important biomarker for tissue oxygenation and pressure ischaemia[42–47]. Although prior work has demonstrated separate wearable electrocardiogram and lactate sensors, these devices were fabricated on separate platforms and thus mandate applying multiple patches on the human body, which is inconvenient and can deter long-term use. By combining a lactate biosensor and an electrocardiogram sensor, the new Chem–Phys hybrid wearable patch represents a powerful platform capable of simultaneously tracking both physicochemical and electrophysiological attributes, thus providing a more comprehensive view of a person's health status than current wearable fitness monitors.

The Chem–Phys hybrid patch was fabricated by leveraging screen-printing technology on a thin, highly flexible polyester sheet that conforms well with the complex three-dimensional (3D) morphology of human skin to provide a low-noise signal. The working electrode of the lactate biosensor was functionalized and coated with a biocompatible biocatalytic layer (lactate oxidase (LOx)-modified prussian blue). The three amperometric electrodes were separated from the Ag/AgCl electrocardiogram electrodes via a printed hydrophobic layer to maximize sensor stability and signal-to-noise ratio even in the presence of significant perspiration. The dimensions of the electrodes and the inter-electrode distances have been optimized based on the human trials to acquire a clean electrocardiogram signal and lactate response with minimal interference between the two sensors. The two sensors were interfaced to a custom-printed circuit board (PCB) featuring a potentiostat, an electrocardiogram analogue front-end (AFE), and a bluetooth low-energy (BLE) radio for wireless telemetry of the results to a mobile platform, such as a smartphone or laptop. The hybrid sensing system was tested on three human subjects during exercise on a stationary bicycle, showing that lactate and electrocardiogram can be measured simultaneously with negligible co-interference. Electrocardiogram data was found to be similar to the data collected from standard electrode types, and extracted heart rate correlated well to commercial heart rate detectors. A control experiment, where an enzyme-free amperometric sensor was applied to a perspiring human subject, corroborated the lactate sensor's sensitivity and selectivity towards on-body detection of physiologic lactate levels. The promising data obtained in this work thus supports the possibility of developing more advanced hybrid wearable sensors that involve complex integration of several physical and chemical sensors on the same platform for monitoring many relevant modalities.

Results

Hybrid patch design rationale and fabrication. The Chem–Phys hybrid multi-sensor system must be compact and easy to wear in a location that offers adequate access to both electrocardiogram signals and perspiration for lactate measurements. The design must also minimize sensor-body motion, must minimize co-interference between the sensing modalities and be low-cost. These requirements motivate a flexible epidermal electronic design that can be worn on the chest and fabricated using screen-printing technology.

The design of the sensing system is shown in Fig. 1. The biosensors were fabricated via conventional low-cost screen-printing technique (conceptually illustrated in Fig. 1a) utilizing custom-designed stencils (photograph in Fig. 1b). The biosensing patches were printed onto a highly flexible, thin polyester sheet (50 μm thickness) for realizing highly conformable sensor patch that adheres well to the human skin without causing any discomfort. An array of fabricated sensors is shown in Fig. 1c. The total patch size was dictated by the bipolar electrocardiogram electrodes, which must be separated by a minimum distance to attain a high-quality signal[48]. Typically, single-lead monitoring systems, such as the present design, are used for basic heart monitoring, arrhythmias diagnosis or studying the effect of exercise on the heart, and are placed in the vicinity of the conventional V1–V6 chest-lead locations. Electrode size, separation and placement parameters were determined through a series of experiments involving the placement of Ag/AgCl-based electrocardiogram electrodes of various sizes (1×1, 1.5×1.5, 1.5×1 and $2 \times 2\,cm^2$) and separation distances (1–6 cm) on

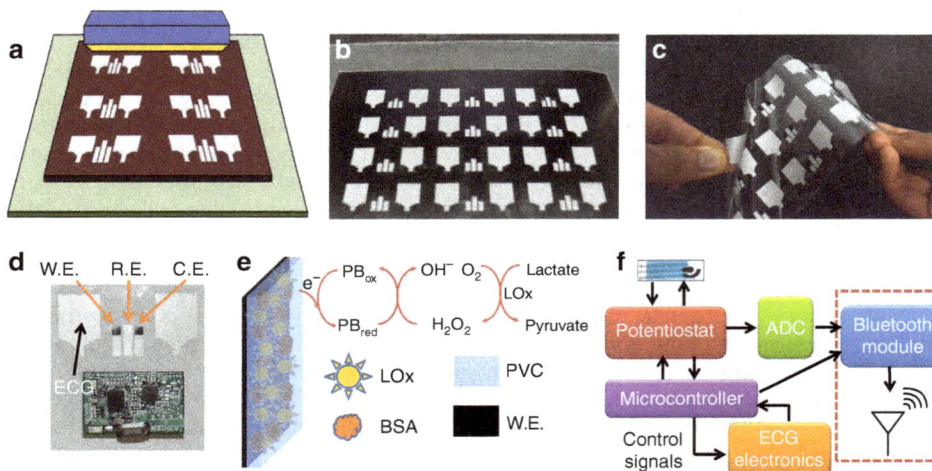

Figure 1 | Fabrication and function of the Chem–Phys hybrid sensor patch. (**a**) Schematic showing the screen-printing process. (**b**) Image of the Chem–Phys printing stencil. (**c**) An array of printed Chem–Phys flexible patches. (**d**) Image of a Chem–Phys patch along with the wireless electronics. (**e**) Schematic showing the LOx-based lactate biosensor along with the enzymatic and detection reactions. (**f**) Block diagram of the wireless readout circuit.

subjects with different chest sizes, and observing the resulting electrocardiogram waveforms. The study revealed that a compact patch that provides favourable electrocardiogram signal could be realized by placing $1.5 \times 1.5\,cm^2$ electrocardiogram electrodes across the V1 and V2 lead sites with an inter-electrode distance of 4 cm, thereby measuring from the vantage point of the septal surface of the heart as suitable for diagnostics of arrhythmias and the effects of exercise on the heart. This sets an upper-end size of the patch to be $7 \times 2\,cm^2$. The chest region is not only convenient for measurement of electrocardiogram, but also has a high-sweat rate during physical excursion[49,50], and can thus serve as an appropriate location to also measure lactate levels in human perspiration. In addition, the epidermis and muscle tissues over these locations do not experience complex 3D strains and remain fairly stable even during intense physical activities, making measurements here especially convenient. Since the performance of amperometric lactate electrodes is not compromised by reducing their dimension, they were fabricated between the two electrocardiogram electrodes, as shown in Fig. 1d. Each of the three electrodes have an active area of $3 \times 2.5\,mm$. The working electrodes were printed using prussian blue ink due to the high selectivity of prussian blue towards hydrogen peroxide, a byproduct of the enzymatic oxidation of lactate[38,51] (Fig. 1e). The reference electrode was printed using Ag/AgCl. Since sweat can provide an alternate electrically conductive pathway between the electrocardiogram electrodes and also between the electrocardiogram and amperometric electrodes, thus leading to a potential distortion of the recorded electrocardiogram signal, a printed hydrophobic layer of Ecoflex was used to separate the amperometric biosensor from the electrocardiogram electrodes. This effectively increases the impedance between the electrocardiogram and amperometric electrodes via sweat, thus minimizing the cross-talk between the two sensors. The entire Chem–Phys patch is highly flexible and can be smoothly mated on curved surfaces. Such flexibility is crucial for achieving unobtrusive wearable devices that cause no hindrance or irritation to the wearer.

The Chem–Phys patch was interfaced to a custom-PCB featuring a potentiostat and analogue-to-digital converter for amperometric data acquisition, an AFE for electrocardiogram data acquisition, and a BLE chip for wireless transmission (Fig. 1d,f).

In-vitro characterization of the lactate biosensor. Lactate concentration in human sweat depends on a person's metabolism and level of exertion, and typically ranges from 0 to 25 mM (ref. 52). A wide linear-detection range, coupled with a fast response time is thus essential for continuous epidermal monitoring of lactate. The operating potential of $-0.1\,V$ (versus to Ag/AgCl) was selected based on the onset potential for electro-oxidation of lactate by the fabricated biosensor, obtained during cyclic voltammetry studies. When the biosensor comes in contact with lactate, the immobilized LOx enzyme catalyses the oxidation of lactate to generate pyruvate and H_2O_2. The prussian blue transducer then selectively reduces the H_2O_2 to generate electrons to quantify the lactate concentration (Fig. 1e). Figure 2a shows the amperometric response of the lactate biosensor to increasing lactate concentrations in the physiological range of 0–28 mM. It is evidenced from this figure that the biosensor responds linearly to the lactate concentrations in this range with a sensitivity of 96 nA/mM.

On-body characterization of the electrocardiogram electrodes. The ability of the printed electrocardiogram electrodes to record electrocardiogram signals was validated by comparing recordings from the fabricated electrodes with commercially-available 3M Red Dot electrocardiogram electrodes. As illustrated in Fig. 2b, on-body electrocardiogram signals recorded for the same subject using commercial and fabricated electrodes at the same location have similar morphologies when acquired using the same AFE circuitry. All on-body experiments were performed in strict compliance with the guidelines of Institutional Review Boards (IRB) and were approved by the Human Research Protections Program at University of California, San Diego (Project name: Epidermal Electrochemical Sensors and Biosensors. Project number: 130,003).

Epidermal evaluation of the Chem–Phys patch. The Chem–Phys hybrid patch (Fig. 3a) was fabricated and applied to three healthy male subjects on the fourth intercostal space of the chest (Fig. 3b). Dynamic changes in sweat-lactate levels and electrocardiogram signals were measured continuously during a bout of intense cycling. To ensure that the anaerobic metabolism was invoked, subjects were asked to mount a stationary cycle and

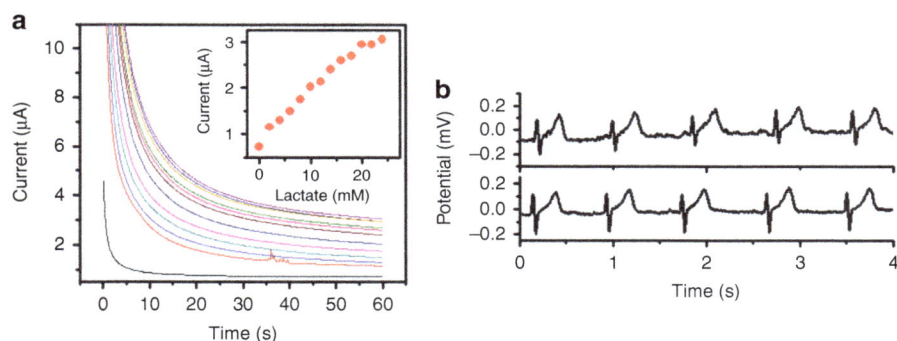

Figure 2 | *In-vitro* characterization of Chem–Phys hybrid patch. (**a**) Amperometric response to increasing lactate concentration from 0 to 28 with 2 mM additions in phosphate buffer (pH 7.0). Applied voltage $= -0.1$ V versus Ag/AgCl. (**b**) Electrocardiogram signals using 3M Red Dot electrodes (top), and printed electrocardiogram sensor (bottom).

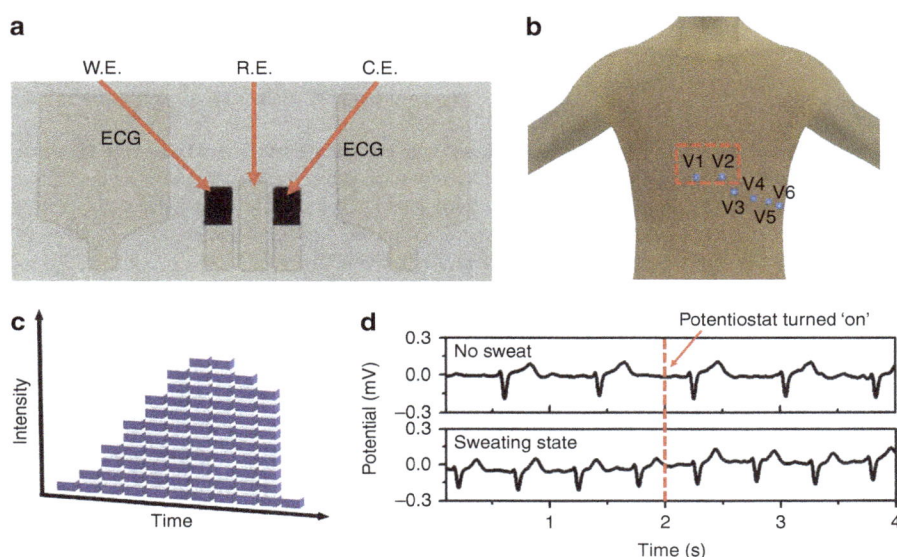

Figure 3 | On-body test configuration. (**a**) A photograph of Chem–Phys hybrid patch. (**b**) Location of the Chem–Phys patch for mounting on the human body—the fourth intercostal space of the chest. (**c**) Cycling resistance profile for on-body tests. (**d**) Effect of amperometric measurement on the electrocardiogram signal before cycling (no sweat state) and during cycling (sweating state).

maintain a steady cycling cadence while the cycling resistance increased periodically as illustrated in Fig. 3c.

Since electrocardiogram measurements were made via bipolar high-impedance electrodes, and lactate measurements were made by applying a constant potential via a low-impedance potentiostat output and measuring current, there is a possibility that a change in the applied potentiostat voltage (for example, during start-up) could interfere with electrocardiogram measurements during the settling time of the potentiostat. At the same time, sweat consists of many ions and could thus act as an electrically conductive medium that can shunt the lactate and electrocardiogram sensors, or the two electrocardiogram electrodes together. Co-sensor interference and shunting effects were mitigated by geometrically separating the lactate and electrocardiogram electrodes and printing two vertically oriented hydrophobic layers were next to the lactate biosensor, thereby facilitating flux of new perspiration across the biosensor itself, while minimizing shunting between the lactate and electrocardiogram sensors. To validate performance under concurrent hybrid sensing scenarios, the Chem–Phys sensor was mounted on a human subject and set to continuously record electrocardiogram before, during, and immediately after turning on the -0.1 V potentiostat output. Experimental results, obtained via a wireless Bluetooth link as

shown in Fig. 3d, reveal that the potentiostat has a negligible effect on the morphology of the electrocardiogram signals, irrespective of whether the subject was in a resting or cycling state.

To validate performance under realistic conditions, the Chem–Phys patch was tested on three subjects during 15–30 min of intense cycling activity; continuous time-series results during each experiment are shown in Fig. 4. At the commencement of the cycling activity, each subject's heart rate, extracted from electrocardiogram data, was within the normal resting range of 60 to 120 beats per minute (b.p.m.)[53]. At the same time, a negligible current response was measured by the lactate biosensor due to the lack of perspiration. With time, the resistance for cycling was increased, causing the subjects to exert increasing levels of effort to maintain constant cycling speed. This resulted in increasing heart rate and generation of sweat. At the onset of perspiration, lactate is released from the epidermis, and is selectively detected by the LOx-based biosensor. As the resistance increases, the sweat-lactate concentration too increases, as illustrated in Fig. 4a–c, showing a correlation between physical exertion, heart rate and, after a physiologic time delay, lactate generation. As the cycling continued, the sweat rate for each subject increased, leading to the well-documented phenomenon of dilution factor

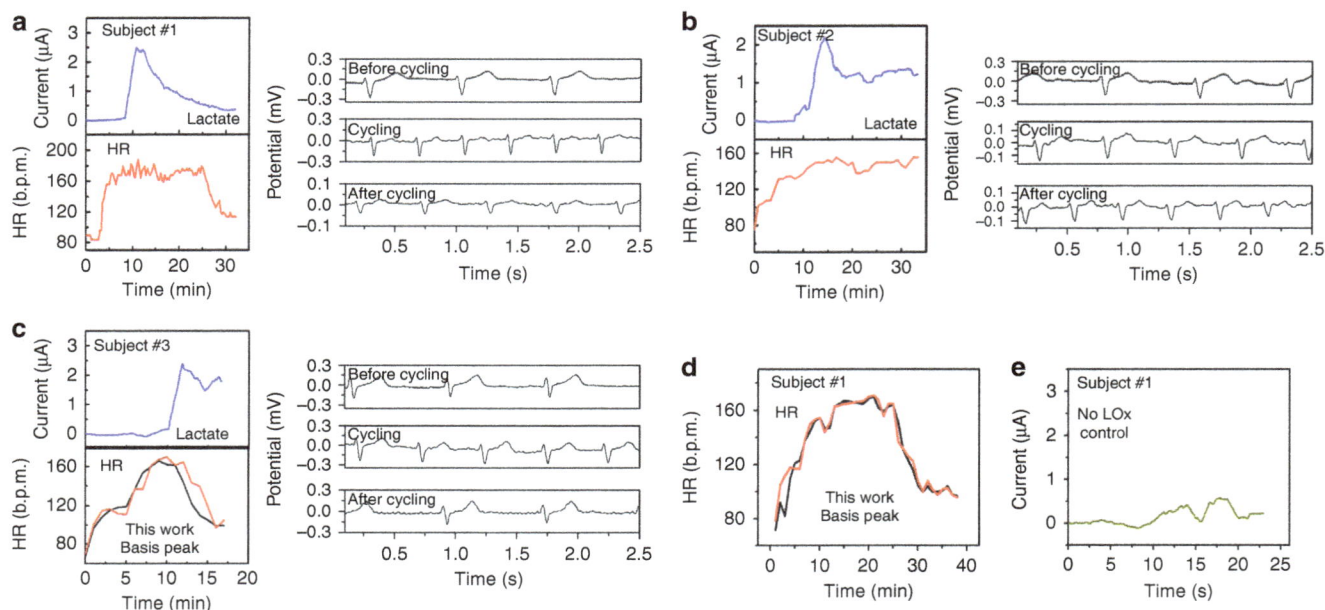

Figure 4 | Real-time on-body evaluation of the Chem–Phys hybrid patch showing the lactate levels and heart rate for three human subjects. (a–c) The corresponding blue plots represent the real-time lactate concentration profiles for each subject, while, the red plots depict the heart rate data obtained by the electrocardiogram electrodes of the Chem–Phys patch. The black plots correspond to the heart rate data recorded by the Basis Peak heart rate monitor. Typical real-time electrocardiogram data obtained before, during and after the cycling bout for each subject is also shown. **(d)** Additional heart rate data from subject #1. **(e)** Response of the control amperometric sensor (without LOx enzyme) for subject #1.

that causes decrease in the lactate concentration[42]. The final stage of the cycling bout involved a 3 min cool-down period. During this phase, as expected, the heart rate normalized back near to the normal resting heart rate. At the same time, the lactate concentration measured by the lactate biosensor continued to decrease.

The lactate biosensor data for each subject resembles the expected sweat-lactate profile for increasing intensity workouts[37]. To validate that lactate, not other sweat constituents, was specifically measured, a control experiment in which an unmodified (LOx-free) amperometric biosensor was used under the same experimental conditions as above to subject #1. As shown in Fig. 4e, the control biosensor leads to a negligible current response without the presence of LOx, confirming the high selectivity of the lactate biosensor. To validate electrocardiogram data over long time series, even under the presence of experimentally-induced motion, heart rate as extracted from the electrocardiogram data is benchmarked against a commercial wristband heat rate monitor (BASIS) for subjects 1 and 3. Extracted heart rate data matched the wrist-worn device with a Pearson's correlation coefficient of $r = 0.975$. These on-body studies illustrate that the hybrid patch could monitor sweat lactate and electrocardiogram in a continuous and simultaneous manner, and that the hydrophobic barrier between the sensors assisted in minimizing potential cross-talk between the two sensing modalities. The data also demonstrates that such a barrier had minimal effect on the supply of oxygen to the enzyme electrode required for biocatalytic detection of lactate.

Discussion

The Chem–Phys sensor patch described in this study represents a hybrid system that fuses the monitoring of electrophysiology with on-body chemical sensing into single fully printable wearable platform. On-body epidermal testing in a realistic fitness environment revealed that electrocardiogram sensing is in-line

with existing wearable devices, and is not adversely affected by simultaneous measurement of lactate via constant-potential amperometry. The lactate control study using an enzyme-free amperometric sensor and correlation of the heart rate data of the hybrid patch to that recorded by a commercial heart rate monitor underscore the promise of the Chem–Phys patch to simultaneously monitor electrocardiogram signals and sweat-lactate levels for tracking the wearer's physicochemical and electrophysiological status. This device represents an important first step in the research and development of multimodal wearable sensors that fuse chemical, electrophysiological and physical sensors for more comprehensive monitoring of human physiology.

Methods

Reagents and materials. Chitosan, acetic acid, polyvinyl chloride, tetrahydrofuran, bovine serum albumin, L-lactic acid, sodium phosphate monobasic and sodium phosphate dibasic were obtained from Sigma-Aldrich (St Louis, MO). L-LOx (activity, 101 U mg^{-1}) was procured from Toyobo Corp. (Osaka, Japan). All reagents were used without further purification. Prussian blue conductive carbon (C2070424P2), Ag/AgCl (E2414) and insulator (Dupont 5036) inks were procured from Gwent Group (Pontypool, UK), Ercon Inc. (Wareham, MA) and Dupont (Wilmington, DE). Electrocardiogram hydrogel conductive adhesive (RG63B, 35-mil thick) was purchased from Covidien. Polyester sheets (MELINEX 453, 50-μm thick) were provided by Tekra Inc. (New Berlin, WI).

Instrumentation. The Chem–Phys patch was printed by using an MPM-SPM semiautomatic screen printer (Speedline Technologies, Franklin, MA). Sensor patterns were designed in AutoCAD (Autodesk, San Rafael, CA) and outsourced for fabrication on stainless steel through-hole 12 × 12 in framed stencils (Metal Etch Services, San Marcos, CA). Electrochemical characterization was performed at room temperature using a CH Instruments electrochemical analyser (model 630C, Austin, TX). A CONTEC MS400 Multi-parameter Patient Simulator, electrocardiogram simulator has been utilized for testing of electrocardiogram instrumentation circuits. 3M Red Dot multi-purpose monitoring electrodes are used for the verification of collected signal using the fabricated electrocardiogram sensors.

Fabrication of Chem–Phys hybrid device. The Chem–Phys hybrid patch was fabricated via screen-printing technology, while the wearable electronic board was realized by relying on standard 4-layer PCB fabrication and assembly protocols.

Printing and functionalization of Chem–Phys hybrid patch. The Chem–Phys patch was fabricated in-house by printing a sequence of Ag/AgCl, Prussian blue and insulator inks were patterned on the highly flexible transparent polyester substrate by using the custom-designed stencils and screen printer. The Ag/AgCl and insulator ink was cured at 90 °C for 10 min, while the Prussian blue ink was cured at 80 °C for 10 min in a convection oven.

On printing of the hybrid patch, the working electrode of the amperometric sensor was functionalized with LOx enzyme. The LOx solution (40 mg ml^{-1} containing 10 mg ml^{-1} bovine serum albumin stabilizer) was mixed with a chitosan solution (0.5 wt% in 1 M acetic acid) in a 1:1 v/v ratio. Subsequently, an 8-μl droplet of the above solution was casted on the electrode and dried under ambient conditions. Thereafter, 4 μl of polyvinyl chloride solution (3 wt% in tetrahydrofuran) was drop casted and allowed to dry under ambient conditions for at least 3 h before use. The electrocardiogram electrodes were covered with conductive hydrogel adhesives. The patch was then affixed to a medical-grade adhesive sheet required for applying to human skin. The patch was stored at 4 °C when not in use.

PCB fabrication. The 4-layer Bluetooth-enabled PCB used a Texas Instrument (TI) CC2541 BLE System-on-Chip for communication and processing. An ADS1293 AFE chip was used for biopotential measurements to record the electrocardiogram (electrocardiogram) signals from the fabricated electro-cardiogram electrodes. An LMP91000 AFE, programmable through an I2C interface driven by the CC2541, was used as the on-board potentiostat for lactate concentration determination. The data from each sensor was collected by the CC2541 and transmitted to a Bluetooth 4.0-enabled receiver. A graphical interface was developed using Python to demonstrate measurement results on a PC. A Johanson Technology 2.45 GHz chip antenna (2450AT42A100) and impedance-matched balun (2450BM15A0002) were used for wireless transmission. A CR2032 button cell lithium battery (3 V, 220 mAh) was utilized as a power source, regulated for the electronics via a TPS61220 boost converter. In the 'active mode', the board consumed, on average, 5 mA from a 3 V supply (15 mW).

***In-vitro* studies.** *Characterization of amperometric lactate sensor.* These studies were performed using a 0.1 M phosphate-buffered solution (pH 7.0). The operating potential for the lactate sensor was selected by using cyclic voltammetry. The amperometric response was recorded after 1 min incubation in the sample solution, using a potential step to −0.1 V (versus Ag/AgCl) for 60 s.

Characterization of the electrocardiogram sensor. Electrocardiogram monitoring has been performed using both commercial 3M Red Dot Multi-Purpose monitoring electrodes, as well as fabricated electrocardiogram electrodes to verify the functionality of the printed Ag/AgCl electrocardiogram sensors.

On-body characterization of Chem–Phys patch. All experiments were performed in strict compliance with the guidelines of IRB and were approved by Human Research Protections Program at University of California, San Diego (Project name: Epidermal Electrochemical Sensors and Biosensors. Project number: 130,003). The study was deemed by the IRB as posing 'no greater than minimal risk' to the prescreened subjects who were recruited for the investigation. A total of 3 healthy male volunteers (recruited in response to follow-up from flyers) with no prior medical history of heart conditions, diabetes or chronic skeletomuscular pain were recruited for participation in the study, and informed, signed consent was obtained from each individual following a rigorous prescreening procedure. A typical study comprised of applying the Chem–Phys hybrid patch on fourth intercostal space of a subject's chest to record the electrocardiogram signal between V1 and V2 positions.

Subjects were then asked to mount a stationary cycle and begin cycling at a steady, comfortable cadence. Subjects were instructed to maintain their cadence while an increasing resistance was applied at 3 min intervals. The absolute resistance level and duration was selected according to subject's fitness level while the same intensity profile was used throughout the human studies. This ensured that the anaerobic metabolism was invoked at similar time scales, hence augmenting the excretion of lactate in the perspiration in a controlled manner. Following the intense fitness bout, the volunteers were asked to gradually reduce their cadence during a 3 min 'cool-down' period whereby the resistance was reduced from maximal levels.

Characterization of instrumentation circuits. The printed circuit board was assembled and tested *in-vitro* to validate both functionality and performance. The potentiostat circuit was verified together with the lactate biosensor through an *in-vitro* amperometric experiment. The electrocardiogram AFE was characterized using a CONTEC MS400 Multi-parameter Patient Simulator (electrocardiogram simulator). The output signal of the electrocardiogram simulator was read using ADS1293 AFE chip, and transferred through BLE link to a BLE-enabled device.

References

1. Kim, D.-H., Ghaffari, R., Lu, N. & Rogers, J. A. Flexible and stretchable electronics for biointegrated devices. *Annu. Rev. Biomed. Eng.* **14**, 113–128 (2012).
2. Bandodkar, A. J. & Wang, J. Non-invasive wearable electrochemical sensors: a review. *Trends Biotechnol.* **32**, 363–371 (2014).
3. Chuang, M. C. *et al.* Flexible thick-film glucose biosensor: influence of mechanical bending on the performance. *Talanta* **81**, 15–19 (2010).
4. Kim, D.-H. *et al.* Epidermal electronics. *Science* **333**, 838–844 (2011).
5. Merritt, C. R., Nagle, H. T. & Grant, E. Textile-based capacitive sensors for respiration monitoring. *IEEE Sens. J.* **9**, 71–78 (2009).
6. Rovira, C. *et al.* Integration of textile-based sensors and Shimmer for breathing rate and volume measurement. *5th Int. Conf. Pervasive Comput. Technol. Healthc. Work* 238–241 (2011).
7. Jeong, J. W., Jang, Y. W., Lee, I., Shin, S. & Kim, S. Wearable respiratory rate monitoring using Piezo-resistive fabric sensor. *World Congr. Med. Phys. Biomed. Eng.* **25**, 282–284 (2009).
8. Chiarugi, F. *et al.* Measurement of heart rate and respiratory rate using a textile-based wearable device in heart failure patients. *Comput. Cardiol.* **35**, 901–904 (2008).
9. Di Rienzo, M. *et al.* Textile technology for the vital signs monitoring in telemedicine and extreme environments. *IEEE Trans. Inf. Technol. Biomed.* **14**, 711–717 (2010).
10. Lee, Y.-D. & Chung, W.-Y. Wireless sensor network based wearable smart shirt for ubiquitous health and activity monitoring. *Sensor Actuat B Chem.* **140**, 390–395 (2009).
11. Lee, S. M. *et al.* Self-adhesive epidermal carbon nanotube electronics for tether-free long-term continuous recording of biosignals. *Sci. Rep.* **4**, 6074 (2014).
12. Pandian, P. S. *et al.* Smart vest: wearable multi-parameter remote physiological monitoring system. *Med. Eng. Phys.* **30**, 466–477 (2008).
13. Rothmaier, M., Selm, B., Spichtig, S., Haensse, D. & Wolf, M. Photonic textiles for pulse oximetry. *Opt. Express* **16**, 12973–12986 (2008).
14. Jung, S., Ji, T. & Varadan, V. K. Point-of-care temperature and respiration monitoring sensors for smart fabric applications. *Smart Mater. Struct.* **15**, 1872–1876 (2006).
15. Bian, Z. G. *et al.* Thermal analysis of ultrathin, compliant sensors for characterization of the human skin. *RSC Adv.* **4**, 5694–5697 (2014).
16. Lorussi, F., Scilingo, E. P., Tesconi, M., Tognetti, A. & De Rossi, D. Strain sensing fabric for hand posture and gesture monitoring. *IEEE Trans. Inf. Technol. Biomed.* **9**, 372–381 (2005).
17. Hasegawa, Y., Shikida, M., Ogura, D., Suzuki, Y. & Sato, K. Fabrication of a wearable fabric tactile sensor produced by artificial hollow fiber. *J. Micromech. Microeng.* **18** (2008).
18. Lorussi, F., Rocchia, W., Scilingo, E. P., Tognetti, A. & De Rossi, D. Wearable, redundant fabric-based sensor arrays for reconstruction of body segment posture. *IEEE Sens. J.* **4**, 807–818 (2004).
19. Mattmann, C., Clemens, F. & Tröster, G. Sensor for measuring strain in textile. *Sensors* **8**, 3719–3732 (2008).
20. Son, D. *et al.* Multifunctional wearable devices for diagnosis and therapy of movement disorders. *Nat. Nanotechnol.* **9**, 397–404 (2014).
21. Gargiulo, G. *et al.* A mobile EEG system with dry electrodes. in *IEEE Biomed. Circuits Syst. Conf.* 273–276 (2008).
22. Chi, Y. M., Deiss, S. R. & Cauwenberghs, G. Non-contact low power EEG/ECG electrode for high density wearable biopotential sensor networks. *Int. Work. Wearable Implant. Body Sens. Networks* 246–250 (2009).
23. Löfhede, J., Seoane, F. & Thordstein, M. Textile electrodes for EEG recording— a pilot study. *Sensors* **12**, 16907–16919 (2012).
24. Cong, P., Ko, W. H. & Young, D. J. Wireless batteryless implantable blood pressure monitoring microsystem for small laboratory animals. *IEEE Sens. J.* **10**, 243–254 (2010).
25. Muehlsteff, J., Aubert, X. A. & Morren, G. Continuous cuff-less blood pressure monitoring based on the pulse arrival time approach: the impact of posture. *Conf. Proc. IEEE Eng. Med. Biol. Soc.* **2008**, 1691–1694 (2008).
26. Wasserman, K., van Kessel, A. & Burton, G. Interaction of physiological mechanisms during exercise. *J. Appl. Physiol.* **22**, 71–85 (1967).
27. Rittweger, J., Beller, G. & Felsenberg, D. Acute physiological effects of exhaustive whole-body vibration exercise in man. *Clin. Physiol.* **20**, 134–142 (2000).
28. Pollock, M. L. *et al.* Resistance exercise in individuals with and without cardiovascular disease. *Circulation* **101**, 828–833 (2000).
29. Tatterson, A. J., Hahn, A. G., Martini, D. T. & Febbraio, M. A. Effects of heat stress on physiological responses and exercise performance in elite cyclists. *J. Sci. Med. Sport* **3**, 186–193 (2000).
30. Yarasheski, K. E., Zachwieja, J. J. & Bier, D. M. Acute effects of resistance exercise on muscle protein synthesis rate in young and elderly men and women. *Am. J. Physiol.* **265**, E210–E214 (1993).
31. Spichiger-Kelle, U. E. *Chemical Sensors and Biosensors for Medical and Biological Applications* (John Wiley and Sons, 1998).
32. Soper, S. A. *et al.* Point-of-care biosensor systems for cancer diagnostics/prognostics. *Biosens. Bioelectron.* **21**, 1932–1942 (2006).

33. Tudos, A. J., Besselink, G. J. & Schasfoort, R. B. Trends in miniaturized total analysis systems for point-of-care testing in clinical chemistry. *Lab. Chip.* **1**, 83–95 (2001).

34. Wang, J. Amperometric biosensors for clinical and therapeutic drug monitoring: a review. *J. Pharm. Biomed. Anal.* **19**, 47–53 (1999).

35. Windmiller, J. R. & Wang, J. Wearable electrochemical sensors and biosensors: a review. *Electroanalysis* **25**, 29–46 (2013).

36. Bandodkar, A. J. *et al.* Tattoo-based potentiometric ion-selective sensors for epidermal pH monitoring. *Analyst* **138**, 123–128 (2013).

37. Jia, W. *et al.* Electrochemical tattoo biosensors for real-time noninvasive lactate monitoring in human perspiration. *Anal. Chem.* **85**, 6553–6560 (2013).

38. Bandodkar, A. J. *et al.* Tattoo-based noninvasive glucose monitoring: a proof-of-concept study. *Anal. Chem.* **87**, 394–398 (2015).

39. Kim, J. *et al.* Non-invasive mouthguard biosensor for continuous salivary monitoring of metabolites. *Analyst* **139**, 1632–1636 (2014).

40. Kim, J. *et al.* Wearable salivary uric acid mouthguard biosensor with integrated wireless electronics. *Biosens. Bioelectron.* **74**, 1061–1068 (2015).

41. Gao, W. *et al.* Fully integrated wearable sensor arrays for multiplexed *in situ* perspiration analysis. *Nature* **529**, 509–514 (2016).

42. Buono, M. J., Lee, N. V. L. & Miller, P. W. The relationship between exercise intensity and the sweat lactate excretion rate. *J. Physiol. Sci.* **60**, 103–107 (2010).

43. Pilardeau, P. A. *et al.* Effect of different work-loads on sweat production and composition in man. *Sport. Med. Phys. Fit.* **28**, 247–252 (1988).

44. Pillardeau, P., Vaysse, J., Garnier, M., Joublin, M. & Valeri, L. Secretion of eccrine sweat glands during exercise. *Brit. J. Sport. Med.* **13**, 118–121 (1979).

45. Falk, B. *et al.* Sweat lactate in exercising children and adolescents of varying physical maturity. *J. Appl. Physiol.* **71**, 1735–1740 (1991).

46. Biagi, S., Ghimenti, S., Onor, M. & Bramanti, E. Simultaneous determination of lactate and pyruvate in human sweat using reversed-phase high-performance liquid chromatography: a noninvasive approach. *Biomed. Chromatogr.* **26**, 1408–1415 (2012).

47. Polliack, A., Taylor, R. & Bader, D. Sweat analysis following pressure ischaemia in a group of debilitated subjects. *J. Rehabil. Res. Dev.* **34**, 303–308 (1997).

48. Harrigan, R. A., Chan, T. C. & Brady, W. J. Electrocardiographic electrode misplacement, misconnection, and artifact. *J. Emerg. Med.* **43**, 1038–1044 (2012).

49. Havenith, G., Fogarty, A., Bartlett, R., Smith, C. J. & Ventenat, V. Male and female upper body sweat distribution during running measured with technical absorbents. *Eur. J. Appl. Physiol.* **104**, 245–255 (2008).

50. Patterson, M. J., Galloway, S. D. & Nimmo, M. A. Variations in regional sweat composition in normal human males. *Exp. Physiol.* **85**, 869–875 (2000).

51. Karyakin, A. A. Prussian blue and its analogues: electrochemistry and analytical applications. *Electroanalysis* **13**, 813–819 (2001).

52. Green, J. M., Pritchett, R. C., Crews, T. R., McLester, J. R. & Tucker, D. C. Sweat lactate response between males with high and low aerobic fitness. *Eur. J. Appl. Physiol.* **91**, 1–6 (2004).

53. Kostis, J. B. *et al.* The effect of age on heart rate in subjects free of heart disease. *Circulation* **65**, 141–145 (1982).

Acknowledgements

We acknowledge support from the National Institute of Biomedical Imaging and Bioengineering of NIH (R21EB019698), Samsung and the Arnold and Mabel Beckman Foundation. The views and conclusions contained herein are those of the authors and should not be interpreted as necessarily representing the official policies or endorsements of the sponsors. We also thank Jeng-Hau Lin for his help in providing electrocardiogram test equipment.

Author contributions

P.P.M. and J.W. conceived the project. S.I. and A.J.B. performed the experiments. S.I. and S.Y. designed and implemented the electronics. A.J.B. designed and implemented the biosensors. All the authors designed the experiments, and contributed to writing and editing the manuscript.

Additional information

Competing financial interests: The authors declare no competing financial interests.

Lysosome triggered near-infrared fluorescence imaging of cellular trafficking processes in real time

Marco Grossi[1], Marina Morgunova[1], Shane Cheung[1,2], Dimitri Scholz[3], Emer Conroy[3], Marta Terrile[3], Angela Panarella[4], Jeremy C. Simpson[4], William M. Gallagher[3] & Donal F. O'Shea[1,2]

Bioresponsive NIR-fluorophores offer the possibility for continual visualization of dynamic cellular processes with added potential for direct translation to *in vivo* imaging. Here we show the design, synthesis and lysosome-responsive emission properties of a new NIR fluorophore. The NIR fluorescent probe design differs from typical amine functionalized lysosomotropic stains with off/on fluorescence switching controlled by a reversible phenol/phenolate interconversion. Emission from the probe is shown to be highly selective for the lysosomes in co-imaging experiments using a HeLa cell line expressing the lysosomal-associated membrane protein 1 fused to green fluorescent protein. The responsive probe is capable of real-time continuous imaging of fundamental cellular processes such as endocytosis, lysosomal trafficking and efflux in 3D and 4D. The advantage of the NIR emission allows for direct translation to *in vivo* tumour imaging, which is successfully demonstrated using an MDA-MB-231 subcutaneous tumour model. This bioresponsive NIR fluorophore offers significant potential for use in live cellular and *in vivo* imaging, for which currently there is a deficit of suitable molecular fluorescent tools.

[1] Department of Pharmaceutical and Medicinal Chemistry, Royal College of Surgeons in Ireland, 123 St Stephen's Green, Dublin 2, Ireland. [2] School of Chemistry and Chemical Biology, Conway Institute, University College Dublin, Belfield, Dublin 4, Ireland. [3] School of Biomolecular and Biomedical Science, Conway Institute of Biomolecular and Biomedical Research, University College Dublin, Belfield, Dublin 4, Ireland. [4] School of Biology and Environmental Science, Conway Institute of Biomolecular and Biomedical Research, University College Dublin, Belfield, Dublin 4, Ireland. Correspondence and requests for materials should be addressed to D.F.O. (email: donalfoshea@rcsi.ie).

E hrlich's use of synthetic dyes as a means of staining biological samples can be viewed as one of the foundation stones of modern scientific research. A century later, the use of fluorescence imaging as a technique to visualize specific regions of live cellular[1–4] or whole organisms[5,6] is often central to research programmes, with clinical applications such as fluorescence-guided surgery now emerging[7–11].

The major shortcomings of fluorescence imaging using molecular fluorophores are interference from nonspecific background fluorescence outside the region of interest (ROI), insufficient photostability and cytotoxicity. Poor ROI selectivity necessitates a time delay to allow background fluorophore clearance and/or a washing procedure between fluorophore administration and image acquisition. This can limit imaging to fixed cells or static snapshots, without the possibility of continuous data acquisition throughout the experiment. An innovative approach to enhance target-to-background signal ratio is to exploit a mechanism of selective fluorescence quenching in the background areas, while establishing the emitting potential of the fluorophore only in the ROI[12,13]. Continuous recording of dynamic cellular events in real time may become feasible if the on/off fluorescence switching is reversible.

Developing a responsive fluorophore suitable for real-time live-cell imaging poses a series of challenges. Stringent criteria are required, such as near-perfect response selectivity, exceptional photostability and low dark and light toxicities. Obtaining selective fluorescence responses for intracellular analytes is not trivial, as analyte selectivity observed in a controlled homogeneous environment of a cuvette does not necessarily translate to far more complex in vitro or in vivo settings. Continuous live-cell imaging places very high demands on photostability of the fluorophore, as the same cell(s) are repeatedly imaged over time. Fluorophore dark toxicity must be low, so that cell viability is not compromised and normal cellular processes are unperturbed. To minimize light-induced toxicity, it is preferable to use low-energy wavelengths in the near-infrared (NIR) spectral region ($\lambda = 700$–900 nm). For in vivo imaging, the use of NIR fluorophores is essential. This spectral region is required for effective light transmission through body tissue, as there are reduced levels of absorption and scattering at these longer wavelengths and less intrinsic autofluorescence. In addition, if on/off NIR fluorescence switching could be accomplished in vivo, then similar imaging advantages could be gained as for in-vitro cell imaging.

Currently, there is a small yet growing selection of NIR fluorophore classes but they often suffer from insufficient photostability and lack emission wavelengths above 700 nm[14]. Our recent research focus led to the development of BF$_2$-chelated azadipyrromethene class 1 (Fig. 1)[15–18]. This class is relatively straightforward to synthesize, amenable to structural elaboration and exhibits excellent photophysical properties. For example, the derivative 1 (R = Ph) has an absorption/emission λ_{max} at 696 and 727 nm in aqueous solutions, high fluorescence quantum yields (0.3–0.4) and excellent photostability[17]. Yet, in spite of recent progress, a significant need remains for new, more sophisticated intracellularly responsive molecular NIR fluorophores, which can be used to visualize dynamic cellular processes in real time with the potential for in vivo translation.

The goal of our current work was to develop an NIR fluorophore capable of a lysosomal-induced off-to-on fluorescence response, thereby permitting real-time imaging of cellular uptake, trafficking and efflux without perturbing function[19]. Endocytosis, the process through which cells internalize biomolecules, is common to all cells and represents a crucial area of research interest due to the numerous associated biological processes[20,21]. The participating organelles at each stage in the endocytosis pathway maintain a unique intravesicular/localized pH, to provide appropriate conditions for specific biochemical processes. Although the extracellular and cytosolic regions are at pH ~7.2, the lysosomes are significantly more acidic. Along the endocytic pathway, the pH lowers from ~6.3 in early endosomes through ~5.5 in late endosomes, down to ~4.5 in lysosomes (Fig. 2)[22]. As such, a difference of almost three orders of magnitude in proton concentration exists between the lysosome interior and the outside of a cell, which is sufficient to establish a selective trigger for fluorescence switching[23–26]. However, a major additional response selectivity challenge still remains, in that pH-responsive molecular fluorophores can also be responsive to micro-environmental polarity, which can compromise their use in cellular experiments (vide infra).

Our novel lysosomal responsive probe design is illustrated in Fig. 1 in which functionalization of the fluorophore core (orange box) with an ortho-nitro phenolic group was chosen to impart the pH-responsive feature of the probe. It would be expected that the electron withdrawing o-nitro group would result in the ionized phenolate dominating at pH 7.2, resulting in fluorescence quenching due to a non-emissive intramolecular charge transfer excited state (Fig. 1, grey box). Following cellular uptake via endocytosis and compartmentalization into acidic organelles such as lysosomes, protonation would occur giving the neutral phenol species and the NIR emission signal would be established (Fig. 1, red box). This approach is a significant departure from other lysosomal stains, which rely on an amine protonation to form a positively charged ammonium salt to concentrate the fluorophore in the acidic compartments[19,22]. An important additional design feature includes a covalently linked polyethylene glycol (PEG) polymer to provide aqueous solubility and promote cellular uptake via endocytic pathways (Fig. 1, blue box)[27].

Results

Synthesis and photophysics. The starting point of the synthesis was the previously reported BF$_2$-chelated bis-phenol azadipyrromethene 3, accessible in three synthetic steps from 1-(4-hydroxyphenyl)-3-phenylpropenone (Fig. 1)[28]. Mono-alkylation of 3 was achieved to produce 4 by reaction with t-butyl bromoacetate and CsF in dimethylsulfoxide (DMSO) at 30 °C. After isolation, compound 4 was then subjected to ortho-nitration of the remaining phenol ring with KHSO$_4$/KNO$_3$ to provide 5. Next, hydrolysis of the t-butyl ester of 5 with trifluoroacetic acid (TFA) gave the carboxylic acid 6, which was converted into its activated ester 7 by reaction with N-hydroxysuccinimide and N-(3-dimethylaminopropyl)-N'-ethylcarbodiimide in DMSO. Formation of the activated ester was monitored by ^{1}H NMR via the diagnostic CH$_2$ peaks at 5.48 for 7 and 4.83 p.p.m. for carboxylic acid 6, which showed complete conversion within 2 h (Supplementary Fig. 1). Conjugation of 7 in DMSO with a terminal amine functionalized PEG polymer (average molecular weight of 4,900) was effective, with the final fluorophore 2 obtained in high yield (Fig. 1). Matrix-assisted laser desorption/ionization–time of flight (MALDI–TOF) analysis of 2 showed the expected molecular weight centred at 5,410 Da, indicating that the covalent linkage was effective. Furthermore, ^{1}H NMR was consistent with the product structure and analytical high-performance liquid chromatography (HPLC) showed a single peak for 2 with retention time differing from that of both the acid 6 and ester 7 (Supplementary Fig. 2).

Comparative absorption and fluorescence emission spectra were recorded for the organic soluble fluorophore 5 in chloroform and aqueous soluble 2 in phenol red-free imaging DMEM medium adjusted to pH 2 (Fig. 3a). Only small differences were

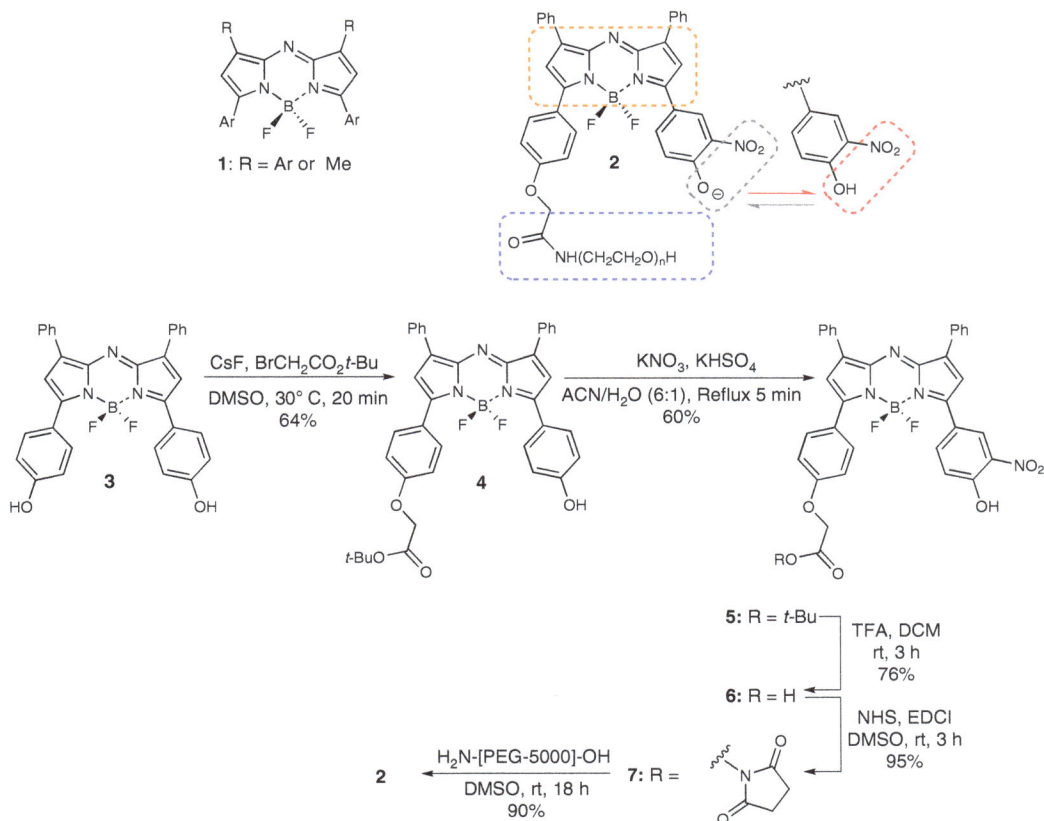

Figure 1 | BF$_2$-azadipyrromethene NIR fluorophores. General structure of BF$_2$-azadipyrromethenes **1**. Design and synthesis of lysosomal responsive BF$_2$-azadipyrromethene NIR fluorophore **2**.

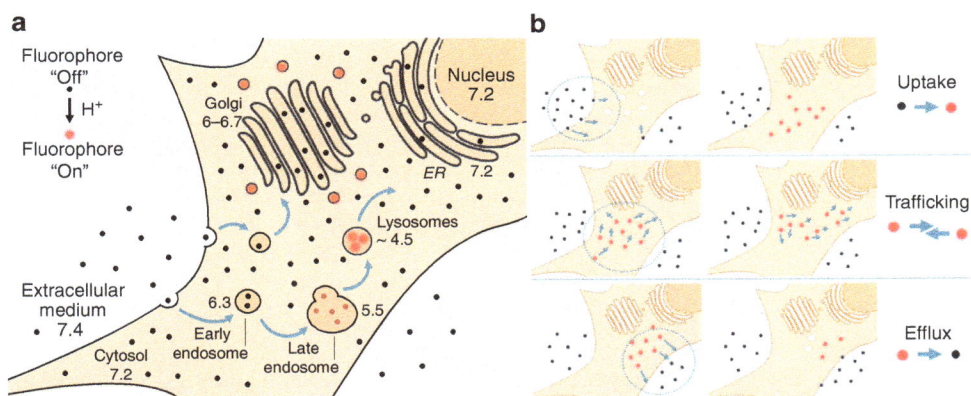

Figure 2 | Cellular uptake responsive NIR-fluorophore. (**a**) Simplified endocytosis of a responsive NIR fluorophore. Numbers represent the approximate pH of the corresponding organelles. (**b**) Three observable stages of the path of the pH-responsive fluorophore in the cellular environment: uptake, trafficking and efflux.

observed between the two fluorophores in the differing organic and aqueous media. Encouragingly, probe **2** had fluorescence λ_{max} at 707 nm with an absorbance λ_{max} at 685 nm. Extinction coefficient and fluorescence quantum yield values for **5** and **2** were similar with polyethylene glycol-substituted **2** having values of 97,000 cm^{-1} M^{-1} and 0.18, respectively (Fig. 3a).

An undesirable feature of some pH-sensitive fluorophores is their strong sensitivity to micro-environmental polarity, which significantly compromises their use in biological settings[29–31]. To test the polarity sensitivity of **2**, its acid/base emission-responsive properties were recorded in toluene, tetrahydrofuran, dimethylformamide and DMSO for both the phenol and phenolate state using 1,8-diazabicyclo[5.4.0]undec-7-ene (DBU)

and TFA to cycle between the two (Fig. 3b). A plot of solvent polarity function $(\Delta f)^{32}$ versus integrated fluorescence intensities in the off states showed highly effective fluorescence quenching as the phenolate irrespective of solvent polarity. A strong fluorescence output was established once protonated to the phenol in all solvents (Fig. 3c). These results predict that the modulation of fluorescence intensity would be selective for pH changes, while remaining unresponsive to differing intracellular micro-environmental polarities, thereby removing the potential for false-positive emissions. An identical study was carried out for fluorophore **5**, giving similar results and indicating that this positive feature is general to the fluorophore class (Supplementary Fig. 3).

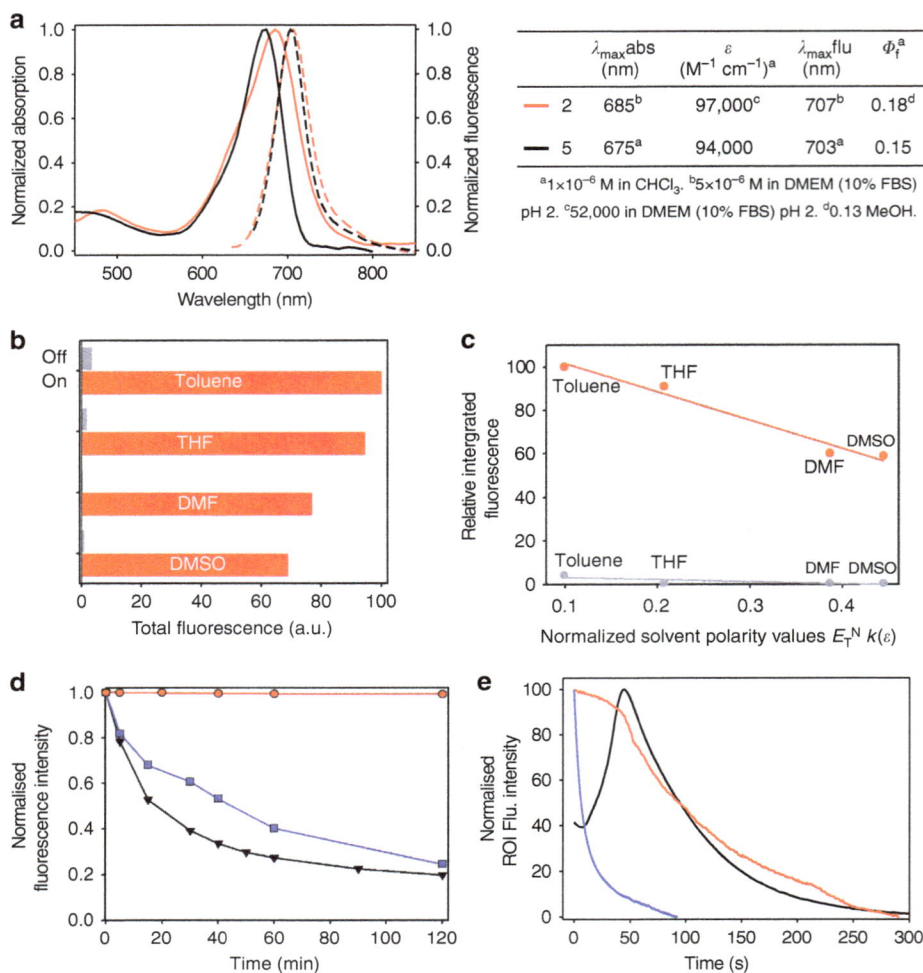

Figure 3 | Photophysical properties of NIR-fluorophores. (a) Light absorption and emission spectra of compounds **2** and **5**, and their photo-physical parameters. **(b)** Integrated off and on fluorescence states of **2** (5×10^{-6} M) in toluene, tetrahydrofuran (THF), dimethylformamide (DMF) and DMSO with TFA (red bars) and DBU (grey bars). **(c)** Plot of relative off and on integrated fluorescence versus solvent polarity values for toluene, THF, DMF and DMSO. **(d)** Comparative photobleaching of 1×10^{-7} M DMEM solutions of **2** (red line), lysotracker red (blue line) and pHrodo red (black line) with 150 W fibre optic delivered light 620(30) nm for **2** and 540(40) nm for lysotracker red and pH-rhodo red at 25 °C. **(e)** *In vitro* photobleaching of **2** (red), lysotracker red (blue line) and pHrodo red (black line) with maximum LED power using excitation filter 640(14) nm for **2** and excitation filter 563(9) nm for lysotracker red and pH-rhodo red.

As sufficient photostability is an essential property for prolonged live-cell imaging, a comparative study of the photo-degradation of **2**, lysotracker red and pH-rhodo red was carried out. DMEM solutions of the three fluorophores were illuminated with light of 620(30) nm for **2** and 540(40) nm for lysotracker red and pH-rhodo red for 2 h, and their fluorescence intensity monitored. Encouragingly, no photobleaching for **2** was observed, whereas both other fluorophores were ~80% degraded within that time frame (Fig. 3d). Comparison of their stabilities in HeLa Kyoto cells using illumination from a solid-state light emitting diode (LED) light source was also examined. Cells stained with **2**, lysotracker red or pH-rhodo red were constantly illuminated with LED power set to a maximum, to promote a fast rate of photobleaching. The same excitation filters used for imaging (640(14) nm for **2** and 563(9) nm for lysotracker red and pH-rhodo red) were used, allowing images to be acquired at various time intervals. Graphing the average cell fluorescence intensity versus time showed that **2** was the most photostable with 50% loss of signal in 94 s and lysotracker red being the least photostable with 50% of signal loss in just 6 s (Fig. 3e, Supplementary Fig. 4 and Supplementary Movies 1 and 2). The behaviour of pH-rhodo

red was more complex, as its intensity first significantly increased throughout the cell followed by photobleaching (Fig. 3e, Supplementary Fig. 4 and Supplementary Movie 3). This response to irradiation is indicative of a photo-conversion occurring for pH-rhodo red but further studies would be required to fully establish the cause for this. Comparison of these results highlights the distinct advantage of **2** for prolonged live-cell imaging in which fluorophore photostability is an essential parameter.

The pH-responsive properties of **2** were investigated in DMEM containing 10% fetal bovine serum (FBS) before its use in imaging studies. Fluorescence output of **2** was negligible at pH 7.4, but became highly fluorescent at acidic pH with its pKa determined as 4.0 (Fig. 4a). Cy5.5 light filter parameters of 690/50 nm were applied to the emission bands at pH 7.4, 5.5 and 4.5, and the integrated fluorescence intensity differences determined. At pH 5.5, as found in late endosomes, the fluorescence enhancement factor (FEF) was 6-fold, while it reached a remarkable 21-fold at lysosomal pH of 4.5 (Fig. 4b). Taken together, these results predict that at a cellular level **2** would remain non-fluorescent in the extracellular environment and become highly NIR fluorescent on uptake and localization in the lysosomes (Fig. 4c).

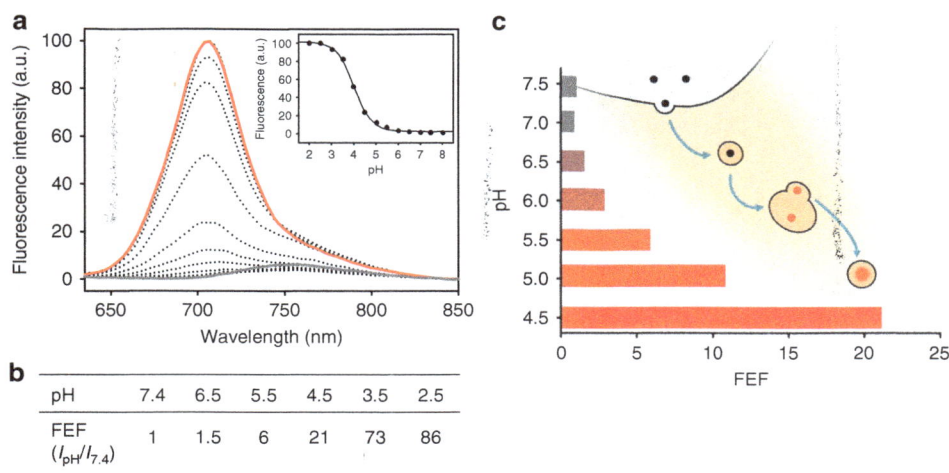

Figure 4 | Cellular uptake responsive NIR-fluorescence. (a) Emission spectra of **2** (5×10^{-6} M) in DMEM (10% FBS) at pH ranging from 8 (grey) to 2 (red). Exc: 625 nm. Inset: fluorescence intensity at $\lambda_{max} = 707$ nm versus pH; sigmoidal plot fit resulted in apparent $pK_a = 4.0$. **(b)** Corresponding FEF values from differing pH solutions applying Cy5.5 filter parameters. **(c)** Diagram represents the pH changes and increasing fluorescence intensity along endocytic path towards lysosomes.

pH	7.4	6.5	5.5	4.5	3.5	2.5
FEF ($I_{pH}/I_{7.4}$)	1	1.5	6	21	73	86

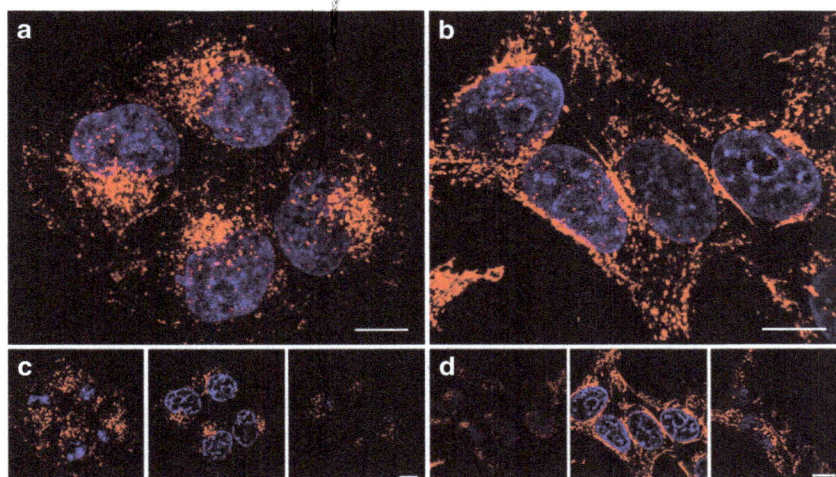

Figure 5 | Intracellular NIR-emission profile. CLSM images showing intracellular localization of pH-responsive compound **2** (10 μM, red) and nuclear counterstain Hoechst 33342 (blue) in fixed (**a,c**) HeLa Kyoto and (**b,d**) HEK293 cell lines. Bottom: three corresponding representative slices of the Z-stack for each cell type. Scale bars, 10 μm.

The difference in emission intensity at pH 5.5 (late endosomes) and 4.5 (lysosomes) suggests that the increased activation of **2** in lysosomes may be sufficient to allow differentiation between these organelles.

***In vitro* fixed and real-time live-cell imaging.** Before testing the imaging capabilities of **2**, its cytotoxicity in HeLa Kyoto (cervical cancer) and HEK (human embryonic kidney) cell lines was determined. Following a 24-h incubation of cells with **2**, an MTT (3-(4,5-dimethylthiazol-2-yl)-2,5-diphenyltetrazolium bromide) assay was performed and EC_{50} values of 0.43 and 0.44 mM, respectively, were obtained (Supplementary Fig. 5). These values were used as the basis to select 10 μM as the concentration for its use in imaging experiments. To establish the ability of **2** to internalize in cells, HeLa Kyoto and HEK293 cells were incubated with **2** for 2 h, followed by fixation, nuclei staining (Hoechst 33342) and imaging with confocal laser scanning microscopy (CLSM) (Fig. 5). These images show that **2** was internalized in both cell lines within 2 h (Supplementary Movie 4). The fluorescence signal was predominately localized in the perinuclear region as would be expected for a lysosomal staining pattern, indicating that the fluorophore had accumulated and become fluorescent in lysosomes[19].

To gain further evidence of selective lysosomal staining, an identical experiment was performed using a HeLa cell line stably expressing the lysosomal-associated membrane protein 1 (LAMP1) fused to green fluorescent protein (GFP)[33]. Following incubation, CLSM-imaged cell images showed very high levels of co-compartmentalization of the red (**2**) and green channel (GFP) emissions to the lysosomes (Fig. 6). Examination of selected focal planes clearly showed a circumferential staining pattern (green) of the LAMP1-GFP in the lysosome membrane and red emission of **2** from within the acidic lumen of the organelles (for Z-stack see Supplementary Movie 5). This ability to resolve the lysosome membrane from the interior is approaching the confocal resolution limit, with ∼1 μm being the average diameter of an individual lysosome (for images from an additional independent experiment, see Supplementary Fig. 6). This was achieved due to the high signal-to-noise ratio, as background emission from **2** is not observed, and high red/green contrast with the genetically

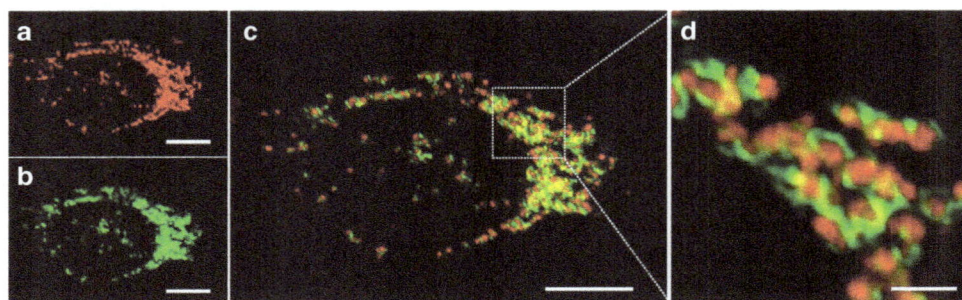

Figure 6 | Identification of subcellular NIR-fluorescent on switch. CLSM fluorescent images showing lysosomal localization of the 'on' state of **2** in LAMP1-GFP-expressing HeLa cells (**a**) Cy5.5 channel; (**b**) GFP channel. (**c**) Three-dimensional image of overlaid Cy5.5 and GFP channels. (**d**) Zoom-in of the dashed box. Scale bars, 10 μm (**a**–**c**) and 2 μm (**d**).

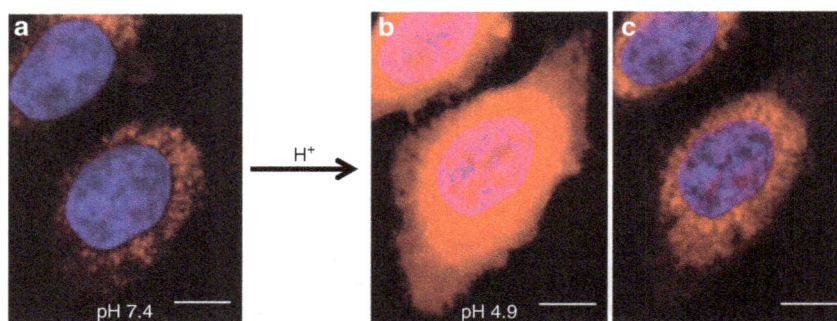

Figure 7 | Illustration of NIR-fluorescence response selectivity. (**a**) CLSM imaging of HeLa Kyoto cells following incubation with **2** (10 μM) for 2 h at 37 °C, DAPI nuclei staining and fixing. (**b**) The same set of cells imaged after buffer changed to pH 4.9, keeping the same laser power and PMT voltage. (**c**) The same set of cells after adjustment of microscope laser power and PMT voltage to obtain a non-saturated image. Red: **2**; blue: DAPI stain. Scale bar, 10 μm.

expressed LAMP1–GFP. Further statistical evidence of co-compartmentalization of the NIR and green emissions was provided by the calculated Manders' coefficients of 0.91 (05) for M_{NIR} and 0.95 (05) for M_{green} and a Pearson's coefficient of 0.79 (09)[34]. The lower Pearson's coefficient value when compared with Manders' may be attributable to the fact that the NIR and green emissions are co-compartmentalized to the lysosomes but not fully co-localized as the green is in the outer membrane and the NIR from within the internal lumen. Golgi co-staining of HeLa cells with **2** showed no significant co-localization (Supplementary Fig. 7, Method 1).

Two possible explanations for the excellent co-compartmentalization of red and green emissions could be envisaged. Either **2** is physically located exclusively in the lysosomes or it is present in other organelles along the endocytic pathway and is only emissive from the lysosomes due to its lower pH, with the remaining **2** being fluorescent silent. To visualize all intracellular **2** in its fluorescent on state, HeLa cells were incubated with **2** in media at pH 7.4 for 2 h, nuclei stained with 4,6-diamidino-2-phenylindole (DAPI) for 15 min, fixed and imaged using CLSM (Fig. 7a and Supplementary Movie 6). The media containing the fixed cells was then changed for media at pH 4.9 (adjusted using HCl), which on penetrating the cell forced on the fluorescence of all **2** within cells not localized in a sufficiently low pH environment. Re-imaging of the same cells (30 min after media change) with identical microscope settings showed a significantly increased fluorescence with saturation of the field of view within the cells (Fig. 7b and Supplementary Movie 7). The mean corrected total fluorescence from the cells (calculated using ImageJ) showed an FEF of 9.5 on lowering of the pH (see Supplementary Fig. 8 for images from additional independent experiment). This clearly illustrates that fluorescence from the majority of **2** was not switched on by the cells and only **2** localized in the lysosomes was emissive under normal pH conditions. The same field of view was imaged for the third time using adjusted microscope laser power and photomultiplier tube (PMT) voltage and it became clear that the additional fluorescence of **2** was predominantly from other cellular organelles of higher pH (Fig. 7c and Supplementary Movie 8). Unfortunately, the adjustment to lower pH caused a loss of the LAMP1–GFP signal; thus, a comparison of Manders' coefficients was not possible.

To show that fluorophore **2** was internalized in cells via endocytosis and not passive diffusion, HeLa Kyoto cells were incubated at 4 °C for 30 min with **2**. It is known that endocytosis is inhibited at 4 °C, whereas passive diffusion can still occur[35]. Following incubation, cells were imaged, the buffer adjusted to pH 4.9 and re-imaged as described above. In contrast to the result shown in Fig. 7, no fluorescence from **2** was detected before or after changing the buffer to pH 4.9 (Supplementary Fig. 9). From these experiments, it was concluded that **2** was internalized via an energy requiring endocytosis rather than passive diffusion.

For the results outlined above, it was anticipated that cellular uptake of **2** could be continuously imaged in real-time without the need for washing or manipulating cells. The first live-cell imaging experiment involved imaging cells in a single focal plane over a 1.5-h time period. Once in focus, HeLa Kyoto cells were treated with **2** and imaged with an epi-fluorescence live-cell microscope, under optimal conditions of temperature and atmosphere for the cells to remain fully active (37 °C and 5% CO_2 humidified environment). Images were continuously acquired every 30 s for 90 min and then combined to form a movie (Supplementary Movie 9). Representative time-lapse images after 1, 30, 60 and 90 min (in black/white for clarity) are shown below in Fig. 8a with the 90 min time point in red

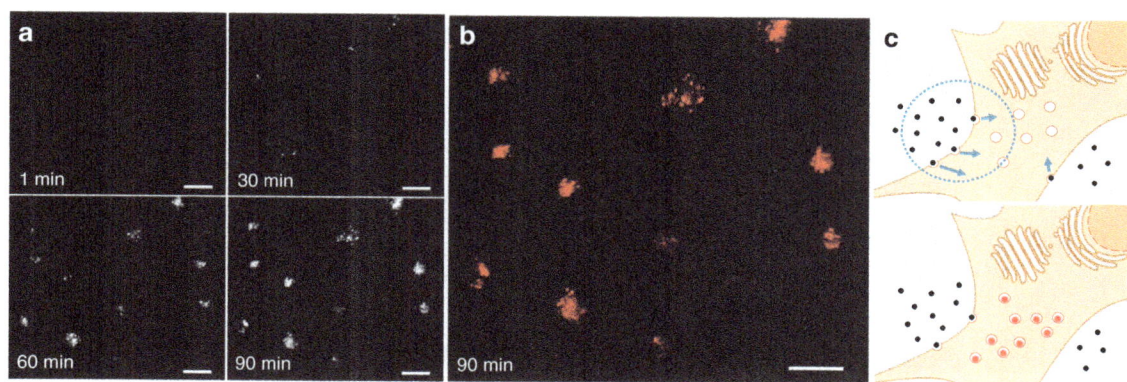

Figure 8 | Widefield live-cell imaging of the uptake of 2 (10 μM) into HeLa Kyoto cells. (a) Time-lapse black and white images are shown 1, 30, 60 and 90 min. (b) Red-coloured image at 90 min. (c) Schematic depiction of the uptake process of responsive fluorophore (c). Scale bars, 20 nm.

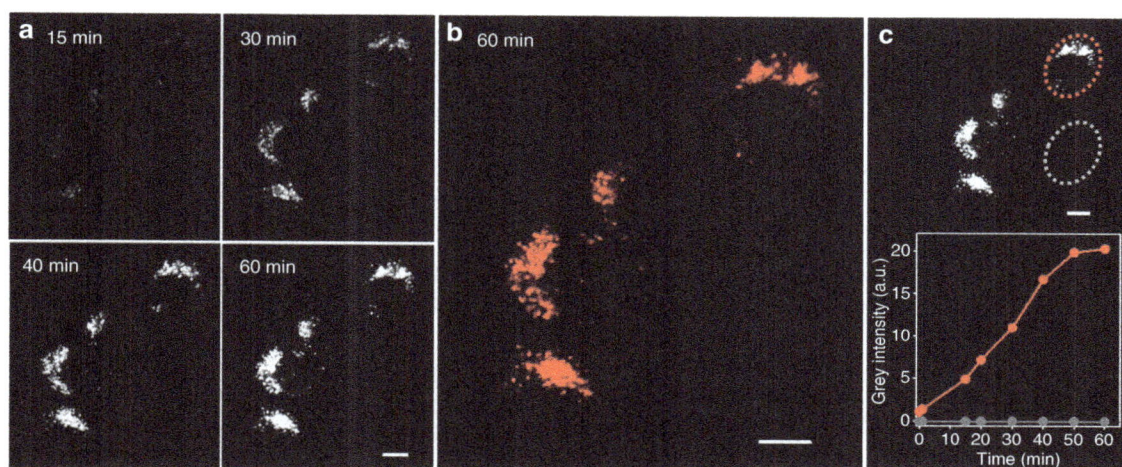

Figure 9 | Z-axis projections of widefield 4D live-cell imaging of the uptake of 2 (10 μM) from HeLa Kyoto cells. Images were acquired in 25 focal planes every 1 min for 60 min. (a) Time lapse b/w images are shown for 15, 30, 40 and 60 min. (b) Red-coloured image at 60 min. (c) Fluorescence intensity quantification in two identical volumes around a selected cell (1) and in the extracellular environment (2). Scale bars, 10 μm.

for comparison (Fig. 8b). The first image acquired at 1 min showed no NIR fluorescence, a signal confirming the effective fluorescence quenching of extracellular **2**. However, over the following 90-min time period, a strong signal arose from point-like organelles as a result of cellular uptake of **2** and transport through the endocytic pathway to the lysosomes (Fig. 8: 60- and 90-min time points). On close inspection, individual lysosomes can be seen emerging into view over the first 30-min time period, following which they increase in intensity and number from 30 to 90 min (Supplementary Movie 9).

The continuous imaging of live cells in the *z*-axis provides the most realistic method for following the progress of biological events over time and is of particular relevance when imaging small mobile organelles. To generate a three-dimensional (3D) representation of the cell in real time, a *z*-stack of 25 focal planes through the cell was acquired every minute[36]. This continual recording of cellular 3D volume over a period of time is termed four-dimensional (4D) imaging as the sample is imaged in the *x*,*y*,*z* and time dimensions, from which a time-lapse video of the 3D cellular volume can be created. Using HeLa cells, a 4D data set of the uptake of **2** over a 60-min time period was acquired followed by deconvolution of the data set to correct for motion of fluorescence objects between focal planes during the 1-min time period required to complete a *z*-stack of the cell. This experiment showed punctuated regions of fluorescence over the 60-min time period starting from a non-fluorescent background at time zero

with individual lysosome movement clearly observable (Fig. 9a,b and Supplementary Movie 10). Quantification of the fluorescence increase was measured by selecting two identical volumes of the imaged area, one overlapping with a chosen cell and the other on the extracellular environment, and applying an image analysis algorithm (ImageJ) to measure the total fluorescence intensity within the volumes (Fig. 9c). A comparative plot of both intensities over time shows how the background remained non-fluorescent, whereas intracellular fluorescence intensity increased over time, reaching a plateau at 60 min at which point a dynamic equilibrium was established between extracellular and intracellular **2**.

This ability to 4D image with **2** provides a tool for tracking lysosomal movements within the cell. Lysosomal staining of HeLa Kyoto cells was achieved by incubation with **2** for 1 h following which they were imaged in 3D for 35 min (without medium replacement) using a widefield microscope. Image analysis software was used to tag individual lysosomes as white spheres to facilitate visualization and the movement of these lysosomes was tracked over the 35-min time period (Fig. 10). The path the lysosome takes through the cell is illustrated by a lengthening white tail, which extends from the lysosome as the video progresses (Fig. 10c,d and Supplementary Movie 11)[37]. Tracking of all lysosomes within a cell (or field of view) was also possible and is shown for Fig. 10 in the Supplementary Information (Supplementary Movie 12).

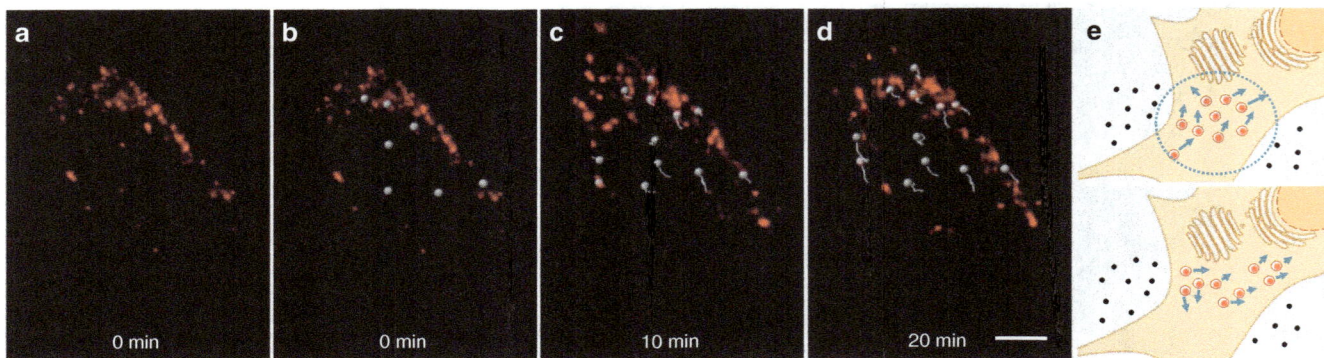

Figure 10 | Lysosome tracking in living HeLa Kyoto cells post 1 h incubation with 2 (10 μM). (**a**) Time-lapse representative snapshot of a single cell chosen for image analysis. (**b**) Lysosome selection at 0 min. (**c,d**) Tracking over time. (**e**) Schematic depiction of tracking intracellular vesicular movements with bioresponsive fluorophore. Scale bar, 5 μm.

Our final *in vitro* experiments with pH-responsive **2** were aimed at studying the efflux of the fluorophore from the cell. HeLa Kyoto cells were pre-treated with **2** for 2 h as previously described, then the medium was replaced with fluorophore-free DMEM. Cells were incubated at 37 °C and imaged at time points of 1, 15, 30 and 120 min using confocal microscopy (Fig. 11). The overall NIR fluorescence detected after 2 h incubation with **2** in the intracellular environment showed to be again arising from point-like organelles (Fig. 11a). The quantity of fluorescent vesicles decreased within 30 min after medium change, with significant clearance of **2** by 120 min (Fig. 11b). Using ImageJ analysis, lysosome number per cell (or field of view) was determined for the different time points and were observed to steadily decrease from time 0 to 120 min (Fig. 11d). In addition, after 60 min cells were fixed and acidified with buffer of pH 4.9 and only a small, 1.6-fold increase in total fluorescence was observed.

***In vivo* imaging**. A distinct advantage of NIR fluorophores is their ability to directly transfer from *in vitro* to *in vivo* imaging due to transparency of biological tissue at these longer wavelengths. To test *in vivo* performance of **2**, the luciferase-expressing human breast cell line MDA-MB-231-luc-D3H1 was chosen to grow subcutaneous tumours of size 100–200 mm^3, which permitted NIR fluorescence imaging with confirmatory bioluminescence. The ability of PEG polymers to act as a drug delivery vehicle has been well established, with several PEGylated drugs in clinical use for over 20 years. PEGylation is known to influence pharmacokinetic properties resulting in prolonged blood circulation times. As such, it was anticipated that the PEG-conjugated **2** may have passive tumour-targeting properties leading to some preferential uptake into tumour cells, thereby generating a distinguishable fluorescent signal. Following an intravenous tail vein injection of **2** (2 mg kg^{-1}), images were acquired at regular intervals over the course of 24 h. A plot of tumour to background NIR fluorescence showed that the fluorescence signal was low in the beginning in both the background and tumour. Emission from the liver peaked at 1 h and subsided over the following 24 h (Fig. 12d dashed red line). In contrast, the tumour fluorescence intensity reached a maximum at 24 h, allowing for good image discrimination at that time point (Fig. 12d solid red line). Bioluminescence and NIR fluorescence imaging at 24 h confirmed that **2** has indeed been taken up into tumour cells and switched on (Fig. 12a–c). No adverse reactions or animal weight loss were observed during or after imaging. These preliminary *in vivo* results represent a unique example of selective NIR *in vivo* tumour imaging using a pH-responsive fluorochrome.

Discussion

Imaging with molecular fluorophores is an indispensable tool for all forms of biological and medical research. Although the full spectrum of colours are available from molecular fluorophores, imaging with lowest energy NIR spectral region offers advantage for prolonged live cellular imaging with the possibility of *in vivo* imaging with the same probe[5,57]. Although most fluorescent markers are permanently fluorescent (unless photobleached), huge imaging advantage can be gained if the fluorescence can be modulated from off to on in a reversible bioresponsive manner[14]. With these goals in mind, we have designed a lysosomal-responsive NIR probe with the potential for real-time visualization of their key cellular operations that can be directly translated for use *in vivo* (Figs 1 and 2). NIR probe **2** is synthesized in five synthetic operations from a bisphenolic substituted member of the BF$_2$-azadipyrromethene fluorophore class (Fig. 1). An *o*-nitro-substituted phenol group on the probe acts as the fluorescent switch with the nitro group tailoring the emission response to the lumen microenvironment of the lysosomes. Photophysical measurements in DMEM shows that **2** remained fluorescent silent at pH 7.2, yet became highly fluorescent in the pH range that corresponds with the acidic micro-environment of the lysosome (Fig. 4). The non-fluorescent state shows little polarity sensitivity, indicating that it could be used in the more complex intracellular environment without resulting in false-positive emissions. Solution and cellular photobleaching experiments indicated high stability, which is an essential feature for imaging over prolonged time periods that is often lacking in both synthetic and genetically expressed probes (Fig. 3).

To illustrate the potential uses of probe **2**, a series of increasingly complex imaging experiments were undertaken in fixed, live cells and *in vivo*. The high fidelity for the switching on of **2** to the lysosomal lumen is observed when imaging with LAMP1-GFP HeLa cells (Fig. 6). We have exploited these responsive emission properties for 3D and 4D real-time live-cell imaging of several fundamental biological events, such as endocytosis, organelle trafficking and efflux. As **2** has extremely low emission in cell media and with fluorescence activated on cellular internalization, a high background-to-noise ratio is achieved, making the continuous acquisition of data as straightforward as just add **2** to cells and image. It could be anticipated that these techniques would be valuable for many types of cellular experiments involving lysosomal response to stimuli.

It is important to note that the mode of action and the use of existing lysosomotropic stains, which are typically amine-functionalized fluorophores to promote retention in acidic

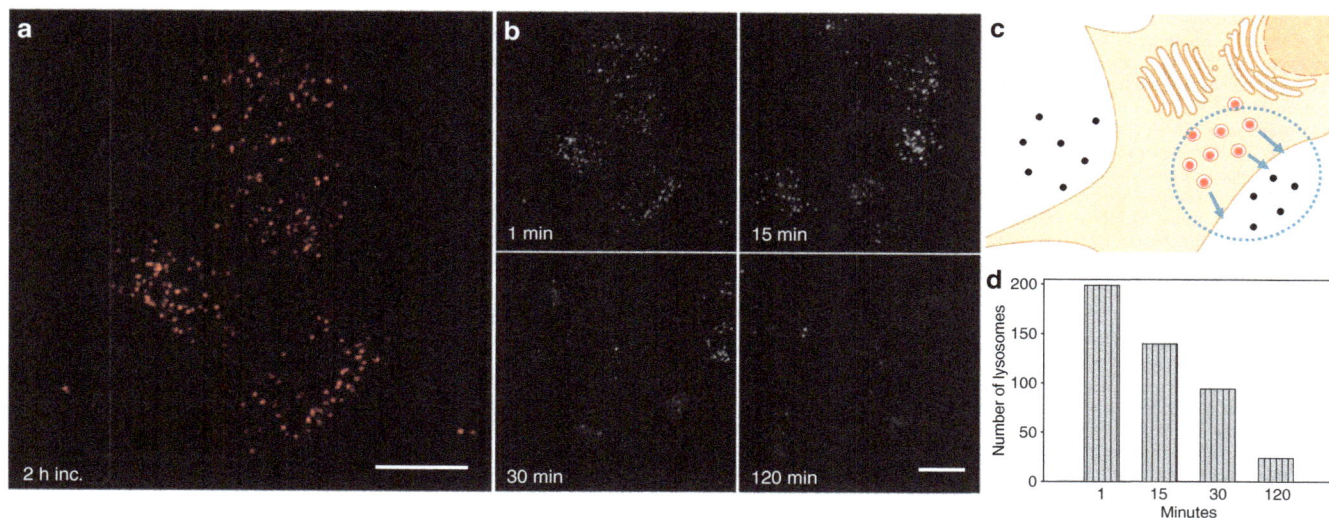

Figure 11 | Imaging of cellular efflux. HeLa Kyoto cells were pre-treated with **2** (10 µM) for 2 h, DMEM replaced with fluorophore-free DMEM and cells fixed at various time points. (**a**) Cells imaged after 2 h incubation. (**b**) Cells imaged after 1, 15, 30 and 120 min post media change. (**c**) Schematic depiction of efflux of bioresponsive fluorophore. (**d**) Decrease in number of NIR fluorescent lysosomes from 1 to 120 min. Scale bars, 20 µm.

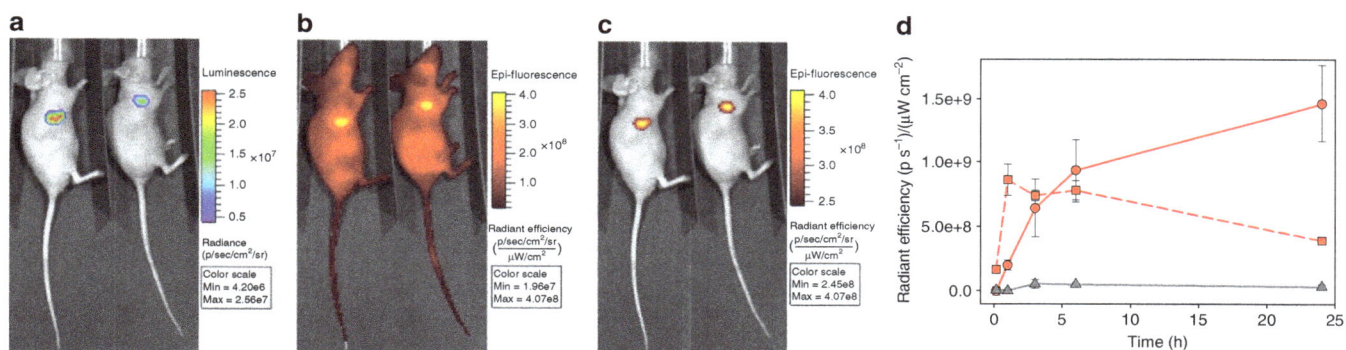

Figure 12 | *In vivo* imaging of 2 using a MDA-MB-231-luc-D3H1 subcutaneous tumour model in two representative mice. (**a**) Bioluminescence imaging confirmation of tumour cells. (**b**) NIR fluorescence imaging 24 h post intravenous (i.v.) administration of **2** (excit. 660–690 nm, emis. 710–730 nm). (**c**) NIR fluorescence imaging 24 h post i.v. administration of **2** with intensity scale adjusted (excit. at 675 nm, emiss. at 720 nm). (**d**) Profile of tumour NIR fluorescence (red solid line) and liver (red dashed line) over time following i.v. tail injection of **2**. Non-injected control tumour NIR fluorescence (grey solid line). Values determined from the same sized ROI from background area and tumour averaged for $n = 3$.

lysosomes on protonation, is significantly different from **2** (refs 38,39). These lysosomotropic fluorophores are pre-incubated for a period of time (typically 30 min), followed by cell washing to remove excess nonspecific fluorophore, before imaging can be carried out for up to a maximum of 1 h (ref. 40). In contrast, **2** in its fluorescent on state is uncharged, showing little cytotoxic effect over long incubation times and no background fluorescence due to the highly selective switching 'on' only in the desired lysosomal ROIs. These characteristics have been demonstrated to be particularly advantageous for continuous real-time live-cell imaging and show significant potential for use in a wider range of complex bio-imaging applications. Overall, the cellular imaging performance, ease of utility and the selectivity for lysosomal staining could be judged as excellent in comparison with the most recently developed probes[41].

To complete the imaging portfolio, we wished to illustrate that the bioresponse of **2** within subcellular compartments could also be visualized at the macroscopic scale of a tumour. This translation to *in vivo* tumour imaging is achievable as shown in

Fig. 12 in which tumour can be clearly distinguished, showing a high potential for targeted responsive fluorescence imaging.

In conclusion, the development of **2** as the first phenol/phenolate controlled molecular NIR lysosome-responsive fluorophore has important implications for the study of intracellular transport mechanisms, lysosome-based diseases and *in vivo* targeting. The use of **2** for further 4D real-time studies of more complex dynamic cell mechanisms involving lysosomes are ongoing. Future studies will include the conjugation of active tumour targeting motifs via **7** (instead of PEG) to the fluorophore, to further broaden the possibilities of *in-vivo* imaging targets with potential applications for fluorescence-guided surgery.

Methods

General information and materials. All commercially available solvents and reagents were used as supplied, unless otherwise stated. All reactions were performed under nitrogen or argon atmosphere in oven-dried glassware. Gel chromatography was performed with Davisil 60 silica (230–400 mesh). On the basis of NMR and reverse-phase HPLC, all final compounds were >95% pure. [1]H and ^{13}C spectra were recorded on 300, 400, 500 or 600 MHz NMR spectrometers and chemical shifts were reported in p.p.m. using solvent residual peak as standard. For spectra of compounds **4**, **5**, **6** and **7**, see Supplementary Figs 10–13, respectively. All ^{19}F NMR chemical shifts are referenced to CFCl$_3$. All ^{11}B NMR chemical shifts are referenced to BF$_3$.Et$_2$O/CDCl$_3$. High-resolution mass spectrometry and tandem mass spectrometry experiments were carried on an electrospray ionisation and MALDI–TOF instruments. Infrared spectra were recorded as a KBr pellet using a Fourier Transform infrared spectrometer. Absorbance spectra were recorded with a Varian Cary 50 Scan ultraviolet–visible spectrometer. Fluorescence spectra were recorded with a Varian Cary Eclipse Fluorescence Spectrometer. Solvents for absorbance and fluorescence experiments were of HPLC quality. SigmaPlot, MestreNova, ChemDraw, Zeiss LSM and ImageJ software were used for data analysis. Phenol red-free imaging DMEM medium was used for all experiments.

Synthesis of compound 4. Compound (ref. 28) (200 mg, 0.38 mmol) and CsF (288 mg, 1.89 mmol) were dissolved in dry DMSO (6 ml) and stirred at 30 °C under a nitrogen atmosphere for 10 min, during which time the colour changed from dark green to dark purple. *t*Butyl bromoacetate (126 mg, 0.64 mmol) was then added via syringe in one go and the solution was stirred at 30 °C for 20 min. The mixture was partitioned between AcOEt (100 ml) and PBS buffer at pH 7 (100 ml). The organic phase was washed with water (3 × 100 ml), brine (50 ml), dried over Na$_2$SO$_4$, filtered and evaporated to dryness. The crude product was purified by silica gel chromatography, eluting with CH$_2$Cl$_2$:AcOEt (99:1 → 90:10) to yield the product **4** as a red metallic solid (151 mg, 64%). mp: 183–187 °C; ^{1}H NMR (400 MHz, DMSO-d_6): δ 10.58 (br s, 1H), 8.19–8.10 (m, 8H), 7.65 (s, 1H), 7.57–7.43 (m, 7H), 7.10 (d, J = 8.7 Hz, 2H), 6.95 (d, J = 8.6 Hz, 2H), 4.81 (s, 2H), 1.46 (s, 9H); ^{13}C NMR (100 MHz, DMSO-d_6): δ 167.5, 161.5, 160.0, 158.9, 155.7, 145.0, 143.8, 142.6, 141.0, 132.5, 132.0, 131.7, 131.4, 129.6, 129.3, 129.1, 129.0, 128.7, 128.6, 124.1, 121.3, 120.3, 119.1, 116.0, 114.8, 81.6, 65.0, 27.7; ^{11}B NMR (128 MHz, DMSO-d_6) δ: 1.00 (t, J = 32.6 Hz); ^{19}F NMR (376 MHz, DMSO-d_6): δ − 130.44 (q, J = 32.6 Hz); ultraviolet–visible: λ_{max} (CHCl$_3$): 680 nm (ε = 85,000 cm^{-1} M^{-1}); emission: λ_{max} (CHCl$_3$): 708 nm, Φ_F (CHCl$_3$) = 0.31; high resolution mass spectrometry (HRMS) (m/z): [M-H]$^-$ calcd. for C$_{38}$H$_{31}$BN$_3$O$_4$F$_2$, 642.2376; found, 642.2357.

Synthesis of compound 5. A solution of **4** (300 mg, 0.48 mmol) in acetonitrile (6 ml) was heated under reflux for 5 min. A solution of KNO$_3$ (53 mg, 0.53 mmol) and KHSO$_4$ (130 mg, 0.96 mmol) in water (1 ml) was added and the mixture was heated under reflux for 5 min. The suspension was cooled to room temperature (rt) and partitioned between AcOEt (200 ml) and water (100 ml). The organic phase was washed with water (100 ml), brine (2 × 100 ml), dried over Na$_2$SO$_4$, filtered and evaporated to dryness. The crude was purified by silica gel chromatography, eluting with CH$_2$Cl$_2$:AcOEt (99: 1) to yield the product **5** as a dark red metallic solid (198 mg, 60%). mp: 192–194 °C; ^{1}H NMR (400 MHz, DMSO-d_6): δ 11.88 (br s, 1H), 8.69 (d, J = 2.0 Hz, 1H), 8.27 (dd, J = 8.9, 2.0, 1H), 8.21 (d, J = 9.0 Hz, 2H), 8.19–8.13 (m, 4H), 7.71 (s, 1H), 7.60 (s, 1H), 7.57–7.43 (m, 6H), 7.27 (d, J = 8.9 Hz, 1H), 7.12 (d, J = 9.0 Hz, 2H), 4.84 (s, 2H), 1.45 (s, 9H); ^{13}C NMR (100 MHz, DMSO-d_6): δ 167.4, 160.9, 159.3, 154.1, 153.9, 145.4, 143.9, 143.3, 141.5, 137.3, 135.6, 132.1, 131.8, 131.5, 129.9, 129.5, 129.2, 129.0, 128.7, 128.7, 126.6, 123.3, 121.8, 120.8, 119.4, 119.1, 115.0, 81.7, 65.1, 27.7; ^{11}B NMR (128 MHz, DMSO-d_6): δ 0.92 (t, J = 32.7 Hz); ^{19}F NMR (376 MHz, DMSO-d_6): δ − 130.31 (q, J = 32.7 Hz); ultraviolet–visible: λ_{max} (CHCl$_3$) 675 nm (ε = 94,000 cm^{-1} M^{-1}); emission: λ_{max} (CHCl$_3$) 703 nm, Φ_F (CHCl$_3$) = 0.15; HRMS (m/z): [M-H]$^-$ calcd. for C$_{38}$H$_{30}$BN$_4$O$_6$F$_2$, 687.2226; found, 687.2229.

Synthesis of compound 6. TFA (1 ml) was added dropwise to a solution of **5** (175 mg, 0.25 mmol) in dichloromethane (DCM) (9 ml) and the solution was stirred at rt for 3 h. The solvent was removed under vacuo and the residual TFA was removed azeotropically with serial additions of DCM and subsequent removal under vacuo. The solid was suspended in DCM, filtered and washed with DCM, to yield the product as a dark purple solid (136 mg, 84%). The product was pure enough to proceed to the next synthetic step. To remove the last trace of starting material, the solid was partitioned between AcOEt (90 ml) and Na$_2$CO$_3$ sat. (180 ml). The organic layer was discarded and the water layer was extracted with AcOEt (90 ml), separated, carefully acidified with 5 M HCl and extracted again with AcOEt (180 ml). The organic layer was separated, dried over anhydrous Na$_2$SO$_4$, filtered and evaporated to dryness. The product **6** was obtained as a dark purple solid (122 mg, 76%). mp: 214–219 °C; ^{1}H NMR (400 MHz, DMSO-d_6): δ 13.15 (br s, 1H), 8.68 (d, J = 2.3 Hz, 1H), 8.28 (dd, J = 8.9, 2.3, 1H), 8.24–8.13 (m, 6H), 7.71 (s, 1H), 7.60 (s, 1H), 7.57–7.44 (m, 6H), 7.28 (d, J = 8.9 Hz, 1H), 7.14 (d, J = 9.0 Hz, 2H), 4.87 (s, 2H); ^{13}C NMR (100 MHz, DMSO-d_6): δ 169.7, 161.0, 159.4, 154.1, 153.8, 145.4, 143.9, 143.4, 141.5, 137.3, 135.5, 132.1, 131.9, 131.5, 129.9, 129.5, 129.2, 129.0, 128.7 (2C), 126.6, 123.2, 121.9, 120.8, 119.5, 119.1, 115.1, 64.6; ^{11}B NMR (128 MHz, DMSO-d_6): δ 0.93 (t, J = 32.8 Hz); ^{19}F NMR (376 MHz, DMSO-d_6): δ − 130.34 (q, J = 32.8 Hz); HRMS (m/z): [M-H]$^-$ calcd. for C$_{34}$H$_{22}$BN$_4$O$_6$F$_2$, 631.1600; found, 631.1603.

^{1}H NMR monitoring of the formation of activated ester 6. A mixture of **5** (45 mg, 0.071 mmol), N-(3-dimethylaminopropyl)-N'-ethylcarbodiimide

hydrochloride (27 mg, 0.14 mmol) and N-hydroxysuccinimide (82 mg, 0.71 mmol) was placed in a sealed dry flask. Anhydrous deuterated DMSO-d_6 (1.2 ml) was added to the mixture and the solution was stirred at rt under N$_2$ atmosphere. Samples (50 μl) were withdrawn at 15, 30, 60, 120 min and 19 h, diluted with DMSO-d_6 in an NMR tube (650 μl) and ^{1}H spectra were recorded at a 600-MHz spectrometer.

Synthesis of compound 7. A mixture of **6** (40 mg, 0.063 mmol), N-(3-dimethylaminopropyl)-N'-ethylcarbodiimide hydrochloride (24 mg, 0.13 mmol) and N-hydroxysuccinimide (73 mg, 0.63 mmol) was dissolved in anhydrous DMSO (1 ml) and stirred at rt for 3 h under N$_2$ atmosphere. The solution was partitioned between with DCM (50 ml) and 0.5 M HCl (50 ml). The organic phase was washed with 0.5 M HCl (50 ml), acidic brine (50 ml), dried over Na$_2$SO$_4$, filtered and evaporated to dryness, keeping the temperature of the bath below 35°C. The product **7** was obtained as a purple metallic solid (44 mg, 95%). m.p.: 177–183 °C; ^{1}H NMR (400 MHz, DMSO-d_6): δ 8.69 (d, J = 2.2 Hz, 1H), 8.29 (dd, J = 8.9, 2.2 Hz, 1H), 8.23 (d, J = 8.9 Hz, 2H), 8.21–8.14 (m, 4H), 7.72 (s, 1H), 7.64 (s, 1H), 7.59–7.45 (m, 6H), 7.28 (d, J = 8.9 Hz, 1H), 7.23 (d, J = 8.9 Hz, 1H), 5.53 (s, 2H), 2.85 (s, 4H); ^{13}C NMR (100 MHz, DMSO-d_6): δ 169.9, 165.2, 159.8, 158.6, 154.5, 145.2, 144.2, 143.1, 142.0, 137.4, 135.5, 132.0, 131.8, 131.6, 129.8, 129.6, 129.2, 129.1, 128.7, 126.8, 124.1, 121.5, 120.6, 119.6, 119.5, 115.2, 63.0, 25.5; ^{11}B NMR (128 MHz, DMSO-d_6): δ 0.93 (t, J = 32.8 Hz); ^{19}F NMR (376 MHz, DMSO-d_6): δ − 130.34 (q, J = 32.8 Hz); HRMS (m/z): [M-H]$^-$ calcd. for C$_{38}$H$_{25}$BF$_2$N$_5$O$_8$, 728.1764; found, 728.1730.

Synthesis of compound 2. A mixture of **7** (6.4 mg, 0.0088 mmol) and O-(2-aminoethyl)polyethylene glycol 5000 (CAS 32130–27–1) (40 mg, 0.008 mmol) was dissolved in anhydrous DMSO (0.88 ml) and stirred at rt for 18 h under a N$_2$ atmosphere. The solvent was removed by short-path distillation at rt overnight and the crude was partitioned between DCM (20 ml) and 1 M Na$_2$CO$_3$ (20 ml). The aqueous phase was extracted with DCM (2 × 20 ml). The organic layers were combined, washed with slightly acidic (HCl) water (20 ml), brine (20 ml), dried over anhydrous Na$_2$SO$_4$, filtered and evaporated to dryness. The residue was dissolved in HPLC grade water (8 ml) and the dark solution was passed through a Sep Pak C18 reverse-phase cartridge, then freeze dried. The product **2** was obtained as a dark green solid (40 mg, 90%). mp: 43–45 °C; ^{1}H NMR (400 MHz, DMSO-d_6): δ 8.74–8.72 (m, 1H), 8.28 (dd, J = 9.0, 2.3 Hz, 1H), 8.24–8.15 (m, 7H), 7.68 (s, 2H), 7.58–7.52 (m, 4H), 7.52–7.45 (m, 2H), 7.25–7.19 (m, 1H), 7.16 (d, J = 9.0 Hz, 2H), 4.66 (s, 2H), 3.70–3.65 (m, 4H), 3.50 (s, 680H); Ultraviolet–visible: λ_{max} (CHCl$_3$) 670 nm, (ε, 97,000 cm^{-1} M^{-1}); emission: λ_{max} (CHCl$_3$) 702 nm, Φ_F (CHCl$_3$) = 0.18; HRMS (m/z): MALDI–TOF distribution maximum centred at 5410.3999 Da.

Fluorescence quantum yields and extinction coefficients. The compound of interest (0.005 mmol) was dissolved in CHCl$_3$ (50 ml) to prepare a stock solution (10^{-4} M). The stock was diluted to concentrations 2, 4, 6, 8 and 10 × 10^{-7} M with CHCl$_3$ and each solution was analysed with an ultraviolet–visible spectrometer and a fluorescence spectrometer against CHCl$_3$ background. Excitation = 640 nm; emission range = 660–900 nm; slit width = 5/5 nm; scan rates = 600 nm min^{-1}. Plots of abs$_{max}$ versus conc and fluorescence area versus abs (640 nm) allowed the calculation of extinction coefficient and fluorescence quantum yield, respectively. Compound **1** (R = Ph, Ar = *p*MeOC$_6$H$_4$) was used as standard for fluorescence quantum yields with Φ_F = 0.36 (refs 17,42).

Fluorescence response of 2 and 5 to addition of DBU/TFA in organic solvents. Compound **2** or **5** was dissolved in toluene, tetrahydrofuran, dimethylformamide and DMSO (25 ml) to a final concentration of 5 μM. A solution of DBU (29.5 mg in 100 ml of CHCl$_3$) was added (64 μl = 1 eq) gradually, and absorbance and fluorescence spectra were recorded before and after each addition. The addition was stopped once spectra remained unchanged. At this stage, an excess of DBU was added and the spectra were recorded. Subsequently, a higher excess of TFA was added and the spectra were recorded. The area below the last two curves was plotted for off/on histogram (shown in Fig. 3). (Note: the toluene solution of **2** contained 1% CHCl$_3$ for solubility).

Fluorescence response of 2 to pH variation in DMEM. Compound **2** (2.8 mg) was dissolved in PBS (500 μl). The stock solution (1 mM) was diluted with DMEM supplemented with 10% FBS to the concentration of 5 μM. The pH of the solution was adjusted with diluted HCl or NaOH, to obtain a range from 8 to 2 at regular intervals, each of which was recorded, and the respective solution analysed by ultraviolet–visible absorption and fluorescence emission. Excitation = 625 nm; emission range = 635–900 nm.

Comparative solution and cellular photobleaching of 2 and lysotracker red and pHrodo red. Entire fluorescence cuvettes contain 1 × 10^{-7} M DMEM solutions at pH 4.0 of **2**; lysotracker red and pH-rodo red were continuously irradiated with light of wavelength 620(30) nm for **2** and 540(40) nm for lysotracker red and pH-rhodo red at 25 °C for 2 h. Filtered light from a 150-W light source used with complete cuvette irradiation via a fibre optic with attached light diffuser. Fluorophore fluorescence intensities were recorded every 20 min. The average fluorescence intensity from three independent experiments were normalized and plotted with sigmaplot 8.

Ten thousand HeLa-Kyoyo cells in DMEM were seeded onto chamber slides and incubated with **2** (20 µM) for 60 min or lysotracker red (150 nM) for 30 min, or pH rhodo red (15 µM) for 30 min. DMEM was replaced with fluorophore-free media and cells constantly irradiated with a Lumencor SPECTRA light engine LED used as the light source set to a maximum power for 400 s. Excitation filter 563/9 nm was used for lysotracker red and pH-rhodo red and excitation filter 640(14) nm was used for **2**. Cells were imaged with the shutter open, a time intervals of either 0.1 or 1.0 or 5 s with exposure of 10 ms and individual frames complied into movie format. The average cellular ROI fluorescence intensities from three independent experiments were plotted. An Olympus × 60 PLANAPO/1.42 objective and Andor iXon 888 ultra were used for signal detection. Acquisition and analysis performed with MetaMorph v7.8.

MTT assay of 2. Compound **2** (4.0 mg) was dissolved in sterile PBS (71 µl) to prepare a stock solution 10 mM. This was serially diluted to prepare samples at 5, 1, 0.5, 0.1 and 0.05 mM. Each of the stock solutions was diluted 1:10 with DMEM medium, which was co-incubated with HeLa or HEK293 cells at 5,000 cells per well on a 96-well plate for 24 h. The solution was removed and substituted with MTT solution (5 mg ml^{-1} in DMEM). The cells were incubated for 3 h. The medium was removed and the wells were treated with DMSO for 10 min. The absorbance of each well was read with a plate reader at 540 nm.

Production and validation of HeLa Kyoto cell line stably expressing LAMP1-GFP fusion protein. An expression plasmid encoding the LAMP1-GFP fusion protein was generated via the complete open-reading frame coded by a I.M.A.G.E. Fully Sequenced cDNA Clone (Source BioScience, I.M.A.G.E. ID: 5019745) of the human LAMP1 (GenBank accession number BC021288) was amplified by PCR using primers designed to append an XhoI site upstream of the translation initiation site and to replace the translation termination site by a segment encoding EcoRI site followed by a linker sequence CTCCTC (single-letter nucleotide code). The PCR product was gel purified and cloned in the XhoI–EcoRI sites of a pEGFP-N1 vector (BD Biosciences Clontech). Constructs were verified by DNA sequencing.

For stable transfection, HeLa cells were grown at 37 °C in complete DMEM supplemented with 10% FBS and 1% glutamine, to 30–40% confluency, and subsequently transfected with the LAMP1–GFP-encoding plasmid using FuGENE 6 (Roche) following the manufacturer's instruction. One day later, 0.6 g l^{-1} G418 was added. The medium was changed every day to remove the G418 non-resistant cells and when the cell number looked stabilized the G418 was lowered to 0.5 g l^{-1}. Cells displaying resistance to G418 and expressing LAMP1-GFP (as judged by fluorescence microscopy) were cloned by limiting dilution and, sorted on a BD FACSAria flow cytometer (Becton Dickinson). The clones were validated by immunostaining, by western blotting and by two functional assays (lysotracker uptake and dextran uptake).

Microscopy. Confocal images (Figs 5–7) were acquired using an Olympus Fluoview FV1000 CLSM and × 60/1.35(oil) UPLSAPO objective with a 635-nm laser at 12%, PMT voltage of ~750 v, pixel dwell time of 4 µs per pixel, pixel size 0.103 µm and image size 1,024 × 1,024. Nuclear staining was performed using Hoescht33342 or DAPI. Hoescht33342 signal was imaged using a 405-nm laser at 10% power and PMT voltage of ~700 v. GFP signal was imaged using a 488 nm laser line at 5% power and PMT voltage of ~600v.

Live-cell images (Figs 8–10) were acquired on a Zeiss AxioVert 200 M epi-fluorescent widefield microscope equipped with a Andor iXon 885 EMCCD, CoolLED pE-2 solid-state LEDs capable of excitation at 445, 488 and 635 nm, and Zeiss Plan-Apochromat × 100/1.40 Oil DIC objective. The microscope was surrounded by an incubation chamber that allowed the temperature and CO$_2$ to be maintained at 37 °C and 5%, respectively. Fluorophore **2** channel was recorded using a 649-nm emission long-pass filter, GFP was imaged using a 520/50 emission bandpass filter.

Fixed cell imaging. Cells were seeded onto an eight-well chambered glass slide and allowed to attach for 24 h. The media was then replaced with 200 µl of **2** (10 µM) in media and incubated for the appropriate time at 37 °C. Cells were counterstained with Hoechst 33342 or DAPI for 15 min. Cells were then washed once with PBS and fixed in 3.7% paraformaldehyde in PBS solution for 3 min and washed thoroughly with PBS. Images were collected by using an Olympus Fluoview 1,000 CLSM. The fluorescence arising from **2** was detected by a Cy5.5 filter. DAPI and GFP channels were used in parallel when cells were counterstained and/or transfected.

Fixed cell imaging at different pH. Cells were seeded onto an eight-well chambered glass slide and allowed to attach for 24 h. The media was then replaced with 200 µl of **2** (10 µM) in media incubated for 2 h at 37 °C. Cells were counterstained with DAPI for 15 min. Cells were then washed once with PBS and fixed in 3.7% paraformaldehyde in PBS solution for 3 min, and washed thoroughly with PBS. A collection of cell were Z-stack imaged using CLSM (PMT voltage = 782v, laser power 12%) and while maintaining focus of the microscope on the same cells the medium was exchanged with medium acidified to pH 4.9 (by addition of HCl (aq)). After allowing 15 min for equilibration the same cells were re-imaged (PMT voltage = 782v) using the same laser power. Following which the same cells were imaged for the third time following the adjustment of the PMT voltage 512v to obtain a non-saturated image. Mean total cell fluorescence was determined from two independent experiments using ImageJ.

Imaging following 4 °C incubation. Cells seeded onto an 8-well chambered glass slide and allowed to attach for 24 h. The media was then replaced with 200 µl of **2** (10 µM) in media incubated for 30 min at 4 °C. Cells were counterstained with DAPI for 15 min. Cells were then washed once with PBS and fixed in 3.7% paraformaldehyde in PBS solution for 3 min and washed thoroughly with PBS. A collection of cell were imaged using CLSM, and while maintaining focus of the microscope on the same cells the medium was exchanged with medium acidified to pH 4.9 (by addition of HCl (aq)). After allowing 15 min for equilibration, the same cells were re-imaged using the same exposure times and laser power.

Real-time live-cell imaging. HeLa Kyoto cells in Dulbecco's cell growth media containing 10% FBS were seeded onto an eight-well chambered glass slide and incubated for 24 h. The slides were placed on the microscope platform and the microscope was focused on a collection of cells. Next, **2** (final concentration 10 µM) was added and fluorescence images (Cy5.5 filter) were acquired at regular intervals. Images were deconvolved and combined in a video format.

Time-dependant efflux of 2. Cells seeded onto an eight8-well chambered glass slide and allowed to attach for 2 h. The media was then replaced with 200 µl of **2** (10 µM) in media incubated for 2 h at 37 °C. Media was replaced with fresh media and the loss of fluorescence monitored over time. Lysosome counting was carried out at 1, 15, 30 and 120 min using ImageJ.

Image processing. Deconvolution of widefield data sets was performed using AutQuant X3 deconvolution software with ten iterations of adaptive point spread function calculations. Lysosome detection and tracking were performed using Imaris 7.7.1 software (Bitplane Scientific). Background subtraction was applied to all images before lysosome detection. The Spots module of Imaris was used to detect lysosomes with an estimated diameter of 1.27 µm. Detected spots were filtered using the 'quality' algorithm. Only spots with values higher than the set threshold value (> 91.76) were analysed. Quality is defined as the intensity at the centre of the spot, Gaussian filtered by the spot radius. The success and accuracy of Spot detection was judged by visual inspection. Tracking lysosome movement over the course of the video was performed using an autoregressive motion algorithm. A maximum search distance of 1 µm was defined to disallow connections between a spot and a candidate match if the distance between the predicted future position of the spot and the candidate position exceeded the maximum distance. A gap-closing algorithm was also implemented to link track segment ends to track segment starts, to recover tracks that were interrupted by the temporary disappearance of particles. The maximum permissible gap length was set equal to three frames. Tracking all the lysosomes in the cell were selected by applying filters, which were based on 'Track Length' (> 0.2 µm) and 'Track Duration' (> 60 s).

Statistical analysis of cell images. Manders' and Pearson's coefficients used to show co-compartmentalization of LAMP1–GFP and **2** emissions were calculated using the Image J plugin 'Coloc2'. Rolling ball background subtraction (50 pixel diameter) and a Gaussian Filter (1 pixel diameter) were applied to all images before running the 'Coloc2' plugin. The ROI surrounding the cell was selected manually using the freeform drawing tool. Analysis was performed on six cells from two independent experiments.

Corrected total cell fluorescence (CTCF) in Fig. 7 was performed on six cells from two different experiments (Supplementary Fig. 8). Z-stack data acquired on the Olympus FLuoview100 was compressed into a single plane using the 'Sum Slice' function in Image J. Individual cells were selected using the freeform drawing tool to create a ROI (ROI). Selecting the 'Measure' function provided the area, the mean grey value and integrated density of the ROI. The mean background level was obtained by measuring the intensity in three different regions outside the cells and averaging the values obtained. The CTCF for each cell was calculated using the formula: CTCF = Integrated density of cell ROI − (Area of ROI × Mean fluorescence of background). The FEF was calculated by dividing the CTCF value of a cell at pH 7 into the CTCF value of the same cell at pH 4.9.

The number of lysosomes per field of view (Fig. 11) after efflux was counted using Z-stack data acquired on the Olympus Fluoview100, which was compressed into a single plane using the 'Sum Slice' function in Image J. A Max Entropy Threshold 15,000–40,000 was applied to each slice followed by use of the 'Despeckle', 'Erode' and 'Dilate' functions to remove noise. To count the number of lysosomes in each image the 'Analyze Particles' function was used to count objects with a circularity of 0.75–1.00 and size from 0 to 200 pixels.

In vivo mouse imaging. MDA-MB-231-luc-D3H1, a luciferase-expressing human breast adenocarcinoma cell line, was obtained from Caliper Life Sciences. Cells were maintained as a monolayer culture in minimum essential medium containing 10% (v/v) FBS and supplemented with 1% (v/v) L-glutamine, 50 U ml^{-1} penicillin, 50 µl ml^{-1} streptomycin, 1% (v/v) sodium pyruvate and 1% (v/v) non-essential amino acids. All cells were maintained in 5% CO_2 (v/v) and 21% O_2 (v/v) at 37 °C. Balb/C nu/nu mice (Harlan) were housed in the Biomedical Facility (UCD) in individually ventilated cages in temperature and humidity controlled rooms with a 12-h light–dark cycle. Two to five million MDA-MB-231-luc-D3H2LN cells in 100 µl of a DPBS:Matrigel (50:50) solution were injected subcutaneously behind the fore limb of the 5-week-old mice using a 25-g needle. Tumours reached an average diameter of 6 mm before injection. All animal protocols were approved by University College Dublin's local Animal Research Ethics Committee and under the licence from the Department of Health and Children. Animals were split into two groups ($n = 4$) and **2** dissolved in PBS (200 µl) was administered through the lateral tail vein at a concentration of 2 mg kg^{-1}. Optical imaging was performed with an IVIS Spectrum small-animal in-vivo imaging system (Caliper LS) with integrated isoflurane anaesthesia. A non-injected control animal was included. Images were acquired at regular intervals post injection of **2** with excitation 675 nm (30 nm band-pass filter) and emission 720 nm (20 nm band-pass filter) narrow band-pass filters and were analysed using Living Image Software v3.0 (Caliper LS).

References

1. Salipalli, S., Singh, P. K. & Borlak, J. Recent advances in live cell imaging of hepatoma cells. *J. BMC Cell Biol.* **15**, 26 (2014).
2. Correa, Jr I. R. Live-cell reporters for fluorescence imaging. *Curr. Opin. Chem. Biol.* **20**, 36–45 (2014).
3. Dean, K. M. & Palmer, A. E. Advances in fluorescence labelling strategies for dynamic cellular imaging. *Nat. Chem. Biol.* **10**, 512–523 (2014).
4. Baker, M. Cellular imaging: taking a long, hard look. *Nature* **466**, 1137–1140 (2010).
5. de Jong, M., Essers, J. & van Weerden, W. M. Imaging preclinical tumour models: improving translational power. *Nat. Rev. Cancer* **14**, 481–493 (2014).
6. Olivo, M., Ho, C. J. H. & Fu, C. Y. Advances in fluorescence diagnosis to track footprints of cancer progression in vivo. *Laser Photon. Rev.* **7**, 646–662 (2013).
7. Vahrmeijer, A. L., Hutteman, M., van der Vorst, J. R., van de Velde, C. J. H. & Frangioni, J. V. Image-guided cancer surgery using near-infrared fluorescence. *Nat. Rev. Clin. Oncol.* **10**, 507–518 (2013).
8. Liu, Y. et al. Near-infrared fluorescence goggle system with complementary metal-oxide-semiconductor imaging sensor and see-through display. *Biomed. Opt.* **18** 101303 1–10 (2013).
9. Sevick-Muraca, E. M. Translation of near-infrared fluorescence imaging technologies: emerging clinical applications. *Annu. Rev. Med.* **63**, 217–231 (2012).
10. van Dam, G. M. et al. Intraoperative tumor-specific fluorescence imaging in ovarian cancer by folate receptor-alpha targeting: first in-human results. *Nat. Med.* **17**, 1315–1319 (2011).
11. Nguyen, Q. T. et al. Surgery with molecular fluorescence imaging using activatable cell-penetrating peptides decreases residual cancer and improves survival. *Proc. Natl Acad. Sci. USA* **107**, 4317–4322 (2010).
12. Li, X., Gao, X., Shi, W. & Ma, H. Design strategies for water-soluble small molecular chromogenic and fluorogenic probes. *Chem. Rev.* **114**, 590–659 (2014).
13. Guo, Z., Park, S., Yoon, J. & Shin, I. Recent progress in the development of near-infrared fluorescent probes for bioimaging applications. *Chem. Soc. Rev.* **43**, 16–29 (2014).
14. Yuan, L., Lin, W., Zheng, K., He, L. & Huang, W. Far-red to near infrared analyte-responsive fluorescent probes based on organic fluorophore platforms for fluorescence imaging. *Chem. Soc. Rev.* **42**, 622–661 (2013).
15. Wu, D. & O'Shea, D. F. Synthesis and properties of BF$_2$-3,3'-dimethyldiaryl azadipyrromethene near-infrared fluorophores. *Org. Lett.* **15**, 3392–3395 (2013).
16. Palma, A. et al. Cellular uptake mediated off/on responsive near-infrared nanoparticles. *J. Am. Chem. Soc.* **133**, 19618–19621 (2011).
17. Batat, P. et al. BF$_2$-azadipyrromethenes: probing the excited-state dynamics of a NIR fluorophore and photodynamic therapy agent. *J. Phys. Chem. A.* **115**, 14034–14039 (2011).
18. Tasior, M. & O'Shea, D. F. BF$_2$-Chelated tetraarylazadipyrromethenes as NIR fluorochromes. *Bioconjugate Chem.* **21**, 1130–1133 (2010).
19. Kilpatrick, B. S., Eden, E. R., Hockey, L. N., Futter, C. E. & Patel, S. Methods for monitoring lysosomal morphology. *Methods Cell Biol.* **126**, 1–19 (2015).
20. Mayor, S. & Pagano, R. E. Pathways of clathrin-independent endocytosis. *Nat. Rev. Mol. Cell Biol.* **8**, 603–612 (2007).
21. Canton, I. & Battaglia, G. Endocytosis at the nanoscale. *Chem. Soc. Rev.* **41**, 2718–2739 (2012).
22. Casey, J. R., Grinstein, S. & Orlowski, J. Sensors and regulators of intracellular pH. *Nat. Rev. Mol. Cell Biol.* **11**, 50–61 (2010).
23. Lee, H. et al. Near-infrared pH-activatable fluorescent probes for imaging primary and metastatic breast tumors. *Bioconjugate Chem.* **22**, 777–784 (2011).
24. Han, J. & Burgess, K. Fluorescent indicators for intracellular pH. *Chem. Rev.* **110**, 2709–2728 (2010).
25. Koide, Y., Urano, Y., Hanaoka, K., Terai, T. & Nagano, T. Evolution of group 14 rhodamines as platforms for near-infrared fluorescence probes utilizing photoinduced electron transfer. *ACS Chem. Biol.* **6**, 600–608 (2011).
26. Urano, Y. et al. Selective molecular imaging of viable cancer cells with pH-activatable fluorescence probes. *Nat. Med.* **15**, 104–109 (2009).
27. Knop, K., Hoogenboom, R., Fischer, D. & Schubert, U. S. Poly(ethylene glycol) in drug delivery: pros and cons as well as potential alternatives. *Angew. Chem. Int. Ed.* **49**, 6288–6308 (2010).
28. Murtagh, J., Frimannsson, D. O. & O'Shea, D. F. Azide conjugatable and pH responsive near-infrared fluorescent imaging probes. *Org. Lett.* **11**, 5386–5389 (2009).
29. Zhang, X.-X. et al. pH-sensitive fluorescent dyes: are they really pH-sensitive in cells? *Mol. Pharm.* **10**, 1910–1917 (2013).
30. Hall, M. J., Allen, L. T. & O'Shea, D. F. PET modulated fluorescent sensing from the BF$_2$ chelated azadipyrromethene platform. *Org. Biomol. Chem.* **4**, 776–780 (2006).
31. Garcia, M. E. D. & Medel, A. S. Dye-surfactant interactions: a review. *Talanta* **33**, 255–264 (1986).
32. Katritzky, A. R., Fara, D. C., Yang, H. & Tamm, K. Quantitative measures of solvent polarity. *Chem. Rev.* **104**, 175–198 (2004).
33. Falcon-Perez, J. M., Nazarian, R., Sabatti, C. & Dell'Angelica, E. C. Distribution and dynamics of Lamp1-containing endocytic organelles in fibroblasts deficient in BLOC-3. *J. Cell Sci.* **118**, 5243–5255 (2005).
34. Bolte, S. & Cordelieres, F. P. A guided tour into subcellular colocalization analysis in light microscopy. *J. Microsc.* **224**, 213–232 (2006).
35. Firdessa, R., Oelschlaeger, T. A. & Moll, H. Identification of multiple cellular uptake pathways of polystyrene nanoparticles and factors affecting the uptake: relevance for drug delivery systems. *Eur. J. Cell Biol.* **93**, 323–337 (2014).
36. De Mey, J. R. et al. Fast 4D Microscopy. *Methods Cell Biol.* **85**, 83–112 (2008).
37. Godley, B. F. et al. Blue light induces mitochondrial DNA damage and free radical production in epithelial cells. *J. Biol. Chem.* **280**, 21061–21066 (2005).
38. Freundt, E. C., Czapiga, M. & Lenardo, M. J. Photoconversion of lysotracker red to a green fluorescent molecule. *Cell Res.* **17**, 956–958 (2007).
39. Galindo, F. et al. Synthetic macrocyclic peptidomimetics as tunable pH probes for the fluorescence imaging of acidic organelles in live cells. *Angew. Chem. Int. Ed.* **44**, 6504–6508 (2005).
40. Chazotte, B. Labeling lysosomes in live cells with lysotracker. *Cold Spring Harb. Protoc.* **2011**, pdb.prot5570 (2011).
41. Zhang, J. et al. Near-infrared fluorescent probes based on piperazine-functionalized BODIPY dyes for sensitive detection of lysosomal pH. *J. Mat. Chem. B* **3**, 2173–2184 (2015).
42. Gorman, A. et al. In vitro demonstration of the heavy-atom effect for photodynamic therapy. *J. Am. Chem. Soc.* **126**, 10619–10631 (2004).

Acknowledgements

D.O.S. gratefully acknowledges Science Foundation Ireland grant number 11/PI/1071(T) for financial support. E.C. and W.M.G. acknowledge the Irish Cancer Society Collaborative Cancer Research Centre BREAST-PREDICT (CCRC13GAL) for financial support. J.C.S. and A.P. acknowledge Science Foundation Ireland grant 09/IN.1/B2604 for financial support.

Author contributions

M.G. carried out all synthetic chemistry and photophysical measurements. M.M. and S.C. did all fixed and live-cell imaging and imaging analysis. D.S. set up and managed all imaging hardware and software used, and provided expertise advice. E.C. and M.T. conducted the in vivo imaging study. A.P. generated LAMP-1 GFP HeLa cells. J.S. provided LAMP-1 GFP HeLa cells and assisted with image and data analysis. W.G. provided expertise on in vivo imaging. D.O.S. wrote the manuscript with input from all the co-authors.

Additional information

Operando NMR spectroscopic analysis of proton transfer in heterogeneous photocatalytic reactions

Xue Lu Wang[1,*], Wenqing Liu[2,*], Yan-Yan Yu[3], Yanhong Song[2], Wen Qi Fang[1], Daxiu Wei[2], Xue-Qing Gong[3], Ye-Feng Yao[2,4] & Hua Gui Yang[1]

Proton transfer (PT) processes in solid–liquid phases play central roles throughout chemistry, biology and materials science. Identification of PT routes deep into the realistic catalytic process is experimentally challenging, thus leaving a gap in our understanding. Here we demonstrate an approach using operando nuclear magnetic resonance (NMR) spectroscopy that allows to quantitatively describe the complex species dynamics of generated H_2/HD gases and liquid intermediates in pmol resolution during photocatalytic hydrogen evolution reaction (HER). In this system, the effective protons for HER are mainly from H_2O, and CH_3OH evidently serves as an outstanding sacrificial agent reacting with holes, further supported by our density functional theory calculations. This results rule out controversy about the complicated proton sources for HER. The operando NMR method provides a direct molecular-level insight with the methodology offering exciting possibilities for the quantitative studies of mechanisms of proton-involved catalytic reactions in solid–liquid phases.

[1] Key Laboratory for Ultrafine Materials of Ministry of Education, School of Materials Science and Engineering, East China University of Science and Technology, Shanghai 200237, China. [2] Department of Physics, Shanghai Key Laboratory of Magnetic Resonance, East China Normal University, Shanghai 200062, China. [3] Key Laboratory for Advanced Materials, Centre for Computational Chemistry, Research Institute of Industrial Catalysis, East China University of Science and Technology, Shanghai 200237, China. [4] NYU-ECNU Institute of Physics at NYU Shanghai, 3663 Zhongshan Road North, Shanghai 200062, China. * These authors contributed equally to this work. Correspondence and requests for materials should be addressed to X.-Q.G. (email: xgong@ecust.edu.cn) or to Y.-F.Y. (email: yfyao@phy.ecnu.edu.cn) or to H.G.Y. (email: hgyang@ecust.edu.cn).

Proton transfer (PT) processes in solid–liquid phases are critically important and can be encountered throughout chemistry, biology and materials science[1-8]. The routes and dynamics of PT processes dictate the efficiency of the key biosynthetic and energy conversion systems, both natural and artificial[1-4]. Hence, tracking the PT trajectories deep into the realistic reaction becomes particularly significant. Molecular hydrogen (H_2) is a clean-burning fuel that can be produced from protons in the reductive half reaction of solar water splitting where the proton-related species are involved[2,9]. Nevertheless, critically important problems including side reactions restrict the productivity of the photocatalytic hydrogen evolution reaction (HER), in which the catalytic steps involve the coupling of light-generated redox equivalents to PT. In these cases, PT in HER is multidirectional. Thus, the addition of specific molecular reagents (also called sacrificial reagents), which is thermodynamically more oxidizable than water (H_2O), into the reaction environment can suppress the recombination of photogenerated charge carriers and the corresponding reverse reactions to some extent[10]. However, from a fundamental standpoint, some molecular reagents themselves contain protons, such as methanol, ethanol and triethanolamine, which means that such molecular agent itself can be a proton source and this undoubtedly increases the complexity of HER. Therefore, the vague and ambiguous proton source, PT routes, as well as the reaction mechanisms greatly limit our understanding and in turn the development of this field.

Recently, a series of methodologies for mechanism study, such as scanning tunnelling microscopy[11-13], time-resolved two-photon photoemission[14], temperature-programmed desorption[15-17], infrared spectroscopy[18-20], solid-state nuclear magnetic resonance[21-24] and density functional theory calculations (DFT)[25,26] have been used to address the central questions in HER. These methods can simulate and observe a rich surface chemistry of the interaction between adsorbed molecules and catalyst substrates, and distinguish various possible HER mechanisms. However, few of these techniques alone permit the direct quantification of specific species in reactions, and, moreover, the vast majority of them usually operate under ideal conditions (solid–gas or in ultrahigh vacuum), albeit with exquisite resolution, and it is still quite tough task to take into account the complexities of practical conditions such as solvent effect, surface specificity and adsorbate–adsorbate interactions. For instance, the adsorption and dissociation of methanol (CH_3OH) gas can occur easily on anatase-TiO_2 (101) surface under continuous ultraviolet illumination; however, in the liquid solution weak interactions such as hydrogen-bonding effects between CH_3OH and H_2O may significantly complicate the reaction and probably impair the conclusion/mechanism derived from the ideal reaction conditions. Furthermore, some possible hydrolysates in HER are out of ability to be observed under the gas–solid vacuum condition, and some changes that occur during the reaction may no longer be apparent when the sample returns to a non-working state. In this regard, there still exists a large gap between over-idealized conclusions and realistic heterogeneous reaction systems. Yet, to the best of our knowledge, no experimental evidence for quantitative insight into the PT mechanisms of HER in solid–liquid phases has been reported.

Here we demonstrate an approach, using operando NMR spectroscopy, that allows to directly quantify the populations of H_2/HD gases and liquid intermediates in pmol resolution, as well as the tracking of the PT routes deep into the realistic heterogeneous HER conditions, which contains Pd/TiO_2 (anatase), H_2O and CH_3OH. In this system, the effective protons for HER are mainly from H_2O, and CH_3OH evidently serves as an outstanding sacrificial agent reacting with holes. Furthermore, it also provides evidence to rule out controversy about the

complicated sources of protons for HER and the role of methanol as sacrificial molecules, with the methodology offering possibilities for the quantitative mechanism studies of proton-related catalytic reactions.

Results

Experimental set-up of operando NMR. Our experimental set-up consists of a micro-reactor system based on a NMR tube, allowing solid–liquid heterogeneous mixture to be directly studied inside the NMR coil. The experimental scheme is illustrated in Fig. 1 and Supplementary Fig. 1. The light source (300 W Xe lamp) beam is directed through a focusing lens assembly to converge on the entrance face of a homemade optical fibre bundle (seven quartz fibres, diameter 3 mm). The NMR study is incident on a suspension containing solid photocatalyst together with the aqueous mixtures, and the signals originate from not only the solid photocatalyst itself but also from the different gas–liquid molecules and chemical species dissolved in the solution[27,28]. This designed scheme has provided excellent feature information among the solid, liquid and gas mixtures without the separation of products. Our NMR studies are carried out on 700-MHz Agilent NMR spectrometers at a magnetic field strength of 16.4 T. An Agilent 5 mm z-axis pulsed field gradient triple resonance probe (^{1}H {^{13}C, ^{15}N}) is used (Detailed information can be found in Supplementary Fig. 2 and Supplementary Note 1).

To operando probe molecular species qualitatively and quantitatively in the solid–liquid phases, we measured some benchmark experiments (Supplementary Figs 3–10). First, the ^{1}H NMR signals of aqueous suspensions containing TiO_2, CH_3OH and D_2O were examined (Supplementary Figs 7 and 8). Two resonance peaks were clearly observed, corresponding to the methyl group (CH_3-) of CH_3OH (Supplementary Fig. 8b) and the

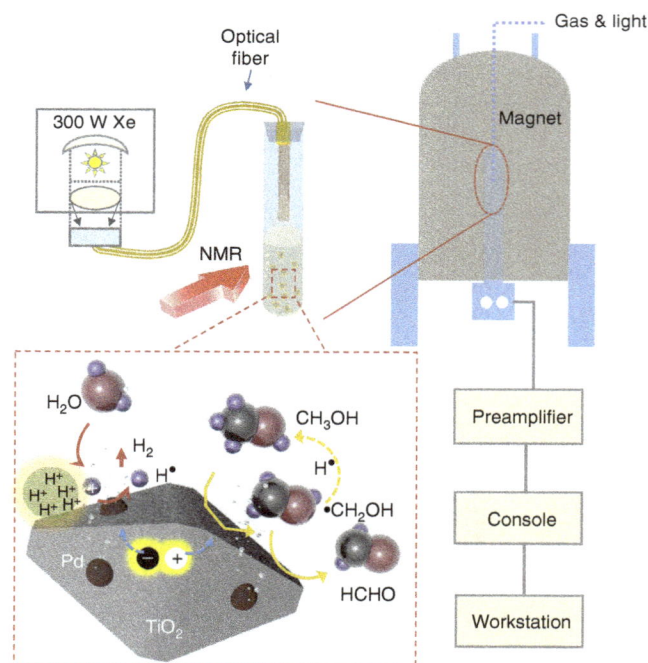

Figure 1 | Schematic layout of set-up for operando NMR studies. The set-up consists of a micro-reactor system based on a homemade NMR tube, allowing solid–liquid heterogeneous mixture to be directly studied inside the NMR coil while maintaining good uniformity. The light source (300-W Xe lamp) beam was directed through a focusing lens assembly to converge on the entrance face of a homemade optical fibre bundle (seven quartz fibres, diameter 3 mm).

residual protons of D_2O (Supplementary Fig. 8a)[29]. Interestingly, the signal of H_2 (singlet, 4.57 p.p.m.) was also remarkably detected in the spectrum of the suspension when the H_2 gas was injected into it (Supplementary Fig. 9). This could provide the way to quantitatively evaluate the amount of H_2 in the aqueous solution by comparing with the signal from the solution of saturated dissolved H_2. Second, the signal sensitivity was evaluated. Take 3-(trimethylsilyl)-1-propanesulfonic acid sodium salt (DSS) for example; when we decreased the content of DSS to nearly 50.7 pmol, the signal still had a relatively good signal-to-noise ratio (Supplementary Fig. 10). Even though by unitary sampling (number of scans is 1) of the heterogeneous mixture, the signals still had a good signal-to-noise ratio and thus could be distinctly distinguished. Therefore, the signals measured by the operando NMR could have pmol sensitivity, indicating a good capability to operando probe the gas–liquid intermediate products quantitatively deep into the solid–liquid working conditions at the very early stage of the reaction.

Operando NMR spectroscopy measurements. In a photocatalysis system, electrons are shuttled to the reductive side where they reduce protons to H_2. However, because of the complexity in building and optimizing such a complete system, when studying the reductive half reaction, it is common for a sacrificial electron donor to be used to optimize catalysts for H_2 production. Thus, for a clear recognition of the PT routes, it is significant to clarify the roles of the sacrificial agents. Widely accepted explanations of the roles for sacrificial agents (for example, methanol) are the thermodynamically more oxidizable ability than water, which work as an external driving force for the surface chemical reactions[12]. Their effects have been attributed to the consumption of holes, with the aim of suppressing the recombination of photogenerated charge carriers and the surface back reaction, which results in an increased H_2 evolution[30]. Yet, no experimental evidence for this mechanism was reported. On the other hand, some groups (for example, Xu et al.[15] and Highfield et al.[19]) pointed out that photo-reforming of methanol (gas) over the photocatalyst may also produce pure H_2, which makes it plausible to assume that the protons from methanol sacrificial agent have participated in the HER. However, these experimental simulations under ideal conditions depend on assumptions and simplifications, and many questions concerning details of the mechanisms in real conditions without the separation of products remain unanswered. As a consequence, the vague and ambiguous roles of sacrificial agents greatly limit our understanding of the PT routes.

To clarify these issues, NMR was applied on two deliberately designed reaction systems, namely System 1 ($H_2O/CD_3OD/$ catalyst) and System 2 ($CH_3OH/D_2O/$catalyst), to provide some qualitative preliminary results. The virtue of those systems lays on the variation of the protonated agent of the similar reaction, which allows selective monitoring of the PT routes in HER. Figure 2a shows ^1H NMR spectra of these two systems before and after the irradiation with ultraviolet/visible (UV/Vis) light. No H_2/HD NMR signal is observed for these samples in the dark. Upon irradiation for 2 h, the H_2 signal (single, 4.56 p.p.m.) and the HD signal (triplet, 4.46, 4.52 and 4.57 p.p.m.) resonances are clearly observed for System 1. Whereas only trace amount of HD gas (triplet, 4.47, 4.53 and 4.58 p.p.m.) is found for System 2, although this system has been irradiated for 24 h before the NMR study. Considering the solution compositions of System 1, we conclude that the proton source of the H_2 gas in the photocatalysis system, $H_2O/CH_3OH/$catalyst, is at least partial from H_2O definitely.

For a deeper understanding, we changed the concentration of H_2O of System 1. As can be seen from Fig. 2b and Supplementary

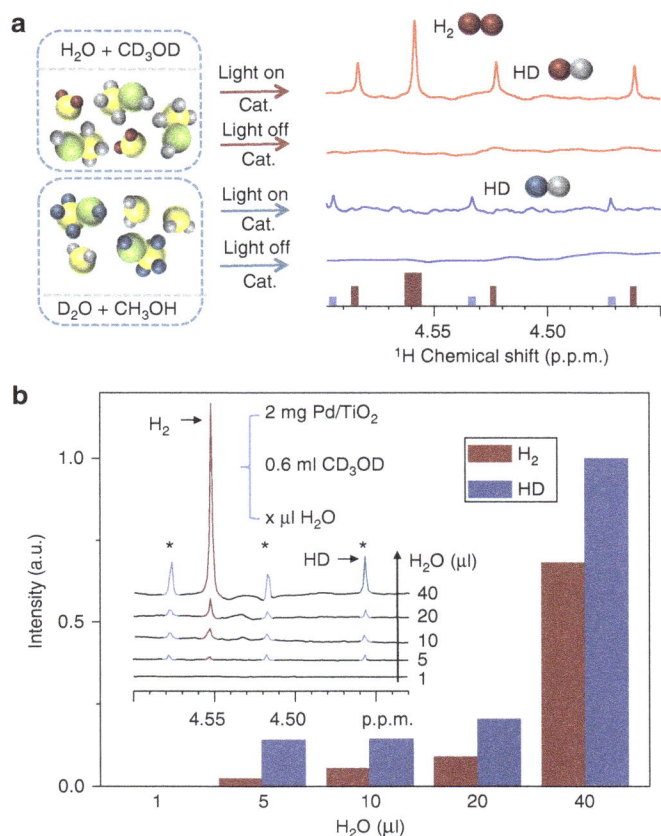

Figure 2 | ^1H NMR spectra of different systems. (a) ^1H NMR spectra of H_2/HD gases produced from heterogeneous systems before and after the irradiation with UV/Vis light. The spectra were acquired from samples containing (A,B) CH_3OH (20 μl), D_2O (500 μl) and Pd/TiO_2 (2 mg) at 298 K with a 5 s recycle delay and 16 scans before (A) and after (B) the UV/ Vis irradiation for 24 h; (C, D) H_2O (20 μl), CD_3OD (500 μl) and Pd/TiO_2 (2 mg) at 298 K with a 5 s recycle delay and 32 scans before (C) and after (D) the UV/Vis irradiation (300-W Xe lamp) for 2 h. The shifts of the H_2/HD peaks are caused by the different deuterated reagents. **(b)** ^1H NMR spectra of the correlations between produced H_2/HD gas and added H_2O. The NMR experiments were acquired from samples containing Pd/TiO_2 (2 mg), CD_3OD (600 μl), D_2O (1% DSS as the internal reference) and H_2O (x μl, x = 1, 5, 10, 20, 40) at 298 K with a 5 s recycle delay and 32 scans after the UV/Vis irradiation for 2 h.

Fig. 11, it is observed that the signal intensity of H_2 gas in the NMR spectra increases with the content of the adding H_2O. This is another direct evidence for its involvement in PT (from H_2O to H_2), which can support our deduction for the origin of H_2 in System 1. Note that the H/D isotope effect has been realized in literature[31,32]. The presence of the H/D isotope effect can decrease the reaction rate of water splitting; however, in principle will not harm our deduction for the origin of H_2 in the studied reaction system.

The origin of HD gas is an intriguing question and also related to the photocatalytic mechanism. In a mixture of H_2O and CD_3OD, few HDO and CD_3OH species can be generated because of the H/D exchange reactions[33], which further increases the complexity of the origin of HD gas. For System 1, there are two possibilities for the origin of HD gas: the product of HDO or the reforming product of CD_3OH. As inspired by the work of Yang et al.[17], the methyl group H of CH_3OH would transfer to the O_{BBO} sites of TiO_2 (anatase) when it is photocatalytically dissociated, while proton reduction reaction prefers to occur on the surface of cocatalyst rather than bulk TiO_2. Consistent with this interpretation, a control experiment without loaded Pd cocatalyst cannot generate any H_2/HD product (Supplementary Fig. 13). Thus, the origin of HD gas is likely from HDO. In order to verify our hypothesis, some compared experiments for the system of $D_2O/CH_3OH/TiO_2$ were carried out (Supplementary Fig. 14). The volume ratios selected for CH_3OH are from 5 to 20%. In these samples, few HDO and CH_3OD species would be generated. After the irradiation with UV/Vis light for 7 h, no H_2/HD NMR signal is observed. Thus, we can safely infer that the reduction process for CH_3OH/CH_3OD to produce H_2/HD is quite limited within 7 h in this system. That is to say, the origin of HD gas in System 1 is mainly from HDO. Similarly, in System 2, we prolonged the illumination time to 24 h; only trace amount of HD gas was found (this might indicate that the concentration of HDO is much more smaller than D_2O in this sample). The results provide a compelling evidence that the dominating proton source of H_2 gas in the HER of CH_3OH/H_2O/catalyst mixture (CH_3OH Vol % < 20%) is H_2O species, and H_2 gas from the photo-reforming process of CH_3OH in the CH_3OH/H_2O mixture solution is negligible.

Note that in a mixture of H_2O and CD_3OD, some H_3O^+ and $CH_3OH_2^+$ (as well as OH^- and CH_3O^-) species may also be generated by autoprotolysis. In this study, the pH value of the system is 7. The influence of the concentration of H_3O^+/OH^- and $CH_3OH_2^+/CH_3O^-$ on the photocatalytic reaction is the ongoing study in our laboratory.

To get a better understanding of the role of methanol, a series of ^{13}C-labelled methanol ($^{13}CH_3OH$) instead of the naturally abundant component was added in the reaction system ($D_2O/^{13}CH_3OH$/catalyst). Figure 3 shows the $^1H/^{13}C$ NMR spectra of methanol intermediates of the reaction system containing Pd/TiO_2, D_2O and CH_3OH. In this system, CH_3OD would be generated because of the fast H/D exchange reactions. Interestingly, after irradiating for 40 h, the ^1H NMR spectrum shows a well-resolved triplet signal with equal intensities (1:1:1 triplet, $J = 1.4$ Hz) at the right side of the CH_3 peak of $CH_3OH(D)$ (Fig. 3a). While similar triplet signals can also be observed from the ^{13}C NMR spectrum (Fig. 3b). According to the chemical shift and the characteristic J coupling value, we have assigned these two signals to CH_2D- of $CH_2DOH(D)$. The possibility of -CHD- can be easily ruled out by the observation in the 1D ^{13}C DEPT-135 (Distortionless Enhancement by Polarizition Transfer) spectrum where the triplet signals show clear negative intensities (Fig. 3b, inset).

For the formation of $CH_2DOH(D)$, two processes can be proposed: one is through an addition reaction of HCHO and HD, while the other is through the reaction of $\bullet CH_2OH(D)$ with D radicals ($\bullet D$). It has been reported that the average adsorption energy of HCHO is much lower than that of CH_3OH; therefore, it may desorb from the TiO_2 surface quickly rather than recombine with an H/D atom[19]. Furthermore, as will be discussed in a further publication, the HCHO may transform into methanediol species quickly in the aqueous solution, which accelerates the desorption rate of HCHO. It is therefore reasonable to conclude that the $CH_2DOH(D)$ is likely formed through a coupling reaction of $\bullet CH_2OH(D)$ and $\bullet D$, providing an indirect evidence of the existence of $\bullet CH_2OH(D)$ radicals during the HER. Meanwhile, $\bullet CH_2OH(D)$ is the inevitable intermediate when $CH_3OH(D)$ reacts with a hole (h^+), which gives a definitive evidence that CH_3OH serves as a sacrificial agent and plays a crucial role in mediating the capture of holes to form $\bullet CH_2OH$ radicals. To our knowledge, $\bullet CH_2OH$ species only exists in the liquid (may not be present in the ideal conditions) and is hard to be captured when it is in an isolated state, while it would also rapidly convert to HCHO (as also confirmed by DFT), thus

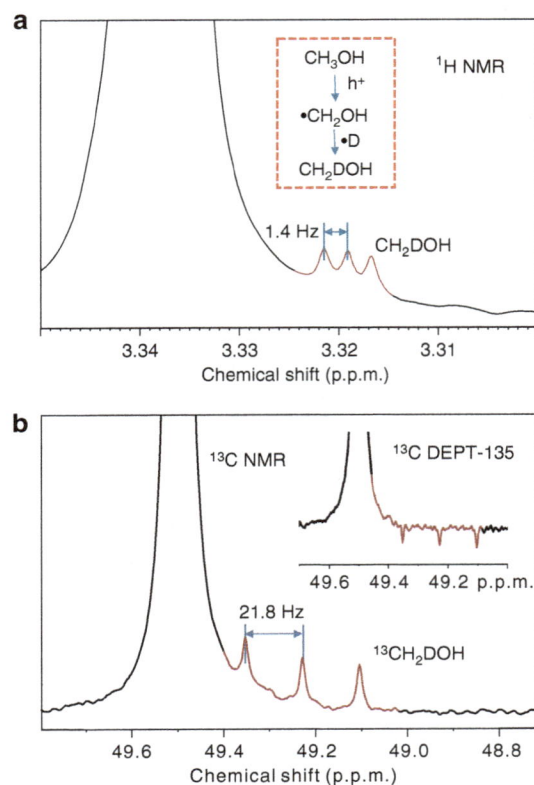

Figure 3 | High resolution NMR spectra of as-prepared samples.
(**a**) ^1H NMR spectrum of the methanol intermediates after the UV/Vis irradiation. The ^1H NMR experiment was acquired from samples containing Pd/TiO_2 (2 mg), CH_3OH (20 µl) and D_2O (500 µl) at 298 K with a 5 s recycle delay and 16 scans after the UV/Vis irradiation for 40 h. (**b**) ^{13}C NMR spectrum of the methanol intermediates after the UV/Vis irradiation. The ^{13}C NMR experiment was acquired from samples containing Pd/TiO_2 (2 mg), $^{13}CH_3OH$ (20 µl) and D_2O (500 µl) at 298 K with a 2 s recycle delay and 32 scans after the UV/Vis irradiation for 40 h. Inset is the DEPT-135 NMR spectrum of the as-prepared samples. The DEPT-135 spectrum was acquired from samples containing Pd/TiO_2 (2 mg), $^{13}CH_3OH$ (20 µl) and D_2O (500 µl) at 298 K with a 1 s recycle delay and 32 scans after the UV/Vis irradiation for 40 h. In the DEPT-135 spectrum, the negative ^{13}C signals indicate that the signals are from the partially deuterated methyl ($-CH_2D$).

resulting in a lacking experimental evidence and unconfirmed mechanisms for this important process.

DFT studies. For a better understanding of the reaction mechanism, we have also conducted systematic DFT calculations concerning the reactions of CH_3OH at both ground and excited states on the anatase-TiO_2 (101) surface[34,35]. The results are illustrated in Fig. 4a (ground state) and Fig. 4b (excited state). As one can see, at the ground state, the dissociation of methanol through breaking of O–H or C–H bond is obviously an endothermic process. Therefore, the CH_3OH may stay intact at the surface under dark conditions. Interestingly, with the help of h^+, both C–H and O–H bond-breaking reactions become exothermic, and the C–H dissociation is more favourable both thermodynamically and dynamically compared with the O–H dissociation. It indicates that CH_3OH may prefer to react with h^+ to form $\bullet CH_2OH$ species under excited conditions. Then, $\bullet CH_2OH$ species may react with $\bullet H$ ($\bullet D$) back to methanol or further react with h^+ to HCHO. Both reactions have also been studied in our calculations, and it turns out that, at the ground

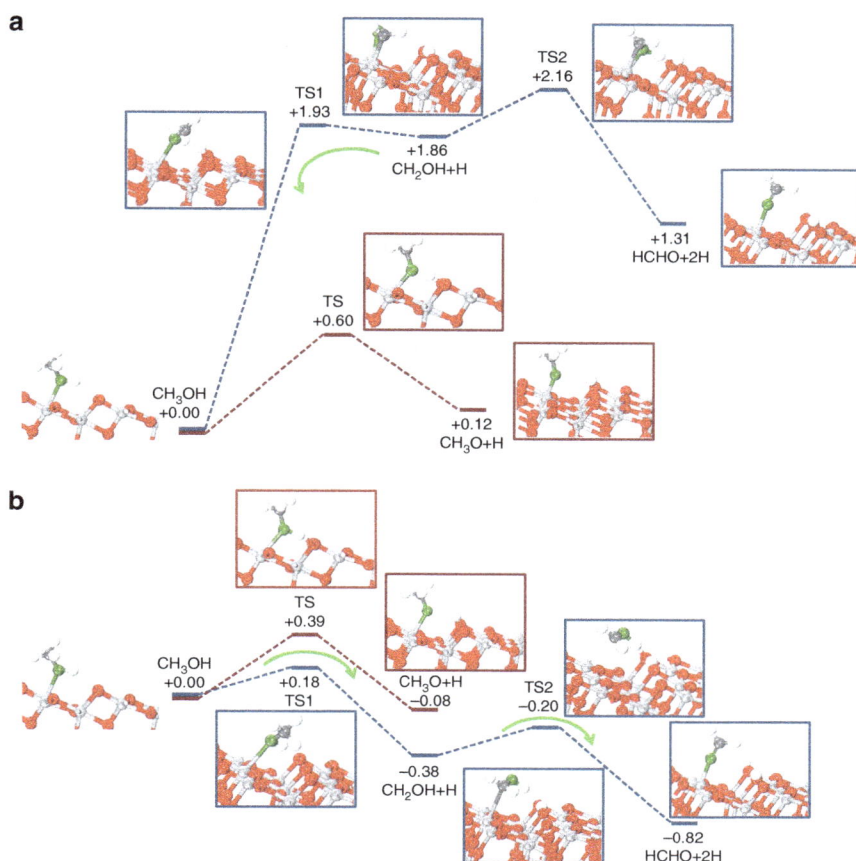

Figure 4 | DFT calculations. (**a**) DFT of the dissociation of CH_3OH at the ground state. (**b**) Photo-oxidation of CH_3OH begins with C–H broken (blue line) and O–H broken (red line). The calculated structures are displayed in the profile. The Ti atoms are in light grey and O in red, while the C atoms are in dark grey and H in white, and O atom of CH_3OH are in green. This notation is used throughout this paper.

state, the $\bullet CH_2OH$ may readily convert back to methanol with the barrier of only 0.07 eV—much lower than the further oxidation to HCHO (0.30 eV). By contrast, at the excited state, CH_2OH prefers the further oxidation to HCHO with a barrier of 0.18 eV rather than returning to methanol, which has a much higher barrier of 0.56 eV.

Quantitative operando NMR spectra. Operando NMR can not only provide qualitative insight into the catalytic mechanism but also be fully quantitative deep into the real solid–liquid phases. In our system, H_2 gas is a very important product, and its variation tendency may reflect some important microcosmic catalytic processes. Although some other characterization techniques such as gas chromatography (GC) and mass spectrometer (MS) can also be used to measure the gas products, they can only detect the gases that escape from the solution (after the dissolved gas reaches the saturation concentration), whereas in fact quite amount of gas molecules might be already present in the solution with partial dissolution and adsorption at the early stage of the reaction. Furthermore, the microcosmic reaction processes and trends before the dissolved gas reaches the saturation concentration will be missing, and the processes could last for more than 10 min for the 40 nmol H_2 dissolving in 100 ml H_2O solution. Meanwhile, it is also impossible for GC or MS to probe some intermediates that cannot escape from the reaction solution. In this context, the operando NMR approach is superior to GC or MS in studying the HER.

In our system, assuming that the H_2 gas does not reach the saturation state at the early stage of the reaction, the observed H_2

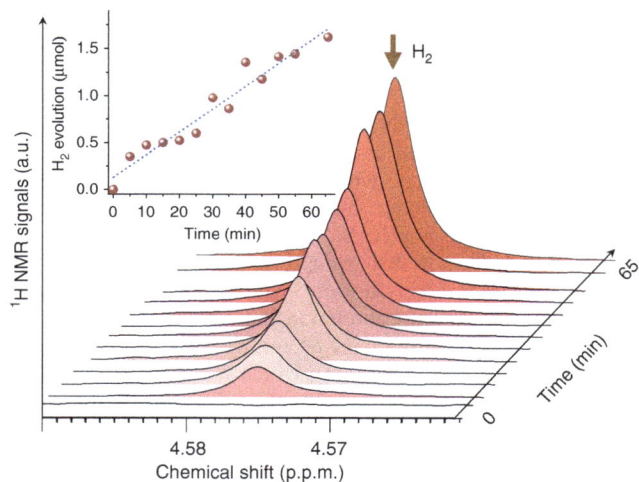

Figure 5 | Quantitative operando ^1H NMR spectra. Quantitative time-related ^1H NMR spectra of Pd/TiO_2 at 298 K in H_2O/CD_3OD mixtures under UV/Vis light irradiation. The operando NMR experiments were acquired from samples containing Pd/TiO_2 (1 mg), CD_3OD (300 µl), D_2O (1% DSS as the internal reference) and H_2O (300 µl) at 298 K with a 5 s recycle delay and eight scans under the UV/Vis irradiation. The first 1,000 points of the FID was cut off to obtain the final separable H_2 signals.

signals in the NMR spectra could reflect the whole H_2 gas from HER. H_2 gas can be quantified by using the intensity of the H_2 signal in the H_2–D_2O saturation solution (estimate 0.018 ml H_2

per 1 ml D_2O) at the same temperature (Supplementary Figs 15 and 16). Through comparison of the H_2 resonance intensities with calibration samples, it is possible to quantify the number of H_2 species in absolute terms at the early stage and gain insight into the PT behaviour properties in working state. Figure 5 shows the quantitative operando ^1H NMR spectrum of $H_2O/CD_3OD/$ catalyst system. For a suitable comparison with the results obtained from the conventional GC detector (Supplementary Fig. 12), the sample was irradiated directly by an Xe light source. The H_2 signal intensity increased linearly on continuous illumination. Under this condition, no HD signal can be detected, and this can be because of the few exchanged HDO products compared with H_2O, consistent with the previous results. The quantification result in Fig. 5 shows that the H_2 evolution rate at the early stage from the NMR measurement is $\sim 1.560\,\mu mol\,h^{-1}$ mg^{-1}. The above result may be far-reaching because it indicates that the widely used methods have the intrinsic shortcoming on the detection of the reaction especially at the early stage. As a consequence, the reaction models derived from the *ex situ* measurement, especially the models for the reaction kinetics, will deviate from the actual reaction mechanisms. Thus, this operando NMR spectroscopic method provides an approach to guide the design of HER systems, offering better measures deep into the realistic reaction.

Discussion

By using an operando NMR method, we successfully quantified the generation of H_2/HD gases and liquid intermediates in pmol resolution without the separation of products, and provided evidence for the PT routes during the HER. On the basis of the photocatalytic system containing anatase-TiO_2, methanol and water, we find that the effective protons for HER are mainly from water, and methanol evidently serves as an outstanding electronic sacrificial agent reacting with holes. However, for the sacrifice reforming process of methanol, a small amount of released methyl group protons still occur, which are likely bound to the surface of TiO_2, and their migration to cocatalyst to take part in the HER would consume more time and energy. This indicates that complicated synergistic effect between water and methanol exists under practical HER conditions. The picture in Supplementary Fig. 17 illustrates the paramount PT processes together with the reaction mechanism of HER. Carrier generation and recombination occur when an electron transfers from the valence band to conduction band in a semiconductor. Once the photoexcited charges reach the surface of the particle, they can take part in surface redox reactions with adsorbed donor or acceptor molecules[34–37]. In our system, the induced electron transfers from the TiO_2 conduction band to the metal particles, and the attached proton would be reduced by an electron to its monoatomic radical species ($\bullet H$), which is the main predecessor of H_2 (Supplementary Fig. 17, (equation 5)). On the other side, the holes would be consumed by CH_3OH, extracting the charges competitively with their recombination and then improving the formation rate of H_2. On the oxidative side, two sequential hole consumptions convert the adjacent CH_3OH species to $\bullet CH_2OH$ (equation 2) and HCHO (equation 3), respectively, accompanying with the release of protons. Subsequent spontaneous (dark) coupling reaction between $\bullet CH_2OH$ and $\bullet H$ radicals results in CH_3OH species re-formation (equation 4). On the reductive side, protons would be reduced to hydrogen by the shuttled electrons on the surface of cocatalyst (equation 5). In H_2O/CH_3OH (CH_3OH vol % $< 20\%$) mixtures, H_2O molecules would provide a vast amount of free protons in solution (equation 1). Although the photocatalytically dissociated protons from CH_3OH are constrained by the O_{BBO} nearby, only a

slim chance for those to be reduced to H_2 exists. As a consequence, the current NMR spectroscopic studies provide a unique viewing angle for us to understand the predominant PT processes of HER in solution.

The results in this work deepen our understanding of interfacial heterogeneous redox reactions, and the operando NMR method will be applicable extensively with the methodology offering exciting possibilities for the microcosmic mechanism study (pmol resolution) in catalysis, photoelectrocatalysis and electrocatalysis of reactions such as CO_2 reduction, waste-water remediation and methanol reforming.

Methods

Synthesis of catalyst. The Pd(1.0 wt %)/TiO_2 photocatalyst powder was prepared by the co-precipitation method. Briefly, appropriate amount of $PdCl_2$ aqueous solution (1 wt%) was added into the TiO_2 (100 mg, Anatase, Sigma) powder and maintained at 80 °C for 1 h. After being dried, the products were calcined at 300 °C for 2 h. Before being characterized and tested, the samples were reduced in 20% H_2/Ar at 300 °C for 1 h.

NMR experiment. All NMR experiments were acquired on 700 MHz Agilent NMR spectrometers at a magnetic field strength of 16.4 T. An Agilent 5 mm z-axis pulsed field gradient triple-resonance static probe was used in all ^1H and ^{13}C NMR experiments. All ^1H NMR measurements were carried out at 298 K with a spectral width of 20 p.p.m., pulse width of 4 μs (45°), a recycle delay of 5 s and 8, 16 or 32 scans. To suppress the water signal, the first 1,000 points of the free induction decay (FID) were cut off before FT. ^{13}C NMR was carried out at 298 K with a spectral width of 253 p.p.m., pulse width of 7.3 μs (45°), a recycle delay of 2 s and 32 scans. All experiments had DSS as the internal reference. To enhance the NMR signals, the examples could also be directly irradiated by an Xe light source during the irradiation accumulation process.

Operando ^1H NMR studies. A 300-W Xe lamp was used as the light source. The light source is outside the magnet. The light source beam was directed through a focusing lens assembly to converge on the entrance face of a homemade optical fibre bundle (seven quartz fibres, diameter 3 mm). The cap of the tube was specially designed to allow the input of the light and the H_2 gas. Note that the fibre may lead to an attenuation of the light intensity and a lower reaction rate (see Fig. 5 and Supplementary Fig. 18). In order to simulate the HER process similar to the conventional GC, the samples might be irradiated directly by an Xe light source.

Calculation details. All the calculations were carried out using the Vienna *ab initio* simulation package[38,39], and the exchange-correlation term was described by the Perdew, Burke and Ernzerhof version within the generalized gradient approximation[40]. The project-augmented wave[41,42] method was used to represent the core–valence electron interaction. The titanium 3s, 3p, 3d, 4s and the carbon and oxygen 2s, 2p electrons were treated as valence electrons and an energy cutoff of 400 eV for the basis-set expansion was used.

The anatase-TiO_2 (101) surface was modelled by a five trilayer slab with only the centre layer fixed, and other atoms were allowed to relax until atomic forces reached below 0.05 eV Å^{-1}. As suggested by Luo and co-workers[43], the hole was introduced by using the triplet state to mimic the singlet excited state[43–47]. A 3 × 1 surface cell and a > 15 Å vacuum gap was used. Different k-point meshes were tested, and it was found the k-point sampling restricted to the Γ point only can already provide reliable results regarding adsorption energies.

The transition states in reactions were located with a constrained optimization scheme[48], and were verified when (i) all forces on atoms vanish and (ii) the when total energy is maximum along the reaction coordinate but minimum with respect to the rest of the degrees of freedom.

References

1. Han, Z., Qiu, F., Eisenberg, R., Holland, P. L. & Krauss, T. D. Robust photogeneration of H_2 in water using semiconductor nanocrystals and a nickel catalyst. *Science* **338**, 1321–1324 (2012).
2. Liu, J. *et al.* Metal-free efficient photocatalyst for stable visible water splitting via a two-electron pathway. *Science* **347**, 970–974 (2015).
3. Hundt, P. M., Jiang, B., Reijzen, M. E., Guo, H. & Beck, R. D. Vibrationally promoted dissociation of water on Ni (111). *Science* **334**, 504–507 (2014).
4. Nielsen, M. *et al.* Low-temperature aqueous-phase methanol dehydrogenation to hydrogen and carbon dioxide. *Nature* **495**, 85–90 (2013).
5. McLaren, A. D. The beckmann rearrangement of aliphatic ketoximes. *Science* **103**, 503 (1946).
6. Corma, A. & García, H. Lewis acids: from conventional homogeneous to green homogeneous and heterogeneous catalysis. *Chem. Rev.* **103**, 4307–4365 (2003).

7. Clarke, H. T., Gillespie, H. B. & Weisshaus, S. Z. The action of formaldehyde on amines and amino acids. *J. Am. Chem. Soc.* **446**, 4571–4587 (1933).

8. Albrecht, Ł. *et al.* Asymmetric organocatalytic formal [2 + 2]-cycloadditions via bifunctional H-bond directing dienamine catalysis. *J. Am. Chem. Soc.* **134**, 2543–2546 (2012).

9. Heyduk, A. F. & Nocera, D. G. Hydrogen produced from hydrohalic acid solutions by a two-electron mixed-valence photocatalyst. *Science* **293**, 1639–1641 (2001).

10. Chen, X., Shen, S., Guo, L. & Mao, S. S. Semiconductor-based photocatalytic hydrogen generation. *Chem. Rev.* **110**, 6503–6570 (2010).

11. Onishi, H. & Iwasawa, Y. Dynamic visualization of a metal-oxide-surface/gas-phase reaction: time-resolved observation by scanning tunneling microscopy at 800 K. *Phys. Rev. Lett.* **76**, 791–794 (1996).

12. Scheiber, P., Riss, A., Schmid, M., Varga, P. & Diebold, U. Observation and destruction of an elusive adsorbate with STM: O$_2$/TiO$_2$ (110). *Phys. Rev. Lett.* **105**, 216101 (2010).

13. Zhang, Z., Bondarchuk, O., White, J. M., Kay, B. D. & Dohnálek, Z. Imaging adsorbate O-H bond cleavage: methanol on TiO$_2$ (110). *J. Am. Chem. Soc.* **128**, 4198–4199 (2006).

14. Li, B. Ultrafast interfacial proton-coupled electron transfer. *Science* **311**, 1436–1440 (2006).

15. Xu, C. *et al.* Molecular hydrogen formation from photocatalysis of methanol on anatase-TiO$_2$ (101). *J. Am. Chem. Soc.* **136**, 602–605 (2014).

16. Xu, C. *et al.* Molecular hydrogen formation from photocatalysis of methanol on TiO$_2$ (110). *J. Am. Chem. Soc.* **135**, 19039–19045 (2013).

17. Guo, Q. *et al.* Stepwies photocatalytic dissociation of methanol and water on TiO$_2$ (110). *J. Am. Chem. Soc.* **134**, 13366–13373 (2012).

18. Zhang, M., Respinis, M. & Frei, H. Time-resolved observations of water oxidation intermediates on a cobalt oxide nanoparticle catalyst. *Nat. Chem.* **6**, 362–367 (2014).

19. Highfield, J. G., Chen, M. H., Nguyen, P. T. & Chen, Z. Mechanistic investigations of photo-driven processes over TiO$_2$ by in-situ DRIFTS-MS: part 1. Platinization and methanol reforming. *Energ. Environ. Sci.* **2**, 991–1002 (2009).

20. Chen, T. *et al.* Mechanistic studies of photocatalytic reaction of methanol for hydrogen production on Pt/TiO$_2$ by *in situ* fourier transform IR and time-resolved IR spectroscopy. *J. Phys. Chem. C* **111**, 8005–8014 (2007).

21. Blanc, F., Leskes, M. & Grey, C. P. In situ solid-state NMR spectroscopy of electrochemical cells: batteries, supercapacitors, and fuel cells. *Acc. Chem. Res.* **46**, 1952–1963 (2013).

22. Cattaneo, A. S. *et al.* Operando electrochemical NMR microscopy of polymer fuel cells. *Energy Environ. Sci.* **8**, 2383–2388 (2015).

23. Chan, K. W. H. & Wieckowski, A. Probing adsorbates on Pt electrode surfaces by the use of ^{13}C spin-echo NMR. Studies of CO generated from methanol electrosorption. *J. Electrochem. Soc.* **137**, 367–368 (1990).

24. Tong, Y. Y., Wieckowski, A. & Oldfield, E. NMR of electrocatalysts. *J. Phys. Chem. B* **106**, 2434–2446 (2002).

25. Sánchez, V. M., Cojulun, J. A. & Scherlis, D. A. Dissociation free energy profiles for water and methanol on TiO$_2$ surfaces. *J. Phys. Chem. C* **114**, 11522–11526 (2010).

26. Setvin, M. *et al.* Reaction of O$_2$ with subsurface oxygen vacancies on TiO$_2$ anatase (101). *Science* **341**, 988–991 (2013).

27. Huang, J., Jiang, Y., Vegten, N., Hunger, M. & Baiker, A. Tuning the support acidity of flame-made Pd/SiO$_2$-Al$_2$O$_3$ catalysts for chemoselective hydrogenation. *J. Catal.* **281**, 352–360 (2011).

28. Kolmer, A. *et al.* The influence of electronic modifications on rotational barriers of bis-NHC-complexes as observed by dynamic NMR spectroscopy. *Magn. Reson. Chem.* **51**, 695–700 (2013).

29. Suleimanov, N. M. *et al.* In situ muSR and NMR investigation of methanol dissociation on carbon-supported nanoscaled Pt-Ru catalyst. *J. Solid State Electrochem.* **17**, 2115–2121 (2013).

30. Schrauben, J. N. *et al.* Titanium and zinc oxide nanoparticles are proton-coupled electron transfer agents. *Science* **336**, 1298–1301 (2012).

31. Krishtalik, L. I. Kinetic isotope effect in the hydrogen evolution reaction. *Electrochim. Acta* **46**, 2949–2960 (2001).

32. Urey, H. C., Brickwedde, F. G. & Murphy, G. M. A hydrogen isotope of mass 2 and its concentration. *Phy. Rev.* **40**, 1–15 (1932).

33. Fenby, D. V. & Chand, A. Thermodynamic study of deuterium exchange in water + methanol systems. *J. Chem. Soc. Faraday Trans.* **74**, 1768–1775 (1978).

34. Lin, K., Zhou, X., Luo, Y. & Liu, S. The microscopic structure of liquid methanol from raman spectroscopy. *J. Phys. Chem. B* **114**, 3567–3573 (2010).

35. Valero, M. C., Raybaud, P. & Sautet, P. Nucleation of Pd$_n$ (n = 1-5) clusters and wetting of Pd particles on γ-Al$_2$O$_3$ surfaces: a density functional theory study. *Phys. Rev. B* **75**, 045427 (2007).

36. Gerritzen, D. & Limbach, H. H. Kinetic and equilibrium isotope effects of proton exchange and autoprotolysis of pure methanol studies by dynamic NMR spectroscopy. *Ber. Bunsenges. Phys. Chem.* **85**, 527–535 (1981).

37. Galińska, A. & Walendziewski, J. Photocatalytic water splitting over Pt-TiO$_2$ in the presence of sacrificial reagents. *Energy and Fuels* **19**, 1143–1147 (2005).

38. Kresse, G. & Furthmüller, J. Efficiency of *ab-initio* total energy calculations for metals and semiconductors using a plane-wave basis set. *Comput. Mater. Sci.* **6**, 15–50 (1996).

39. Kresse, G. & Furthmüller, J. Efficient iterative schemes for *ab initio* total-energy calculations using a plane-wave basis set. *Phys. Rev. B* **54**, 11169–11186 (1996).

40. Perdew, J. P., Burke, K. & Ernzerhof, M. Generalized gradient approximation made simple. *Phys. Rev. Lett.* **77**, 3865–3868 (1996).

41. Kresse, G. & Joubert, D. From ultrasoft pseudopotentials to the projector augmented-wave method. *Phys. Rev. B* **56**, 1758–1775 (1999).

42. Blöchl, P. E. Projector augmented-wave method. *Phys. Rev. B* **50**, 17953–17979 (1994).

43. Ji, Y., Wang, B. & Luo, Y. A comparative theoretical study of proton-coupled hole transfer for H$_2$O and small organic molecules (CH$_3$OH, HCOOH, H$_2$CO) on the anatase TiO$_2$ (101) Surface. *J. Phys. Chem. C* **118**, 21457–21462 (2014).

44. Ji, Y., Wang, B. & Luo, Y. Location of trapped hole on rutile-TiO$_2$ (110) surface and its role in water oxidation. *J. Phys. Chem. C* **116**, 7863–7866 (2012).

45. Di, V. C. & Fittipaldi, D. Hole scavenging by organic adsorbates on the TiO$_2$ surface: a DFT model study. *J. Phys. Chem. Lett.* **4**, 1901–1906 (2013).

46. Ji, Y., Wang, B. & Luo, Y. First principles study of O$_2$ adsorption on reduced rutile TiO$_2$ (110) surface under UV illumination and its role on CO oxidation. *J. Phys. Chem. C* **117**, 956–961 (2013).

47. Jedidi, A., Markovits, A., Minot, C., Bouzriba, S. & Abderraba, M. Modeling localized photoinduced electrons in rutile-TiO$_2$ using periodic DFT + U methodology. *Langmuir* **26**, 16232–16238 (2010).

48. Alavi, A., Hu, P., Deutsch, T., Silvestrelli, P. L. & Hutter, J. CO oxidation on Pt (111): an *ab initio* density functional theory study. *Phys. Rev. Lett.* **80**, 3650–3653 (1998).

Acknowledgements

This work was financially supported by the National Natural Science Foundation of China (21373083, 21421004, 21322307 and 21574043), the Fundamental Research Funds for the Central Universities (WD1313009 and WD1514303), SRF for ROCS, SEM, SRFDP, Program of Shanghai Subject Chief Scientist (15XD1501300), National High-tech R&D Program of China (863 Program; 2014AA123400, 2014AA123401).

Author contributions

H.G.Y. conceived the project and contributed to the design of the experiments and analysis of the data. X.L.W. performed the catalyst preparation, characterizations and wrote the paper. Y.-F.Y., Y.S., W.Q.F., D.W. and W.L. conducted the NMR examination and contributed to writing the NMR section. X.-Q.G. and Y.-Y.Y. conducted DFT calculations and wrote part of the paper (calculations). All the authors discussed the results and commented on the manuscript.

Additional information

Permissions

List of Contributors

Akinori Kuzuya, Yusuke Sakai, Takahiro Yamazaki, Yan Xu and Makoto Komiyama
Research Center for Advanced Science and Technology, The University of Tokyo, 4-6-1 Komaba, Meguro, Tokyo 153-8904, Japan

Seong W. Choi and Graham S. Timmins
Department of Pharmaceutical Sciences, College of Pharmacy, University of New Mexico, Albuquerque, New Mexico 87131, USA

Mamoudou Maiga, Mariama C. Maiga and William R. Bishai
Department of Medicine, Center for Tuberculosis Research, Johns Hopkins University, Baltimore, Maryland 21231, USA

Viorel Atudorei and Zachary D. Sharp
Department of Earth and Planetary Sciences, College of Pharmacy, University of New Mexico, Albuquerque, New Mexico 87131, USA

Wei Ma, Hua Kuang, Liguang Xu and Chuanlai Xu
State Key Lab of Food Science and Technology, School of Food Science and Technology, Jiangnan University, Wuxi, Jiangsu 214122, China

Li Ding
State Key Lab of Food Safety Test (Hunan), Changsha, Hunan 410004, China

Nicholas A. Kotov
Department of Chemical Engineering, University of Michigan, Ann Arbor, Michigan 48109, USA
Department of Materials Science, University of Michigan, Ann Arbor, Michigan 48109, USA
Department of Biomedical Engineering, University of Michigan, Ann Arbor, Michigan 48109, USA
Biointerface Institute, University of Michigan, Ann Arbor, Michigan 48109, USA

Libing Wang
State Key Lab of Food Science and Technology, School of Food Science and Technology, Jiangnan University, Wuxi, Jiangsu 214122, China
State Key Lab of Food Safety Test (Hunan), Changsha, Hunan 410004, China

Katsuro Hayashi
Center for Secure Materials, Materials and Structures Laboratory, Tokyo Institute of Technology, R3-34, 4259 Nagatsuta, Yokohama 226-8503, Japan

Peter V. Sushko and Alexander L. Shluger
Department of Physics and Astronomy, University College London, London WC1E 6BT, UK

Yasuhiro Hashimoto
Department of New Business Development, Asahi Kasei Corporation, 1-105 Kanda-Jinbocho, Tokyo 101-8101, Japan

Hideo Hosono
Frontier Research Center, Tokyo Institute of Technology, S2-13, 4259 Nagatsuta, Yokohama 226-8503, Japan

Yurui Xue, Xun Li and Wenke Zhang
State Key Laboratory of Supramolecular Structure and Materials, College of Chemistry, Jilin University, 2699 Qianjin Street, Changchun 130012, China

Hongbin Li
Department of Chemistry, University of British Columbia, Vancouver, British Columbia, Canada V6T 1Z1

Igor Dolamic, Stefan Knoppe and Thomas Bürgi
Département de Chimie Physique, Université de Genève, 30 Quai Ernest-Ansermet, 1211 Genève 4, Switzerland

Amala Dass
Department of Chemistry and Biochemistry, University of Mississippi, 352 Coulter Hall, University, Mississippi 38677, USA

Yuanmu Yang
Interdisciplinary Materials Science Program, Vanderbilt University, Nashville, Tennessee 37212, USA

Ivan I. Kravchenko and Dayrl P. Briggs
Center for Nanophase Materials Sciences, Oak Ridge National Laboratory, Oak Ridge, Tennessee 37831, USA

Jason Valentine
Department of Mechanical Engineering, Vanderbilt University, Nashville, Tennessee 37212, USA

Joosub Lee and Minkyeong Pyo
Department of Chemical Engineering, Hanyang University, Seoul 133-791, Korea

Sang-hwa Lee
Department of Physics, Hanyang University, Seoul 133-791, Korea

Jaeyong Kim
Department of Physics, Hanyang University, Seoul
133-791, Korea
Institute of Nano Science and Technology, Hanyang
University, Seoul 133-791, Korea

Moonsoo Ra and Whoi-Yul Kim
Department of Electronic Engineering, Hanyang
University, Seoul 133-791, Korea

Bum Jun Park
Department of Chemical Engineering, Kyung Hee
University, Youngin-Si, Gyeonggi-do 446-701, Korea

Chan Woo Lee
Institute of Nano Science and Technology, Hanyang
University, Seoul 133-791, Korea

Jong-Man Kim
Department of Chemical Engineering, Hanyang
University, Seoul 133-791, Korea. Institute of Nano
Science and Technology, Hanyang University, Seoul
133-791, Korea

**Nicoló Maccaferri, Keith E. Gregorczyk and Thales
V.A.G. de Oliveira**
CIC nanoGUNE, 20018 Donostia-San Sebastián, Spain

Mikko Kataja and Sebastiaan van Dijken
NanoSpin, Department of Applied Physics, Aalto
University School of Science, 00076 Aalto, Finland

Zhaleh Pirzadeh and Alexandre Dmitriev
Department of Applied Physics, Chalmers University
of Technology, 41296 Gothenburg, Sweden

Johan Åkerman
Materials Physics, KTH Royal Institute of Technology,
Electrum 229, 16440 Kista, Sweden
Department of Physics, University of Gothenburg,
41296 Gothenburg, Sweden

Mato Knez and Paolo Vavassori
CIC nanoGUNE, 20018 Donostia-San Sebastián, Spain
IKERBASQUE, Basque Foundation for Science, 48011
Bilbao, Spain

**Iban Amenabar, Simon Poly, Wiwat Nuansing, Roman
Krutokhvostov, Lianbing Zhang and Alexander A.
Govyadinov**
CIC nanoGUNE Consolider, 20018 Donostia—San
Sebastián, Spain

Elmar H. Hubrich and Joachim Heberle
Experimental Molecular Biophysics, Department of
Physics, Freie Universität Berlin, 14195 Berlin, Germany

Florian Huth
CIC nanoGUNE Consolider, 20018 Donostia—San
Sebastián, Spain
Neaspec GmbH, 82152 Martinsried, Germany

**Mato Knez, Alexander M. Bittner and Rainer
Hillenbrand**
CIC nanoGUNE Consolider, 20018 Donostia—San
Sebastián, Spain
IKERBASQUE, Basque Foundation for Science, 48011
Bilbao, Spain

**Kohki Okabe, Takashi Funatsu, Seiichi Uchiyama
and Chie Gota**
Graduate School of Pharmaceutical Sciences, The
University of Tokyo, 7-3-1 Hongo Bunkyo-ku, Tokyo
113-0033, Japan

Noriko Inada
The Graduate School of Biological Sciences, Nara
Institute of Science and Technology, 8916-5 Takayama-
Cho Ikoma-shi, Nara 630-0101, Japan

Yoshie Harada
Institute for Integrated Cell-Material Sciences (WPI-
iCeMS), Kyoto University, Yoshida-Honmachi Sakyo-
ku, Kyoto 606-8501, Japan

Jonathan R. Felts
Mechanical Engineering Department, Texas A&M
University, 3123 TAMU, College Station, Texas 77843,
USA

Andrew J. Oyer and Keith E. Whitener Jr
National Research Council, US Naval Research
Laboratory, 4555 Overlook Avenue SW, Washington,
District Of Columbia 20375, USA

Sandra C. Hernández and Scott G. Walton
Plasma Physics Division, US Naval Research
Laboratory, 4555 Overlook Avenue SW, Washington,
District Of Columbia 20375, USA

Jeremy T. Robinson
Electronics Science and Technology Division, US
Naval Research Laboratory, 4555 Overlook Avenue
SW, Washington, District Of Columbia 20375, USA

Paul E. Sheehan
Chemistry Division, US Naval Research Laboratory,
Washington, District Of Columbia 20375, USA

Chularat Wattanakit and Sudarat Yadnum
Univ. de Bordeaux, CNRS, ISM, UMR 5255, ENSCBP,
16 Avenue Pey Berland, Pessac FR-33607, France

Yémima Bon Saint Côme, Veronique Lapeyre, Matthias Heim and Alexander Kuhn
Univ. de Bordeaux, CNRS, ISM, UMR 5255, ENSCBP, 16 Avenue Pey Berland, Pessac FR-33607, France

Somkiat Nokbin and Chompunuch Warakulwit
Department of Chemistry and NANOTEC Center for Nanoscale Materials Design for Green Nanotechnology, Kasetsart University, Bangkok 10900, Thailand

Jumras Limtrakul
Department of Chemistry and NANOTEC Center for Nanoscale Materials Design for Green Nanotechnology, Kasetsart University, Bangkok 10900, Thailand
PTT Group Frontier Research Center, PTT Public Company Limited, 555 Vibhavadi Rangsit Road, Chatuchak, Bangkok 10900, Thailand

Philippe A. Bopp
Univ. de Bordeaux, CNRS, ISM, UMR 5255, 351 cours de la Libération, Talence FR-33405, France

Brian Lam
Department of Chemistry, Faculty of Arts and Sciences, University of Toronto, Toronto, Ontario, Canada M5S 3M2

Jagotamoy Das, Ludovic Live and Andrew Sage
Department of Pharmaceutical Sciences, Leslie Dan Faculty of Pharmacy, University of Toronto, Toronto, Ontario, Canada M5S 3M2

Richard D. Holmes
Institute for Biomaterials and Biomedical Engineering, University of Toronto, Toronto, Ontario, Canada M5S 3M2

Edward H. Sargent
Department of Electrical and Computer Engineering, Faculty of Engineering, University of Toronto, Toronto, Ontario, Canada M5S 3M2

Shana O. Kelley
Department of Chemistry, Faculty of Arts and Sciences, University of Toronto, Toronto, Ontario, Canada M5S 3M2
Department of Pharmaceutical Sciences, Leslie Dan Faculty of Pharmacy, University of Toronto, Toronto, Ontario, Canada M5S 3M2
Institute for Biomaterials and Biomedical Engineering, University of Toronto, Toronto, Ontario, Canada M5S 3M2
Department of Biochemistry, Faculty of Medicine, University of Toronto, Toronto, Ontario, Canada M5S 3M2

Yu Zhang and Yu-Rong Zhen
Department of Physics and Astronomy, Rice University, Houston, Texas 77005, USA
Laboratory for Nanophotonics, Rice University, Houston, Texas77005, USA

Oara Neumann and Jared K. Day
Laboratory for Nanophotonics, Rice University, Houston, Texas 77005, USA
Department of Electrical and Computer Engineering, Rice University, Houston, Texas 77005, USA

Peter Nordlander and Naomi J. Halas
Department of Physics and Astronomy, Rice University, Houston, Texas 77005, USA
Laboratory for Nanophotonics, Rice University, Houston, Texas 77005, USA
Department of Electrical and Computer Engineering, Rice University, Houston, Texas 77005, USA

Hiroshi Nonaka, Ryunosuke Hata, Tomohiro Doura and Tatsuya Nishihara
INAMORI Frontier Research Center, Kyushu University, 744 Motooka, Nishi-ku, Fukuoka 819 0395, Japan

Keiko Kumagai and Mai Akakabe
Science Research Center, Kochi University, Kochi 783 8506, Japan

Masashi Tsuda
Center for Advanced Marine Core Research, Kochi University, Kochi 783 8502, Japan

Kazuhiro Ichikawa and Shinsuke Sando
Innovation Center for Medical Redox Navigation, Kyushu University, Fukuoka 812 8582, Japan

Daniel Ortiz and Sophie Le Caër
Institut Rayonnement Matière de Saclay, LIDyL et Service Interdisciplinaire sur les Systèmes Moléculaires et les Matériaux UMR 3299 CNRS/CEA SIS2M Laboratoire de Radiolyse, Bâtiment 546, F-91191 Gif-sur-Yvette, France

Vincent Steinmetz and Philippe Maître
Laboratoire de Chimie-Physique, UMR 8000 CNRS Université Paris Sud, Faculté des Sciences, Bâtiment 349, F-91405 Orsay, France

Delphine Durand, Solène Legand and Vincent Dauvois
CEA/Saclay, DEN/DANS/DPC/SECR/LRMO, F-91191 Gif-sur-Yvette, France

Elizaveta Pustovgar, Marta Palacios and Robert J. Flatt
Institute for Building Materials, Department of Civil, Environmental and Geomatic Engineering, ETH Zürich 8093, Switzerland

Rahul P. Sangodkar and Bradley F. Chmelka
Department of Chemical Engineering, University of California, Santa Barbara, California 93106, USA

Andrey S. Andreev
Soft Matter Science and Engineering Laboratory, UMR CNRS 7615, ESPCI Paris, PSL Research University, 10 rue Vauquelin, Paris 75005, France

Jean-Baptiste d'Espinose de Lacaillerie
Institute for Building Materials, Department of Civil, Environmental and Geomatic Engineering, ETH Zürich 8093, Switzerland
Soft Matter Science and Engineering Laboratory, UMR CNRS 7615, ESPCI Paris, PSL Research University, 10 rue Vauquelin, Paris 75005, France

Shan Cong, Zhigang Chen, Mei Yang, Yongyi Zhang, Qingwen Li and Zhigang Zhao
Key Lab of Nanodevices and Applications, Suzhou Institute of Nano-Tech and Nano-Bionics, Chinese Academy of Sciences (CAS), Suzhou 215123, China

Yinyin Yuan
Key Lab of Nanodevices and Applications, Suzhou Institute of Nano-Tech and Nano-Bionics, Chinese Academy of Sciences (CAS), Suzhou 215123, China
Key Laboratory for Ultrafine Materials of Ministry of Education, School of Materials Science and Engineering, East China University of Science and Technology, Shanghai 200237, China

Junyu Hou, Yanli Su and Fengxia Geng
College of Chemistry, Chemical Engineering and Materials Science, Soochow University, Suzhou 215123, Chinaz

Liang Li
Key Laboratory for Ultrafine Materials of Ministry of Education, School of Materials Science and Engineering, East China University of Science and Technology, Shanghai 200237, China

Alex Belianinov, Mikhail Kravchenko, Stephen Jesse and Sergei V. Kalinin
Institute for Functional Imaging of Materials, Oak Ridge National Laboratory, Oak Ridge, Tennessee 37831, USA
The Center for Nanophase Materials Sciences, Oak Ridge National Laboratory, Oak Ridge, Tennessee 37831, USA

Qian He
Materials Sciences and Technology Division, Oak Ridge National Laboratory, Oak Ridge, Tennessee 37831, USA

Albina Borisevich
Institute for Functional Imaging of Materials, Oak Ridge National Laboratory, Oak Ridge, Tennessee 37831, USA
Materials Sciences and Technology Division, Oak Ridge National Laboratory, Oak Ridge, Tennessee 37831, USA

Clinton Yu and Lan Huang
Department of Physiology and Biophysics, University of California, Irvine, California 92697, USA

Haibin Mao, Xiaobo Tang and Ning Zheng
Department of Pharmacology and Howard Hughes Medical Institute, University of Washington, Seattle, Washington 98195, USA

Eric J. Novitsky and Scott D. Rychnovsky
Department of Chemistry, University of California, Irvine, California 92697, USA

Wei Xie and Sebastian Schlücker
Physical Chemistry I, Faculty of Chemistry and Center for Nanointegration Duisburg-Essen (CENIDE), University of Duisburg-Essen, Universitätsstr. 5, Essen 45141, Germany

Somayeh Imani, Patrick P. Mercier and Shengfei Yu
Department of Electrical and Computer Engineering, University of California, San Diego, La Jolla, California 92093-0407, USA

Amay J. Bandodkar, A.M. Vinu Mohan, Rajan Kumar and Joseph Wang
Department of Nanoengineering, University of California, San Diego, La Jolla, California 92093-0448, USA

Marco Grossi and Marina Morgunova
Department of Pharmaceutical and Medicinal Chemistry, Royal College of Surgeons in Ireland, 123 St Stephen's Green, Dublin 2, Ireland

Dimitri Scholz, Emer Conroy, Marta Terrile and William M. Gallagher
School of Biomolecular and Biomedical Science, Conway Institute of Biomolecular and Biomedical Research, University College Dublin, Belfield, Dublin 4, Ireland

Shane Cheung and Donal F. O'Shea
Department of Pharmaceutical and Medicinal Chemistry, Royal College of Surgeons in Ireland, 123 St Stephen's Green, Dublin 2, Ireland
School of Chemistry and Chemical Biology, Conway Institute, University College Dublin, Belfield, Dublin 4, Ireland

Angela Panarella and Jeremy C. Simpson
School of Biology and Environmental Science, Conway Institute of Biomolecular and Biomedical Research, University College Dublin, Belfield, Dublin 4, Ireland

Xue Lu Wang, Wen Qi Fang and Hua Gui Yang
Key Laboratory for Ultrafine Materials of Ministry of Education, School of Materials Science and Engineering, East China University of Science and Technology, Shanghai 200237, China

Wenqing Liu, Yanhong Song and Daxiu Wei
Department of Physics, Shanghai Key Laboratory of Magnetic Resonance, East China Normal University, Shanghai 200062, China

Yan-Yan Yu and Xue-Qing Gong
Key Laboratory for Advanced Materials, Centre for Computational Chemistry, Research Institute of Industrial Catalysis, East China University of Science and Technology, Shanghai 200237, China

Ye-Feng Yao
Department of Physics, Shanghai Key Laboratory of Magnetic Resonance, East China Normal University, Shanghai 200062, China
NYU-ECNU Institute of Physics at NYU Shanghai, 3663 Zhongshan Road North, Shanghai 200062, China

Index

A

Accelerated Ageing Studies, 126
Achiral Ligands, 22, 40, 43, 102
Anti-stokes Raman Scattering, 78, 112-113, 118
Antibiotic-resistance Profiling, 104
Artificial Reproduction, 96
Atomic Force Microscopy, 1, 8, 31-32, 38-39, 79
Attomolar Dna Detection, 15

B

Breath Test, 9, 11-12, 14

C

Calcium Ions, 119, 122, 138
Cellular Functions, 80-81, 158
Cellular Trafficking, 183, 195
Chiral Metal Surfaces, 96-97
Chiral Nanorod Assemblies, 15
Circular Dichroism Spectra, 15, 22, 40, 102
Circular Polarization, 15
Colorimetric Transition, 53-56
Covalent Bond, 31-32, 34-38, 93
Cycle-inhibiting Factor (cif), 158

D

Deamidation Blocks, 158
Density Functional Theory, 91-93, 129, 138-139, 157, 196-197, 202
Dielectric Metasurface Analogue, 46
Diethyl Carbonate, 126, 133

E

Efficient Strength Control, 31
Electromagnetically Induced Transparency (eit), 46
Electron-induced Reduction, 170, 173-175
Enantioselective Recognition, 44, 96-97, 99, 101
Enantioseparation, 22, 40, 42-43, 102
Environmental Monitoring, 15

F

Fitness Monitoring, 176
Flexible Biosensors, 176
Fluorescence Imaging, 61, 87, 183-184, 190-191, 194-195
Fluorescence Lifetime Imaging Microscopy, 80
Fluorescent Polymeric Thermometer, 80, 87
Fourier Transform, 71-72, 78-79, 132, 153-154, 192, 202
Functional Groups, 89-94, 103

H

Heterogeneous Catalysis, 40, 135, 171, 201
Heterogeneous Photocatalytic Reactions, 196
Hybrid Biosensing System, 176
Hydride Ions, 23, 29
Hydrochromic Conjugated Polymers, 53
Hydroxide Ions, 23, 137
Hyperpolarized Magnetic Resonance, 119

I

In Vivo Detection, 9
Individual Protein Complexes, 71-73
Infectious Disease Speciation, 104
Infrared Nanospectroscopy, 71-73, 76
Intracellular Temperature Mapping, 80-82, 85

L

Light Phase Control, 63
Lithium-ion Batteries, 29, 126, 132
Local Crystallography, 150
Lysosome, 87, 183-184, 187, 189-191, 193, 195

M

Magnetoplasmonic Nanoantennas, 63, 66-67, 69
Mechanical Stress, 89-90
Mechanochemical Cleavage, 89-90
Metal Oxides, 143
Molecular-level Detection, 63
Multivariate Algorithms, 150
Mycobacterium Tuberculosis, 9, 14

N

Nanomechanical Dna Origami, 1-3, 7
Natural Systems, 96
Nuclear Magnetic Resonance (nmr), 26, 28, 196

O

Octahedral Tilts, 156
Oligonucleotides, 15, 19
Oxide Hosts, 23
Oxygen Vacancies, 143-147, 202

P

Pathogen Detection, 104, 107-108
Photorecycling Metal Surfaces, 170
Plasmonic Fano Resonance, 112
Plasmonic Nanostructures, 51, 64, 69, 112-113, 144

Polymerization, 54, 61, 133-135, 137-138, 140, 169
Proton Transfer, 78, 130, 196

Q
Quantitative Analyses, 134, 136

R
Radiolysis, 126-127, 129-133
Reactive Oxygen Species, 119
Real-time Health, 176
Real-time Monitoring, 2, 111, 119, 176
Refractive Index, 22, 46-48, 50, 64-66, 69-70
Refractometric Sensing, 63, 69

S
Self-assembled Monolayer (sam), 171
Shape Transition, 1-3
Silicate Hydration, 134-137, 140-141
Silicon-based Metasurfaces, 46

Single-molecule Beacons, 1-2
Single-molecule Sensitivity, 112-113, 117-118
Small Molecules, 115, 131, 170
Solution-based Circuits, 104-109
Spectroscopic Analysis, 61, 196
Structural Analysis, 71
Structural Reconfiguration, 158
Surface Vacancy, 146, 148
Surface-enhanced Raman Spectroscopy (sers), 143
Sweat Pore Mapping, 53, 58-60

T
Therapeutic Efficacy, 12
Thiol-gold Interactions, 31-33, 35-36
Tricalcium Silicates, 134
Tumour Imaging, 183, 190-191

U
Ubiquitin Ligase, 158-160, 168-169